T0250169

General Motors

Cadillac Eldorado, Seville, Deville and Fleetwood (FWD) Oldsmobile Toronado Buick Riviera

Automotive Repair Manual

by Robert Maddox and John H Haynes

Member of the Guild of Motoring Writers

Models covered:

Cadillac Eldorado (1986 thru 1991)
Cadillac Seville (1986 thru 1991)
Cadillac Deville (FWD) (1986 thru 1993)
Cadillac Fleetwood (FWD) (1986 thru 1992)
Oldsmobile Toronado (1986 thru 1992)
Buick Riviera (1986 thru 1993)

(38031-7V3)

Haynes Group Limited
Haynes North America, Inc.
www.haynes.com

About this manual

Its purpose

The purpose of this manual is to help you get the best value from your vehicle. It can do so in several ways. It can help you decide what work must be done, even if you choose to have it done by a dealer service department or a repair shop; it provides information and procedures for routine maintenance and servicing; and it offers diagnostic and repair procedures to follow when trouble occurs.

We hope you use the manual to tackle the work yourself. For many simpler jobs, doing it yourself may be quicker than arranging an appointment to get the vehicle into a shop and making the trips to leave it and pick it up. More importantly, a lot of money can be saved by avoiding the expense the shop must pass on to you to cover its labor and overhead costs. An added benefit is the sense of satisfaction and accomplishment that you feel after doing the job yourself.

Using the manual

The manual is divided into Chapters. Each Chapter is divided into numbered Sections, which are headed in bold type between horizontal lines. Each Section consists of consecutively numbered paragraphs.

At the beginning of each numbered Section you will be referred to any illustrations which apply to the procedures in that Section. The reference numbers used in illustration captions pinpoint the pertinent Section and the Step within that Section. That is, illustration 3.2 means the illustration refers to Section 3 and Step (or paragraph) 2 within that Section.

Procedures, once described in the text, are not normally repeated. When it's necessary to refer to another Chapter, the reference will be given as Chapter and Section number. Cross references given without use of the word "Chapter" apply to Sections and/or paragraphs in the same Chapter. For example, "see Section 8" means in the same Chapter.

References to the left or right side of the vehicle assume you are sitting in the driver's seat, facing forward.

Even though we have prepared this manual with extreme care, neither the publisher nor the author can accept responsibility for any errors in, or omissions from, the information given.

NOTE

A **Note** provides information necessary to properly complete a procedure or information which will make the procedure easier to understand.

CAUTION

A **Caution** provides a special procedure or special steps which must be taken while completing the procedure where the Caution is found. Not heeding a Caution can result in damage to the assembly being worked on.

WARNING

A **Warning** provides a special procedure or special steps which must be taken while completing the procedure where the Warning is found. Not heeding a Warning can result in personal injury.

Acknowledgements

Wiring diagrams and certain illustrations originated exclusively for Haynes North America, Inc. by Valley Forge Technical Information Systems. Technical writers who contributed to this project include Jeff Kibler, Jay Storer and Larry Warren.

A book in the Haynes Automotive Repair Manual Series

ISBN-10: 1-56392-347-5

ISBN-13: 978-1-56392-347-0

Library of Congress Catalog Card Number 99-63083

Contents

Haynes mechanic, author and photographer with 1991 Cadillac Deville

Introduction to the Cadillac Eldorado, Seville, Deville and Fleetwood, Oldsmobile Toronado, and Buick Riviera front-wheel drive models

Cadillac Eldorado and Fleetwood Coupe, Oldsmobile Toronado and Buick Riviera are two-door models, while the Cadillac Seville, Deville and Fleetwood models are four-door sedans.

Cadillac models have 4.1L, 4.5L and 4.9L V8 engines while Buick Riviera and Oldsmobile Toronado models are powered by 3.8L (3800) V6 engines. All models are equipped with fuel injection.

The transversely mounted engine trans- mits power to the front wheels through a four-speed automatic transaxle via independent driveaxles.

Suspension is independent in the front, utilizing coil springs with struts and lower control arms to locate the knuckle assembly at each wheel. The rear suspension on Cadillac Eldorado and Seville, Oldsmobile Toronado and Buick Riviera models uses a transverse plastic composite spring, lower control arms and suspension knuckle and shock absorbers. The rear suspension on Cadillac Deville and Fleetwood models features strut/coil spring assemblies, trailing arms and lateral link rods.

The rack-and-pinion steering unit is mounted behind the engine with power-assist as standard equipment.

The brakes are disc at the front and drum at the rear, depending on model, with power assist standard. An Anti-lock Braking System (ABS) is standard on most models.

Vehicle identification numbers

Vehicle Identification Number (VIN)

This very important identification number is stamped on a plate attached to the left side of the dashboard and is visible through the driver's side of the windshield (see illustration). The VIN also appears on the Vehicle Certificate of Title and Registration. It contains information such as where and when the vehicle was manufactured, the model year and the body style.

VIN engine and model year codes

Two particularly important pieces of information found in the VIN are the engine code and the model year code. Counting from the left, the engine code letter designation is the 8th digit and the model year code letter designation is the 10th digit.

On the models covered by this manual the engine codes are:

Buick and Oldsmobile

1986 through 1988
B.. 3.8L V6
1989 through 1991
C 3.8L V6
1992
L 3.8L V6

Cadillac

1986 and 1987
8.. 4.1L V8
1988
5.. 4.5L V8
1989
B.. 4.5L V8
1990
3.. 4.5L V8
1991 and 1992
B.. 4.9L V8
1993
B.. 4.9L V8

On the models covered by this manual the model year codes are:

G 1986
H 1987
J 1988
K... 1989
L 1990
M... 1991
N 1992
P... 1993

Vehicle Certification label

The Vehicle Certification Plate (VC label) is affixed to the rear of the left front door (see illustration). The plate contains the name of the manufacturer, the month and year of production, the Gross Vehicle Weight Rating (GVWR) and the certification statement.

The vehicle Identification Number (VIN) is on a plate attached to the top of the dashboard on the driver's side of the vehicle - it can be seen through the windshield

Body identification plate

The body identification plate is located in the engine compartment on the upper surface of the radiator support (see illustration). Like the VIN, it contains valuable information concerning the production of the vehicle, as well as information on the options with which

The Vehicle Certification label is located on the end of the driver side door, above the tire loading information label

The body identification plate can be found on the radiator support

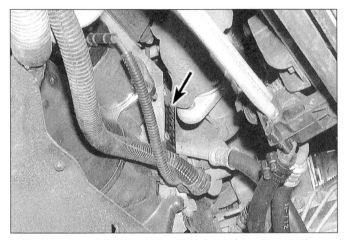

The V8 engine identification number (arrow) can be seen from below the vehicle

The V8 engine unit number label is on the right valve cover

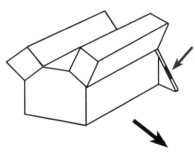

FRONT OF VEHICLE

24032-00D-HAYNES

On V6 engines the engine identification number is in one of two places: At the end of the block adjacent to the water pump, or at the front side of the block adjacent to the starter

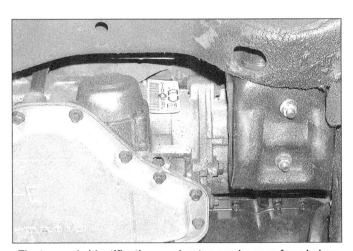

The transaxle identification number tag can be seen from below the vehicle

it is equipped. This plate is especially useful for matching the color and type of paint for repair work.

Engine identification number

The ID number on V8 engines is located on a tag at the left hand side at the rear of the engine block (see illustration). The V8 engine unit number label is on the rear (firewall) side valve cover (see illustration). On V6 engines the ID number can be found adjacent to the water pump, on a pad at the front surface of the block at the transaxle left (driver side) end (see illustration), adjacent to the starter or on the cylinder head below the valve cover at the front (radiator) side.

Service parts identification label

This label is located inside the trunk on the spare tire cover. It lists the VIN number, wheelbase, paint number, options and other information specific to your vehicle. Always refer to this label when ordering parts.

Transaxle identification number

The transaxle identification number is located on the right rear side of the transaxle (see illustration).

Vehicle Emissions Control Information label

The Vehicle Emissions Control Information label is under the hood, often attached to the left shock tower (see Chapter 6 for more information and an illustration of the label).

Buying parts

Replacement parts are available from many sources, which generally fall into one of two categories - authorized dealer parts departments and independent retail auto parts stores. Our advice concerning these parts is as follows:

Retail auto parts stores: Good auto parts stores will stock frequently needed components which wear out relatively fast, such as clutch components, exhaust systems, brake parts, tune-up parts, etc. These stores often supply new or reconditioned parts on an exchange basis, which can save a considerable amount of money. Discount auto parts stores are often very good places to buy materials and parts needed for general vehicle maintenance such as oil, grease, filters, spark plugs, belts, touch-up paint, bulbs, etc. They also usually sell tools and general accessories, have convenient hours, charge lower prices and can often be found not far from home.

Authorized dealer parts department: This is the best source for parts which are unique to the vehicle and not generally available elsewhere (such as major engine parts, transmission parts, trim pieces, etc.).

Warranty information: If the vehicle is still covered under warranty, be sure that any replacement parts purchased - regardless of the source - do not invalidate the warranty!

To be sure of obtaining the correct parts, have engine and chassis numbers available and, if possible, take the old parts along for positive identification.

Maintenance techniques, tools and working facilities

Maintenance techniques

There are a number of techniques involved in maintenance and repair that will be referred to throughout this manual. Application of these techniques will enable the home mechanic to be more efficient, better organized and capable of performing the various tasks properly, which will ensure that the repair job is thorough and complete.

Fasteners

Fasteners are nuts, bolts, studs and screws used to hold two or more parts together. There are a few things to keep in mind when working with fasteners. Almost all of them use a locking device of some type, either a lockwasher, locknut, locking tab or thread adhesive. All threaded fasteners should be clean and straight, with undamaged threads and undamaged corners on the hex head where the wrench fits. Develop the habit of replacing all damaged nuts and bolts with new ones. Special locknuts with nylon or fiber inserts can only be used once. If they are removed, they lose their locking ability and must be replaced with new ones.

Rusted nuts and bolts should be treated with a penetrating fluid to ease removal and prevent breakage. Some mechanics use turpentine in a spout-type oil can, which works quite well. After applying the rust penetrant, let it work for a few minutes before trying to loosen the nut or bolt. Badly rusted fasteners may have to be chiseled or sawed off or removed with a special nut breaker, available at tool stores.

If a bolt or stud breaks off in an assembly, it can be drilled and removed with a special tool commonly available for this purpose. Most automotive machine shops can perform this task, as well as other repair procedures, such as the repair of threaded holes that have been stripped out.

Flat washers and lockwashers, when removed from an assembly, should always be replaced exactly as removed. Replace any damaged washers with new ones. Never use a lockwasher on any soft metal surface (such as aluminum), thin sheet metal or plastic.

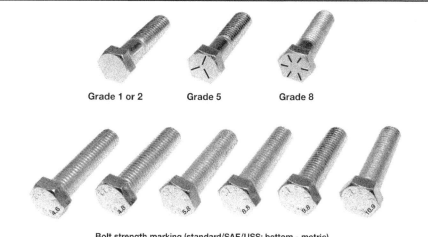

Grade 1 or 2   Grade 5   Grade 8

Bolt strength marking (standard/SAE/USS; bottom - metric)

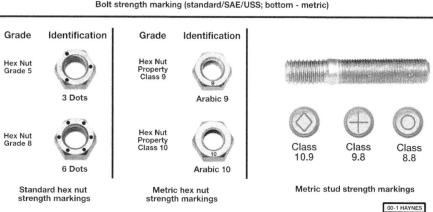

Grade	Identification	Grade	Identification
Hex Nut Grade 5	3 Dots	Hex Nut Property Class 9	Arabic 9
Hex Nut Grade 8	6 Dots	Hex Nut Property Class 10	Arabic 10

Standard hex nut strength markings

Metric hex nut strength markings

Class 10.9   Class 9.8   Class 8.8

Metric stud strength markings

00-1 HAYNES

Fastener sizes

For a number of reasons, automobile manufacturers are making wider and wider use of metric fasteners. Therefore, it is important to be able to tell the difference between standard (sometimes called U.S. or SAE) and metric hardware, since they cannot be interchanged.

All bolts, whether standard or metric, are sized according to diameter, thread pitch and length. For example, a standard 1/2 - 13 x 1 bolt is 1/2 inch in diameter, has 13 threads per inch and is 1 inch long. An M12 - 1.75 x 25 metric bolt is 12 mm in diameter, has a thread pitch of 1.75 mm (the distance between threads) and is 25 mm long. The two bolts are nearly identical, and easily confused, but they are not interchangeable.

In addition to the differences in diameter, thread pitch and length, metric and standard bolts can also be distinguished by examining the bolt heads. To begin with, the distance across the flats on a standard bolt head is measured in inches, while the same dimension on a metric bolt is sized in millimeters (the same is true for nuts). As a result, a standard wrench should not be used on a metric bolt and a metric wrench should not be used on a standard bolt. Also, most standard bolts have slashes radiating out from the center of the head to denote the grade or strength of the bolt, which is an indication of the amount of torque that can be applied to it. The greater the number of slashes, the greater the strength of the bolt. Grades 0 through 5 are commonly used on automobiles. Metric bolts have a property class (grade) number, rather than a slash, molded into their heads to indicate bolt strength. In this case, the higher the number, the stronger the bolt. Property class numbers 8.8, 9.8 and 10.9 are commonly used on automobiles.

Strength markings can also be used to distinguish standard hex nuts from metric hex nuts. Many standard nuts have dots stamped into one side, while metric nuts are marked with a number. The greater the number of dots, or the higher the number, the greater the strength of the nut.

Metric studs are also marked on their ends according to property class (grade). Larger studs are numbered (the same as metric bolts), while smaller studs carry a geometric code to denote grade.

It should be noted that many fasteners, especially Grades 0 through 2, have no distinguishing marks on them. When such is the case, the only way to determine whether it is standard or metric is to measure the thread pitch or compare it to a known fastener of the same size.

Standard fasteners are often referred to as SAE, as opposed to metric. However, it should be noted that SAE technically refers to a non-metric fine thread fastener only. Coarse thread non-metric fasteners are referred to as USS sizes.

Since fasteners of the same size (both standard and metric) may have different strength ratings, be sure to reinstall any bolts,

Metric thread sizes	Ft-lbs	Nm
M-6	6 to 9	9 to 12
M-8	14 to 21	19 to 28
M-10	28 to 40	38 to 54
M-12	50 to 71	68 to 96
M-14	80 to 140	109 to 154

Pipe thread sizes		
1/8	5 to 8	7 to 10
1/4	12 to 18	17 to 24
3/8	22 to 33	30 to 44
1/2	25 to 35	34 to 47

U.S. thread sizes		
1/4 - 20	6 to 9	9 to 12
5/16 - 18	12 to 18	17 to 24
5/16 - 24	14 to 20	19 to 27
3/8 - 16	22 to 32	30 to 43
3/8 - 24	27 to 38	37 to 51
7/16 - 14	40 to 55	55 to 74
7/16 - 20	40 to 60	55 to 81
1/2 - 13	55 to 80	75 to 108

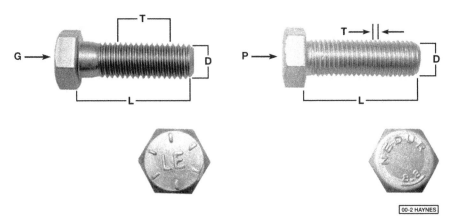

Standard (SAE and USS) bolt dimensions/grade marks

G   Grade marks (bolt strength)
L   Length (in inches)
T   Thread pitch (number of threads per inch)
D   Nominal diameter (in inches)

Metric bolt dimensions/grade marks

P   Property class (bolt strength)
L   Length (in millimeters)
T   Thread pitch (distance between threads in millimeters)
D   Diameter

studs or nuts removed from your vehicle in their original locations. Also, when replacing a fastener with a new one, make sure that the new one has a strength rating equal to or greater than the original.

Tightening sequences and procedures

Most threaded fasteners should be tightened to a specific torque value (torque is the twisting force applied to a threaded component such as a nut or bolt). Overtightening the fastener can weaken it and cause it to break, while undertightening can cause it to eventually come loose. Bolts, screws and studs, depending on the material they are made of and their thread diameters, have specific torque values, many of which are noted in the Specifications at the beginning of each Chapter. Be sure to follow the torque recommendations closely. For fasteners not assigned a specific torque, a general torque value chart is presented here as a guide. These torque values are for dry (unlubricated) fasteners threaded into steel or cast iron (not aluminum). As was previously mentioned, the size and grade of a fastener determine the amount of torque that can safely be applied to it. The figures listed here are approximate for Grade 2 and Grade 3 fasteners. Higher grades can tolerate higher torque values.

Fasteners laid out in a pattern, such as cylinder head bolts, oil pan bolts, differential cover bolts, etc., must be loosened or tightened in sequence to avoid warping the component. This sequence will normally be

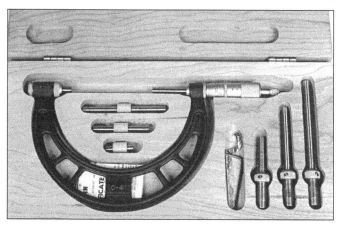

Micrometer set

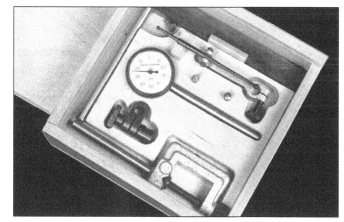

Dial indicator set

shown in the appropriate Chapter. If a specific pattern is not given, the following procedures can be used to prevent warping.

Initially, the bolts or nuts should be assembled finger-tight only. Next, they should be tightened one full turn each, in a criss-cross or diagonal pattern. After each one has been tightened one full turn, return to the first one and tighten them all one-half turn, following the same pattern. Finally, tighten each of them one-quarter turn at a time until each fastener has been tightened to the proper torque. To loosen and remove the fasteners, the procedure would be reversed.

Component disassembly

Component disassembly should be done with care and purpose to help ensure that the parts go back together properly. Always keep track of the sequence in which parts are removed. Make note of special characteristics or marks on parts that can be installed more than one way, such as a grooved thrust washer on a shaft. It is a good idea to lay the disassembled parts out on a clean surface in the order that they were removed. It may also be helpful to make sketches or take instant photos of components before removal.

When removing fasteners from a component, keep track of their locations. Sometimes threading a bolt back in a part, or putting the washers and nut back on a stud, can prevent mix-ups later. If nuts and bolts cannot be returned to their original locations, they should be kept in a compartmented box or a series of small boxes. A cupcake or muffin tin is ideal for this purpose, since each cavity can hold the bolts and nuts from a particular area (i.e. oil pan bolts, valve cover bolts, engine mount bolts, etc.). A pan of this type is especially helpful when working on assemblies with very small parts, such as the carburetor, alternator, valve train or interior dash and trim pieces. The cavities can be marked with paint or tape to identify the contents.

Whenever wiring looms, harnesses or connectors are separated, it is a good idea to identify the two halves with numbered pieces of masking tape so they can be easily reconnected.

Gasket sealing surfaces

Throughout any vehicle, gaskets are used to seal the mating surfaces between two parts and keep lubricants, fluids, vacuum or pressure contained in an assembly.

Many times these gaskets are coated with a liquid or paste-type gasket sealing compound before assembly. Age, heat and pressure can sometimes cause the two parts to stick together so tightly that they are very difficult to separate. Often, the assembly can be loosened by striking it with a soft-face hammer near the mating surfaces. A regular hammer can be used if a block of wood is placed between the hammer and the part. Do not hammer on cast parts or parts that could be easily damaged. With any particularly stubborn part, always recheck to make sure that every fastener has been removed.

Avoid using a screwdriver or bar to pry apart an assembly, as they can easily mar the gasket sealing surfaces of the parts, which must remain smooth. If prying is absolutely necessary, use an old broom handle, but keep in mind that extra clean up will be necessary if the wood splinters.

After the parts are separated, the old gasket must be carefully scraped off and the gasket surfaces cleaned. Stubborn gasket material can be soaked with rust penetrant or treated with a special chemical to soften it so it can be easily scraped off. A scraper can be fashioned from a piece of copper tubing by flattening and sharpening one end. Copper is recommended because it is usually softer than the surfaces to be scraped, which reduces the chance of gouging the part. Some gaskets can be removed with a wire brush, but regardless of the method used, the mating surfaces must be left clean and smooth. If for some reason the gasket surface is gouged, then a gasket sealer thick enough to fill scratches will have to be used during reassembly of the components. For most applications, a non-drying (or semi-drying) gasket sealer should be used.

Hose removal tips

Warning: *If the vehicle is equipped with air conditioning, do not disconnect any of the A/C hoses without first having the system depressurized by a dealer service department or a service station.*

Hose removal precautions closely parallel gasket removal precautions. Avoid scratching or gouging the surface that the hose mates against or the connection may leak. This is especially true for radiator hoses. Because of various chemical reactions, the rubber in hoses can bond itself to the metal spigot that the hose fits over. To remove a hose, first loosen the hose clamps that secure it to the spigot. Then, with slip-joint pliers, grab the hose at the clamp and rotate it around the spigot. Work it back and forth until it is completely free, then pull it off. Silicone or other lubricants will ease removal if they can be applied between the hose and the outside of the spigot. Apply the same lubricant to the inside of the hose and the outside of the spigot to simplify installation.

As a last resort (and if the hose is to be replaced with a new one anyway), the rubber can be slit with a knife and the hose peeled from the spigot. If this must be done, be careful that the metal connection is not damaged.

If a hose clamp is broken or damaged, do not reuse it. Wire-type clamps usually weaken with age, so it is a good idea to replace them with screw-type clamps whenever a hose is removed.

Tools

A selection of good tools is a basic requirement for anyone who plans to maintain and repair his or her own vehicle. For the owner who has few tools, the initial investment might seem high, but when compared to the spiraling costs of professional auto maintenance and repair, it is a wise one.

To help the owner decide which tools are needed to perform the tasks detailed in this manual, the following tool lists are offered: *Maintenance and minor repair, Repair/overhaul* and *Special*.

The newcomer to practical mechanics

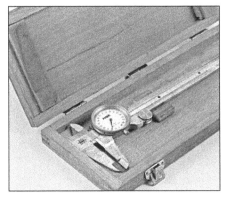

Dial caliper

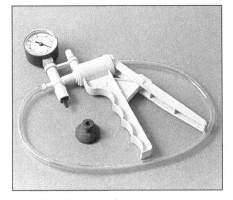

Hand-operated vacuum pump

Timing light

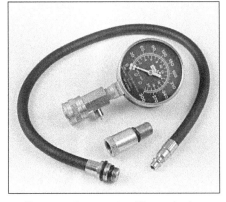

Compression gauge with spark plug
hole adapter

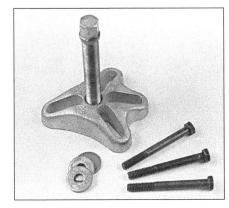

Damper/steering wheel puller

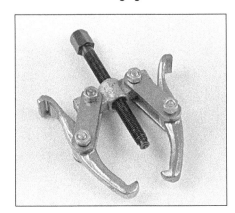

General purpose puller

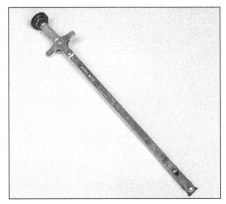

Hydraulic lifter removal tool

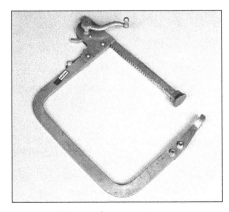

Valve spring compressor

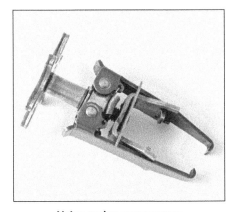

Valve spring compressor

Ridge reamer

Piston ring groove cleaning tool

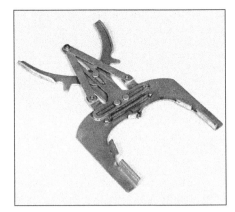

Ring removal/installation tool

Ring compressor

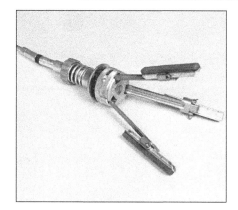

Cylinder hone

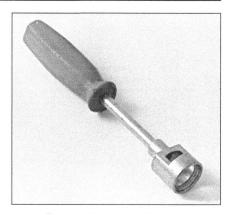

Brake hold-down spring tool

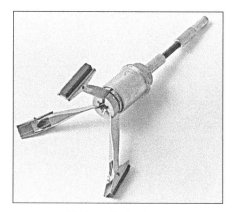

Brake cylinder hone

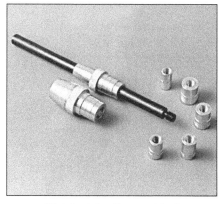

Clutch plate alignment tool

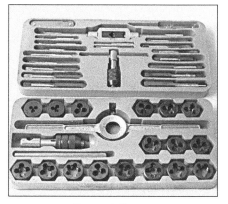

Tap and die set

should start off with the *maintenance and minor repair* tool kit, which is adequate for the simpler jobs performed on a vehicle. Then, as confidence and experience grow, the owner can tackle more difficult tasks, buying additional tools as they are needed. Eventually the basic kit will be expanded into the *repair and overhaul* tool set. Over a period of time, the experienced do-it-yourselfer will assemble a tool set complete enough for most repair and overhaul procedures and will add tools from the special category when it is felt that the expense is justified by the frequency of use.

Maintenance and minor repair tool kit

The tools in this list should be considered the minimum required for performance of routine maintenance, servicing and minor repair work. We recommend the purchase of combination wrenches (box-end and open-end combined in one wrench). While more expensive than open end wrenches, they offer the advantages of both types of wrench.

Combination wrench set (1/4-inch to 1 inch or 6 mm to 19 mm)
Adjustable wrench, 8 inch
Spark plug wrench with rubber insert
Spark plug gap adjusting tool
Feeler gauge set
Brake bleeder wrench

Standard screwdriver (5/16-inch x 6 inch)
Phillips screwdriver (No. 2 x 6 inch)
Combination pliers - 6 inch
Hacksaw and assortment of blades
Tire pressure gauge
Grease gun
Oil can
Fine emery cloth
Wire brush
Battery post and cable cleaning tool
Oil filter wrench
Funnel (medium size)
Safety goggles
Jackstands (2)
Drain pan

Note: *If basic tune-ups are going to be part of routine maintenance, it will be necessary to purchase a good quality stroboscopic timing light and combination tachometer/dwell meter. Although they are included in the list of special tools, it is mentioned here because they are absolutely necessary for tuning most vehicles properly.*

Repair and overhaul tool set

These tools are essential for anyone who plans to perform major repairs and are in addition to those in the maintenance and minor repair tool kit. Included is a comprehensive set of sockets which, though expensive, are invaluable because of their versatil-

ity, especially when various extensions and drives are available. We recommend the 1/2-inch drive over the 3/8-inch drive. Although the larger drive is bulky and more expensive, it has the capacity of accepting a very wide range of large sockets. Ideally, however, the mechanic should have a 3/8-inch drive set and a 1/2-inch drive set.

Socket set(s)
Reversible ratchet
Extension - 10 inch
Universal joint
Torque wrench (same size drive as sockets)
Ball peen hammer - 8 ounce
Soft-face hammer (plastic/rubber)
Standard screwdriver (1/4-inch x 6 inch)
Standard screwdriver (stubby - 5/16-inch)
Phillips screwdriver (No. 3 x 8 inch)
Phillips screwdriver (stubby - No. 2)
Pliers - vise grip
Pliers - lineman's
Pliers - needle nose
Pliers - snap-ring (internal and external)
Cold chisel - 1/2-inch
Scribe
Scraper (made from flattened copper tubing)
Centerpunch
Pin punches (1/16, 1/8, 3/16-inch)
Steel rule/straightedge - 12 inch

Allen wrench set (1/8 to 3/8-inch or
     4 mm to 10 mm)
A selection of files
Wire brush (large)
Jackstands (second set)
Jack (scissor or hydraulic type)

Note: Another tool which is often useful is an electric drill with a chuck capacity of 3/8-inch and a set of good quality drill bits.

Special tools

The tools in this list include those which are not used regularly, are expensive to buy, or which need to be used in accordance with their manufacturer's instructions. Unless these tools will be used frequently, it is not very economical to purchase many of them. A consideration would be to split the cost and use between yourself and a friend or friends. In addition, most of these tools can be obtained from a tool rental shop on a temporary basis.

This list primarily contains only those tools and instruments widely available to the public, and not those special tools produced by the vehicle manufacturer for distribution to dealer service departments. Occasionally, references to the manufacturer's special tools are included in the text of this manual. Generally, an alternative method of doing the job without the special tool is offered. However, sometimes there is no alternative to their use. Where this is the case, and the tool cannot be purchased or borrowed, the work should be turned over to the dealer service department or an automotive repair shop.

Valve spring compressor
Piston ring groove cleaning tool
Piston ring compressor
Piston ring installation tool
Cylinder compression gauge
Cylinder ridge reamer
Cylinder surfacing hone
Cylinder bore gauge
Micrometers and/or dial calipers
Hydraulic lifter removal tool
Balljoint separator
Universal-type puller
Impact screwdriver
Dial indicator set
Stroboscopic timing light (inductive
     pick-up)
Hand operated vacuum/pressure pump
Tachometer/dwell meter
Universal electrical multimeter
Cable hoist
Brake spring removal and installation
     tools
Floor jack

Buying tools

For the do-it-yourselfer who is just starting to get involved in vehicle maintenance and repair, there are a number of options available when purchasing tools. If maintenance and minor repair is the extent of the work to be done, the purchase of individual tools is satisfactory. If, on the other hand, extensive work is planned, it would be a good idea to purchase a modest tool set from one

of the large retail chain stores. A set can usually be bought at a substantial savings over the individual tool prices, and they often come with a tool box. As additional tools are needed, add-on sets, individual tools and a larger tool box can be purchased to expand the tool selection. Building a tool set gradually allows the cost of the tools to be spread over a longer period of time and gives the mechanic the freedom to choose only those tools that will actually be used.

Tool stores will often be the only source of some of the special tools that are needed, but regardless of where tools are bought, try to avoid cheap ones, especially when buying screwdrivers and sockets, because they won't last very long. The expense involved in replacing cheap tools will eventually be greater than the initial cost of quality tools.

Care and maintenance of tools

Good tools are expensive, so it makes sense to treat them with respect. Keep them clean and in usable condition and store them properly when not in use. Always wipe off any dirt, grease or metal chips before putting them away. Never leave tools lying around in the work area. Upon completion of a job, always check closely under the hood for tools that may have been left there so they won't get lost during a test drive.

Some tools, such as screwdrivers, pliers, wrenches and sockets, can be hung on a panel mounted on the garage or workshop wall, while others should be kept in a tool box or tray. Measuring instruments, gauges, meters, etc. must be carefully stored where they cannot be damaged by weather or impact from other tools.

When tools are used with care and stored properly, they will last a very long time. Even with the best of care, though, tools will wear out if used frequently. When a tool is damaged or worn out, replace it. Subsequent jobs will be safer and more enjoyable if you do.

How to repair damaged threads

Sometimes, the internal threads of a nut or bolt hole can become stripped, usually from overtightening. Stripping threads is an all-too-common occurrence, especially when working with aluminum parts, because aluminum is so soft that it easily strips out.

Usually, external or internal threads are only partially stripped. After they've been cleaned up with a tap or die, they'll still work. Sometimes, however, threads are badly damaged. When this happens, you've got three choices:

1) Drill and tap the hole to the next suitable oversize and install a larger diameter bolt, screw or stud.
2) Drill and tap the hole to accept a threaded plug, then drill and tap the plug to the original screw size. You can also buy a plug already threaded to the original size. Then you simply drill a hole to

the specified size, then run the threaded plug into the hole with a bolt and jam nut. Once the plug is fully seated, remove the jam nut and bolt.
3) The third method uses a patented thread repair kit like Heli-Coil or Slimsert. These easy-to-use kits are designed to repair damaged threads in straight-through holes and blind holes. Both are available as kits which can handle a variety of sizes and thread patterns. Drill the hole, then tap it with the special included tap. Install the Heli-Coil and the hole is back to its original diameter and thread pitch.

Regardless of which method you use, be sure to proceed calmly and carefully. A little impatience or carelessness during one of these relatively simple procedures can ruin your whole day's work and cost you a bundle if you wreck an expensive part.

Working facilities

Not to be overlooked when discussing tools is the workshop. If anything more than routine maintenance is to be carried out, some sort of suitable work area is essential.

It is understood, and appreciated, that many home mechanics do not have a good workshop or garage available, and end up removing an engine or doing major repairs outside. It is recommended, however, that the overhaul or repair be completed under the cover of a roof.

A clean, flat workbench or table of comfortable working height is an absolute necessity. The workbench should be equipped with a vise that has a jaw opening of at least four inches.

As mentioned previously, some clean, dry storage space is also required for tools, as well as the lubricants, fluids, cleaning solvents, etc. which soon become necessary.

Sometimes waste oil and fluids, drained from the engine or cooling system during normal maintenance or repairs, present a disposal problem. To avoid pouring them on the ground or into a sewage system, pour the used fluids into large containers, seal them with caps and take them to an authorized disposal site or recycling center. Plastic jugs, such as old antifreeze containers, are ideal for this purpose.

Always keep a supply of old newspapers and clean rags available. Old towels are excellent for mopping up spills. Many mechanics use rolls of paper towels for most work because they are readily available and disposable. To help keep the area under the vehicle clean, a large cardboard box can be cut open and flattened to protect the garage or shop floor.

Whenever working over a painted surface, such as when leaning over a fender to service something under the hood, always cover it with an old blanket or bedspread to protect the finish. Vinyl covered pads, made especially for this purpose, are available at auto parts stores.

Jacking and towing

Jacking

Warning: *The jack supplied with the vehicle should only be used for raising the vehicle when changing a tire or placing jackstands under the frame. Never work under the vehicle or start the engine while the jack is being used as the only means of support.*

The vehicle must be on a level surface with the wheels blocked and the transaxle in Park. Apply the parking brake if the front of the vehicle must be raised. Make sure no one is in the vehicle as it's being raised with the jack.

Remove the jack, lug nut wrench and spare tire (if needed) from the vehicle. If a tire is being replaced, use the lug wrench to remove the wheel cover. **Warning:** *Wheel covers may have sharp edges - be very careful not to cut yourself.* Loosen the lug nuts one-half turn, but leave them in place until the tire is raised off the ground. Position the jack under the vehicle at the indicated jacking point. There's a front and rear jacking point on each side of the vehicle **(see illustration)**.

Turn the jack handle clockwise until the tire clears the ground. Remove the lug nuts, pull the tire off and replace it with the spare. Replace the lug nuts with the beveled edges facing in and tighten them snugly. Don't attempt to tighten them completely until the vehicle is lowered or it could slip off the jack.

Turn the jack handle counterclockwise to lower the vehicle. Remove the jack and

The head of the jack should engage securely on the rocker panel flange at either the front or rear of the vehicle

tighten the lug nuts in a criss-cross pattern. If possible, tighten the nuts with a torque wrench (see Chapter 1 for the torque figures). If you don't have access to a torque wrench, have the nuts checked by a service station or repair shop as soon as possible.

Stow the tire, jack and wrench and unblock the wheels.

Towing

As a general rule, these vehicles should be towed with the front (drive) wheels off the ground. The vehicle may be towed with the rear end raised and the front wheels on the ground for distances up to 500 miles provided speed does not exceed 55 mph. These vehicles should not be towed with all four wheels on the ground.

Be sure to release the parking brake. If the vehicle is being towed with the front wheels on the ground, place the transaxle in Neutral. Also, the ignition key must be in the ACC position, since the steering lock mechanism isn't strong enough to hold the front wheels straight while towing.

Equipment specifically designed for towing should be used. It must be attached to the main structural members of the vehicle, not the bumpers or brackets.

Safety is a major consideration when towing and all applicable state and local laws must be obeyed. A safety chain must be used at all times. Remember that power steering and brakes won't work with the engine off.

Booster battery (jump) starting

Observe these precautions when using a booster battery to start a vehicle:

a) *Before connecting the booster battery, make sure the ignition switch is in the Off position.*
b) *Turn off the lights, heater and other electrical loads.*
c) *Your eyes should be shielded. Safety goggles are a good idea.*
d) *Make sure the booster battery is the same voltage as the dead one in the vehicle.*
e) *The two vehicles MUST NOT TOUCH each other!*
f) *Make sure the transaxle is in Neutral (manual) or Park (automatic).*
g) *If the booster battery is not a maintenance-free type, remove the vent caps and lay a cloth over the vent holes.*

Connect the red jumper cable to the positive (+) terminals of each battery **(see illustration)**.

Connect one end of the black jumper cable to the negative (-) terminal of the booster battery. The other end of this cable should be connected to a good ground on the vehicle to be started, such as a bolt or bracket on the body.

Start the engine using the booster battery, then, with the engine running at idle speed, disconnect the jumper cables in the reverse order of connection.

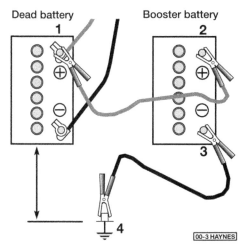

Make the booster battery cable connections in the numerical order shown (note that the negative cable of the booster battery is **NOT** attached to the negative terminal of the dead battery)

Automotive chemicals and lubricants

A number of automotive chemicals and lubricants are available for use during vehicle maintenance and repair. They include a wide variety of products ranging from cleaning solvents and degreasers to lubricants and protective sprays for rubber, plastic and vinyl.

Cleaners

Carburetor cleaner and choke cleaner is a strong solvent for gum, varnish and carbon. Most carburetor cleaners leave a dry-type lubricant film which will not harden or gum up. Because of this film it is not recommended for use on electrical components.

Brake system cleaner is used to remove grease and brake fluid from the brake system, where clean surfaces are absolutely necessary. It leaves no residue and often eliminates brake squeal caused by contaminants.

Electrical cleaner removes oxidation, corrosion and carbon deposits from electrical contacts, restoring full current flow. It can also be used to clean spark plugs, carburetor jets, voltage regulators and other parts where an oil-free surface is desired.

Demoisturants remove water and moisture from electrical components such as alternators, voltage regulators, electrical connectors and fuse blocks. They are non-conductive, non-corrosive and non-flammable.

Degreasers are heavy-duty solvents used to remove grease from the outside of the engine and from chassis components. They can be sprayed or brushed on and, depending on the type, are rinsed off either with water or solvent.

Lubricants

Motor oil is the lubricant formulated for use in engines. It normally contains a wide variety of additives to prevent corrosion and reduce foaming and wear. Motor oil comes in various weights (viscosity ratings) from 5 to 80. The recommended weight of the oil depends on the season, temperature and the demands on the engine. Light oil is used in cold climates and under light load conditions. Heavy oil is used in hot climates and where high loads are encountered. Multi-viscosity oils are designed to have characteristics of both light and heavy oils and are available in a number of weights from 5W-20 to 20W-50.

Gear oil is designed to be used in differentials, manual transmissions and other areas where high-temperature lubrication is required.

Chassis and wheel bearing grease is a heavy grease used where increased loads and friction are encountered, such as for wheel bearings, balljoints, tie-rod ends and universal joints.

High-temperature wheel bearing grease is designed to withstand the extreme temperatures encountered by wheel bearings in disc brake equipped vehicles. It usually contains molybdenum disulfide (moly), which is a dry-type lubricant.

White grease is a heavy grease for metal-to-metal applications where water is a problem. White grease stays soft under both low and high temperatures (usually from -100 to +190-degrees F), and will not wash off or dilute in the presence of water.

Assembly lube is a special extreme pressure lubricant, usually containing moly, used to lubricate high-load parts (such as main and rod bearings and cam lobes) for initial start-up of a new engine. The assembly lube lubricates the parts without being squeezed out or washed away until the engine oiling system begins to function.

Silicone lubricants are used to protect rubber, plastic, vinyl and nylon parts.

Graphite lubricants are used where oils cannot be used due to contamination problems, such as in locks. The dry graphite will lubricate metal parts while remaining uncontaminated by dirt, water, oil or acids. It is electrically conductive and will not foul electrical contacts in locks such as the ignition switch.

Moly penetrants loosen and lubricate frozen, rusted and corroded fasteners and prevent future rusting or freezing.

Heat-sink grease is a special electrically non-conductive grease that is used for mounting electronic ignition modules where it is essential that heat is transferred away from the module.

Sealants

RTV sealant is one of the most widely used gasket compounds. Made from silicone, RTV is air curing, it seals, bonds, waterproofs, fills surface irregularities, remains flexible, doesn't shrink, is relatively easy to remove, and is used as a supplementary sealer with almost all low and medium temperature gaskets.

Anaerobic sealant is much like RTV in that it can be used either to seal gaskets or to form gaskets by itself. It remains flexible, is solvent resistant and fills surface imperfections. The difference between an anaerobic sealant and an RTV-type sealant is in the curing. RTV cures when exposed to air, while an anaerobic sealant cures only in the absence of air. This means that an anaerobic sealant cures only after the assembly of parts, sealing them together.

Thread and pipe sealant is used for sealing hydraulic and pneumatic fittings and vacuum lines. It is usually made from a Teflon compound, and comes in a spray, a paint-on liquid and as a wrap-around tape.

Chemicals

Anti-seize compound prevents seizing, galling, cold welding, rust and corrosion in fasteners. High-temperature anti-seize, usually made with copper and graphite lubricants, is used for exhaust system and exhaust manifold bolts.

Anaerobic locking compounds are used to keep fasteners from vibrating or working loose and cure only after installation, in the absence of air. Medium strength locking compound is used for small nuts, bolts and screws that may be removed later. High-strength locking compound is for large nuts, bolts and studs which aren't removed on a regular basis.

Oil additives range from viscosity index improvers to chemical treatments that claim to reduce internal engine friction. It should be noted that most oil manufacturers caution against using additives with their oils.

Gas additives perform several functions, depending on their chemical makeup. They usually contain solvents that help dissolve gum and varnish that build up on carburetor, fuel injection and intake parts. They also serve to break down carbon deposits that form on the inside surfaces of the combustion chambers. Some additives contain upper cylinder lubricants for valves and piston rings, and others contain chemicals to remove condensation from the gas tank.

Miscellaneous

Brake fluid is specially formulated hydraulic fluid that can withstand the heat and pressure encountered in brake systems. Care must be taken so this fluid does not come in contact with painted surfaces or plastics. An opened container should always be resealed to prevent contamination by water or dirt.

Weatherstrip adhesive is used to bond weatherstripping around doors, windows and trunk lids. It is sometimes used to attach trim pieces.

Undercoating is a petroleum-based, tar-like substance that is designed to protect metal surfaces on the underside of the vehicle from corrosion. It also acts as a sound-deadening agent by insulating the bottom of the vehicle.

Waxes and polishes are used to help protect painted and plated surfaces from the weather. Different types of paint may require the use of different types of wax and polish. Some polishes utilize a chemical or abrasive cleaner to help remove the top layer of oxidized (dull) paint on older vehicles. In recent years many non-wax polishes that contain a wide variety of chemicals such as polymers and silicones have been introduced. These non-wax polishes are usually easier to apply and last longer than conventional waxes and polishes.

Conversion factors

Length (distance)
Inches (in)	X	25.4	= Millimetres (mm)	X 0.0394	= Inches (in)
Feet (ft)	X	0.305	= Metres (m)	X 3.281	= Feet (ft)
Miles	X	1.609	= Kilometres (km)	X 0.621	= Miles

Volume (capacity)
Cubic inches (cu in; in^3)	X	16.387	= Cubic centimetres (cc; cm^3)	X 0.061	= Cubic inches (cu in; in^3)
Imperial pints (Imp pt)	X	0.568	= Litres (l)	X 1.76	= Imperial pints (Imp pt)
Imperial quarts (Imp qt)	X	1.137	= Litres (l)	X 0.88	= Imperial quarts (Imp qt)
Imperial quarts (Imp qt)	X	1.201	= US quarts (US qt)	X 0.833	= Imperial quarts (Imp qt)
US quarts (US qt)	X	0.946	= Litres (l)	X 1.057	= US quarts (US qt)
Imperial gallons (Imp gal)	X	4.546	= Litres (l)	X 0.22	= Imperial gallons (Imp gal)
Imperial gallons (Imp gal)	X	1.201	= US gallons (US gal)	X 0.833	= Imperial gallons (Imp gal)
US gallons (US gal)	X	3.785	= Litres (l)	X 0.264	= US gallons (US gal)

Mass (weight)
Ounces (oz)	X	28.35	= Grams (g)	X 0.035	= Ounces (oz)
Pounds (lb)	X	0.454	= Kilograms (kg)	X 2.205	= Pounds (lb)

Force
Ounces-force (ozf; oz)	X	0.278	= Newtons (N)	X 3.6	= Ounces-force (ozf; oz)
Pounds-force (lbf; lb)	X	4.448	= Newtons (N)	X 0.225	= Pounds-force (lbf; lb)
Newtons (N)	X	0.1	= Kilograms-force (kgf; kg)	X 9.81	= Newtons (N)

Pressure
Pounds-force per square inch (psi; lbf/in^2; lb/in^2)	X	0.070	= Kilograms-force per square centimetre (kgf/cm^2; kg/cm^2)	X 14.223	= Pounds-force per square inch (psi; lbf/in^2; lb/in^2)
Pounds-force per square inch (psi; lbf/in^2; lb/in^2)	X	0.068	= Atmospheres (atm)	X 14.696	= Pounds-force per square inch (psi; lbf/in^2; lb/in^2)
Pounds-force per square inch (psi; lbf/in^2; lb/in^2)	X	0.069	= Bars	X 14.5	= Pounds-force per square inch (psi; lbf/in^2; lb/in^2)
Pounds-force per square inch (psi; lbf/in^2; lb/in^2)	X	6.895	= Kilopascals (kPa)	X 0.145	= Pounds-force per square inch (psi; lbf/in^2; lb/in^2)
Kilopascals (kPa)	X	0.01	= Kilograms-force per square centimetre (kgf/cm^2; kg/cm^2)	X 98.1	= Kilopascals (kPa)

Torque (moment of force)
Pounds-force inches (lbf in; lb in)	X	1.152	= Kilograms-force centimetre (kgf cm; kg cm)	X 0.868	= Pounds-force inches (lbf in; lb in)
Pounds-force inches (lbf in; lb in)	X	0.113	= Newton metres (Nm)	X 8.85	= Pounds-force inches (lbf in; lb in)
Pounds-force inches (lbf in; lb in)	X	0.083	= Pounds-force feet (lbf ft; lb ft)	X 12	= Pounds-force inches (lbf in; lb in)
Pounds-force feet (lbf ft; lb ft)	X	0.138	= Kilograms-force metres (kgf m; kg m)	X 7.233	= Pounds-force feet (lbf ft; lb ft)
Pounds-force feet (lbf ft; lb ft)	X	1.356	= Newton metres (Nm)	X 0.738	= Pounds-force feet (lbf ft; lb ft)
Newton metres (Nm)	X	0.102	= Kilograms-force metres (kgf m; kg m)	X 9.804	= Newton metres (Nm)

Vacuum
Inches mercury (in. Hg)	X	3.377	= Kilopascals (kPa)	X 0.2961	= Inches mercury
Inches mercury (in. Hg)	X	25.4	= Millimeters mercury (mm Hg)	X 0.0394	= Inches mercury

Power
Horsepower (hp)	X	745.7	= Watts (W)	X 0.0013	= Horsepower (hp)

Velocity (speed)
Miles per hour (miles/hr; mph)	X	1.609	= Kilometres per hour (km/hr; kph)	X 0.621	= Miles per hour (miles/hr; mph)

Fuel consumption*
Miles per gallon, Imperial (mpg)	X	0.354	= Kilometres per litre (km/l)	X 2.825	= Miles per gallon, Imperial (mpg)
Miles per gallon, US (mpg)	X	0.425	= Kilometres per litre (km/l)	X 2.352	= Miles per gallon, US (mpg)

Temperature
Degrees Fahrenheit = (°C x 1.8) + 32   Degrees Celsius (Degrees Centigrade; °C) = (°F - 32) x 0.56

*It is common practice to convert from miles per gallon (mpg) to litres/100 kilometres (l/100km),
where mpg (Imperial) x l/100 km = 282 and mpg (US) x l/100 km = 235

Safety first!

Regardless of how enthusiastic you may be about getting on with the job at hand, take the time to ensure that your safety is not jeopardized. A moment's lack of attention can result in an accident, as can failure to observe certain simple safety precautions. The possibility of an accident will always exist, and the following points should not be considered a comprehensive list of all dangers. Rather, they are intended to make you aware of the risks and to encourage a safety conscious approach to all work you carry out on your vehicle.

Essential DOs and DON'Ts

DON'T rely on a jack when working under the vehicle. Always use approved jackstands to support the weight of the vehicle and place them under the recommended lift or support points.

DON'T attempt to loosen extremely tight fasteners (i.e. wheel lug nuts) while the vehicle is on a jack - it may fall.

DON'T start the engine without first making sure that the transmission is in Neutral (or Park where applicable) and the parking brake is set.

DON'T remove the radiator cap from a hot cooling system - let it cool or cover it with a cloth and release the pressure gradually.

DON'T attempt to drain the engine oil until you are sure it has cooled to the point that it will not burn you.

DON'T touch any part of the engine or exhaust system until it has cooled sufficiently to avoid burns.

DON'T siphon toxic liquids such as gasoline, antifreeze and brake fluid by mouth, or allow them to remain on your skin.

DON'T inhale brake lining dust - it is potentially hazardous (see Asbestos below).

DON'T use loose fitting wrenches or other tools which may slip and cause injury.

DON'T push on wrenches when loosening or tightening nuts or bolts. Always try to pull the wrench toward you. If the situation calls for pushing the wrench away, push with an open hand to avoid scraped knuckles if the wrench should slip.

DON'T attempt to lift a heavy component alone - get someone to help you.

DON'T rush or take unsafe shortcuts to finish a job.

DON'T allow children or animals in or around the vehicle while you are working on it.

DO wear eye protection when using power tools such as a drill, sander, bench grinder, etc. and when working under a vehicle.

DO keep loose clothing and long hair well out of the way of moving parts.

DO make sure that any hoist used has a safe working load rating adequate for the job.

DO get someone to check on you periodically when working alone on a vehicle.

DO carry out work in a logical sequence and make sure that everything is correctly assembled and tightened.

DO keep chemicals and fluids tightly capped and out of the reach of children and pets.

DO remember that your vehicle's safety affects that of yourself and others. If in doubt on any point, get professional advice.

Steering, suspension and brakes

These systems are essential to driving safety, so make sure you have a qualified shop or individual check your work. Also, compressed suspension springs can cause injury if released suddenly - be sure to use a spring compressor.

Airbags

Airbags are explosive devices that can **CAUSE** injury if they deploy while you're working on the vehicle. Follow the manufacturer's instructions to disable the airbag whenever you're working in the vicinity of airbag components.

Asbestos

Certain friction, insulating, sealing, and other products - such as brake linings, brake bands, clutch linings, torque converters, gaskets, etc. - may contain asbestos or other hazardous friction material. Extreme care must be taken to avoid inhalation of dust from such products, since it is hazardous to health. If in doubt, assume that they do contain asbestos.

Fire

Remember at all times that gasoline is highly flammable. Never smoke or have any kind of open flame around when working on a vehicle. But the risk does not end there. A spark caused by an electrical short circuit, by two metal surfaces contacting each other, or even by static electricity built up in your body under certain conditions, can ignite gasoline vapors, which in a confined space are highly explosive. Do not, under any circumstances, use gasoline for cleaning parts. Use an approved safety solvent.

Always disconnect the battery ground (-) cable at the battery before working on any part of the fuel system or electrical system. Never risk spilling fuel on a hot engine or exhaust component. It is strongly recommended that a fire extinguisher suitable for use on fuel and electrical fires be kept handy in the garage or workshop at all times. Never try to extinguish a fuel or electrical fire with water.

Fumes

Certain fumes are highly toxic and can quickly cause unconsciousness and even death if inhaled to any extent. Gasoline vapor falls into this category, as do the vapors from some cleaning solvents. Any draining or pouring of such volatile fluids should be done in a well ventilated area.

When using cleaning fluids and solvents, read the instructions on the container carefully. Never use materials from unmarked containers.

Never run the engine in an enclosed space, such as a garage. Exhaust fumes contain carbon monoxide, which is extremely poisonous. If you need to run the engine, always do so in the open air, or at least have the rear of the vehicle outside the work area.

The battery

Never create a spark or allow a bare light bulb near a battery. They normally give off a certain amount of hydrogen gas, which is highly explosive.

Always disconnect the battery ground (-) cable at the battery before working on the fuel or electrical systems.

If possible, loosen the filler caps or cover when charging the battery from an external source (this does not apply to sealed or maintenance-free batteries). Do not charge at an excessive rate or the battery may burst.

Take care when adding water to a non maintenance-free battery and when carrying a battery. The electrolyte, even when diluted, is very corrosive and should not be allowed to contact clothing or skin.

Always wear eye protection when cleaning the battery to prevent the caustic deposits from entering your eyes.

Household current

When using an electric power tool, inspection light, etc., which operates on household current, always make sure that the tool is correctly connected to its plug and that, where necessary, it is properly grounded. Do not use such items in damp conditions and, again, do not create a spark or apply excessive heat in the vicinity of fuel or fuel vapor.

Secondary ignition system voltage

A severe electric shock can result from touching certain parts of the ignition system (such as the spark plug wires) when the engine is running or being cranked, particularly if components are damp or the insulation is defective. In the case of an electronic ignition system, the secondary system voltage is much higher and could prove fatal.

Hydrofluoric acid

This extremely corrosive acid is formed when certain types of synthetic rubber, found in some O-rings, oil seals, fuel hoses, etc. are exposed to temperatures above 750-degrees F (400-degrees C). The rubber changes into a charred or sticky substance containing the acid. Once formed, the acid remains dangerous for years. If it gets onto the skin, it may be necessary to amputate the limb concerned.

When dealing with a vehicle which has suffered a fire, or with components salvaged from such a vehicle, wear protective gloves and discard them after use.

Troubleshooting

Contents

This section provides an easy reference guide to the more common problems which may occur during the operation of your vehicle. Various symptoms and their possible causes are grouped under headings denoting components or systems, such as *Engine, Cooling system,* etc. They also refer to the Chapter and/or Section that deals with the problem.

Remember that successful troubleshooting isn't a mysterious "black art" practiced only by professional mechanics. It's simply the result of knowledge combined with an intelligent, systematic approach to a problem. Always use a process of elimination, starting with the simplest solution and working through to the most complex - and never overlook the obvious. Anyone can run the gas tank dry or leave the lights on overnight, so don't assume that you're exempt from such oversights.

Finally, always establish a clear idea why a problem has occurred and take steps to ensure that it doesn't happen again. If the electrical system fails because of a poor connection, check all other connections in the system to make sure they don't fail as well. If a particular fuse continues to blow, find out why - don't just go on replacing fuses. Remember, failure of a small component can often be indicative of potential failure or incorrect functioning of a more important component or system.

Engine and performance

1 Engine will not rotate when attempting to start

1   Battery terminal connections loose or corroded (Chapter 1).
2   Battery discharged or faulty (Chapter 1).
3   Automatic transaxle not completely engaged in Park (Chapter 7).
4   Broken, loose or disconnected wiring in the starting circuit (Chapters 5 and 12).
5   Starter motor pinion jammed in flywheel ring gear (Chapter 5).
6   Starter solenoid faulty (Chapter 5).
7   Starter motor faulty (Chapter 5).
8   Ignition switch faulty (Chapter 12).
9   Neutral start switch faulty (Chapter 7).
10   Starter pinion or driveplate teeth worn or broken (Chapter 5).
11   PASS-Key theft deterrent system malfunctioning. **Note:** *Sometimes the receptor in the ignition key lock cylinder becomes dirty and won't "read" the resistance value of the key. This problem can sometimes be rectified by spraying some electrical contact cleaner into the lock cylinder and allowing it to dry.*

2 Engine rotates but will not start

1   Fuel tank empty.

2   Battery discharged (engine rotates slowly) (Chapter 5).
3   Battery terminal connections loose or corroded (Chapter 1).
4   Leaking fuel injector(s), fuel pump, pressure regulator, etc. (Chapter 4).
5   Fuel not reaching fuel injection system (Chapter 4).
6   Ignition components damp or damaged (Chapter 5).
7   Worn, faulty or incorrectly gapped spark plugs (Chapter 1).
8   Broken, loose or disconnected wiring in the starting circuit (Chapter 5).
9   Broken, loose or disconnected wires at the ignition coil(s) or faulty coil(s) (Chapter 5).

3 Engine hard to start when cold

1   Battery discharged or low (Chapter 1).
2   Fuel system malfunctioning (Chapter 4).
3   Injector(s) leaking (Chapter 4).

4 Engine hard to start when hot

1   Air filter clogged (Chapter 1).
2   Fuel not reaching the fuel injection system (Chapter 4).
3   Corroded battery connections, especially ground (Chapter 1).

5 Starter motor noisy or excessively rough in engagement

1   Pinion or driveplate gear teeth worn or broken (Chapter 5).
2   Starter motor mounting bolts loose or missing (Chapter 5).

6 Engine starts but stops immediately

1   Loose or faulty electrical connections at coil pack or alternator (Chapter 5).
2   Insufficient fuel reaching the fuel injectors (Chapter 4).
3   Vacuum leak at the gasket between the intake manifold/plenum and throttle body (Chapters 1 and 4).
4   Restricted exhaust system (most likely the catalytic converter) (Chapters 4 and 6).

7 Oil puddle under engine

1   Oil pan gasket and/or oil pan drain bolt seal leaking (Chapters 1 and 2).
2   Oil pressure sending unit leaking (Chapter 2).
3   Rocker arm cover gaskets leaking (Chapter 2).

4   Engine oil seals leaking (Chapter 2).
5   Timing cover sealant or sealing flange leaking (Chapter 2).

8 Engine lopes while idling or idles erratically

1   Vacuum leakage (Chapter 4).
2   Leaking EGR valve or plugged PCV valve (Chapters 1 and 6).
3   Air filter clogged (Chapter 1).
4   Fuel pump not delivering sufficient fuel to the fuel injection system (Chapter 4).
5   Leaking head gasket (Chapter 2).
6   Timing chain worn (Chapter 2).
7   Camshaft lobes worn (Chapter 2).

9 Engine misses at idle speed

1   Spark plugs worn or not gapped properly (Chapter 1).
2   Faulty spark plug wires (Chapter 1).
3   Vacuum leaks (Chapters 1 and 4).
4   Incorrect ignition timing (Chapter 5).
5   Uneven or low compression (Chapter 2).

10 Engine misses throughout driving speed range

1   Fuel filter clogged and/or impurities in the fuel system (Chapters 1 and 4).
2   Low fuel output at the injector (Chapter 4).
3   Faulty or incorrectly gapped spark plugs (Chapter 1).
4   Incorrect ignition timing (Chapter 5).
5   Leaking spark plug wires (Chapter 1).
6   Faulty emission system components (Chapter 6).
7   Low or uneven cylinder compression pressures (Chapter 2).
8   Weak or faulty ignition system (Chapter 5).
9   Vacuum leak in fuel injection system, intake manifold or vacuum hoses (Chapter 4).

11 Engine stumbles on acceleration

1   Spark plugs fouled (Chapter 1).
2   Fuel injection system needs adjustment or repair (Chapter 4).
3   Fuel filter clogged (Chapter 1).
4   Incorrect ignition timing (Chapter 5).
5   Intake manifold air leak (Chapter 4).

12 Engine surges while holding accelerator steady

1   Intake air leak (Chapter 4).
2   Fuel pump faulty (Chapter 4).

3   Loose fuel injector harness connections (Chapter 4).
4   Defective ECM (Chapter 6).

13 Engine stalls

1   Idle speed incorrect (Chapters 1 and 4).
2   Fuel filter clogged and/or water and impurities in the fuel system (Chapters 1 and 4).
3   Ignition components damp or damaged (Chapter 5).
4   Faulty emissions system components (Chapter 6).
5   Faulty or incorrectly gapped spark plugs (Chapter 1).
6   Faulty spark plug wires (Chapter 1).
7   Vacuum leak in the fuel injection system, intake manifold or vacuum hoses (Chapter 4).

14 Engine lacks power

1   Incorrect ignition timing (Chapter 5).
2   Faulty or incorrectly gapped spark plugs (Chapter 1).
3   Fuel injection system malfunctioning (Chapter 4).
4   Faulty coil(s) (Chapter 5).
5   Brakes binding (Chapter 1).
6   Automatic transaxle fluid level incorrect (Chapter 1).
7   Fuel filter clogged and/or impurities in the fuel system (Chapter 1).
8   Emission control system not functioning properly (Chapter 6).
9   Low or uneven cylinder compression pressures (Chapter 2).
10   Restricted exhaust system (most likely the catalytic converter (Chapters 4 and 6).

15 Engine backfires

1   Emissions system not functioning properly (Chapter 6).
2   Ignition timing incorrect (Chapter 5).
3   Faulty secondary ignition system (Chapter 5).
4   Fuel injection system malfunctioning (Chapter 4).
5   Vacuum leak at fuel injectors, intake manifold or vacuum hoses (Chapter 4).
6   Valves sticking (Chapter 2).

16 Pinging or knocking engine sounds during acceleration or uphill

1   Incorrect grade of fuel.
2   Ignition timing incorrect (Chapter 5).
3   Fuel injection system malfunctioning Chapter 4).
4   Improper or damaged spark plugs or wires (Chapter 1).
5   Worn or damaged ignition components (Chapter 5).

6   Faulty emissions system (Chapter 6).
7   Vacuum leak (Chapter 4).

17 Engine runs with oil pressure light on

1   Low oil level (Chapter 1).
2   Idle rpm below specification (Chapter 1).
3   Short in wiring circuit (Chapter 12).
4   Faulty oil pressure sender (Chapter 2).
5   Oil viscosity too low or oil diluted.
6   Worn engine bearings and/or oil pump (Chapter 2).

18 Engine diesels (continues to run) after switching off

Leaking fuel injector(s).

Engine electrical system

19 Battery will not hold a charge

1   Alternator drivebelt defective or not adjusted properly (Chapter 1).
2   Battery terminals loose or corroded (Chapter 1).
3   Alternator not charging properly (Chapter 5).
4   Loose, broken or faulty wiring in the charging circuit (Chapter 5).
5   Short in vehicle wiring (Chapters 5 and 12).
6   Internally defective battery (Chapters 1 and 5).

20 Voltage warning light fails to go out

1   Faulty alternator or charging circuit (Chapter 5).
2   Alternator drivebelt defective or out of adjustment (Chapter 1).
3   Alternator voltage regulator inoperative (Chapter 5).

21 Voltage warning light fails to come on when key is turned on

1   Warning light bulb defective (Chapter 12).
2   Fault in the printed circuit, dash wiring or bulb holder (Chapter 12).

Fuel system

22 Excessive fuel consumption

1   Dirty or clogged air filter element (Chap-

ter 1).
2   Incorrectly set ignition timing (Chapter 5).
3   Emissions system not functioning properly (Chapter 6).
4   Fuel injection system malfunctioning (Chapter 4).
5   Low tire pressure or incorrect tire size (Chapter 1).

23 Fuel leakage and/or fuel odor

1   Leak in a fuel feed or vent line (Chapter 4).
2   Tank overfilled.
3   Evaporative emissions control canister filter clogged (Chapters 1 and 6).
4   Fuel injector internal parts excessively worn (Chapter 4).

Cooling system

24 Overheating

1   Insufficient coolant in system (Chapter 1).
2   Water pump drivebelt defective or out of adjustment (Chapter 1).
3   Radiator core blocked or grille restricted (Chapter 3).
4   Thermostat faulty (Chapter 3).
5   Electric cooling fan blades broken or cracked (Chapter 3).
6   Radiator cap not maintaining proper pressure (Chapter 3).
7   Ignition timing incorrect (Chapter 5).

25 Overcooling

Incorrect (opening temperature too low) or faulty thermostat (Chapter 3).

26 External coolant leakage

1   Deteriorated/damaged hoses or loose clamps (Chapters 1 and 3).
2   Water pump seal defective (Chapters 1 and 3).
3   Leakage from radiator core or header tank (Chapter 3).
4   Engine drain or water jacket core plugs leaking (Chapter 2).

27 Internal coolant leakage

1   Leaking cylinder head gasket (Chapter 2).
2   Cracked cylinder bore or cylinder head (Chapter 2).

28 Coolant loss

1   Too much coolant in system (Chapter 1).
2   Coolant boiling away because of over-heating (Chapter 3).
3   Internal or external leakage (Chapter 3).
4   Faulty radiator cap (Chapter 3).

29 Poor coolant circulation

1   Inoperative water pump (Chapter 3).
2   Restriction in cooling system (Chapters 1 and 3).
3   Water pump drivebelt defective or out of adjustment (Chapter 1).
4   Thermostat sticking (Chapter 3).

Automatic transaxle

Note: *Due to the complexity of the automatic transaxle, it's difficult for the home mechanic to properly diagnose and service this component. For problems other than the following, the vehicle should be taken to a dealer service department or a transmission shop.*

30 Fluid leakage

1   Automatic transmission fluid is a deep red color. Fluid leaks should not be confused with engine oil, which can easily be blown by airflow to the transaxle.
2   To pinpoint a leak, first remove all built-up dirt and grime from the transaxle housing with degreasing agents and/or steam cleaning. Drive the vehicle at low speeds so air flow will not blow the leak far from its source. Raise the vehicle and determine where the leak is coming from. Common areas of leakage are:

a) *Pan (Chapters 1 and 7)*
b) *Filler pipe (Chapter 7)*
c) *Fluid cooler lines (Chapter 7)*
d) *Speedometer gear or sensor (Chapter 7)*

31 Transaxle fluid brown or has a burned smell

Transaxle overheated. Change fluid (Chapter 1).

32 General shift mechanism problems

1   Chapter 7 deals with checking and adjusting the shift linkage on automatic transaxles. Common problems which may be attributed to a poorly adjusted linkage are:

a) *Engine starting in gears other than Park or Neutral.*
b) *Indicator on shifter pointing to a gear other than the one actually being used.*
c) *Vehicle moves when in Park.*
2   Refer to Chapter 7 for the shift linkage adjustment procedure.

33 Transaxle will not downshift with accelerator pedal pressed to the floor

Throttle valve (TV) cable out of adjustment (Chapter 7).

34 Engine will start in gears other than Park or Neutral

Starter safety switch malfunctioning (Chapter 7).

35 Transaxle slips, shifts roughly, is noisy or has no drive in forward or reverse gears

There are many probable causes for the above problems, but the home mechanic should be concerned with only one possibility - fluid level. Before taking the vehicle to a repair shop, check the level and condition of the fluid as described in Chapter 1.
Correct the fluid level as necessary or change the fluid and filter if needed. If the problem persists, have a professional diagnose the probable cause.

Driveaxles

36 Clicking noise in turns

Worn or damaged outer CV joint. Check for cut or damaged boots (Chapter 1). Repair as necessary (Chapter 8).

37 Knock or clunk when accelerating after coasting

Worn or damaged outer CV joint. Check for cut or damaged boots (Chapter 1). Repair as necessary (Chapter 8).

38 Shudder or vibration during acceleration

1   Worn or damaged CV joints. Repair or replace as necessary (Chapter 8).
2   Sticking inner joint assembly. Correct or replace as necessary (Chapter 8).

Brakes

Note: *Before assuming that a brake problem exists, make sure . . .*
a) *The tires are in good condition and properly inflated (Chapter 1).*
b) *The front end alignment is correct (Chapter 10).*
c) *The vehicle isn't loaded with weight in an unequal manner.*

39 Vehicle pulls to one side during braking

1   Incorrect tire pressures (Chapter 1).
2   Front end out of line (have the front end aligned).
3   Unmatched tires on same axle.
4   Restricted brake lines or hoses (Chapter 9).
5   Malfunctioning brake assembly (Chapter 9).
6   Loose suspension parts (Chapter 10).
7   Loose brake calipers (Chapter 9).

40 Noise (high-pitched squeal when the brakes are applied)

Front disc brake pads worn out. The noise comes from the wear sensor rubbing against the disc. Replace pads with new ones immediately (Chapter 9).

41 Brake roughness or chatter (pedal pulsates)

1   Excessive front brake disc lateral runout (Chapter 9).
2   Parallelism of disc not within specifications (Chapter 9).
3   Uneven pad wear caused by caliper not sliding due to improper clearance or dirt (Chapter 9).
4   Defective brake disc (Chapter 9).

42 Excessive pedal effort required to stop vehicle

1   Malfunctioning power brake booster (Chapter 9).
2   Partial system failure (Chapter 9).
3   Excessively worn pads or shoes (Chapter 9).
4   One or more caliper pistons seized or sticking (Chapter 9).
5   Brake pads or shoes contaminated with oil or grease (Chapter 9).
6   New pads or shoes installed and not yet seated. It will take a while for the new material to seat.

43 Excessive brake pedal travel

1   Partial brake system failure (Chapter 9).
2   Insufficient fluid in master cylinder (Chapters 1 and 9).
3   Air trapped in system (Chapters 1 and 9).
4   Faulty master cylinder (Chapter 9).

44 Dragging brakes

1   Master cylinder pistons not returning correctly (Chapter 9).
2   Restricted brakes lines or hoses (Chapters 1 and 9).
3   Incorrect parking brake adjustment (Chapter 9).

45 Grabbing or uneven braking action

1   Malfunction of proportioning valve (Chapter 9).
2   Malfunction of power brake booster unit (Chapter 9).
3   Binding brake pedal mechanism (Chapter 9).
4   Contaminated brake linings (Chapter 9).

46 Brake pedal feels spongy when depressed

1   Air in hydraulic lines (Chapter 9).
2   Master cylinder mounting bolts loose (Chapter 9).
3   Master cylinder defective (Chapter 9).

47 Brake pedal travels to the floor with little resistance

Little or no fluid in the master cylinder reservoir caused by leaking caliper, loose, damaged or disconnected brake lines (Chapter 9).

48 Parking brake does not hold

Parking brake linkage improperly adjusted (Chapter 9).

Suspension and steering systems

Note: *Before attempting to diagnose the suspension and steering systems, perform the following preliminary checks:*
   a) *Check the tire pressures and look for uneven wear.*

   b) *Check the steering universal joints or coupling from the column to the steering gear for loose fasteners and wear.*
   c) *Check the front and rear suspension and the steering gear assembly for loose and damaged parts.*
   d) *Look for out-of-round or out-of-balance tires, bent rims and loose and/or rough wheel bearings.*

49 Vehicle pulls to one side

1   Mismatched or uneven tires (Chapter 10).
2   Broken or sagging springs (Chapter 10).
3   Wheel alignment incorrect (Chapter 10).
4   Front brakes dragging (Chapter 9).

50 Abnormal or excessive tire wear

1   Front wheel alignment incorrect (Chapter 10).
2   Sagging or broken springs (Chapter 10).
3   Tire out-of-balance (Chapter 10).
4   Worn shock absorber (Chapter 10).
5   Overloaded vehicle.
6   Tires not rotated regularly.

51 Wheel makes a "thumping" noise

1   Blister or bump on tire (Chapter 1).
2   Improper shock absorber action (Chapter 10).

52 Shimmy, shake or vibration

1   Tire or wheel out-of-balance or out-of-round (Chapter 10).
2   Loose or worn wheel bearings (Chapter 10).
3   Worn tie-rod ends (Chapter 10).
4   Worn balljoints (Chapter 10).
5   Excessive wheel runout (Chapter 10).
6   Blister or bump on tire (Chapter 1).

53 Hard steering

1   Lack of lubrication at balljoints, tie-rod ends and steering gear assembly (Chapter 10).
2   Front wheel alignment incorrect (Chapter 10).
3   Low tire pressure (Chapter 1).

54 Steering wheel does not return to center position correctly

1   Lack of lubrication at balljoints and tie-rod ends (Chapter 10).

2   Binding in steering column (Chapter 10).
3   Defective rack-and-pinion assembly (Chapter 10).
4   Front wheel alignment problem (Chapter 10).

55 Abnormal noise at the front end

1   Lack of lubrication at balljoints and tie-rod ends (Chapter 1).
2   Loose upper strut mount (Chapter 10).
3   Worn tie-rod ends (Chapter 10).
4   Loose stabilizer bar (Chapter 10).
5   Loose wheel lug nuts (Chapter 1).
6   Loose suspension bolts (Chapter 10).

56 Wander or poor steering stability

1   Mismatched or uneven tires (Chapter 10).
2   Lack of lubrication at balljoints or tie-rod ends (Chapters 1 and 10).
3   Worn shock absorbers (Chapter 10).
4   Loose stabilizer bar (Chapter 10).
5   Broken or sagging springs (Chapter 10).
6   Front wheel alignment incorrect (Chapter 10).
7   Worn steering gear clamp bushings (Chapter 10).

57 Erratic steering when braking

1   Wheel bearings worn (Chapters 8 and 10).
2   Broken or sagging springs (Chapter 10).
3   Leaking caliper (Chapter 9).
4   Warped rotors (Chapter 9).
5   Worn steering gear clamp bushings (Chapter 10).

58 Excessive pitching and/or rolling around corners or during braking

1   Loose stabilizer bar (Chapter 10).
2   Worn shock absorbers or mounts (Chapter 10).
3   Broken or sagging springs (Chapter 10).
4   Overloaded vehicle.

59 Suspension bottoms

1   Overloaded vehicle.
2   Worn shock absorbers (Chapter 10).
3   Incorrect, broken or sagging springs (Chapter 10).

60 Cupped tires

1   Front wheel alignment incorrect (Chapter 10).

2   Worn shock absorbers (Chapter 10).
3   Wheel bearings worn (Chapters 8 and 10).
4   Excessive tire or wheel runout (Chapter 10).
5   Worn balljoints (Chapter 10).

61 Excessive tire wear on outside edge

1   Inflation pressures incorrect (Chapter 1).
2   Excessive speed in turns.
3   Wheel alignment incorrect (excessive toe-in or positive camber). Have professionally aligned.
4   Suspension arm bent or twisted (Chapter 10).

62 Excessive tire wear on inside edge

1   Inflation pressures incorrect (Chapter 1).
2   Wheel alignment incorrect (toe-out or excessive negative camber). Have professionally aligned.
3   Loose or damaged steering components (Chapter 10).

63 Tire tread worn in one place

1   Tires out-of-balance.
2   Damaged or buckled wheel. Inspect and replace if necessary.
3   Defective tire (Chapter 1).

64 Excessive play or looseness in steering system

1   Wheel bearings worn (Chapter 10).
2   Tie-rod end loose or worn (Chapter 10).
3   Steering gear loose (Chapter 10).

65 Rattling or clicking noise in rack and pinion

Steering gear clamps loose (Chapter 10).

Chapter 1
Tune-up and routine maintenance

Contents

Specifications

Recommended lubricants and fluids

Note: *Listed here are manufacturer recommendations at the time this manual was written. Manufacturers occasionally upgrade their fluid and lubricant specifications, so check with you local auto parts store for current recommendations.*

Engine oil
 Type .... API grade SH or SH/CC multigrade and fuel efficient oil
 Viscosity .... See accompanying chart

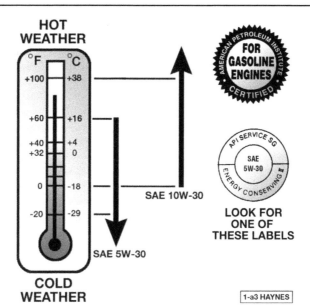

For best fuel economy and cold starting, select the lowest SAE viscosity grade oil for the expected temperature range

LOOK FOR ONE OF THESE LABELS

1-a3 HAYNES

Recommended lubricants and fluids

Automatic transaxle fluid	Dexron II, Dexron IIE or Dexron III ATF
Engine coolant	50/50 mixture of water and ethylene glycol-based antifreeze (and cooling system sealant on Cadillac V8 engines)
Brake fluid...	DOT 3 brake fluid
Power steering fluid	GM power steering fluid or equivalent
Chassis lubrication	Multi-purpose, lithium-base chassis grease

Capacities*

Engine oil
V6	0 qts (3.8 liters)
V8	5.5 qts (5.2 liters)
Cooling system..	13.2 qts (12.5 liters)

Automatic transaxle (when draining pan and replacing filter only)
Buick/Olds...	6.0 qts (5.7 liters)
Cadillac	6.5 qts (6.2 liters)

All capacities approximate. Add as necessary to bring to the appropriate level.

Ignition system

Spark plug type and gap
Buick Riviera
1985 through 1989	AC R44LTS or equivalent @ 0.045 inch
1990 and 1991	AC R45LTS6 or equivalent @ 0.060 inch
1992 and 1993	AC 41-600 or equivalent @ 0.060 inch

Oldsmobile Toronado
1986 and 1987	AC R44LTS or equivalent @ 0.045 inch
1988 through 1991	AC R45LTS6 or equivalent @ 0.060 inch
1992	AC 41-600 or equivalent @ 0.060 inch

Cadillac
1986 through 1989 (all)...	AC R44LTS6 or equivalent @ 0.060 inch

1990 and 1991
Eldorado	AC R44LTS6K or equivalent @ 0.060 inch

Deville, Fleetwood and Seville
1990 and 1991	AC R45LTSK or equivalent @ 0.060
1992 and 1993	AC 41-602 or equivalent @ 0.060 inch

Ignition timing
V6 engines...	Not adjustable

V8 engines (Cadillac)
1986 to 1990	10-degrees BTDC, below 800 rpm
1991 on	6 to 10-degrees BTDC, below 800 rpm

Firing order
V6	1-6-5-4-3-2
V8	1-8-4-3-6-5-7-2

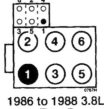
1986 to 1988 3.8L
Type I

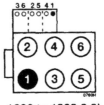
1986 to 1988 3.8L
Type II

Cylinder location, distributor
rotation and coil terminal
identification

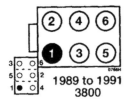
1989 to 1991
3800

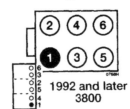
1992 and later
3800

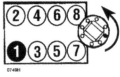
V8 Engine

Radiator cap pressure rating .... 15 psi

Brakes
Brake pad wear limit.. 1/8 inch
Brake shoe wear limit .... 1/16 inch

Torque specifications   **Ft-lbs** (unless otherwise indicated)
Spark plugs
    V6 20
    V8 .... 132 in-lbs
Engine oil drain plug
    V6 30
    V8 22
Automatic transaxle oil pan bolts .... 96 to 120 in-lbs
Wheel lug nuts 100

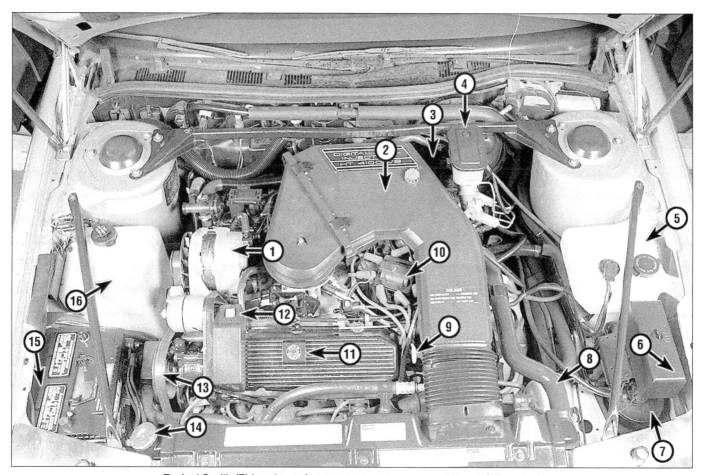

Typical Seville/Eldorado engine compartment component layout (V8 engine)

1	Alternator	6	Fuse block	12	Power steering fluid dipstick
2	Air cleaner assembly	7	EVAP system canister	13	Drivebelt
3	Automatic transaxle fluid dipstick location	8	Upper radiator hose	14	Radiator cap
4	Brake fluid reservoir	9	Engine oil dipstick	15	Battery
5	Windshield washer fluid reservoir	10	Distributor	16	Coolant reservoir
		11	Engine oil filler cap		

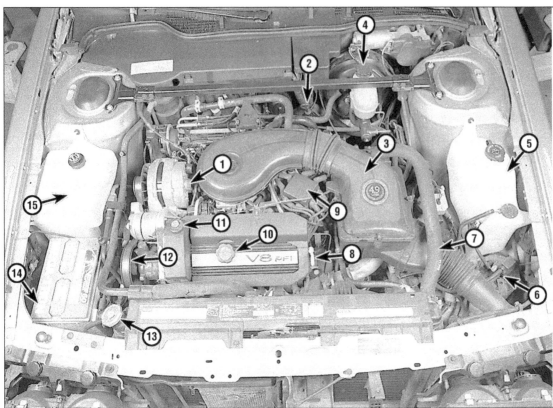

Typical Deville/Fleetwood engine compartment component layout (V8 engine)

1   Alternator
2   Automatic transaxle fluid dipstick location
3   Air cleaner assembly
4   Brake fluid reservoir
5   Windshield washer fluid reservoir
6   EVAP system canister
7   Upper radiator hose
8   Engine oil dipstick
9   Distributor
10 Engine oil filler cap
11 Power steering fluid dipstick
12 Drivebelt
13 Radiator cap
14 Battery
15 Coolant reservoir

Typical Toronado/Riviera engine compartment component layout (V6 engine)

1   Power steering fluid dipstick
2   Automatic transaxle fluid dipstick
3   Brake fluid reservoir
4   Air cleaner assembly
5   Engine compartment fuse block
6   Windshield washer fluid reservoir
7   Engine oil filler cap
8   Engine oil dipstick
9   Alternator
10 Ignition coils
11 Radiator cap
12 Battery
13 Coolant reservoir

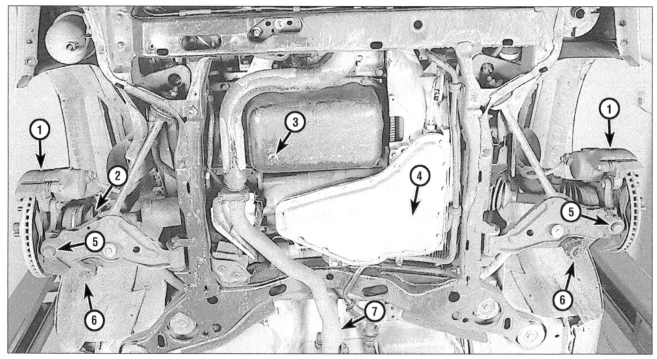

Typical Seville/Eldorado front underside components

1	Disc brake caliper	4	Automatic transaxle fluid pan	6	Tie-rod end
2	CV joint boot	5	Balljoint	7	Exhaust pipe
3	Engine oil drain plug				

**Typical
Deville/Fleetwood
front underside
components**

1   CV joint boot
2   Automatic
    transaxle fluid
    pan
3   Exhaust pipe
4   Front
    suspension
    strut
5   Engine oil
    drain plug
6   Disc brake
    caliper
7   Balljoint
8   Tie-rod end

Typical rear underside component layout (all except Deville/Fleetwood)

| 1 | Muffler | 3 | Fuel tank | 5 | Fuel tank filler hose and | 6 | Rear disc brake caliper |
| 2 | Parking brake cable | 4 | Rear leaf spring | | pipe | | |

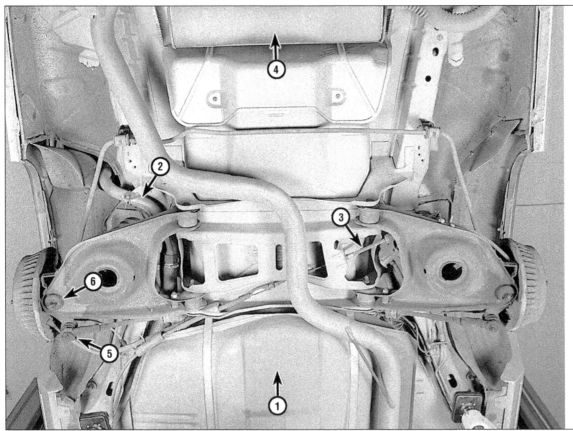

**Typical
Deville/Fleetwood
rear underside
component layout**

1   Fuel tank
2   Fuel tank filler
    hose and pipe
3   Parking brake
    cable
4   Muffler
5   Toe-link end
    lube fitting
6   Balljoint lube
    fitting

1 Maintenance schedule

The following maintenance intervals are based on the assumption that the vehicle owner will be doing the maintenance or service work, as opposed to having a dealer service department do the work. Although the time/mileage intervals are loosely based on factory recommendations, most have been shortened to ensure, for example, that such items as lubricants and fluids are checked/changed at intervals that promote maximum engine/driveline service life. Also, subject to the preference of the individual owner interested in keeping his or her vehicle in peak condition at all times, and with the vehicle's ultimate resale in mind, many of the maintenance procedures may be performed more often than recommended in the following schedule. We encourage such owner initiative.

When the vehicle is new it should be serviced initially by a factory authorized dealer service department to protect the factory warranty. In many cases the initial maintenance check is done at no cost to the owner (check with your dealer service department for more information).

Every 250 miles or weekly, whichever comes first

Check the engine oil level (Section 4)
Check the engine coolant level (Section 4)
Check the windshield washer fluid level (Section 4)
Check the brake fluid level (Section 4)
Check the tires and tire pressures (Section 5)

Every 3000 miles or 3 months, whichever comes first

All items listed above plus:
Check the power steering fluid level (Section 6)*
Check the automatic transaxle fluid level (Section 7)*
Change the engine oil and filter (Section 8)*
Check the seat belts (Section 9)
Inspect and replace, if necessary, the windshield wiper blades (Section 10)
Check and service the battery (Section 11)

Every 6000 miles or 6 months, whichever comes first

All items listed above plus:
Check the engine drivebelts (Section 12)

Inspect and replace, if necessary, all underhood hoses (Section 13)
Check the cooling system (Section 14)
Check the park/neutral switch (Section 15)
Rotate the tires (Section 16)
Lubricate the chassis components (Section 17)
Inspect the suspension, steering and driveaxle boots (Section 18)*
Inspect the exhaust system (Section 19)*
Check the brakes (Section 20)*

Every 30,000 miles or 24 months, whichever comes first

All items listed above plus:
Inspect the fuel system (Section 21)
Replace the fuel filter (Section 22)
Change the automatic transaxle fluid (Section 23)**
Replace the brake fluid (Section 24)
Replace the air filter (Section 25)*
Service the cooling system (drain, flush and refill) (Section 26)
Inspect and replace, if necessary, the PCV valve (Section 27)
Check the EGR system (Section 28)
Replace the spark plugs (Section 29)
Inspect the spark plug wires (Section 30)
Inspect and replace, if necessary, the distributor cap and rotor (Cadillac models only) (Section 31)
Check and adjust, if necessary, the ignition timing (Cadillac models only) (Section 32)

This item is affected by "severe" operating conditions as described below. If the vehicle is operated under severe conditions, perform all maintenance indicated with an asterisk (*) at 3000 mile/3 month intervals. Severe conditions are indicated if the vehicle is operated mainly . . .

in dusty areas
while towing a trailer
when allowed to idle for extended periods and/or at low speeds when outside temperatures remain below freezing and most trips are less than four miles long

**If operated under one or more of the following conditions, change the automatic transaxle fluid every 15,000 miles.*

In heavy city traffic where the outside temperature regularly reaches 90-degrees F or higher
In hilly or mountainous terrain
Frequent trailer pulling

2   Introduction

This Chapter is designed to help the home mechanic maintain the GM Cadillac full-size front wheel drive models with the goals of maximum performance, economy, safety and reliability in mind.

Included is a master maintenance schedule, followed by procedures dealing specifically with each item on the schedule. Visual checks, adjustments, component replacement and other helpful items are included. Refer to the **accompanying illustrations** of the engine compartment and the underside of the vehicle for the locations of various components.

Servicing your vehicle in accordance with the mileage/time maintenance schedule and the step-by-step procedures will result in a planned maintenance program that should produce a long and reliable service life. Keep in mind that it's a comprehensive plan, so maintaining some items but not others at the specified intervals will not produce the same results.

As you service your vehicle, you'll discover that many of the procedures can - and should - be grouped together because of the nature of the particular procedure you're performing or because of the close proximity of two otherwise unrelated components to one another.

For example, if the vehicle is raised, you should inspect the exhaust, suspension, steering and fuel systems while you're under the vehicle. When you're rotating the tires, it makes good sense to check the brakes since the wheels are already removed. Finally, let's suppose you have to borrow or rent a torque wrench. Even if you only need it to tighten the spark plugs, you might as well check the torque of as many critical fasteners as time allows.

The first step in this maintenance program is to prepare yourself before the actual work begins. Read through all the procedures you're planning to do, then gather up all the parts and tools needed. If it looks like you might run into problems during a particular job, seek advice from a mechanic or an experienced do-it-yourselfer.

3   Tune-up general information

The term tune-up is used in this manual to represent a combination of individual operations rather than one specific procedure.

If, from the time the vehicle is new, the routine maintenance schedule is followed closely and frequent checks are made of fluid levels and high wear items, as suggested throughout this manual, the engine will be kept in relatively good running condition and the need for additional work will be minimized.

More likely than not, however, there will be times when the engine is running poorly

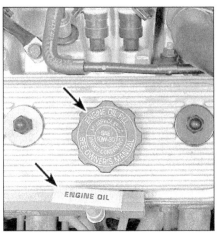

4.2a On most V6 models, the engine oil dipstick is clearly marked "ENGINE OIL" (arrow), as is the oil filler cap, which threads into the valve cover (arrow)

due to lack of regular maintenance. This is even more likely if a used vehicle, which has not received regular and frequent maintenance checks, is purchased. In such cases, an engine tune-up will be needed outside of the regular routine maintenance intervals.

The first step in any tune-up or diagnostic procedure to help correct a poor running engine is a cylinder compression check. A compression check (see Chapter 2C) will help determine the condition of internal engine components and should be used as a guide for tune-up and repair procedures. If, for instance, a compression check indicates serious internal engine wear, a conventional tune-up won't improve the performance of the engine and would be a waste of time and money. Because of its importance, the compression check should be done by someone with the right equipment and the knowledge to use it properly.

The following procedures are those most often needed to bring a generally poor running engine back into a proper state of tune.

Minor tune-up

Check all engine related fluids (Section 4)
Clean, inspect and test the battery (Section 11)
Check the cooling system (Section 14)
Check all underhood hoses (Section 13)
Check the air filter (Section 25)
Check the drivebelt (Section 12)
Check the PCV valve (Section 27)
Replace the spark plugs (Section 29)
Inspect the spark plug wires (Section 30)
Inspect the distributor cap and rotor (Cadillac models only) (Section 31)
Check and adjust, if necessary, the ignition timing (Cadillac models only) (Section 32)

Major tune-up

All items listed under Minor tune-up plus . . .
Check the fuel system (Section 21)

4.2b V8 engine oil dipstick and filler cap locations (arrows)

Replace the air filter (Section 25)
Check the EGR system (Section 28)
Check the ignition system (Section 29 and Chapter 5)
Check the charging system (Chapter 5)
Replace the spark plug wires (Section 30)
Replace the distributor cap and rotor (Cadillac models only) (Section 31)

4   Fluid level checks (every 250 miles or weekly)

Note: *The following are fluid level checks to be done on a 250 mile or weekly basis. Additional fluid level checks can be found in specific maintenance procedures which follow. Regardless of intervals, be alert to fluid leaks under the vehicle which would indicate a problem to be corrected immediately.*

1   Fluids are an essential part of the lubrication, cooling, brake and windshield washer systems. Because the fluids gradually become depleted and/or contaminated during normal operation of the vehicle, they must be periodically replenished. See Recommended lubricants and fluids at the beginning of this Chapter before adding fluid to any of the following components. **Note:** The vehicle must be on level ground when fluid levels are checked.

Engine oil

Refer to illustrations 4.2a, 4.2b, 4.4 and 4.6

2   The engine oil level is checked with a dipstick **(see illustrations)**. The dipstick extends through a metal tube down into the oil pan.

3   The oil level should be checked before the vehicle has been driven, or about 15 minutes after the engine has been shut off. If the oil is checked immediately after driving the vehicle, some of the oil will remain in the upper part of the engine, resulting in an inaccurate reading on the dipstick.

4   Pull the dipstick from the tube and wipe all the oil from the end with a clean rag or paper towel. Insert the clean dipstick all the

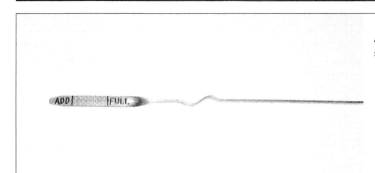

4.4 The oil level should be in the cross-hatched area - if it's below the ADD line, add enough oil to bring the level into the cross-hatched area

4.6 Unscrew the oil cap from the valve cover to add oil

way back into the tube and pull it out again. Note the oil at the end of the dipstick. Add oil as necessary to keep the level above the ADD mark in the cross-hatched area of the dipstick **(see illustration)**.

5   Do not overfill the engine by adding too much oil since this may result in oil fouled spark plugs, oil leaks or oil seal failures.

6   Oil is added to the engine after removing a twist off cap located on the valve cover **(see illustration)**. An oil can spout or funnel may help to reduce spills.

7   Checking the oil level is an important preventive maintenance step. A consistently low oil level indicates oil leakage through damaged seals, defective gaskets or past worn rings or valve guides. If the oil looks milky in color or has water droplets in it, the cylinder head gasket may be blown or the head or block may be cracked. The engine should be checked immediately. The condition of the oil should also be checked. Whenever you check the oil level, slide your thumb and index finger up the dipstick before wiping off the oil. If you see small dirt or metal particles clinging to the dipstick, the oil should be changed (see Section 8).

Engine coolant

Refer to illustration 4.9

Warning: *Do not allow antifreeze to come in contact with your skin or painted surfaces of the vehicle. Flush contaminated areas immediately with plenty of water. Do not store new coolant or leave old coolant lying around where it's accessible to children or pets - they're attracted by its sweet taste. Ingestion of even a small amount of coolant can be fatal! Wipe up garage floor and drip pan coolant spills immediately. Keep antifreeze containers covered and repair leaks in the cooling system immediately.*

8   All vehicles covered by this manual are equipped with a pressurized coolant recovery system. A plastic coolant reservoir located in the right front corner of the engine compartment is connected by a hose to the radiator filler neck. If the engine overheats, coolant escapes through a valve in the radiator cap and travels through the hose into the reservoir. As the engine cools, the coolant is automatically drawn back into the cooling system to maintain the correct level.

9   The coolant level in the reservoir should be checked regularly. **Warning:** *Do not remove*

the radiator cap to check the coolant level when the engine is warm. The level in the reservoir varies with the temperature of the engine. When the engine is cold, the coolant level should be at or slightly above the FULL COLD mark on the reservoir. Once the engine has warmed up, the level should be at or near the FULL HOT mark **(see illustration)**. If it isn't, allow the engine to cool, then unscrew the cap from the reservoir and add a 50/50 mixture of ethylene glycol based antifreeze and water. The coolant and windshield washer reservoirs are similar-looking, so be sure to add the correct fluids; the caps are clearly marked.

10   Drive the vehicle and recheck the coolant level. If only a small amount of coolant is required to bring the system up to the proper level, water can be used. However, repeated additions of water will dilute the antifreeze and water solution. In order to maintain the proper ratio of antifreeze and water, always top up the coolant level with the correct mixture. An empty plastic milk jug or bleach bottle makes an excellent container for mixing coolant. Do not use rust inhibitors or additives.

11   If the coolant level drops consistently, there may be a leak in the system. Inspect the radiator, hoses, filler cap, drain plugs and water pump (see Section 14). If no leaks are noted, have the radiator cap pressure tested by a service station.

12   If you have to remove the radiator cap, wait until the engine has cooled completely,

then wrap a thick cloth around the cap and turn it to the first stop. If coolant or steam escapes, let the engine cool down longer, then remove the cap.

13   Check the condition of the coolant as well. It should be relatively clear. If it is brown or rust colored, the system should be drained, flushed and refilled. Even if the coolant appears to be normal, the corrosion inhibitors wear out, so it must be replaced at the specified intervals.

Windshield washer fluid

Refer to illustration 4.14

14   Fluid for the windshield washer system is located in a plastic reservoir on the left side of the engine compartment **(see illustration)**. In milder climates, plain water can be used in the reservoir, but it should be kept no more than two-thirds full to allow for expansion if the water freezes. In colder climates, use windshield washer system antifreeze, available at any auto parts store, to lower the freezing point of the fluid. Mix the antifreeze with water in accordance with the manufacturer's directions on the container. **Caution:** *Do not use cooling system antifreeze - it will damage the vehicle's paint.*

4.9 The coolant level must be maintained between the FULL HOT and FULL COLD marks on the reservoir (arrow)

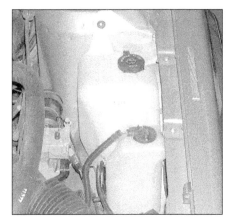

4.14 The reservoir for the windshield washer is located on the left side of the engine compartment

4.18a On white plastic translucent reservoirs, make sure the fluid level is even with the MAX mark (arrow)

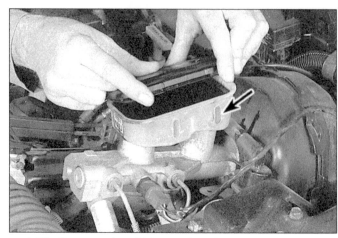

4.18b On reservoirs with inspection windows, the fluid level inside should be kept near the top of the window (arrow) - when adding fluid, grasp the tabs and rotate the cover up

15   To help prevent icing in cold weather, warm the windshield with the defroster before using the washer.

Battery electrolyte

16   All vehicles covered by this manual are equipped with a battery which is permanently sealed (except for vent holes) and has no filler caps. Water does not have to be added to these batteries at any time.

Brake fluid

Refer to illustrations 4.18a and 4.18b

17   The brake master cylinder is mounted on the front of the power booster unit in the engine compartment.

18   The fluid level is readily visible through the translucent reservoir - the level should be near the top at the MAX mark **(see illustrations)**. If a low level is indicated, be sure to clean the reservoir cap, to prevent contamination of the brake system, before removing it.

19   When adding fluid, pour it carefully into the reservoir to avoid spilling it on surrounding painted surfaces. Be sure the specified fluid is used, since mixing different types of brake fluid can cause damage to the system. See *Recommended lubricants and fluids* at the front of this Chapter or your owner's manual. **Warning:** *Brake fluid can harm your eyes and damage painted surfaces, so use extreme caution when handling or pouring it. Do not use brake fluid that has been standing open or is more than one year old. Brake fluid absorbs moisture from the air. Excess moisture can cause a dangerous loss of braking effectiveness.*

20   At this time the fluid and master cylinder can be inspected for contamination. The system should be drained and refilled if deposits, dirt particles or contamination are seen in the fluid.

21   After filling the reservoir to the proper level, make sure the cap is on tight to prevent fluid leakage.

22   The brake fluid level in the master cylin-

der will drop slightly as the pads at each wheel wear down during normal operation. If the master cylinder requires repeated replenishing to keep it at the proper level, this is an indication of leakage in the brake system, which should be corrected immediately. Check all brake lines and connections (see Section 20 for more information).

23   If, when checking the master cylinder fluid level, you discover one or both reservoirs empty or nearly empty, the brake system should be bled (see Chapter 9) and the cause of the fluid loss found.

---

5   Tire and tire pressure checks (250 miles or weekly)

Refer to illustrations 5.2, 5.3, 5.4a, 5.4b and 5.8

1   Periodic inspection of the tires may spare you the inconvenience of being stranded with a flat tire. It can also provide you with vital information regarding possible problems in the steering and suspension systems before major damage occurs.

2   The original tires on this vehicle are equipped with 1/2-inch wide bands that appear when tread depth reaches 1/16-inch, at which point the tires can be considered worn out. Tread wear can be monitored with a simple, inexpensive device known as a tread depth indicator **(see illustration)**.

3   Note any abnormal tread wear **(see illustration)**. Tread pattern irregularities such as cupping, flat spots and more wear on one side than the other are indications of front end alignment and/or balance problems. If any of these conditions are noted, take the vehicle to a tire shop or service station to correct the problem.

4   Look closely for cuts, punctures and embedded nails or tacks. Sometimes a tire will hold air pressure for a short time or leak down very slowly after a nail has embedded itself in the tread. If a slow leak persists, check the valve stem core to make sure it's

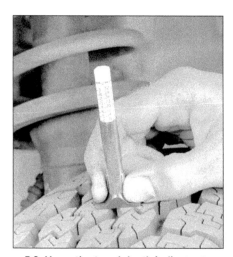

5.2 Use a tire tread depth indicator to monitor tire wear - they are available at auto parts stores and service stations and cost very little

tight **(see illustration)**. Examine the tread for an object that may have embedded itself in the tire or for a "plug" that may have begun to leak (radial tire punctures are repaired with a plug that's installed in the hole). If a puncture is suspected, it can be easily verified by spraying a solution of soapy water onto the suspected area **(see illustration)**. The soapy solution will bubble if there's a leak. Unless the puncture is unusually large, a tire shop or service station can usually repair the tire.

5   Carefully inspect the inner sidewall of each tire for evidence of brake fluid. If you see any, inspect the brakes immediately.

6   Correct air pressure adds miles to the lifespan of the tires, improves mileage and enhances overall ride quality. Tire pressure cannot be accurately estimated by looking at a tire, especially if it's a radial. A tire pressure gauge is essential. Keep an accurate gauge in the vehicle. The pressure gauges attached to the nozzles of air hoses at gas stations are often inaccurate.

UNDERINFLATION

CUPPING

Cupping may be caused by:
- Underinflation and/or mechanical irregularities such as out-of-balance condition of wheel and/or tire, and bent or damaged wheel.
- Loose or worn steering tie-rod or steering idler arm.
- Loose, damaged or worn front suspension parts.

OVERINFLATION

INCORRECT TOE-IN OR EXTREME CAMBER

FEATHERING DUE TO MISALIGNMENT

5.3 This chart will help you determine the condition of the tires, the probable cause(s) of abnormal wear and the corrective action necessary

5.4a If a tire loses air on a steady basis, check the valve core first to make sure it's snug (special inexpensive wrenches are commonly available at auto parts stores)

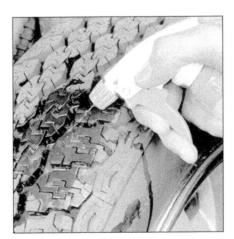

5.4b If the valve core is tight, raise the corner of the vehicle with the low tire and spray a soapy water solution onto the tread as the tire is turned slowly - leaks will cause small bubbles to appear

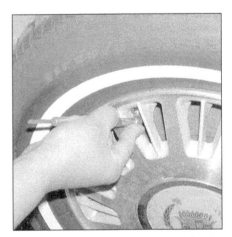

5.8 To extend the life of the tires, check the air pressure at least once a week with an accurate gauge (don't forget the spare!)

7   Always check tire pressure when the tires are cold. Cold, in this case, means the vehicle has not been driven over a mile in the three hours preceding a tire pressure check. A pressure rise of four to eight pounds is not uncommon once the tires are warm.

8   Unscrew the valve cap protruding from the wheel or hubcap and push the gauge firmly onto the valve stem **(see illustration)**. Note the reading on the gauge and compare the figure to the recommended tire pressure shown on the label attached to the rear edge of the driver's door. Be sure to reinstall the valve cap to keep dirt and moisture out of the valve stem mechanism. Check all four tires and, if necessary, add enough air to bring them up to the recommended pressure.

9   Don't forget to keep the spare tire inflated to the specified pressure (refer to your owner's manual or the tire sidewall).

6.2 The power steering fluid reservoir (arrow) is located near the front (drivebelt end) of the engine

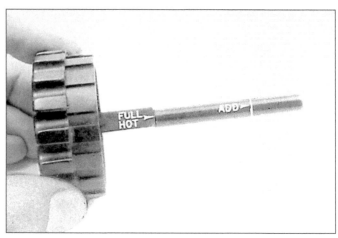

6.6 The marks on the power steering fluid dipstick indicate the safe fluid level range

6   Power steering fluid level check (every 3000 miles or 3 months)

Refer to illustrations 6.2 and 6.6

1   The power steering system relies on fluid which may, over a period of time, require replenishing.

2   The fluid reservoir for the power steering pump is located behind the radiator near the front (drivebelt end) of the engine **(see illustration)**.

3   For the check, the front wheels should be pointed straight ahead and the engine should be off.

4   Use a clean rag to wipe off the reservoir cap and the area around the cap. This will help prevent any foreign matter from entering the reservoir during the check.

5   Twist off the cap and check the temperature of the fluid at the end of the dipstick with your finger.

6   Wipe off the fluid with a clean rag, reinsert it, then withdraw it and read the fluid level. The level should be at the HOT mark if the fluid was hot to the touch **(see illustra-**

tion). It should be at the COLD mark if the fluid was cool to the touch. Note that on some models the marks (FULL HOT and COLD) are on opposite sides of the dipstick. At no time should the fluid level drop below the ADD mark.

7   If additional fluid is required, pour the specified type directly into the reservoir, using a funnel to prevent spills.

8   If the reservoir requires frequent fluid additions, all power steering hoses, hose connections, the power steering pump and the rack-and-pinion assembly should be carefully checked for leaks.

7   Automatic transaxle fluid level check (every 3000 miles or 6 months)

Refer to illustrations 7.3 and 7.6

1   The automatic transaxle fluid level should be carefully maintained. Low fluid level can lead to slipping or loss of drive, while overfilling can cause foaming (which can lead to burned clutch discs in the transaxle) and loss of fluid.

2   With the parking brake set, start the engine, then move the shift lever through all the gear ranges, ending in Park. The fluid level must be checked with the vehicle level and the engine running at idle. **Note:** *Incorrect fluid level readings will result if the vehicle has just been driven at high speeds for an*

extended period, in hot weather in city traffic, or if it has been pulling a trailer. If any of these conditions apply, wait until the fluid has cooled (about 30 minutes).

3   With the transaxle at normal operating temperature, remove the dipstick from the filler tube. The dipstick is located at the rear of the engine compartment **(see illustration)**.

4   Carefully touch the fluid at the end of the dipstick to determine if the fluid is cool, warm or hot. Wipe the fluid from the dipstick with a clean rag and push it back into the filler tube until the cap seats.

5   Pull the dipstick out again and note the fluid level.

6   If the fluid felt cool, the level should be about 1/8-to-3/8 inch below the "ADD 1 PT" mark **(see illustration)**. If it felt warm, the level should be close to the "ADD 1 PT" mark. If the fluid was hot, the level should be within the cross-hatched area. If additional fluid is required, pour it directly into the tube using a funnel. It takes about one pint to raise the level from the ADD mark to the upper edge of the cross-hatched area with a hot transaxle, so add the fluid a little at a time and keep checking the level until it's correct.

7   The condition of the fluid should also be checked along with the level. If the fluid at the end of the dipstick is a dark reddish-brown color, or if the fluid has a burned smell, the fluid should be changed. If you're in doubt about the condition of the fluid, purchase some new fluid and compare the two for color and smell.

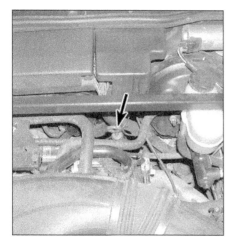

7.3 The automatic transaxle fluid dipstick (arrow) is located at the rear of the engine compartment

7.6 The automatic transaxle fluid level must be maintained within the cross-hatched area of the dipstick

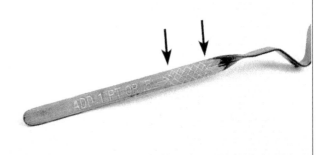

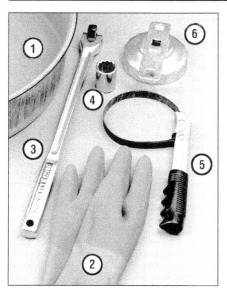

8.2 These tools are required when changing the engine oil and filter

1   **Drain pan** - It should be fairly shallow in depth, but wide to prevent spills
2   **Rubber gloves** - When removing the drain plug and filter, you will get oil on your hands (the gloves will prevent burns)
3   **Breaker bar** - Sometimes the oil drain plug is tight, and a long breaker bar is needed to loosen it
4   **Socket** - To be used with the breaker bar or a ratchet (must be the correct size to fit the drain plug - six-point preferred)
5   **Filter wrench** - This is a metal band-type wrench, which requires clearance around the filter to be effective
6   **Filter wrench** - This type fits on the bottom of the filter and can be turned with a ratchet or breaker bar (different-size wrenches are available for different types of filters)

8   Engine oil and filter change (every 3000 miles or 3 months)

Refer to illustrations 8.2, 8.7, 8.12 and 8.14

1   Frequent oil changes are the best preventive maintenance the home mechanic can give the engine, because aging oil becomes diluted and contaminated, which leads to premature engine wear.
2   Make sure you have all the necessary tools before you begin this procedure **(see illustration)**. You should also have plenty of rags or newspapers handy for mopping up any spills.
3   Access to the underside of the vehicle is greatly improved if the vehicle can be lifted on a hoist, driven onto ramps or supported by jackstands. **Warning:** *Do not work under a vehicle which is supported only by a hydraulic or scissors-type jack.*
4   If this is your first oil change, get under the vehicle and familiarize yourself with the

8.7 The engine oil drain plug is located at the rear of the oil pan - it's usually very tight, so use a box-end wrench to avoid rounding off the hex

locations of the oil drain plug and the oil filter. The engine and exhaust components will be warm during the actual work, so try to anticipate any potential problems before the engine and accessories are hot.
5   Park the vehicle on a level spot. Start the engine and allow it to reach its normal operating temperature. Warm oil and sludge will flow out more easily. Turn off the engine when it's warmed up. Remove the filler cap from the valve cover.
6   Raise the vehicle and support it securely on jackstands. **Warning:** *Never get beneath the vehicle when it is supported only by a jack. The jack provided with your vehicle is designed solely for raising the vehicle to remove and replace the wheels. Always use jackstands to support the vehicle when it becomes necessary to place your body underneath the vehicle.*
7   Being careful not to touch the hot exhaust components, place the drain pan under the drain plug in the bottom of the pan and remove the plug **(see illustration)**. You may want to wear gloves while unscrewing the plug the final few turns if the engine is hot.
8   Allow the old oil to drain into the pan. It may be necessary to move the pan farther under the engine as the oil flow slows to a trickle. Inspect the old oil for the presence of metal shavings and chips.
9   After all the oil has drained, wipe off the drain plug with a clean rag. Even minute metal particles clinging to the plug would immediately contaminate the new oil.
10   Clean the area around the drain plug opening, reinstall the plug and tighten it to the torque listed in the Specifications at the beginning of this Chapter.
11   Move the drain pan into position under the oil filter.
12   Loosen the oil filter **(see illustration)** by turning it counterclockwise with the filter wrench. **Note:** *Oil filters on V8 engines are located on the rear end of the engine block below the brake master cylinder, while the oil filter on V6 engines is located on the rear side of the engine block next the crankshaft pulley.*

8.12 The oil filter is usually on very tight as well and will require a special wrench for removal - DO NOT use the wrench to tighten the new filter!

8.14 Lubricate the oil filter gasket with clean engine oil before installing the filter on the engine

Use a quality filter wrench of the correct size and be careful not to collapse the canister as you apply pressure. Once the filter is loose, use your hands to unscrew it from the block. Just as the filter is detached from the block, immediately tilt the open end up to prevent the oil inside the filter from spilling out. **Warning:** *The exhaust system may still be hot, so be careful.*
13   With a clean rag, wipe off the mounting surface on the block. If a residue of old oil is allowed to remain, it will smoke when the block is heated up. Also make sure that none of the old gasket remains stuck to the mounting surface. It can be removed with a scraper if necessary.
14   Compare the old filter with the new one to make sure they are the same type. Smear some clean engine oil on the rubber gasket of the new filter and screw it into place **(see illustration)**. Because overtightening the filter will damage the gasket, do not use a filter wrench to tighten the filter. Tighten it by hand until the gasket contacts the seating surface. Then seat the filter by giving it an additional 3/4-turn.
15   Remove all tools, rags, etc. from under

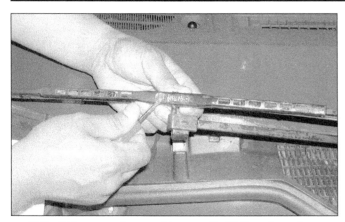

10.5a Using a small screwdriver, gently lift the release lever

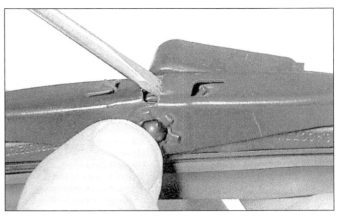

10.5b . . . or pry on the spring at the center of the windshield wiper arm . . .

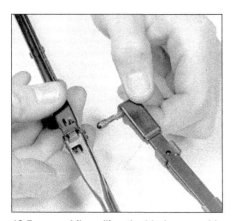

10.5c . . . while pulling the blade assembly away from the arm

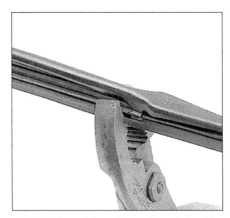

10.7a If the rubber element is retained in the blade by small clips, the metal backing of the rubber element can be compressed at one end with pliers, allowing the element to slide out of the clips

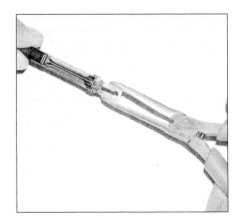

10.7b If the element is retained by a clip at the end, use needle nose pliers to compress it and pull it out

the vehicle, being careful not to spill the oil in the drain pan, then lower the vehicle.

16   Add new oil to the engine through the oil filler cap in the valve cover. Use a funnel, if necessary, to prevent oil from spilling onto the top of the engine. Pour three quarts of fresh oil into the engine. Wait a few minutes to allow the oil to drain into the pan, then check the level on the oil dipstick (see Section 4 if necessary). If the oil level is at or near the upper hole on the dipstick, install the filler cap hand tight, start the engine and allow the new oil to circulate.

17   Allow the engine to run for about a minute. While the engine is running, look under the vehicle and check for leaks at the oil pan drain plug and around the oil filter. If either is leaking, stop the engine and tighten the plug or filter.

18   Wait a few minutes to allow the oil to trickle down into the pan, then recheck the level on the dipstick and, if necessary, add enough oil to bring the level to the upper hole.

19   If the vehicle is equipped with an oil life indicator light, reset it now (see Section 33).

20   During the first few trips after an oil change, make it a point to check frequently for leaks and proper oil level.

21   The old oil drained from the engine cannot be re-used in its present state and should be discarded. Check with your local refuse

disposal company, disposal facility or environmental agency to see if they will accept the oil for recycling. Don't pour used oil into drains or onto the ground. After the oil has cooled, it can be drained into a suitable container (capped plastic jugs, topped bottles, milk cartons, etc.) for transport to one of these disposal sites.

9   Seat belt check (every 6000 miles or 6 months)

1   Check the seat belts, buckles, latch plates and guide loops for obvious damage and signs of wear.

2   See if the seat belt reminder light comes on when the key is turned to the Run or Start position. A chime should also sound.

3   The seat belts are designed to lock up during a sudden stop or impact, yet allow free movement during normal driving. Make sure the retractors return the belt against your chest while driving and rewind the belt fully when the buckle is unlatched.

4   If any of the above checks reveal problems with the seat belt system, replace parts as necessary.

10   Wiper blade inspection and replacement (every 6000 miles or 6 months)

Refer to illustrations 10.5a, 10.5b, 10.5c, 10.7a and 10.7b

1   The windshield wiper and blade assembly should be inspected periodically for damage, loose components and cracked or worn blade elements.

2   Road film can build up on the wiper blades and affect their efficiency, so they should be washed regularly with a mild detergent solution.

3   The action of the wiping mechanism can loosen the bolts, nuts and fasteners, so they should be checked and tightened, as necessary, at the same time the wiper blades are checked.

4   If the wiper blade elements (sometimes called inserts) are cracked, worn or warped, they should be replaced with new ones.

5   Remove the wiper blade assembly from the wiper arm by inserting a small screwdriver into the spring under the release lever while pulling on the blade to release it **(see illustrations)**.

6   With the blade removed from the vehicle, you can remove the rubber element from

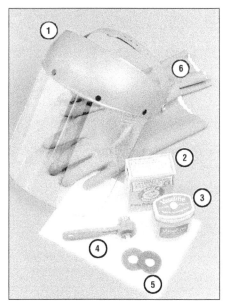

11.1 Tools and materials required for battery maintenance

1   **Face shield/safety goggles** - When removing corrosion with a brush, the acidic particles can easily fly up into your eyes
2   **Baking soda** - A solution of baking soda and water can be used to neutralize corrosion
3   **Petroleum jelly** - A layer of this on the battery terminals will help prevent corrosion
4   **Battery post/cable cleaner** - This wire brush cleaning tool will remove all traces of corrosion from the battery and cable terminals
5   **Treated felt washers** - Placing one of these on each terminal will help prevent corrosion
6   **Rubber gloves** - Another safety item to consider when servicing the battery; remember that's acid inside the battery!

the blade.
7   Using pliers, pinch the metal backing or pull out the retaining clip of the element **(see illustrations)**, then slide the element out of the blade assembly.
8   Compare the new element with the old for length, design, etc.
9   Slide the new element into place. It will automatically lock at the correct location.
10   Reinstall the blade assembly on the arm, wet the windshield glass and test for proper operation.

11   Battery check and maintenance (every 6000 miles or 6 months)

Refer to illustrations 11.1, 11.7a, 11.7b and 11.7c

Warning: *Certain precautions must be followed when checking and servicing the battery. Hydrogen gas, which is highly flammable,*

11.7a A tool like this one (available at auto parts stores) is used to clean the side terminal type battery contact area

is always present in the battery cells, so keep lighted tobacco and all other open flames and sparks away from the battery. The electrolyte inside the battery is actually dilute sulfuric acid, which will cause injury if splashed on your skin or in your eyes. It will also ruin clothes and painted surfaces. When removing the battery cables, always detach the negative cable first and hook it up last!

1   A routine preventive maintenance program for the battery in your vehicle is the only way to ensure quick and reliable starts. But before performing any battery maintenance, make sure that you have the proper equipment necessary to work safely around the battery **(see illustration)**.
2   There are also several precautions that should be taken whenever battery maintenance is performed. Before servicing the battery, always turn the engine and all accessories off and disconnect the cable from the negative terminal of the battery.
3   The battery produces hydrogen gas, which is both flammable and explosive. Never create a spark, smoke or light a match around the battery. Always charge the battery in a ventilated area.
4   Electrolyte contains poisonous and corrosive sulfuric acid. Do not allow it to get in your eyes, on your skin on your clothes. Never ingest it. Wear protective safety glasses when working near the battery. Keep children away from the battery.
5   Note the external condition of the battery. Look for any corroded or loose connections, cracks in the case or cover or loose hold-down clamps. Also check the entire length of each cable for cracks and frayed conductors.
6   If corrosion, which looks like white, fluffy deposits is evident, particularly around the terminals, the battery should be removed for cleaning. Loosen the cable clamp bolts with a wrench, being careful to remove the ground cable first, and detach the cables from the battery. Then disconnect the hold-down clamp bolt, remove the clamp and lift the battery from the engine compartment.

11.7b Use the brush to finish the cleaning job

11.7c The result should be a clean, shiny terminal area

7   Clean the cable terminals thoroughly with a battery terminal tool and brush or a terminal cleaner and a solution of warm water and baking soda. Wash the terminals and the top of the battery case with the same solution but make sure that the solution doesn't get into the battery. When cleaning the cables, terminals and battery top, wear safety goggles and rubber gloves to prevent any solution from coming in contact with your eyes or hands. Wear old clothes too - even diluted, sulfuric acid splashed onto clothes will burn holes in them. If the terminals have been extensively corroded, clean them up with a terminal cleaner tool **(see illustrations)**. Thoroughly wash all cleaned areas with plain water.
8   Make sure that the battery tray is in good condition and the clamp bolts are tight. If the battery is removed from the tray, make sure no parts remain in the bottom of the tray when the battery is reinstalled. When reinstalling the hold-down clamp bolts, do not overtighten them.
9   Information on removing and installing the battery can be found in Chapter 5. Information on jump starting can be found at the front of this manual. For more detailed battery checking procedures, refer to the *Haynes Automotive Electrical Manual*.

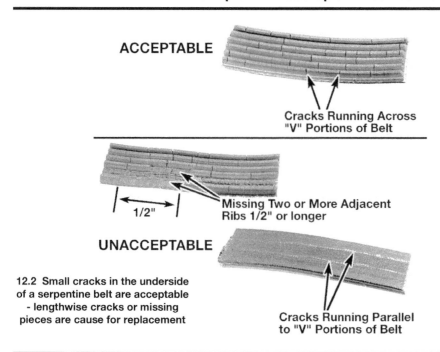

ACCEPTABLE

Cracks Running Across
"V" Portions of Belt

1/2"

Missing Two or More Adjacent
Ribs 1/2" or longer

UNACCEPTABLE

Cracks Running Parallel
to "V" Portions of Belt

**12.2 Small cracks in the underside
of a serpentine belt are acceptable
- lengthwise cracks or missing
pieces are cause for replacement**

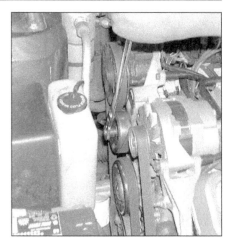

**12.5a Using a wrench or socket and ratchet,
rotate the V6 engine drivebelt tensioner
counterclockwise to release tension
on the belt**

Cleaning

10   Corrosion on the hold-down components, battery case and surrounding areas
can be removed with a solution of water and
baking soda. Thoroughly rinse all cleaned
areas with plain water.

11   Any metal parts of the vehicle damaged
by corrosion should be covered with a zinc-
based primer, then painted.

Charging

Warning: *When batteries are being charged,
hydrogen gas, which is very explosive and
flammable, is produced. Do not smoke or
allow open flames near a charging or a
recently charged battery. Wear eye protection
when near the battery during charging. Also,
make sure the charger is unplugged before
connecting or disconnecting the battery from
the charger.*

12   Slow-rate charging is the best way to
restore a battery that's discharged to the
point where it will not start the engine. It's
also a good way to maintain the battery
charge in a vehicle that's only driven a few
miles between starts. Maintaining the battery
charge is particularly important in the winter
when the battery must work harder to start
the engine and electrical accessories that
drain the battery are in greater use.

13   It's best to use a one or two-amp battery charger (sometimes called a "trickle"
charger). They are the safest and put the
least strain on the battery. They are also the
least expensive. For a faster charge, you can
use a higher amperage charger, but don't use
one rated more than 1/10th the amp/hour rating of the battery. Rapid boost charges that
claim to restore the power of the battery in
one to two hours are hardest on the battery
and can damage batteries not in good condi-

tion. This type of charging should only be
used in emergency situations.

14   The average time necessary to charge a
battery should be listed in the instructions
that come with the charger. As a general rule,
a trickle charger will charge a battery in 12 to
16 hours.

12   Drivebelt check and replacement (every 6000 miles or 6 months)

Drivebelt

*Refer to illustrations 12.2, 12.5a, 12.5b and
12.5c*

1   A single serpentine drivebelt is located at
the front of the engine and plays an important
role in the overall operation of the engine and its
components. Due to its function and material
make up, the belt is prone to wear and should
be periodically inspected. The serpentine belt
drives the alternator, power steering pump,
water pump and air conditioning compressor.

**12.5b V8 engine drivebelt tensioner location - rotate it clockwise
to release belt tension**

**12.5c Most models have a drivebelt routing decal on the upper
radiator panel**

2   With the engine off, open the hood and use your fingers (and a flashlight, if necessary), to move along the belt checking for cracks and separation of the belt plies. Also check for fraying and glazing, which gives the belt a shiny appearance **(see illustration)**. Both sides of the belt should be inspected, which means you will have to twist the belt to check the underside.

3   Check the ribs on the underside of the belt. They should all be the same depth, with none of the surface uneven.

4   The tension of the belt is maintained by the tensioner assembly and isn't adjustable. The belt should be checked at the mileage specified in the maintenance schedule at the front of this Chapter, if the belt shows noticeable damage or wear during these checks it should be replaced.

5   Rotate the tensioner away from the belt to release belt tension **(see illustrations)**. **Note:** *These models have a drivebelt routing decal on the engine to help during drivebelt installation* **(see illustration)**. *If the decal is missing, make a sketch.*

6   Remove the belt from the pulleys components and slowly release the tensioner.

7   Route the new belt over the various pulleys, again rotating the tensioner to allow the belt to be installed, then release the belt tensioner. Make sure the belt is positioned properly on all of the pulleys.

13   Underhood hose check and replacement (every 6000 miles or 6 months)

General

Refer to illustration 13.1

1   **Caution:** *Replacement of air conditioning hoses must be left to a dealer service department or air conditioning shop that has the equipment to depressurize the system safely. Never remove air conditioning components or hoses* **(see illustration)** *until the system has been depressurized.*

2   High temperatures under the hood can cause the deterioration of the rubber and plastic hoses used for engine, accessory and emission systems operation. Periodic inspection should be made for cracks, loose clamps, material hardening and leaks. Information specific to the cooling system hoses can be found in Section 14.

3   Some, but not all, hoses are secured to the fittings with clamps. Where clamps are used, check to be sure they haven't lost their tension, allowing the hose to leak. If clamps aren't used, make sure the hose hasn't expanded and/or hardened where it slips over the fitting, allowing it to leak.

Vacuum hoses

4   It's quite common for vacuum hoses, especially those in the emissions system, to be color coded or identified by colored stripes molded into each hose. Various systems require hoses with different wall thicknesses, collapse resistance and temperature resistance. When replacing hoses, be sure the new ones are made of the same material.

5   Often the only effective way to check a hose is to remove it completely from the vehicle. If more than one hose is removed, be sure to label the hoses and fittings to ensure correct installation.

6   When checking vacuum hoses, be sure to include any plastic T-fittings in the check. Inspect the fittings for cracks and the hose where it fits over the fitting for distortion, which could cause leakage.

7   A small piece of vacuum hose (1/4-inch inside diameter) can be used as a stethoscope to detect vacuum leaks. Hold one end of the hose to your ear and probe around vacuum hoses and fittings, listening for the "hissing" sound characteristic of a vacuum leak. **Warning:** *When probing with the vacuum hose stethoscope, be careful not to allow your body or the hose to come into contact with moving engine components such as the drivebelt, cooling fan, etc.*

Fuel hose

Warning: *There are certain precautions which must be taken when inspecting or servicing fuel system components. Work in a* well ventilated area and do not allow open flames (cigarettes, appliance pilot lights, etc.) or bare light bulbs near the work area. Mop up any spills immediately and do not store fuel soaked rags where they could ignite. The fuel system is under pressure, so if any fuel lines must be disconnected, the pressure in the system must be relieved first (see Chapter 4 for more information).

8   Check all rubber fuel lines for deterioration and chafing. Check especially for cracks in areas where the hose bends and just before fittings, such as where a hose attaches to the fuel filter and fuel injection unit.

9   High quality fuel line meeting original equipment specifications must be used for fuel line replacement. Never, under any circumstances, use unreinforced vacuum line, clear plastic tubing or water hose for fuel lines.

10   Spring-type clamps are commonly used on fuel lines. These clamps often lose their tension over a period of time, and can be "sprung" during the removal process. Therefore, spring-type clamps should be replaced with screw-type clamps whenever a hose is replaced.

Metal lines

11   Sections of steel tubing are often used for fuel line between the fuel pump and engine compartment. Check carefully for cracks, kinks and flat spots in the line.

12   If a section of metal fuel line must be replaced, only seamless steel tubing should be used, since copper and aluminum tubing do not have the strength necessary to withstand normal engine vibration.

13   Check the metal brake lines where they enter the master cylinder and brake proportioning unit (if used) for cracks in the lines and loose fittings. Any sign of brake fluid leakage calls for an immediate thorough inspection of the brake system.

14   Cooling system check (every 6000 miles or 6 months)

Refer to illustration 14.4

1   Many major engine failures can be attributed to a faulty cooling system. The cooling system also cools the transaxle fluid and plays an important role in prolonging transaxle life.

2   The cooling system should be checked with the engine cold. Do this before the vehicle is driven for the day or after the engine has been shut off for at least three hours.

3   Remove the radiator cap by turning it to the left until it reaches a stop. If you hear any hissing sounds (indicating there is still pressure in the system), wait until it stops. Now press down on the cap with the palm of your hand and continue turning to the left until the cap can be removed. Thoroughly clean the cap, inside and out, with clean water. Also clean the filler neck on the radiator. All traces

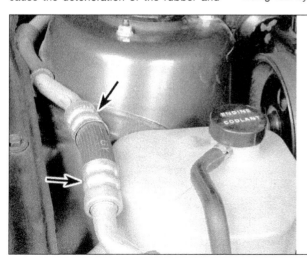

13.1 Air conditioning hoses are easily identified by the swaged connections (arrows) - DO NOT disconnect or accidentally damage the air conditioning hoses (the system is under high pressure)

Check for a chafed area that could fail prematurely.

Check for a soft area indicating the hose has deteriorated inside.

Overtightening the clamp on a hardened hose will damage the hose and cause a leak.

Check each hose for swelling and oil-soaked ends. Cracks and breaks can be located by squeezing the hose

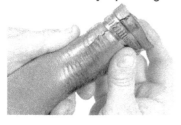

14.4 Hoses, like drivebelts, have a habit of failing at the worst possible time - to prevent the inconvenience of a blown radiator or heater hose, inspect them carefully as shown here

of corrosion should be removed. The coolant inside the radiator should be relatively transparent. If it is rust colored, the system should be drained and refilled (Section 26). If the coolant level is not up to the top, add additional antifreeze/coolant mixture (Section 4).

4   Carefully check the large upper and lower radiator hoses along with any smaller diameter heater hoses which run from the engine to the firewall. Inspect each hose along its entire length, replacing any hose that is cracked, swollen or shows signs of deterioration. Cracks may become more apparent if the hose is squeezed **(see illustration)**.

5   Make sure all hose connections are tight. A leak in the cooling system will usually show up as white or rust colored deposits on the areas adjoining the leak. If wire-type clamps are used at the ends of the hoses, it

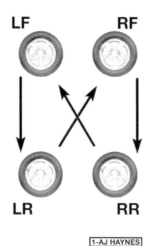

16.2 Tire rotation diagram

may be wise to replace them with more secure screw-type clamps.

6   Use compressed air or a soft brush to remove bugs, leaves, etc. from the front of the radiator or air conditioning condenser. Be careful not to damage the delicate cooling fins or cut yourself on them.

7   Every other inspection, or at the first indication of cooling system problems, have the cap and system pressure tested. If you don't have a pressure tester, most gas stations and repair shops will do this for a minimal charge.

15   Park/neutral switch check (every 6000 miles or 6 months)

Warning: *During the following checks there's a chance the vehicle could lunge forward, possibly causing damage or injuries. Allow plenty of room around the vehicle, apply the parking brake and hold down the regular brake pedal during the checks.*

1   Try to start the engine in each gear. The engine should crank only in Park or Neutral. If it starts in any other position, replace the switch (see Chapter 7).

2   Also make sure the steering column lock allows the key to go into the Lock position only when the shift lever is in Park.

3   The ignition key should come out only in the Lock position.

16   Tire rotation (every 6000 miles or 6 months)

Refer to illustration 16.2

1   The tires should be rotated at the specified intervals and whenever uneven wear is noticed.

2   Refer to the **accompanying illustration** for the preferred tire rotation pattern.

3   Refer to the information in *Jacking and*

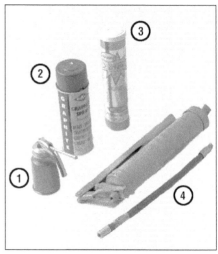

17.1 Materials required for chassis and body lubrication

1   **Engine oil** - *Light engine oil in a can like this can be used for door and hood hinges*

2   **Graphite spray** - *Used to lubricate lock cylinders*

3   **Grease** - *Grease, in a variety of types and weights, is available for use in a grease gun. Check the Specifications for your requirements*

4   **Grease gun** - *A common grease gun, shown here with a detachable hose and nozzle, is needed for chassis lubrication. After use, clean it thoroughly*

towing at the front of this manual for the proper procedures to follow when raising the vehicle and changing a tire. If the brakes are to be checked, don't apply the parking brake as stated. Make sure the tires are blocked to prevent the vehicle from rolling as it's raised. **Note:** *Loosen the wheel lug nuts 1/2-turn before raising the vehicle.*

4   Preferably, the entire vehicle should be raised at the same time. This can be done on a hoist or by jacking up each corner and then lowering the vehicle onto jackstands placed under the frame rails. Always use four jackstands and make sure the vehicle is safely supported.

5   After rotation, check and adjust the tire pressures as necessary and be sure to properly tighten the lug nuts.

17   Chassis lubrication (every 6000 miles or 6 months)

Refer to illustrations 17.1 and 17.6

1   Refer to *Recommended lubricants and fluids* at the front of this Chapter to obtain the necessary lubricants. You'll also need a grease gun **(see illustration)**. Occasionally plugs will be installed rather than grease fittings. If so, grease fittings will have to be purchased and installed.

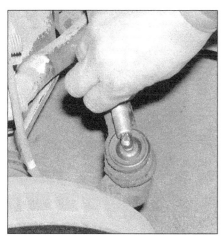

17.6 After cleaning the grease fitting, push the gun nozzle firmly into place and pump the grease into the component (usually about two pumps will be sufficient)

2   Look under the vehicle and see if grease fittings or plugs are installed in the balljoints and tie-rod ends. If there are plugs, remove them and buy grease fittings, which will thread into the component. An auto parts store will be able to supply the correct fittings. Straight, as well as angled, fittings are available.

3   For easier access under the vehicle, raise it with a jack and place jackstands under the frame. Make sure it's securely supported by the stands. If the wheels are being removed at this interval for rotation or brake inspection, loosen the lug nuts slightly while the vehicle is still on the ground.

4   Before beginning, force a little grease out of the nozzle to remove any dirt from the end of the gun. Wipe the nozzle clean with a rag.

5   With the grease gun and plenty of clean rags, crawl under the vehicle and begin lubricating the components **(see the under-vehicle photos at the beginning of this Chapter).**

6   Wipe off the grease fitting and push the nozzle firmly over it **(see illustration)**. Squeeze the trigger on the grease gun to force grease into the component. The balljoints and tie-rod ends should be lubricated until each rubber seal is firm to the touch. Do not pump too much grease into the fitting or it could rupture the seal. If the grease escapes around the grease gun nozzle, the fitting is clogged or the nozzle isn't completely seated on the fitting. Resecure the gun nozzle to the fitting and try again. If necessary, replace the fitting with a new one.

7   Wipe the excess grease off the components and the grease fitting. Repeat the procedure for the remaining fittings.

8   Lubricate the shift linkage with a little multi-purpose grease. While you are under the vehicle, clean and lubricate the parking brake cable along with the cable guides and levers. This can be done by smearing some of the chassis grease onto the cable and its related parts with your fingers.

9   Open the hood and smear a little chassis grease on the hood latch mechanism. Have an assistant pull the hood release lever from inside the vehicle as you lubricate the cable at the latch.

10   Lubricate all the hinges (door, hood, etc.) with engine oil.

11   The key lock cylinders can be lubricated with spray-on graphite or silicone lubricant, which is available at auto parts stores. **Caution:** *The manufacturer doesn't recommend using oil in black plastic lock cylinders - it could damage them by washing out the factory-applied lubricant.*

12   Lubricate the door weatherstripping with silicone spray. This will reduce chafing and retard wear.

---

18   Steering, suspension and driveaxle boot check (every 6000 miles or 6 months)

Note: *The steering linkage and suspension components should be checked periodically.*

Worn or damaged suspension and steering linkage components can result in excessive and abnormal tire wear, poor ride quality and vehicle handling, and reduced fuel economy. For detailed illustrations of the steering and suspension components, refer to Chapter 10.

Shock absorber check

1   Park the vehicle on level ground, turn the engine off and set the parking brake. Check the tire pressures.

2   Push down at one corner of the vehicle, then release it while noting the movement of the body. It should stop moving and come to rest in a level position within one or two bounces.

3   If the vehicle continues to move up-and-down or if it fails to return to its original position, a worn or weak shock absorber is probably the reason.

4   Repeat the above check at each of the three remaining corners of the vehicle.

5   Raise the vehicle and support it securely on jackstands.

6   Check the front struts and rear shock absorbers for evidence of fluid leakage. A light film of fluid is no cause for concern. Make sure that any fluid noted is from the shocks and not from some other source. If leakage is noted, replace the shocks as a set.

7   Check the shocks to be sure that they are securely mounted and undamaged. Check the upper mounts for damage and wear. If damage or wear is noted, replace the shocks as a set (front or rear).

8   If the shocks must be replaced, refer to Chapter 10 for the procedure.

Steering and suspension check

Refer to illustrations 18.9a, 18.9b, 18.9c, 18.9d and 18.11

9   Visually inspect the steering and suspension components for damage and distortion. Look for damaged seals, boots and bushings and leaks of any kind **(see illustrations).**

18.9a Inspect the lower control arm bushings and balljoint (arrows) . . .

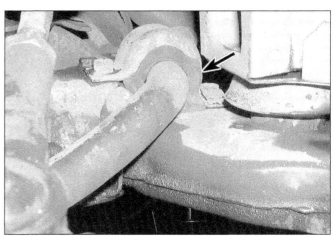

18.9b . . . the front and rear stabilizer bar bushings (arrow) . . .

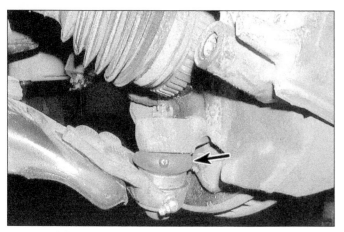

18.9c ... and the tie-rod ends and balljoint boots

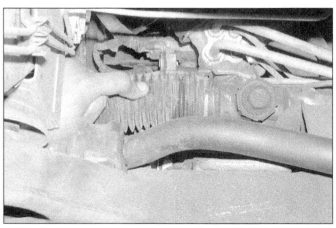

18.9d Check the steering gear boots for cracks and leaking steering fluid (if fluid is present, the rack seals are leaking)

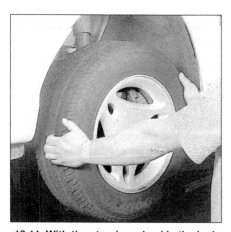

18.11 With the steering wheel in the lock position and the vehicle raised, grasp the front tire as shown and try to move it back-and-forth - if any play is noted, check the steering gear mounts and tie rod ends for looseness

18.14 Check the driveaxle boots for cracks and/or leaking grease

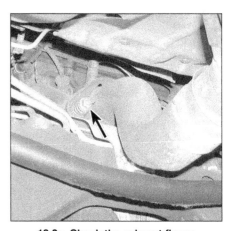

19.2a Check the exhaust flange (arrow) connections ...

10 Clean the lower end of the steering knuckle. Have an assistant grasp the lower edge of the tire and move the wheel in-and-out while you look for movement at the steering knuckle-to-control arm balljoint. If there is any movement the suspension balljoint(s) must be replaced.

11 Grasp each front tire at the front and rear edges, push in at the front, pull out at the rear and feel for play in the steering system components. If any freeplay is noted, check the steering gear mounts and the tie-rod ends for looseness **(see illustration)**.

12 Additional steering and suspension system information and illustrations can be found in Chapter 10.

Driveaxle boot check)

Refer to illustration 18.14

13 The driveaxle boots are very important because they prevent dirt, water and foreign material from entering and damaging the constant velocity (CV) joints. Oil and grease can cause the boot material to deteriorate prematurely, so it's a good idea to wash the

boots with soap and water. Because it constantly pivots back and forth following the steering action of the front hub, the outer CV boot wears out sooner and should be inspected regularly.

14 Inspect the boots for tears and cracks as well as loose clamps **(see illustration)**. If there is any evidence of cracks or leaking lubricant, they must be replaced as described in Chapter 8.

19 Exhaust system check (every 6000 miles or 6 months)

Refer to illustrations 19.2a, 19.2b and 19.2c

1 With the engine cold (at least three hours after the vehicle has been driven), check the complete exhaust system from the engine to the end of the tailpipe. Ideally, the inspection should be done with the vehicle on a hoist to permit unrestricted access. If a hoist isn't available, raise the vehicle and support it securely on jackstands.

2 Check the exhaust pipes and connections for evidence of leaks, severe corrosion and damage. Make sure that all brackets and hangers are tight and in good condition **(see illustrations)**.

3 At the same time, inspect the underside of the body for holes, corrosion, open seams, etc. which may allow exhaust gases to enter the passenger compartment. Seal all body openings with silicone or body putty.

4 Rattles and other noises can often be traced to the exhaust system, especially the mounts and hangers. Try to move the pipes, muffler and catalytic converter. If the components can come in contact with the body or

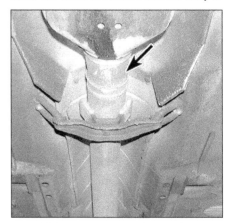

19.2b ... and the exhaust pipe connections (arrow) for exhaust leaks

19.2c Check each exhaust system hanger (arrow) for damage and cracks

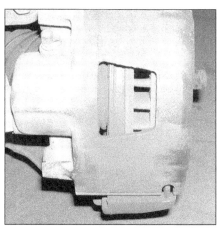

20.6 You will find an inspection hole like this in each caliper - placing a ruler across the hole should enable you to determine the thickness of remaining pad material on the inner pad

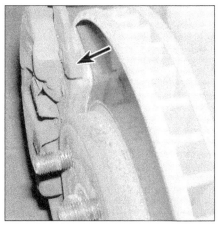

20.7 Check the end of the outer pad to determine its thickness - most models have a pad wear indicator (arrow) that makes a squealing noise when the pad is worn

suspension parts, secure the exhaust system with new mounts.

5   Check the running condition of the engine by inspecting inside the end of the tailpipe. The exhaust deposits here are an indication of engine state-of-tune. If the pipe is black and sooty or coated with white deposits, the engine may need a tune-up, including a thorough fuel system inspection and adjustment.

20   Brake check (every 6000 miles or 6 months)

Warning: *The dust created by the brake system may contain asbestos, which is harmful to your health. Never blow it out with compressed air and don't inhale any of it. An approved filtering mask should be worn when working on the brakes. Do not, under any circumstances, use petroleum-based solvents to clean brake parts. Use brake system cleaner only! Try to use non-asbestos replacement parts whenever possible.*
Note: *For detailed photographs of the brake system, refer to Chapter 9.*
1   In addition to the specified intervals, the brakes should be inspected every time the wheels are removed or whenever a defect is

suspected.
2   Any of the following symptoms could indicate a potential brake system defect: The vehicle pulls to one side when the brake pedal is depressed; the brakes make squealing or dragging noises when applied; brake pedal travel is excessive; the pedal pulsates; brake fluid leaks, usually onto the inside of the tire or wheel.
3   Loosen the wheel lug nuts.
4   Raise the vehicle and place it securely on jackstands.
5   Remove the wheels (see *Jacking and towing* at the front of this manual, or your owner's manual, if necessary).

Disc brakes
Refer to illustrations 20.6, 20.7, 20.9 and 20.11
6   There are two pads (an outer and an inner) in each caliper. The pads are visible through inspection holes in each caliper **(see illustration)**.
7   Check the pad thickness by looking at each end of the caliper and through the inspection hole in the caliper body **(see illustration)**. If the lining material is less than the

thickness listed in this Chapter's Specifications, replace the pads. **Note:** *Keep in mind that the lining material is riveted or bonded to a metal backing plate and the metal portion is not included in this measurement.*
8   If it is difficult to determine the exact thickness of the remaining pad material by the above method, or if you are at all concerned about the condition of the pads, remove the caliper(s), then remove the pads from the calipers for further inspection (refer to Chapter 9).
9   Once the pads are removed from the calipers, clean them with brake cleaner and re-measure them with a ruler or a vernier caliper **(see illustration)**.
10   Measure the disc thickness with a micrometer to make sure that it still has service life remaining. If any disc is thinner than the specified minimum thickness, replace it (refer to Chapter 9). Even if the disc has service life remaining, check its condition. Look for scoring, gouging and burned spots. If these conditions exist, remove the disc and have it resurfaced (see Chapter 9).
11   Before installing the wheels, check all brake lines and hoses for damage, wear,

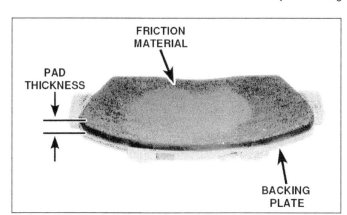

20.9 Measure the pad thickness to determine how much friction material remains on the brake backing plate

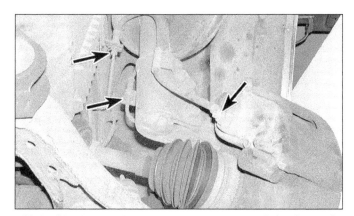

20.11 Check along the brake hoses and at each fitting (arrows) for deterioration and cracks

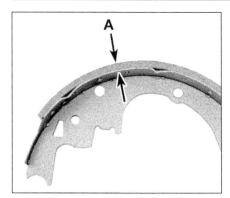

20.15 If the lining is bonded to the brake shoe, measure the lining thickness from the outer surface to the metal shoe, as shown here. If the lining is riveted to the shoe, measure from the lining outer surface to the rivet head

20.16 Typical assembled view of a rear drum brake (left side shown, right side is the exact opposite)

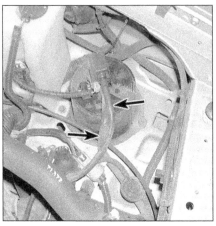

21.9 Inspect the charcoal canister and hoses (arrows) for damage and leaks

deformation, cracks, corrosion, leakage, bends and twists, particularly in the vicinity of the rubber hoses at the calipers **(see illustration)**. Check the clamps for tightness and the connections for leakage. Make sure that all hoses and lines are clear of sharp edges, moving parts and the exhaust system. If any of the above conditions are noted, repair, reroute or replace the lines and/or fittings as necessary (see Chapter 9).

Drum brakes

Refer to illustrations 20.15 and 20.16

12   On rear drum brakes, make sure the parking brake is off then proceed to tap on the outside of the drum with a rubber mallet to loosen it.

13   Remove the brake drums.

14   With the drums removed, carefully clean the brake assembly with brake system cleaner. **Warning:** *Don't blow the dust out with compressed air and don't inhale any of it (it may contain asbestos, which is harmful to your health).*

15   Note the thickness of the lining material on both front and rear brake shoes. If the material has worn away to within 1/16-inch of the recessed rivets or 1/8-inch of the metal backing on bonded type shoes, the shoes should be replaced **(see illustration)**. The shoes should also be replaced if they're cracked, glazed (shiny areas), or covered with brake fluid.

16   Make sure all the brake assembly springs are connected and in good condition **(see illustration)**.

17   Check the brake components for signs of fluid leakage. With your finger or a small screwdriver, carefully pry back the rubber cups on the wheel cylinder located at the top of the brake shoes. Any leakage here is an indication that the wheel cylinders should be overhauled immediately (see Chapter 9). Also, check all hoses and connections for signs of leakage.

18   Clean the inside of the drum with brake system cleaner. Again, be careful not to breathe the dangerous asbestos dust.

19   Check the inside of the drum for cracks, score marks, deep scratches and "hard spots" which will appear as small discolored areas. If imperfections cannot be removed with fine emery cloth, the drum must be taken to an automotive machine shop for resurfacing.

20   Repeat the procedure for the remaining wheel. If the inspection reveals that all parts are in good condition, reinstall the brake drums, install the wheels and lower the vehicle to the ground.

Brake booster check

21   Sit in the driver's seat and perform the following sequence of tests.

22   With the brake fully depressed, start the engine - the pedal should move down a little when the engine starts.

23   With the engine running, depress the brake pedal several times - the travel distance should not change.

24   Depress the brake, stop the engine and hold the pedal in for about 30 seconds - the pedal should neither sink nor rise.

25   Restart the engine, run it for about a minute and turn it off. Then firmly depress the brake several times - the pedal travel should decrease with each application.

26   If your brakes do not operate as described, the brake booster has failed. Refer to Chapter 9 for the replacement procedure.

Parking brake

27   Rear drum brakes utilize a self-adjusting parking brake mechanism and do not require regular scheduled maintenance or routine adjustment. For more detailed information on the parking brake assembly see Chapter 9.

---

21   Fuel system check (every 6000 miles or 6 months)

Refer to illustration 21.9

Warning: *Gasoline is extremely flammable, so take extra precautions when you work on any part of the fuel system. Don't smoke or allow open flames or bare light bulbs near the work area, and don't work in a garage where a natural gas-type appliance (such as a water heater or clothes dryer) with a pilot light is present. Since gasoline is carcinogenic, wear latex gloves when there's a possibility of being exposed to fuel, and, if you spill any fuel on your skin, rinse it off immediately with soap and water. Mop up any spills immediately and do not store fuel-soaked rags where they could ignite. When you perform any kind of work on the fuel system, wear safety glasses and have a Class B type fire extinguisher on hand. The fuel system is under constant pressure, so, before any lines are disconnected, the fuel system pressure must be relieved. See Chapter 4.*

1   If you smell gasoline while driving or after the vehicle has been sitting in the sun, inspect the fuel system immediately.

2   Remove the gas filler cap and inspect if for damage and corrosion. The gasket should have an unbroken sealing imprint. If the gasket is damaged or corroded, install a new cap.

3   Inspect the fuel feed and return lines for cracks. Make sure that the connections between the fuel lines and the fuel injection system and between the fuel lines and the in-line fuel filter are tight. **Warning:** *Your vehicle is fuel injected, so you must relieve the fuel system pressure before servicing fuel system components. The fuel system pressure-relief procedure is outlined in Chapter 4.*

4   If the fuel injectors are visible, look for signs of fuel leakage (wet spots) around any of the injectors, they may need new O-rings (see Chapter 4).

5   Since some components of the fuel system - the fuel tank and part of the fuel feed and return lines, for example - are underneath the vehicle, they can be inspected more easily with the vehicle raised on a hoist. If that's not possible, raise the vehicle and support it on jackstands.

6   With the vehicle raised and safely supported, inspect the gas tank and filler neck for punctures, cracks and other damage. The

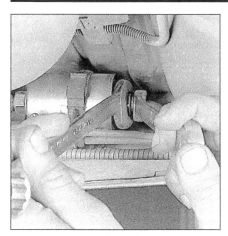

22.3 To remove the fuel filter, use two wrenches to unscrew the fuel line fittings (early models)

connection between the filler neck and the tank is particularly critical. Sometimes a rubber filler neck will leak because of loose clamps or deteriorated rubber. Inspect all fuel tank mounting brackets and straps to be sure that the tank is securely attached to the vehicle. **Warning:** *Do not, under any circumstances, try to repair a fuel tank (except rubber components). A welding torch or any open flame can easily cause fuel vapors inside the tank to explode.*

7   Carefully check all rubber hoses and metal lines leading away from the fuel tank. Check for loose connections, deteriorated hoses, crimped lines and other damage. Repair or replace damaged sections as necessary (see Chapter 4).

8   The evaporative emissions control system can also be a source of fuel odors. The function of the system is to store fuel vapors from the fuel tank in a charcoal canister until they can be routed to the intake manifold where they mix with incoming air before being burned in the combustion chambers.

9   The most common symptom of a faulty evaporative emissions system is a strong odor of fuel in the engine compartment. If a fuel odor has been detected, and you have already checked the areas described above, check the charcoal canister, located in the engine compartment, and the hoses connected to it **(see illustration)**.

22   Fuel filter replacement (every 30,000 miles or 24 months)

Refer to illustration 22.3
Warning: *Gasoline is extremely flammable, so extra precautions must be taken when working on any part of the fuel system. See the* **Warning** *in Section 21.*

Earlier models

1   Relieve the fuel system pressure (see Chapter 4).
2   Raise the vehicle and support it securely on jackstands.

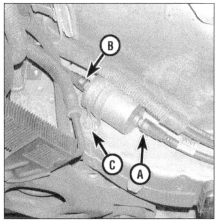

22.11 Squeeze the white plastic quick-disconnect tabs together and pull the inlet line (A) away from the filter, then detach the outlet line (B) from the fuel filter - unbolt the filter mounting bracket (C) to remove the fuel filter (later models)

3   Unscrew the fuel line-to-fuel filter fittings **(see illustration)**. Use a back-up wrench on the filter to keep from twisting the line.
4   Remove the filter from the clip or mount.
5   Install the new filter securely into the clip or mount. Make sure the arrow on the filter points toward the engine.
6   Reattach the fuel lines to the filter, using new O-ring seals. **Note:** *The fuel line and/or filter must be replaced with new ones if they are scratched or damaged during installation.*
7   Start the engine and check for fuel leaks.

Later models

8   Relieve the fuel system pressure (see Chapter 4).
9   Raise the vehicle and support it securely on jackstands.
10   Twist the connector 1/4 turn in each direction to loosen any debris within the connector, then blow out any dirt or debris from the connector. Squeeze the plastic tabs on the male end of the connector and pull the connector apart. Repeat for the other connector.
11   Remove the filter mounting bolt, unhook the tab from the receptacle in the frame and remove the filter **(see illustration)**.
12   Wipe off the male end of both connec-

tors, then apply a drop of clean engine oil onto the fuel filter male connectors where they are inserted into the female connector. Position the new filter so the arrow on the filter points toward the engine.
13   Insert the fuel line female connectors onto the fuel filter's male connectors and push them on until the plastic tabs snap and lock into place.
14   Pull on the female end of each connector to make sure it is securely locked into place.
15   Move the filter into place and install the bolt.
16   Start the engine and check for fuel leaks.

23   Automatic transaxle fluid and filter change (every 30,000 miles or 24 months)

Refer to illustrations 23.7, 23.10a, 23.10b and 23.12
1   At the specified time intervals, the transaxle fluid should be drained and replaced. Since the fluid will remain hot long after driving, perform this procedure only after everything has cooled down completely.
2   Before beginning work, purchase the specified transaxle fluid (see *Recommended lubricants and fluids* at the front of this Chapter) and a new filter.
3   Other tools necessary for this job include jackstands to support the vehicle in a raised position, a drain pan capable of holding several quarts, newspapers and clean rags.
4   Raise and support the vehicle on jackstands.
5   With a drain pan in place, remove the front and side transaxle pan mounting bolts.
6   Loosen the rear pan bolts one turn.
7   Carefully pry the transaxle pan loose with a screwdriver, allowing the fluid to drain **(see illustration)**.
8   Remove the remaining bolts, pan and gasket. Carefully clean the gasket surface of the transaxle to remove all traces of the old gasket and sealant.
9   Drain the fluid from the transaxle pan, clean the pan with solvent and dry it with compressed air. Be careful not to lose the magnet.

23.7 After removing the front and side pan bolts, loosen the rear bolts and allow the fluid to drain, then remove the bolts and lower the pan from the vehicle

23.10a Pull the filter straight down to remove it

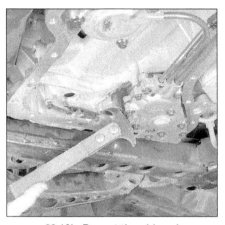

23.10b Pry out the old seal

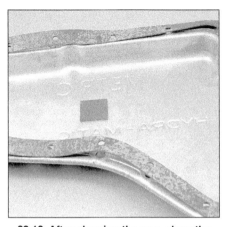

23.12 After cleaning the pan, place the magnet in position and install the gasket

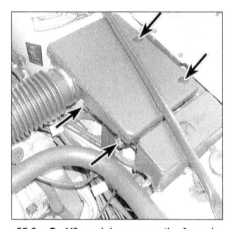

25.3a On V6 models, remove the four air cleaner cover screws (arrows) and detach the cover . . .

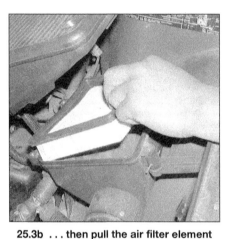

25.3b . . . then pull the air filter element out of the housing

25.4 Detach the clip (upper arrow), remove the nuts (lower arrows) and lift out the air filter element

10 Remove the filter and pry out the seal **(see illustrations)**.

11 Push a new filter seal fully into its bore, then install the new filter.

12 Make sure the gasket surface on the transaxle pan is clean, then install the magnet and a new gasket **(see illustration)**. Put the pan in place against the transaxle and install the bolts. Working around the pan, tighten each bolt a little at a time until the final torque figure is reached.

13 Lower the vehicle and add the specified amount of automatic transmission fluid through the filler tube (see Section 7).

14 With the shift lever in Park and the parking brake set, run the engine at a fast idle, but don't race it.

15 Move the shift lever through each gear and back to Park. Check the fluid level.

16 Check under the vehicle for leaks during the first few trips.

24 Brake fluid change (every 30,000 miles or 24 months)

Warning: *Brake fluid can harm your eyes and damage painted surfaces, so use extreme caution when handling or pouring it. Do not*

use brake fluid that has been standing open or is more than one year old. Brake fluid absorbs moisture from the air. Excess moisture can cause a dangerous loss of braking effectiveness.

1 At the specified intervals, the brake fluid should be drained and replaced. Since the brake fluid may drip or splash when pouring it, place plenty of rags around the master cylinder to protect any surrounding painted surfaces.

2 Before beginning work, purchase the specified brake fluid (see *Recommended lubricants and fluids* at the beginning of this Chapter).

3 Remove the cap from the master cylinder reservoir.

4 Using a hand-held suction pump or similar device, withdraw the fluid from the master cylinder reservoir.

5 Add new fluid to the master cylinder until it rises to the base of the filler neck.

6 Bleed the brake system as described in Chapter 9 at all four brakes until new and uncontaminated fluid is expelled from the bleeder screw. Be sure to maintain the fluid level in the master cylinder as you perform the bleeding process. If you allow the master cylinder to run dry, air will enter the system.

7 Refill the master cylinder with fluid and check the operation of the brakes. The pedal should feel solid when depressed, with no sponginess. **Warning:** *Do not operate the vehicle if you are in doubt about the effectiveness of the brake system.*

25 Air filter replacement (every 30,000 miles or 24 months)

Refer to illustrations 25.3a, 25.3b and 25.4

1 At the specified intervals, the air filter should be replaced with a new one On some V8 models, the PCV filter should be replaced as well.

V6 models

2 It may be necessary to loosen the engine compartment brace rear bolt, remove the front bolt and move the brace out of the way for access to the air cleaner. Be sure to reinstall the bolt and tighten both bolts securely after replacing the filter element.

3 Remove the screws, lift the top cover off and withdraw the air filter **(see illustrations)**.

V8 models

4 Detach the hose clamp and unhook the

26.5 Push the radiator cap downward and rotate it counterclockwise to remove it - never remove it when the engine is hot!

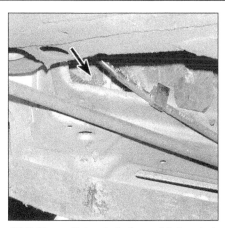

26.6 The radiator drain (arrow) is located at the lower corner of the radiator (viewed from below)

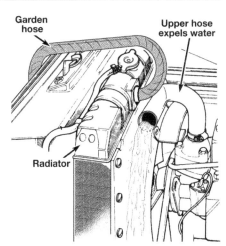

26.13 With the thermostat removed, disconnect the upper radiator hose and flush the radiator and engine block with a garden hose

air intake duct from the throttle body. Remove the nuts, release the clips, then remove the air cleaner cover **(see illustration)**. Withdraw the air filter.

5   On HT4100 models with a PCV filter, remove the air filter housing assembly. Detach the hose and remove the PCV filter. Install a new PCV filter, connect the hose and install the air filter housing.

6   While the filter housing cover is off, be careful not to drop anything down into the air duct or air filter assembly.

7   Wipe out the inside of the air filter housing with a clean rag.

8   Place a new air filter in the air filter housing. Make sure it seats properly in the bottom of the housing.

9   Install the cover and retaining nuts and connect the clips securely.

---

26 Cooling system servicing (draining, flushing and refilling) (every 30,000 miles or 24 months)

Refer to illustrations 26.5, 26.6 and 26.13
Warning: *Do not allow antifreeze to come in contact with your skin or painted surfaces of the vehicle. Rinse off spills immediately with plenty of water. Antifreeze is highly toxic if ingested. Never leave antifreeze lying around in an open container or in puddles on the floor; children and pets are attracted by it's sweet smell and may drink it. Check with local authorities about disposing of used antifreeze. Many communities have collection centers which will see that antifreeze is disposed of safely.*

1   Periodically, the cooling system should be drained, flushed and refilled to replenish the antifreeze mixture and prevent formation of rust and corrosion, which can impair the performance of the cooling system and cause engine damage.

2   At the same time the cooling system is serviced, all hoses and the radiator cap should be inspected and replaced if defective (see Section 14).

3   Since antifreeze is poisonous, be careful not to spill any of the coolant mixture on your skin. Also, antifreeze will damage paint. If antifreeze contacts your skin or the vehicle's paint, rinse it off immediately with plenty of clean water. Consult local authorities about the dumping of antifreeze before draining the cooling system. In many areas, reclamation centers have been set up to collect automobile oil and drained antifreeze/water mixtures, rather than allowing them to be added to the sewage system.

Draining

Warning: *Wait until the engine is completely cool before beginning this procedure.*

4   Apply the parking brake and block the wheels. If the vehicle has just been driven, wait several hours to allow the engine to cool down before beginning this procedure.

5   Once the engine is completely cool, remove the radiator cap and the reservoir cap **(see illustration)**.

6   Drain the radiator by opening the drain valve at the bottom of the radiator **(see illustration)**. If the valve is corroded and can't be turned easily, or if the radiator isn't equipped with a drain valve, disconnect the lower radiator hose to allow the coolant to drain. Be careful not to get antifreeze on your skin or in your eyes.

7   After the coolant stops flowing out of the radiator, remove the lower radiator hose (if not done already) and allow the remaining coolant in the engine block to drain.

8   While the coolant is draining from the engine block, disconnect the hose from the coolant reservoir and remove the reservoir (see Chapter 3 if necessary). Flush the reservoir out with water until it's clean, and if necessary, wash the inside with soapy water and a brush to make reading the fluid level easier.

9   While the coolant is draining, check the condition of the radiator hoses, heater hoses and clamps (refer to Section 14 if necessary).

10   Replace any damaged clamps or hoses.

Flushing

11   Once the system is completely drained, remove the thermostat from the engine (see

Chapter 3). Then reinstall the thermostat housing without the thermostat. This will allow the system to be flushed.

12   Reinstall the radiator hoses and tighten the radiator drain plug.

13   Disconnect the upper radiator hose from the radiator, then place a garden hose in the upper radiator inlet and flush the system until the water runs clear out of the upper radiator hose **(see illustration)**.

14   In severe cases of contamination or clogging of the radiator, remove the radiator (see Chapter 3) and have a radiator repair facility clean and repair it if necessary.

15   Many deposits can be removed by the chemical action of a cleaner available at auto parts stores. Follow the procedure outlined in the manufacturer's instructions. **Note:** *When the coolant is regularly drained and the system refilled with the correct antifreeze/water mixture, there should be no need to use chemical cleaners or descalers.*

Refilling

16   To refill the system, install the thermostat, reconnect any radiator hoses and install the reservoir and the overflow hose.

17   Place the heater temperature control in the maximum heat position.

18   Fill the cooling system with a 50/50 mixture of antifreeze and water through the radiator filler neck. **Note:** *A special cooling system sealant must be added to Cadillac V8 engines. Check with your local auto parts store or dealer parts department for the proper sealant to use.* Slowly fill the radiator to the base of the filler neck. Add coolant to the reservoir until it reaches the FULL COLD mark. Wait five minutes and recheck the coolant level in the radiator, adding if necessary.

19   Leave the radiator cap off and run the engine in a well-ventilated area until the thermostat opens (coolant will begin flowing through the radiator hose will become hot).

20   Turn the engine off and let it cool. Add

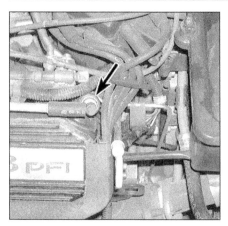

27.1 On Cadillac V8 engines the PCV valve plugs into the valve cover (arrow)

27.2 Remove the PCV valve, then feel for suction at the end of the valve and shake it, listening for a clicking sound

27.9 Remove the plastic nut (arrow), grasp the fuel injector cover and lift up to detach the retaining clip

more coolant mixture to bring the level up to the base of the filler neck.

21   Squeeze the upper radiator hose to expel air, then add more coolant mixture if necessary. Reinstall the radiator cap.

22   Start the engine and allow it to reach normal operating temperature and check for leaks.

---

27   Positive Crankcase Ventilation (PCV) valve check and replacement (every 30,000 miles or 24 months)

V8 engine and 1990 and earlier V6 engines

Refer to illustrations 27.1 and 27.2

1   With the engine idling at normal operating temperature, pull the valve (with hose attached) out of the rubber grommet in the valve cover or intake manifold **(see illustration)**.

2   Place your finger over the end of the valve. If there is no vacuum at the valve, check for a plugged hose, manifold port, or the valve itself. Replace any plugged or dete-

riorated hoses **(see illustration)**.

3   Turn off the engine and shake the PCV valve, listening for a rattle. If the valve doesn't rattle, replace it with a new one.

4   To replace the valve, pull it out of the end of the hose, noting its installed position and direction.

5   When purchasing a replacement PCV valve, make sure it's for your particular vehicle, model year and engine size. Compare the old valve with the new one to make sure they are the same.

6   Push the valve into the end of the hose until it's seated.

7   Inspect the rubber grommet for damage and replace it with a new one if necessary.

8   Push the PCV valve and hose securely into position in the valve cover or intake manifold.

1991 and later V6 engines

Refer to illustrations 27.9 and 27.10

9   Remove the plastic nut and detach the fuel injector cover **(see illustration)**.

10   Push down on the PCV valve access cover (there is a spring under the cover holding the PCV valve in place) and remove the fasteners **(see illustration)**.

11   Remove the spring, O-ring and PCV valve from the intake manifold noting the installed position and direction of the valve. Shake the PCV valve, listening for a rattle. If it doesn't rattle, replace it with a new one.

12   When purchasing a replacement PCV valve, make sure it's for your particular vehicle, model year and engine size. Compare the old valve with the new one to make sure they are the same and replace the O-ring with a new one if it's damaged.

13   Installation of the valve is the reverse of removal.

---

28   Exhaust Gas Recirculation (EGR) system check (every 30,000 miles or 24 months)

Refer to illustration 28.2

1   The EGR valve is located on the intake manifold. Most of the time, when a problem develops in the emissions system, it's due to a stuck or corroded EGR valve.

2   With the engine cold to prevent burns, reach under the valve and push on the diaphragm **(see illustration)**. Using moderate pressure, you should be able to move the diaphragm.

3   If the diaphragm doesn't move or moves only with much effort, replace the EGR valve with a new one. If in doubt about the condition of the valve, compare the free movement with a new valve.

4   Refer to Chapter 6 for more information on the EGR system.

---

29   Spark plug replacement (every 30,000 miles or 24 months)

Refer to illustrations 29.2, 29.5a, 29.5b, 29.6, 29.9 and 29.10

1   All vehicles covered by this manual are equipped with transversely mounted engines which locate the spark plugs on the side of the engine at the front and the rear of the

27.10 Press down on the PCV valve access cover and remove the fasteners (arrows)

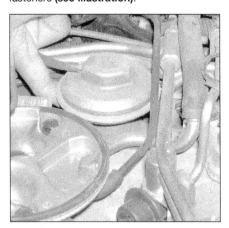

28.2 The diaphragm, reached from under the EGR valve, should move with finger pressure

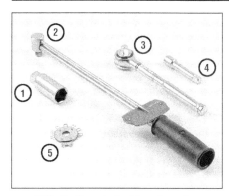

29.2 Tools required for changing spark plugs

1   **Spark plug socket** - *This will have special padding inside to protect the spark plug's porcelain insulator*
2   **Torque wrench** - *Although not mandatory, using this tool is the best way to ensure the plugs are tightened properly*
3   **Ratchet** - *Standard hand tool to fit the spark plug socket*
4   **Extension** - *Depending on model and accessories, you may need special extensions and universal joints to reach one or more of the plugs*
5   **Spark plug gap gauge** - *This gauge for checking the gap comes in a variety of styles. Make sure the gap for your engine is included*

29.5a Spark plug manufacturers recommend using a wire type gauge when checking the gap - if the wire does not slide between the electrodes with a slight drag, adjustment is required

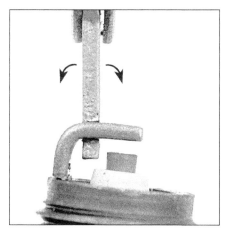

29.5b To change the gap, bend the *side* electrode only, as indicated by the arrows, and be very careful not to crack or chip the porcelain insulator surrounding the center electrode

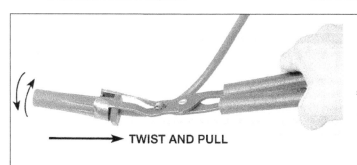

TWIST AND PULL

29.6 When removing the spark plug wires, pull only on the boot and twist it back-and-forth

engine compartment. The left side (front) spark plugs can be reached from the front of the vehicle, but the right side (rear) spark plugs are located between the engine and the firewall. Removal of the right side (rear) spark plugs is best accomplished by raising the vehicle, supporting it securely on jackstands and working from below.

2   In most cases, the tools necessary for spark plug replacement include a spark plug socket which fits onto a ratchet (spark plug sockets are padded inside to prevent damage to the porcelain insulators on the new plugs), various extensions and a gap gauge to check and adjust the gaps on the new plugs **(see illustration)**. A special plug wire removal tool is available for separating the wire boots from the spark plugs, and is a good idea on these models because the boots fit very tightly. A torque wrench should be used to tighten the new plugs. It is a good idea to allow the engine to cool before removing or installing the spark plugs.

3   The best approach when replacing the spark plugs is to purchase the new ones in advance, adjust them to the proper gap and replace the plugs one at a time. When buying the new spark plugs, be sure to obtain the correct plug type for your particular engine. The plug type can be found in the Specifications at the front of this Chapter and on the *Emission Control Information* label located under the hood. If these two sources list different plug types, consider the emission control label correct.

4   Allow the engine to cool completely before attempting to remove any of the plugs. While you are waiting for the engine to cool, check the new plugs for defects and adjust the gaps.

5   Check the gap by inserting the proper thickness gauge between the electrodes at the tip of the plug **(see illustration)**. The gap between the electrodes should be the same as the one specified on the *Emissions Control Information* label or as listed in this Chapter's Specifications. The wire should slide between the electrodes with a slight amount of drag. If the gap is incorrect, use the adjuster on the gauge body to bend the curved side electrode slightly until the proper gap is obtained **(see illustration)**. If the side electrode is not exactly over the center electrode, bend it with the adjuster until it is. Check for cracks in the porcelain insulator (if any are found, the plug should not be used).

6   With the engine cool, remove the spark plug wire from one spark plug. Pull only on the boot at the end of the wire - do not pull on the wire. A plug wire removal tool should be used if available **(see illustration)**.

7   If compressed air is available, use it to blow any dirt or foreign material away from the spark plug hole. The idea here is to eliminate the possibility of debris falling into the cylinder as the spark plug is removed.

8   The spark plugs on these models are, for the most part, difficult to reach so a spark

plug socket incorporating a universal joint will probably be necessary. Place the spark plug socket over the plug and remove it from the engine by turning it in a counterclockwise direction.

9   Compare the spark plug with the chart shown on the inside back cover of this manual to get an indication of the general running condition of the engine. Before installing the new plugs, it is a good idea to apply a thin coat of anti-seize compound to the threads **(see illustration)**.

29.9 Apply a thin coat of anti-seize compound to the spark plug threads

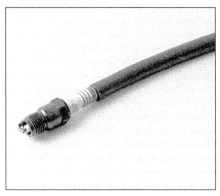

29.10 A length of snug-fitting rubber hose will save time and prevent damaged threads when installing the spark plugs

10   Thread one of the new plugs into the hole until you can no longer turn it with your fingers, then tighten it with a torque wrench (if available) or the ratchet. It's a good idea to slip a short length of rubber hose over the end of the plug to use as a tool to thread it into place **(see illustration)**. The hose will grip the plug well enough to turn it, but will start to slip if the plug begins to cross-thread in the hole - this will prevent damaged threads and the accompanying repair costs.

11   Before pushing the spark plug wire onto the end of the plug, inspect it following the procedures outlined in Section 30.

12   Attach the plug wire to the new spark plug, again using a twisting motion on the boot until it's seated on the spark plug.

13   Repeat the procedure for the remaining spark plugs, replacing them one at a time to prevent mixing up the spark plug wires.

30   Spark plug wire check and replacement (every 30,000 miles or 24 months)

1   The spark plug wires should be checked at the recommended intervals and whenever new spark plugs are installed in the engine.

2   The wires should be inspected one at a time to prevent mixing up the order, which is essential for proper engine operation.

3   Disconnect the plug wire from the spark plug. To do this, grab the rubber boot, twist slightly and pull the wire off. Do not pull on the wire itself, only on the rubber boot.

4   Check inside the boot for corrosion, which will look like a white crusty powder. Push the wire and boot back onto the end of the spark plug. It should be a tight fit on the plug. If it isn't, remove the wire and use pliers to carefully crimp the metal connector inside the boot until it fits securely on the end of the spark plug.

5   Using a clean rag, wipe the entire length of the wire to remove any built-up dirt and grease. Once the wire is clean, check for burns, cracks and other damage. Do not bend the wire excessively or pull the wire lengthwise - the conductor inside might break.

6   Disconnect the wire from the coil or distributor. Again, pull only on the rubber boot. Check for corrosion and a tight fit in the same manner as the spark plug end. Replace the wire at the coil or distributor.

7   Check the remaining spark plug wires one at a time, making sure they are securely fastened at the ignition coil or distributor and the spark plug when the check is complete.

8   If new spark plug wires are required, purchase a set for your specific engine model. Wire sets are available pre-cut, with the rubber boots already installed. Remove and replace the wires one at a time to avoid mix-ups in the firing order.

31   Distributor cap and rotor check and replacement (Cadillac models only) (every 30,000 miles or 24 months)

Refer to illustrations 31.2 and 31.3

1   At the distributor cap, disconnect the electrical connectors.

2   Remove the distributor cap. On some models it is secured by four screws **(see illustration)**. On other models it may be secured by four spring-loaded latches which are disengaged by placing a screwdriver on the slotted head of each latch, press down and turn each latch 90-degrees to release the hooks. When all the screws are removed or the latches are disengaged, separate the cap from the distributor with the spark plug wires still attached.

3   Inspect the cap for cracks and other damage. Closely examine the terminals on the inside of the cap for excessive corrosion **(see illustration)**. Slight pitting is normal. Deposits on the terminals may be removed with a small file.

4   The rotor is visible, with the cap removed, at the top of the distributor shaft. Remove the two retaining screws and remove the rotor.

5   Inspect the rotor for cracks and other damage. Carefully check the condition of the rotor tip and the metal contact at the top of the rotor for excessive burning or pitting. Check the top of the rotor for carbon tracks, replace the rotor if carbon tracks are visible.

6   Install the rotor (either new or original). Note that the slot in the rotor engages the tab on the distributor shaft when the rotor is properly seated. Tighten the screws securely.

7   If a new cap is being installed, release the locking tabs to the spark plug wire retaining ring and remove the ring with the wires attached (this will prevent mixing the wires). If the ring is missing, or the wires have been previously replaced with a type that does not attach to the ring, make sure you label the wires and mark the cap before removing so they can be reinstalled in their original locations. Transfer the ignition coil to the new cap (see Chapter 5).

8   Install the cap to the distributor. Note

that the tab on the cap engages the slot in the distributor housing when the cap is properly seated. Tighten the screws securely or engage the latches by pushing down and rotating 90 degrees. Connect the electrical connectors to the coil.

32   Ignition timing check and adjustment (Cadillac models only) (every 30,000 miles or 24 months)

Refer to illustrations 32.1 and 32.2
Note: *It is imperative that the procedures included on the Vehicle Emission Control Information label be followed when adjusting the ignition timing. The label will include all information concerning preliminary steps to be performed before adjusting the timing as well as the timing specifications.*

1   Locate the VECI label under the hood and read through and perform the preliminary instructions, if any. Some special tools will be needed for this procedure **(see illustration)**.

2   Locate the timing scale, near the crankshaft pulley, and the notch on the crankshaft pulley. Clean the timing scale and mark the notch with a dab of white paint **(see illustration)**.

3   Connect the pick-up lead of the timing light to the number one spark plug wire and the power leads to the battery, according to the manufactures instructions.

4   On 1988 and later models, connect terminals A and B of the ALDL connector with a jumper wire while NOT in the diagnostic mode. **Note:** *On some models it may be necessary to remove the air cleaner cover for access to the ALDL connector.*

5   Start the engine and allow it to reach normal operating temperature. Aim the timing light at the timing scale and note at what point on the scale the notch is lined up with. The numbers on the scale represent degrees of advance (BTDC) or retard (ATDC), with the zero representing TDC.

6   If the notch is not lined up at the correct

31.2 Distributor cap retaining screw locations (arrows)

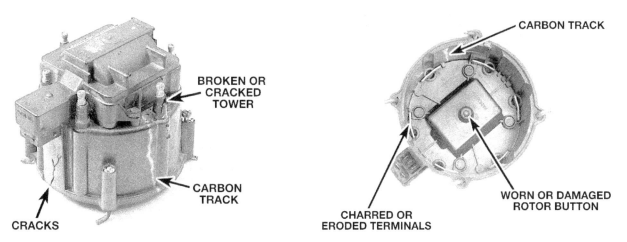

31.3 Inspect the distributor cap for cracks and carbon tracks and the terminals for corrosion and damage

point, according to the specifications on the VECI label, Loosen the distributor hold-down bolt slightly, and while watching the scale rotate the distributor until the notch lines up with the correct mark.

7   Tighten the distributor hold-down bolt and recheck the timing.

8   Turn off the engine and remove the timing light. On 1988 and later models, be sure to remove the jumper wire from the ALDL connector.

33 Oil life indicator lamp - resetting

Refer to illustration 33.2

1   Some models have an oil life indicator lamp on the instrument panel that automatically lights when an oil change is required.

2   Once the oil has been changed, press the reset button to turn the lamp off **(see illustration)**.

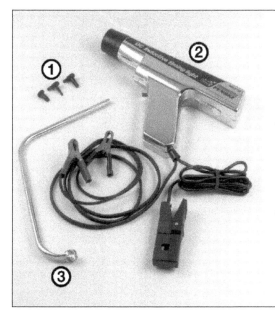

32.1 Tools needed to check and adjust the ignition timing

1   *Vacuum plugs - Vacuum plugs may have to be disconnected and plugged. Molded plugs in various shapes and sizes are available for this*

2   *Inductive pick-up timing light - flashes a bright, concentrated beam of light when the number one spark plug fires. Connect the leads according to the instructions supplied with the light*

3   *Distributor wrench - On some models, the hold-down bolt for the distributor is difficult to reach and turn with conventional wrenches or sockets. A special wrench like this must be used*

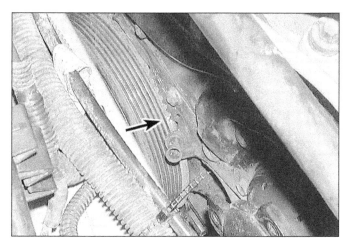

32.2 The ignition timing scale is located at the front of the engine

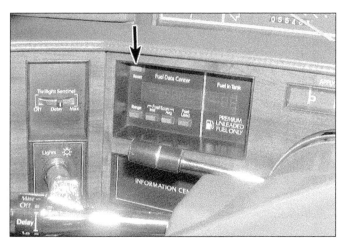

33.2 Press the reset button (arrow) to turn off the oil life indicator lamp

Notes

Chapter 2 Part A
V6 Engine

Contents

Specifications

General

Cylinder numbers (drivebelt end-to-transaxle end)	
Front bank (radiator side)	1-3-5
Rear bank	2-4-6
Firing order	1-6-5-4-3-2
Displacement	3.8 liters (231 cubic inches)

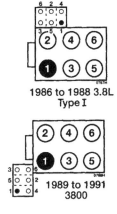

1986 to 1988 3.8L
Type I

1986 to 1988 3.8L
Type II

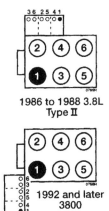

1989 to 1991
3800

1992 and later
3800

**Cylinder location and
coil terminal identification**

General (continued)

Oil pump

Outer gear-to-housing clearance	0.008 to 0.015 inch
Inner gear tip-to-outer gear tip clearance	0.006 inch
Gear end clearance	0.001 to 0.0035 inch
Pump cover warpage limit	0.002 inch

Torque specifications

Ft-lbs (unless otherwise indicated)

Camshaft sprocket bolts

1986 through 1991	27
1992	
Step one	74
Step two	Tighten an additional 105-degrees
1993	
Step one	74
Step two	Tighten an additional 90-degrees

Cylinder head bolts

1986 through 1989	
Step one	25
Step two	Tighten an additional 90-degrees*
Step three	Tighten an additional 90-degrees*

In Steps 2 and 3, if 60 ft-lbs is reached, do not turn further

1990 through 1993	
Step one	35
Step two	Tighten an additional 130-degrees
Step three (four center bolts only)	Tighten an additional 30-degrees

Driveplate-to-crankshaft bolts

1986 through 1991	60
1992 and 1993	
Step one	132 in-lbs
Step two	Tighten an additional 50-degrees

Exhaust manifold-to-cylinder head bolts

1986	20
1987 through 1993	38
Exhaust crossover bolts/studs	15

Intake manifold-to-cylinder head bolts/nuts

1986	45
1987	32
1988 through 1990 (tighten twice, in sequence)	88 in-lbs
1991 through 1993	
Upper intake plenum to lower intake	22
Lower intake-to-heads	88 in-lbs
Oil pan bolts	124 in-lbs
Oil filter adapter-to-timing chain cover bolts	22

Oil pump

Cover-to-timing chain cover bolts	97 in-lbs
Pickup tube and screen assembly bolts	132 in-lbs

Rocker arm shaft/pivot bolts

1986 through 1992	25
1993	
Step one	18
Step two	Tighten an additional 70-degrees
Timing chain cover bolts	22
Timing chain damper bolt	16
Valve cover nuts/bolts	88 in-lbs

Crankshaft balancer-to-crankshaft bolt

1986	200
1987 through 1990	219
1991	
Step one	105
Step two	Tighten an additional 56-degrees
1992	
Step one	110
Step two	Tighten an additional 76-degrees

Right-side engine mounts

Mount-to-engine bracket nuts	30
Bracket-to-engine bolts	66
Mount-to-frame nuts	30

Left-side engine mounts	**Ft-lbs** (unless otherwise indicated)
1986 through 1990	
Mount-to-engine bracket nuts	30
Bracket-to-engine bolts	66
Mount-to-frame nuts..	30
1991 through 1993 cradle mount	
Cradle-to-transaxle bolts	50
Cradle-to-mount nuts	30
Mount-to-chassis nuts	30
Engine torque strut	
Through-bolts...	42
Bracket bolts	17
Bracket nuts	20

1 General information

This Part of Chapter 2 is devoted to in-vehicle repair procedures for the 3800 V6 engine in Buick and Oldsmobile models.

Information concerning camshaft, lifters, balance shaft and engine removal and installation, as well as engine block and cylinder head overhaul, is in Part C of this Chapter. The camshaft cannot be removed with the engine in-vehicle.

The following repair procedures are based on the assumption the engine is installed in the vehicle. If the engine has been removed from the vehicle and mounted on a stand, many of the steps included in this Part of Chapter 2 will not apply.

The Specifications included in this Part of Chapter 2 apply only to the procedures in this Part. The Specifications necessary for rebuilding the block and cylinder heads are found in Part C.

2 Repair operations possible with the engine in the vehicle

Many major repair operations can be accomplished without removing the engine from the vehicle.

Clean the engine compartment and the exterior of the engine with some type of pressure washer before any work is done. A clean engine will make the job easier and will help keep dirt out of the internal areas of the engine.

Depending on the components involved, it may be a good idea to remove the hood to improve access to the engine as repairs are performed (refer to Chapter 11 if necessary).

If vacuum, exhaust, oil or coolant leaks develop, indicating a need for gasket or seal replacement, the repairs can generally be made with the engine in the vehicle. The intake and exhaust manifold gaskets, oil pan gasket and cylinder head gaskets are all accessible with the engine in place.

Exterior engine components such as the intake and exhaust manifolds, the oil pan, the oil pump, the water pump, the starter motor, the alternator and the fuel injection system can be removed for repair with the engine in

place. The timing chain and sprockets can also be replaced with the engine in the vehicle, but the camshaft and balance shaft cannot be removed with the engine in place.

Since the cylinder heads can be removed without pulling the engine, valve component servicing can also be accomplished with the engine in the vehicle.

In extreme cases caused by a lack of necessary equipment, repair or replacement of piston rings, pistons, connecting rods and rod bearings is possible with the engine in the vehicle. However, this practice is not recommended because of the cleaning and preparation work that must be done to the components involved.

3 Top Dead Center (TDC) for number one piston - locating

Refer to illustration 3.8

1   Top Dead Center (TDC) is the highest point in the cylinder that each piston reaches as it travels up and down when the crankshaft turns. Each piston reaches TDC on the compression stroke and again on the exhaust stroke, but TDC generally refers to piston position on the compression stroke. The timing marks on the vibration damper installed on the front of the crankshaft are referenced to the number one piston at TDC on the compression stroke.

2   Positioning the piston(s) at TDC is an essential part of many procedures such as valve seal replacement and timing chain and sprocket replacement.

3   Before beginning this procedure, be sure to place the transaxle in Park, apply the parking brake and block the rear wheels.

4   Remove the spark plugs (see Chapter 1).

5   When looking at the drivebelt end of the engine, normal crankshaft rotation is clockwise. In order to bring any piston to TDC, the crankshaft must be turned with a socket and ratchet attached to the bolt threaded into the center of the vibration damper on the crankshaft.

6   Have an assistant turn the crankshaft with a socket and ratchet while you hold a finger over the number 1 spark plug hole. **Note:** *See the cylinder numbering sequence*

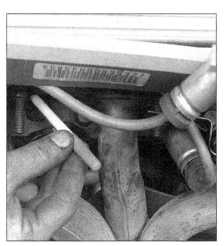

3.8 A plastic pen inserted into the spark plug hole can be used to determine the highest point reached by that piston

in the Specifications for this Chapter.

7   When the piston approaches TDC, air pressure will be felt at the spark plug hole. Instruct your assistant to turn the crankshaft slowly.

8   Insert a plastic pen into the spark plug hole **(see illustration)**. As the piston rises, the pen will be pushed out. Note the point at which the pen stops moving, this is TDC on the compression stroke for that piston.

9   After the number one piston has been positioned at TDC on the compression stroke, TDC for any of the remaining cylinders can be located by turning the crankshaft 120-degrees at a time and following the firing order (refer to the Specifications).

4 Valve covers - removal and installation

Removal

Front cover

Refer to illustration 4.4

1   Disconnect the negative battery cable from the battery.

2   Remove the spark plug wires from the spark plugs. Number each wire before removal to ensure correct reinstallation.

4.4 Remove the valve cover bolts (arrows) - 1993 model shown

5.2a Loosen the three rocker arm shafts bolts (arrows) and remove the assembly - 1986 model shown

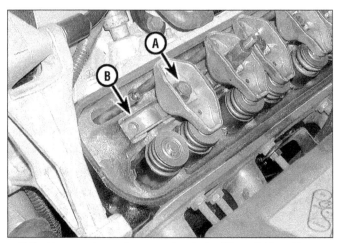

5.2b Remove the rocker arm pivot bolts (A) on 1988 and later models - (B) is the rocker arm pivot support

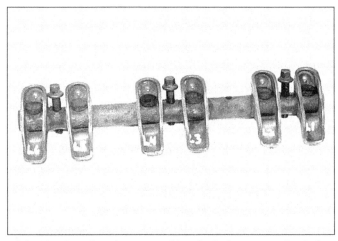

5.2c Number all rocker arms with paint before removal - shaft-type assembly shown

3   Remove the serpentine drivebelt (see Chapter 1) and the alternator brace (see Chapter 5).

4   Remove the valve cover mounting bolts/nuts **(see illustration)**. Some 1986 models have three top-mounted nuts holding the valve cover to studs, while all later engines have mounting bolts around the perimeter of the valve cover.

5   Detach the valve cover. **Note:** *If the cover sticks to the cylinder head, use a soft-face hammer to dislodge it.*

Rear cover

6   Disconnect the negative battery cable from the battery.

7   Remove the spark plug wires from the spark plugs and remove the wire holder. Be sure each wire is labeled before removal to ensure correct reinstallation.

8   Remove the serpentine drivebelt (see Chapter 1).

9   Loosen the power steering pump-to-bracket bolts and move the power steering pump toward the right fenderwell to clear the valve cover. Refer to Chapter 10 if necessary.

10   Refer to Chapter 6 and remove the EGR

tube (between the EGR valve and the rear exhaust manifold) and EGR valve (1988 through 1993 models).

11   Remove the valve cover mounting bolts/nuts.

12   Detach the valve cover. **Note:** *If the cover sticks to the cylinder head, use a soft-face hammer to dislodge it.*

Installation

13   The mating surfaces of the cylinder head and valve cover must be perfectly clean when the covers are installed. Use a gasket scraper to remove all traces of sealant or old gasket, then clean the mating surfaces with lacquer thinner or acetone (if there's sealant or oil on the mating surfaces when the cover is installed, oil leaks may develop). The valve covers are made of aluminum, so be extra careful not to nick or gouge the mating surfaces with the scraper.

14   Clean the mounting bolt threads and apply thread-locking compound. Place the valve cover and new gasket in position, then install the bolts.

15   Tighten the bolts/nuts in several steps to

the torque listed in this Chapter's specifications.

16   Complete the installation by reversing the removal procedure.

17   Start the engine and check for oil leaks at the valve cover-to-cylinder head joints.

---

5   Rocker arms and pushrods - removal, inspection and installation

Removal

Refer to illustrations 5.2a, 5.2b, 5.2c, 5.3 and 5.4

1   Refer to Section 4 and detach the valve covers from the cylinder heads. The rocker arm assembly in 1986 and 1987 models mounts the rocker arms on a shaft. On all later models, the rocker arms are mounted on individual studs with pivots.

2   Loosen the rocker arm pivot bolts (shaft bolts on 1986 and 1987 models) one at a time and detach the rocker arms, bolts, pivots and pivot retainers **(see illustrations)**. Keep track

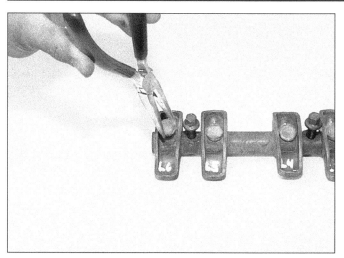

5.3 Remove the rocker arm retainer buttons only if you have to inspect the shaft - the buttons can break easily, so make sure you have new ones on hand

5.4 If more than one pushrod is being removed, store them in a perforated cardboard box to prevent mix-ups during installation - note the label indicating the front of the engine

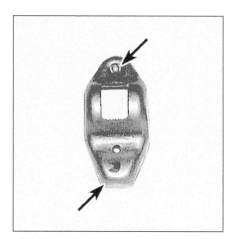

5.5 Check the bottom of the rocker arm for wear at these points (arrows)

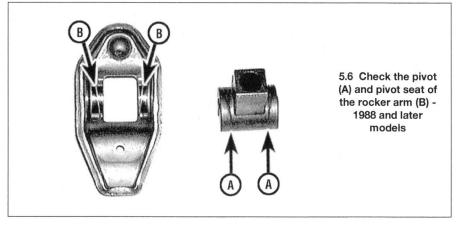

5.6 Check the pivot (A) and pivot seat of the rocker arm (B) - 1988 and later models

of the rocker arm positions, since they must be returned to the same location **(see illustration)**.

3   Store each set of rocker components separately in a marked plastic bag to ensure that they're reinstalled in their original locations. If you are removing the rocker arms from the shaft on 1986 and 1987 models, for rocker or shaft inspection or replacement, use pliers to remove the plastic buttons that retain them to the shaft **(see illustration)**. **Note:** *After years of use, these buttons become brittle and may break during removal. Order new retainer buttons before reassembly.*

4   Remove the pushrods and store them separately to make sure they don't get mixed up during installation **(see illustration)**.

Inspection

Refer to illustrations 5.5 and 5.6

5   Check each rocker arm for wear, cracks and other damage, especially where the pushrods and valve stems contact the rocker arm **(see illustration)**.

6   Check the pivot seat in each rocker arm and the pivot faces **(see illustration)**. Look for galling, stress cracks and unusual wear patterns. If the rocker arms are worn or damaged, replace them with new ones and install new pivots or shafts as well.

7   Make sure the hole at the pushrod end of each rocker arm is open.

8   Inspect the pushrods for cracks and excessive wear at the ends. Roll each pushrod across a piece of plate glass to see if it's bent (if it wobbles, it's bent).

Installation

9   Lubricate the lower end of each pushrod with clean engine oil or moly-base grease and install them in their original locations. Make sure each pushrod seats completely in the lifter socket.

10   Apply moly-base grease to the ends of the valve stems, the upper ends of the pushrods and to the pivot faces to prevent damage to the mating surfaces before engine oil pressure builds up.

11   Coat the rocker arm pivot bolts (or shaft on 1986 and 1987 models) with moly-base grease. Install the rocker arms, pivots and pivot retainers. Tighten the bolts to the

torque listed in this Chapter's specifications. As the bolts are tightened, make sure the pushrods seat properly in the rocker arms.

12   Install the valve covers (see Section 4).

6   Valve springs, retainers and seals - replacement

Refer to illustrations 6.4, 6.7, 6.8, 6.13 and 6.16

Note: *Broken valve springs and defective valve stem seals can be replaced without removing the cylinder heads. Two special tools and a compressed air source are normally required to perform this operation, so read through this Section carefully and rent or buy the tools before beginning the job.*

1   Refer to Section 4 and remove the valve cover from the affected cylinder head. If all of the valve stem seals are being replaced, remove both valve covers.

2   Remove the spark plug from the cylinder which has the defective component. If all of the valve stem seals are being replaced, all of the spark plugs should be removed.

3   Turn the crankshaft until the piston in the affected cylinder is at Top Dead Center

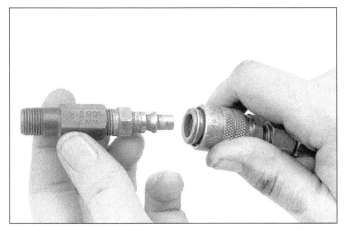

6.4 This is what the typical air hose adapter looks like - they're commonly available in auto parts stores

6.7 A tool that attaches to the rocker arm stud can be used to compress the valve springs

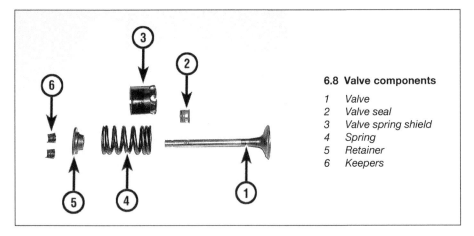

6.8 Valve components

1　Valve
2　Valve seal
3　Valve spring shield
4　Spring
5　Retainer
6　Keepers

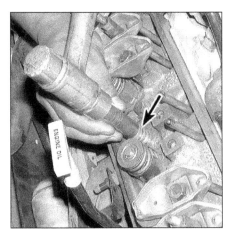

6.13 Tap the new seal (arrow) over the guide, using a deep socket - tap the seal on only until it stops

(TDC) on the compression stroke (refer to Section 3 for instructions). If you are replacing all of the valve stem seals, begin with cylinder number one and work on the valves for one cylinder at a time. Move from cylinder to cylinder following the firing order sequence (see this Chapter's Specifications).

4　Thread an adapter into the spark plug hole and connect an air hose from a compressed air source to it **(see illustration)**. Most auto parts stores can supply the air hose adapter. **Note:** *Many cylinder compression gauges utilize a screw-in fitting that may work with your air hose quick-disconnect fitting.*

5　Remove the rocker arm assembly (see Section 5) and the pushrod for the valve with the defective part. If all of the valve stem seals are being replaced, remove all of the pushrods.

6　Apply compressed air to the cylinder.

7　Stuff shop rags into the cylinder head holes above and below the valves to prevent parts and tools from falling into the engine, then use a valve spring compressor to compress the valve spring. **Note:** *If the keepers are stuck in the retainer (which will cause the valve to open as you compress the spring), place a socket over the retainer and strike it sharply with a hammer. This will break the bond between the retainer and keepers.*

Remove the keepers with small needle-nose pliers or a magnet **(see illustration)**.

8　Remove the spring retainer and valve spring assembly, then remove the guide seal **(see illustration)**. **Note:** *If air pressure fails to hold the valve in the closed position during this operation, the valve face and/or seat is probably damaged. If so, the cylinder head will have to be removed for additional repair operations.*

9　Wrap a rubber band or tape around the top of the valve stem so the valve won't fall into the combustion chamber, then release the air pressure.

10　Inspect the valve stem for damage. Rotate the valve in the guide and check the end for eccentric movement, which would indicate that the valve is bent.

11　Move the valve up and down in the guide and make sure it doesn't bind. If the valve stem binds, either the valve is bent or the guide is damaged. In either case, the head will have to be removed for repair.

12　Reapply air pressure to the cylinder to retain the valve in the closed position, then remove the tape or rubber band from the valve stem.

13　Lubricate the valve stem with engine oil and install a new guide seal **(see illustration)**.

14　Install the valve spring assembly in posi-

tion over the valve. **Note:** *The valve spring shield (if used) goes over the bottom of the spring (the spring end with the closely-wound coils), with the oil flow holes at the bottom.*

15　Install the valve spring retainer or rotator, and compress the valve spring assembly.

16　Position the keepers in the upper groove. Apply a small dab of grease to the

6.16 Apply a small dab of grease to the keepers before installation - this will keep them in place during installation

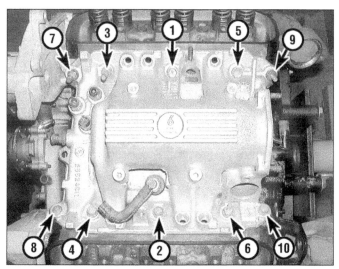

7.5a Intake manifold mounting bolts (arrows) - TIGHTENING sequence shown; reverse this sequence for disassembly (1986 model shown)

7.5b Carefully pry up on a casting boss - don't pry between gasket surfaces

inside of each keeper to hold it in place if necessary **(see illustration)**. Remove the pressure from the spring and make sure the keepers are seated.

17 Disconnect the air hose and remove the adapter from the spark plug hole.

18 Refer to Section 5 and install the pushrods and rocker arm assembly.

19 Refer to Section 4 and install the valve covers.

20 Install the spark plugs and hook up the wires.

21 Start and run the engine, then check for oil leaks and unusual sounds coming from the valve cover area.

7.6 The upper plenum can be removed alone if necessary (arrows indicate the three front bolts), but it can also be removed with the lower manifold

---

7 Intake manifold - removal and installation

Warning 1: *Wait until the engine is completely cool before beginning this procedure.*
Warning 2: *Gasoline is extremely flammable, so take extra precautions when you work on any part of the fuel system. Don't smoke or allow open flames or bare light bulbs near the work area, and don't work in a garage where a natural gas-type appliance (such as a water heater or a clothes dryer) with a pilot light is present. Since gasoline is carcinogenic, wear latex gloves when there's a possibility of being exposed to fuel, and, if you spill any fuel on your skin, rinse it off immediately with soap and water. Mop up any spills immediately and do not store fuel-soaked rags where they could ignite. The fuel system is under constant pressure, so, if any fuel lines are to be disconnected, the fuel pressure in the system must be relieved first. When you perform any kind of work on the fuel system, wear safety glasses and have a Class B type fire extinguisher on hand.*
Caution: *After the fuel pressure has been relieved, it's a good idea to lay a shop towel*

over any fuel connection to be disassembled, to absorb the residual fuel that may leak out when servicing the fuel system.

Removal

1986 through 1990 models
Refer to illustrations 7.5a and 7.5b

1 Disconnect the negative battery cable from the battery. Refer to Chapter 1 and drain the engine coolant.

2 Refer to Chapter 4 to relieve the fuel system pressure and disconnect the fuel hoses, electrical connectors at the injectors, throttle linkage, throttle cable bracket and tag and disconnect all hoses or electrical connectors at the throttle body, including the coolant hoses.

3 Refer to Chapter 6 and remove the EGR tube from the rear exhaust manifold and the EGR valve from the intake manifold (1988 and later models). Refer to Chapter 5 and remove the alternator and bracket.

4 Remove the two braces on the power steering pump (see Chapter 10 if necessary). Remove the brake booster vacuum hose from the intake manifold, the heater hose, bypass hose and the upper radiator hose.

5 Remove the intake manifold mounting

bolts and separate the manifold from the engine **(see illustrations)**. Do not pry between the manifold and heads, as damage to the gasket sealing surfaces may result. If you're installing a new manifold, transfer all fittings and sensors to the new manifold.

1991 and later models
Refer to illustration 7.6

6 The 1991 and later models have a two-piece intake manifold, consisting of a lower manifold and a removable upper plenum **(see illustration)**. The upper plenum can be removed alone, but if the whole intake manifold is to be removed, do not separate the two components.

7 All of the Steps listed above for the earlier engines apply to the later models, but later models have a sight shield over the fuel injectors that must be removed before disconnecting the injectors. Also, removal of the heat shield over the exhaust crossover pipe will facilitate intake manifold removal.

Installation
Refer to illustrations 7.8, 7.10a and 7.10b
Note: *The mating surfaces of the cylinder heads, block and manifold must be perfectly*

7.8 Remove all traces of gasket material, but don't gouge the mating surfaces

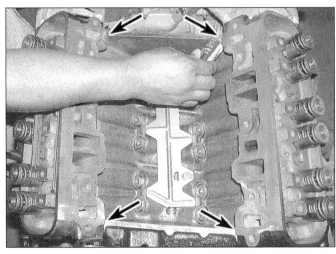

7.10a Seal the corners (arrows) where the intake gaskets meet with RTV sealant - apply a small amount before the end seals are in place, then another dab after the end seals and gaskets are in position

clean when the manifold is installed. Gasket removal solvents in aerosol cans are available at most auto parts stores and may be helpful when removing old gasket material that's stuck to the heads and manifold (since the manifold is made of aluminum, aggressive scraping can cause damage). Be sure to follow the directions printed on the container.

8   Use a gasket scraper to remove all traces of sealant and old gasket material **(see illustration)**, then clean the mating surfaces with lacquer thinner or acetone. If there's old sealant or oil on the mating surfaces when the manifold is installed, oil or vacuum leaks may develop. Use shop towels to protect the lifter valley during gasket removal and a vacuum cleaner to remove any final debris. On 1991 through 1993 models, if the upper plenum was removed, clean the mating surface on the lower manifold with gasket remover only, do not use scraping tools.

9   Use a tap of the correct size to chase the threads in the bolt holes, then use compressed air (if available) to remove the debris from the holes. **Warning:** *Wear safety glasses or a face shield to protect your eyes when using compressed air.*

10   Install the new manifold gaskets, applying RTV sealant to the four corners where the manifold, head and block come together **(see illustrations)**. **Note:** *On some early models, the intake gasket set has one large gasket that covers the whole lifter valley, and two rubber end seals. Later engines all have two side gaskets (between the manifold and the heads) and two rubber end gaskets.*

11   Thoroughly clean the top of the block where the rubber end seals go and glue the seals to the block with contact cement, then apply a bead of RTV sealant where the side gaskets meet the end seals in the corners.

12   Carefully lower the manifold into place. Apply thread-locking compound to the mounting bolt threads and install the bolts finger tight.

13   Tighten the mounting bolts, following

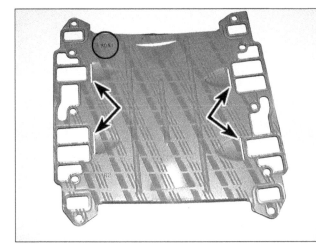

7.10b When using the large, one-piece gasket on early engines, note the marking (circled) for FRONT, and snap the dished cutouts (arrows) under the edge of the cylinder heads

the recommended sequence **(see illustration 7.5a)**, to the torque listed in this Chapter's specifications. Tighten the bolts again, in sequence.

14   Install the remaining components in the reverse order of removal. If the plenum was removed on 1991 through 1993 models, apply a 1/16-inch bead of RTV sealant to the top of the lower manifold, encircling the perimeter and all the bolt holes. Lower the plenum carefully in place and install the bolts. Tighten from the center out to the torque listed in this Chapter's Specifications.

15   Change the oil and filter and fill the cooling system (see Chapter 1). Start the engine and check for oil and vacuum leaks.

---

8   Exhaust manifolds - removal and installation

Warning: *Allow the engine to cool completely before beginning this procedure.*

Note: *Exhaust system fasteners are frequently difficult to remove - they get frozen in place because of the heating/cooling cycle to*

which they're constantly exposed. To ease removal, apply penetrating oil to the threads of all exhaust manifold and exhaust pipe fasteners and allow it to soak in.

Removal

Front manifold

Refer to illustrations 8.4 and 8.5

1   Disconnect the negative battery cable (see Chapter 1).

2   Remove the dipstick, the dipstick tube hold-down nut and work the dipstick tube out of the block.

3   Detach the spark plug wires from the front spark plugs, then remove the spark plugs (see Chapter 1).

4   Unbolt the crossover pipe and heat shield from the manifold **(see illustration)**. **Note:** *If more working room is desired, refer to Chapter 3 and remove the engine cooling fan.*

5   Unbolt and remove the exhaust manifold **(see illustration)**. **Note:** *On 1993 models, the crossover pipe is permanently attached to the front manifold and is attached to the rear manifold with two nuts at a flange.*

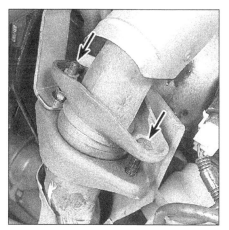

8.4 Remove the bolts/nuts (arrows) where the crossover pipe bolts to the manifold

Rear manifold

Refer to illustration 8.14

6   Disconnect the negative battery cable.

7   On 1991 through 1993 models, remove the cowl-mounted vacuum tank, rear engine lifting eye (bolted to rear exhaust manifold), and the throttle cable bracket at the throttle body.

8   Remove the two nuts attaching the exhaust pipe to the rear exhaust manifold. **Note:** *Soak the nuts with penetrating oil before attempting removal.*

9   Disconnect the spark plug wires from the rear spark plugs and remove the spark plugs (see Chapter 1).

10   Remove the EGR pipe and transaxle dipstick tube if it's in the way (see Chapters 6 and 7).

11   Disconnect the oxygen sensor electrical connector (see Chapter 6).

12   Remove the heat shield attached to the cowl. **Note:** *There are two bolts accessed from below.*

13   Remove the two bolts/nuts holding the crossover pipe to the rear exhaust manifold.

14   Remove the nuts/bolts and detach the manifold from the head **(see illustration)**. Note the location of the studs/nuts for later reference.

Installation

15   Clean the mating surfaces of the manifold and cylinder head, then check the manifold for warpage and cracks. If the manifold was leaking, take it to an automotive machine shop for resurfacing.

16   Place the manifold in position and install the bolts finger tight.

17   Starting in the middle and working out toward the ends, tighten the mounting bolts a little at a time until all of them are at the torque listed in this Chapter's Specifications. **Note:** *Clean the threads of all exhaust bolts/studs and apply a light film of anti-seize compound. This will ease installation and any future disassembly.*

18   Install the remaining components in the reverse order of removal.

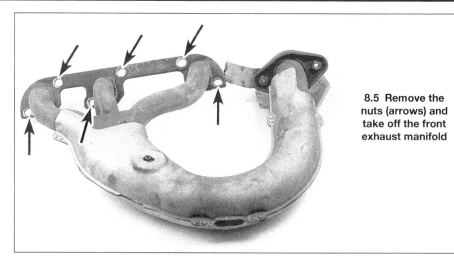

8.5 Remove the nuts (arrows) and take off the front exhaust manifold

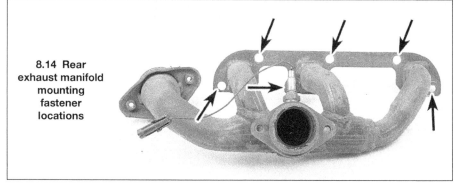

8.14 Rear exhaust manifold mounting fastener locations

19   Start the engine and check for exhaust leaks between the manifold and cylinder head, between the manifold and exhaust pipe, between the manifolds and the crossover pipe, and at the EGR pipe connections.

9   Cylinder heads - removal and installation

Warning: *Wait until the engine is completely cool before beginning this procedure.*

Removal

Refer to illustrations 9.5 and 9.11

1   Disconnect the negative battery cable at the battery. Refer to Chapter 1 and drain the cooling system and remove the drivebelt.

2   Disconnect the spark plug wires and remove the spark plugs (see Chapter 1). Be sure to label the plug wires to simplify reinstallation.

3   Remove the intake manifold as described in Section 7.

4   Refer to Chapter 5 and remove the ignition coil-pack.

5   Disconnect all wires and hoses from the cylinder head(s). Be sure to label them to simplify reinstallation. If removing the rear cylinder head, remove the power steering pump (see Chapter 10), then remove the water bypass hose in front of the belt tensioner and remove the belt tensioner **(see illustration)**.

9.5 Remove the bolts (arrows) holding the drivebelt tensioner assembly to the rear cylinder head (shown with power steering pump already removed)

9.11 Pry carefully - don't force a tool between the gasket surfaces

9.14 Carefully remove all traces of old gasket material

9.17 Look for gasket marks (arrow) to ensure correct installation

6   When removing the front cylinder head, refer to Chapter 5 and remove the alternator and its bracket. One upper bolt will have to be removed from the air conditioning compressor bracket.

7   Detach the exhaust manifold(s) from the cylinder head(s) being removed (see Section 8).

8   Remove the valve cover(s) (see Section 4).

9   Remove the rocker arms, pushrods and guide plates (see Section 5). **Caution:** *Keep the rocker arm components for each cylinder identified and together. They must go back in their original locations.*

10   Loosen the head bolts in 1/4-turn increments until they can be removed by hand. Work in a pattern opposite that of the tightening sequence **(see illustration 9.19)**.

11   Lift the cylinder head off the engine. If resistance is felt, don't pry between the cylinder head and block as damage to the mating surfaces will result. Recheck for cylinder head bolts that may have been overlooked, then use a hammer and block of wood to tap the cylinder head and break the gasket seal. Be careful because there are locating dowels in the block which position each cylinder head. As a last resort, pry each cylinder head up at the rear corner only and be careful not to damage anything **(see illustration)**. After removal, place the cylinder head on blocks of wood on your workbench to prevent damage to the gasket surfaces.

12   Refer to Chapter 2, Part C, for cylinder head disassembly, inspection and valve service procedures.

Installation

Refer to illustrations 9.14, 9.17 and 9.19

13   The mating surfaces of the cylinder heads and block must be perfectly clean when the heads are installed.

14   Use a gasket scraper to remove all traces of carbon and old gasket material **(see illustration)**, then clean the mating surfaces with lacquer thinner or acetone. If there's oil on the mating surfaces when the heads are installed, the gaskets may not

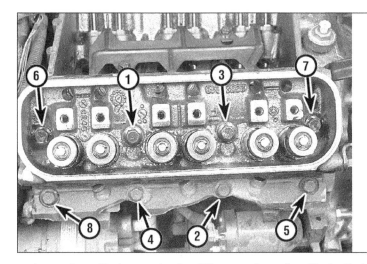

9.19 Cylinder head bolt TIGHTENING sequence

seal correctly and leaks may develop. When working on the block, it's a good idea to cover the lifter valley with shop rags to keep debris out of the engine. Use a shop rag or vacuum cleaner to remove any debris that falls into the cylinders.

15   Check the block and head mating surfaces for nicks, deep scratches and other damage. If damage is slight, it can be removed with a file; if it's excessive, machining may be the only alternative.

16   Use a tap of the correct size to chase the threads in the head bolt holes. Dirt, corrosion, sealant and damaged threads will affect torque readings.

17   Position the new gaskets over the dowel pins in the block. Install the cylinder head gaskets dry (no sealant), unless the manufacturer states otherwise. Most gaskets are marked UP or TOP **(see illustration)** because they must be installed a certain way and they have an arrow that must point to the front (timing chain end) of the engine. The left gasket may have an "L" near the arrow.

18   Carefully position the heads on the block without disturbing the gaskets. They are located by pressing them over the two dowel pins on the exhaust side of the block.

19   Tighten the bolts as directed in this Chapter's specifications in the sequence

shown **(see illustration)**. This must be done in three steps, following the sequence each time. **Note:** *Apply a small amount of RTV sealant to the underside of the bolt heads, and thread-locking compound to the (cleaned) bolt threads.*

20   The remaining installation steps are the reverse of removal.

21   Change the oil and filter (see Chapter 1).

10   Crankshaft front oil seal - replacement

Refer to illustrations 10.6, 10.9, 10.11 and 10.13

Caution: *The crankshaft balancer is serviced as an assembly. Do not attempt to separate the pulley from the balancer hub.*

1   Disconnect the negative cable from the battery.

2   Loosen the lug nuts on the right front wheel.

3   Raise the vehicle and support it securely on jackstands.

4   Remove the right front wheel.

5   Remove the serpentine drivebelt (see Chapter 1). **Note:** *On 1989 and later V6 engines, remove the nuts on top and bottom*

10.6 Remove the bolt in the center of the crankshaft pulley, while using a chain or strap wrench to keep the pulley from turning - if using a chain wrench, protect the pulley's grooves with a leather strap or a piece of old drivebelt wrapped around the pulley

10.9 Pry the crankshaft seal out of the front cover, using a seal removal tool - be careful not to scratch the crankshaft

of the torque-axis mount at the right-front of the timing cover. Remove the upper torque-axis mount from the timing cover before removing the crankshaft pulley.

6   Use a strap wrench or chain wrench to hold the crankshaft pulley while you loosen the crankshaft pulley bolt **(see illustration)**. The bolt is normally quite tight, so use a large breaker bar and a six-point socket. **Note:** *If you are using a chain wrench, use a section of an old leather belt over the pulley to protect the grooves of the crankshaft pulley.*

7   The crankshaft pulley should pull off the crankshaft by hand. If the pulley can't be removed by hand, a puller that bolts to the center hub will be required to remove the pulley. Leave the Woodruff key in place in the end of the crankshaft.

8   If there's a groove worn into the seal contact surface on the crankshaft balancer, sleeves are available that fit over the groove, restoring the contact surface to like-new condition. These sleeves are sometimes included with the seal kit. Check with your parts supplier for details.

9   Pry the old oil seal out with a seal removal tool or a screwdriver **(see illustration)**. Be very careful not to nick or otherwise damage the crankshaft in the process.

10   Apply a thin coat of RTV sealant to the outer edge of the new seal. Lubricate the seal lip with moly-base grease or clean engine oil.

11   Place the seal squarely in position in the bore and drive it into place with a special seal installer (available at auto parts stores). Make sure the seal enters the bore squarely and seats completely **(see illustration)**.

12   If the special tool is unavailable, carefully drive the seal into place with a hammer and large socket or section of pipe. The outer diameter of the socket or pipe should be the same size as the seal outer diameter.

13   Installation is the reverse of removal. Align the keyway of the crankshaft pulley with the key **(see illustration)** and avoid bending the metal tabs. Be sure to apply moly-base

10.11 The new seal can be installed with a seal installation tool, or driven in evenly with a large socket or section of pipe the same diameter as the seal

10.13 Be sure to align the keyway (arrow) with the key on the crankshaft

grease to the seal contact surface on the back side of the crankshaft pulley (if it isn't lubricated, the seal lip could be damaged and oil leakage would result).

14   Apply thread locking compound to the threads and tighten the crankshaft bolt in two steps to the torque listed in this Chapter's Specifications.

15   Reinstall the remaining parts in the reverse order of removal.

16   Start the engine and check for oil leaks at the seal.

11   Timing chain - removal and installation

Warning: *Wait until the engine is completely cool before beginning this procedure.*
Note: *1986 and 1987 models have a "conventional" timing chain with camshaft and crankshaft sprockets. The 1989 and later engines have a balance shaft, which is*

mounted in the block above the camshaft, and is gear-driven by a gear behind the camshaft sprocket. The timing chain procedure is much the same for all of the V6 engines covered by this manual, with or without the balance shaft. The balance shaft is covered in Chapter 2, Part C.

Removal

Refer to illustrations 11.9, 11.11, 11.12, 11.14, 11.15 and 11.17

1   Disconnect the negative battery cable from the battery.

2   Set the parking brake and put the transaxle in Park. Raise the front of the vehicle and support it securely on jackstands. Remove the splash shield from the right inner fender.

3   Drain the oil and coolant (see Chapter 1).

4   Remove the coolant hoses from the timing chain cover and water pump, and loosen the water pump pulley (see Chapter 3).

5   Remove the serpentine drivebelt (see

11.9 Timing chain cover bolt locations (arrows)

11.11 Remove all traces of old gasket material

11.12 The camshaft thrust surface on this
cover is worn away (arrow) which means
that a new cover must be installed

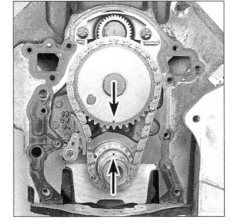

11.14 The marks on the crankshaft and
camshaft sprockets (arrows) must be
aligned adjacent to each other as shown

11.15 Remove the timing chain damper

Chapter 1) and the drivebelt tensioner assembly (see Section 9). Remove the water pump pulley.

6   Remove the crankshaft front pulley (see Section 10).

7   Unplug the connectors from the oil pressure, camshaft and crankshaft sensors (see Chapter 6). Remove the shield, if equipped, over the front of the crankshaft sensor.

8   Remove the oil pan bolts which attach to the front cover, then loosen the rest of the oil pan bolts (see Section 12).

9   Remove the timing chain cover-to-engine block bolts **(see illustration)**. Note that two of the bolts also secure the crankshaft sensor. Lift the sensor off when removing these bolts.

10   Separate the cover from the front of the engine, removing it with the water pump and oil filter adapter still attached.

11   Use a gasket scraper to remove all traces of old gasket material and sealant from the cover and engine block **(see illustration)**. The cover is made of aluminum, so be careful not to nick or gouge it. Clean the gasket sealing surfaces with lacquer thinner or acetone.

12   Check the camshaft thrust surface in the cover for excessive wear **(see illustration)**. If it's worn, a new cover will be required.

13   The timing chain should be replaced with a new one if the total freeplay midway between the sprockets exceeds one inch. Failure to replace the timing chain may result in erratic engine performance, loss of power and lowered fuel mileage.

14   Temporarily install the crankshaft pulley bolt and turn the crankshaft clockwise to align the timing marks on the crankshaft and camshaft sprockets directly opposite each other **(see illustration)**.

15   Detach the spring, then remove the bolt and separate the timing chain damper from the block **(see illustration)**.

16   Remove the camshaft sprocket bolt. Try not to turn the camshaft in the process (if you do, realign the timing marks after the bolts are loosened).

17   Alternately pull the camshaft sprocket and then the crankshaft sprocket forward and remove the sprockets and timing chain as an assembly **(see illustration)**.

18   Clean the timing chain components with

solvent and dry them with compressed air (if available). **Warning:** *Wear eye protection when using compressed air.*

19   Inspect the components for wear and damage. Look for teeth that are deformed, chipped, pitted, polished or discolored.

11.17 Remove the camshaft sprocket,
timing chain and crankshaft sprocket as
an assembly

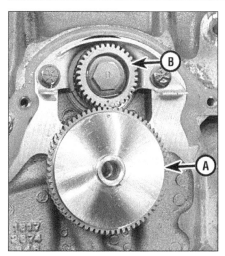

11.21 On 1988 and later engines, the balance shaft drive gear (A) must align with the balance shaft driven gear (B) before installing the timing chain and sprockets - note the dots on both gears are aligned

Installation

Refer to illustration 11.21 and 11.26

20   If the crankshaft has been disturbed, install the crankshaft sprocket temporarily and turn the crankshaft until the mark on the sprocket is exactly at the top. If the camshaft was disturbed, install the camshaft sprocket temporarily and turn the camshaft until the timing mark is at the bottom, opposite the mark on the crankshaft sprocket **(see illustration 11.14)**.

21   Assemble the timing chain on the sprockets, then slide the sprocket and chain assembly onto the shafts with the timing marks aligned as shown in **illustration 11.14**. **Note:** *If it was removed for any reason, be sure to install the balance shaft drive gear, which is behind the camshaft sprocket on 1988 and later engines. It drives the gear on the balance shaft and the two gears must be aligned before installing the timing chain with camshaft sprocket* **(see illustration)**.

22   Install the camshaft sprocket bolt and tighten to the torque listed in this Chapter's Specifications.

23   Attach the timing chain damper assembly to the block and install the spring.

24   Lubricate the chain and sprocket with clean engine oil.

25   Before the cover is installed, the oil pump cover must be removed and the cavity packed with petroleum jelly as described in Section 13.

26   Install a new camshaft thrust button and spring in the center of the camshaft sprocket. Some models may be equipped with a plastic button/spring assembly. This should be replaced with an improved thrust assembly that uses a bearing **(see illustration)**; check with your local auto parts store or dealer parts department for parts availability.

27   Apply a thin layer of RTV sealant to both sides of the new gasket, then position the

11.26 If your engine has a button-type camshaft thrust assembly (at right), replace it with the improved assembly at left that incorporates a bearing

gasket on the engine block (the dowel pins should keep it in place). Apply RTV sealant to the front portion of the oil pan and at the joint between the pan and the block, then attach the cover to the engine, making sure that the oil pump drive engages with the crankshaft.

28   Apply Teflon thread sealant to the bolt threads, then install them finger tight. Follow a criss-cross pattern when tightening the bolts and work up to the torque listed in this Chapter's Specifications in three steps to avoid warping the cover. Don't forget to install the front oil pan bolts, and tighten the rest of the oil pan bolts if they were loosened before.

29   The remainder of installation is the reverse of removal. **Caution:** *Be sure to align the crankshaft sensor as described in Chapter 6 before installing the crankshaft pulley or damage to the sensor may result.*

30   Add oil and coolant, start the engine and check for leaks.

12 Oil pan - removal and installation

Removal

Refer to illustration 12.4

1   Disconnect the cable from the negative battery terminal.

2   Raise the vehicle and place it securely on jackstands. Drain the engine oil and remove the oil filter (refer to Chapter 1 if necessary).

3   Remove the driveplate inspection cover and starter (1986 through 1990 models), if necessary for access (refer to Chapter 5). Disconnect the electrical connector from the oil lever sensor, then remove the oil level sensor from the pan.

4   Remove the oil pan mounting bolts **(see illustration)** and carefully lower the oil pan from the block. Don't pry between the block and the pan or damage to the sealing surfaces may result and oil leaks may develop. Instead, tap the pan with a soft-face hammer to break the gasket seal.

Installation

Refer to illustration 12.9

5   As long as the pan is off, this is a good time to remove and clean the screen and pickup tube with solvent and dry it with compressed air, if available. **Warning:** *Wear eye protection.*

6   If the oil screen is damaged or has metal chips in it, replace it. An abundance of metal chips indicates a major engine problem that must be corrected. **Caution:** *The presence of plastic pieces indicates that the plastic teeth of the camshaft gear (early models) are breaking down. Remove the timing cover and replace the timing chain/sprocket set right away. If the plastic parts get into the oil pump, the engine will be severely damaged.*

7   Make sure the mating surfaces of the pipe flange and the engine block are clean and free of nicks.

8   Clean the pan with solvent and remove all old sealant and gasket material from the block and pan mating surfaces. Clean the mating surfaces with lacquer thinner or acetone and make sure the bolt holes in the block are clear. Check the oil pan flange for distortion, particularly around the bolt holes. If necessary, place the pan on a block of wood and use a hammer to flatten and restore the gasket surface.

9   Always use a new gasket whenever the oil pan is installed. The one-piece gasket is

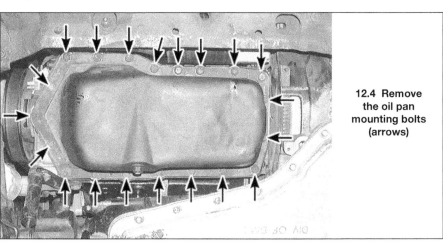

12.4 Remove the oil pan mounting bolts (arrows)

12.9 Before installing the pan gasket, apply a bead of RTV sealant (arrows) where the front cover and rear seal retainer meet the block

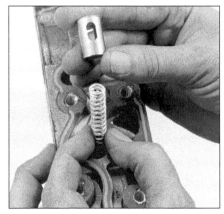

13.4a Remove the pressure regulator valve and spring, then check the valve for wear and damage

13.4b There are two types of oil pressure relief valves; the ventilated type (right) and the later, solid type (left). New timing covers may come with the solid type - do not interchange valves and springs between these types

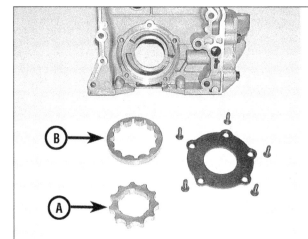

13.5 The oil pump cover is attached to the inside of the timing chain cover - a T-30 Torx driver is required for removal of the screws - (A) is the inner rotor, (B) is the outer rotor

not reusable. Before installing the new gasket, apply a small amount of RTV sealant to the four joints between the front cover/block and rear seal plate/block **(see illustration)**.
10   After the bolts are installed, tighten them to the torque listed in this Chapter's Specifications. Starting at the center, follow a criss-cross pattern and work up to the final torque in three steps.
11   The remaining steps are the reverse of the removal procedure.
12   Refill the engine with oil. Run the engine until normal operating temperature is reached and check for leaks.

---

13   Oil pump - removal, inspection and installation

Removal

Refer to illustrations 13.4a, 13.4b and 13.5
1   Remove the oil filter (see Chapter 1).
2   Remove the timing chain cover (see Section 11).

3   Remove the four bolts holding the oil filter adapter to the timing chain cover. The cover is spring loaded, so remove the bolts while keeping pressure on the cover, then release the spring pressure carefully.
4   Remove the pressure regulator valve and spring **(see illustrations)**. Use a gasket scraper to remove all traces of the old gasket.
5   Remove the oil pump cover-to-timing chain cover bolts **(see illustration)**.
6   Lift out the cover and oil pump gears as an assembly.

Inspection

Refer to illustrations 13.10, 13.11 and 13.12
7   Clean the parts with solvent and dry them with compressed air (if available). **Warning:** *Wear eye protection!*
8   Inspect all components for wear and score marks. Replace any worn out or damaged parts.
9   Reinstall the gears in the timing chain cover.
10   Measure the outer gear-to-housing clearance with a feeler gauge **(see illustration)**.
11   Measure the inner gear-to-outer gear clearance at several points **(see illustration)**.

12   Use a depth micrometer or a straight-edge and feeler gauge to measure the gear end clearance (distance from the gear to the gasket surface of the cover) **(see illustration)**.
13   Check for pump cover warpage by laying a precision straightedge across the cover and trying to slip a feeler gauge between the

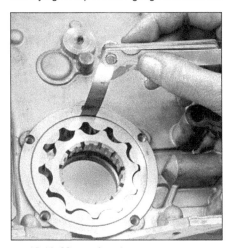

13.10 Measuring the outer gear-to-housing clearance with a feeler gauge

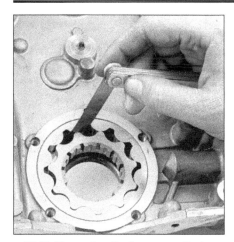

13.11 Measuring the inner gear tip-to-outer gear tip clearance with a feeler gauge

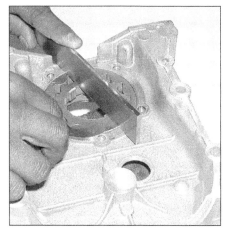

13.12 Measure the clearance between the gasket surface and the outer gear with a straightedge and feeler gauge

cover and straightedge.

14 Compare the measurements to this Chapter's Specifications. Replace all worn or damaged components with new ones.

Installation

15 Remove the gears and pack the pump cavity with petroleum jelly.

16 Install the gears - make sure petroleum jelly is forced into every cavity. Failure to do so could cause the pump to lose its prime when the engine is started, causing damage from lack of oil pressure.

17 Install the pump cover.

18 Install the pressure regulator spring and valve and the oil filter adapter (with a new gasket). **Note:** *There are early and late oil pressure relief valves* **(see illustration 13.4b)**. *Do not interchange them in the same timing chain cover. If a new timing chain cover is installed, use the new style relief valve and spring.*

19 Install the timing chain cover (see Section 11). Be sure to align the crankshaft sensor before installing the crankshaft pulley (see Chapter 6).

20 Install the oil filter and check the oil level. Start and run the engine and check for correct oil pressure, then look carefully for oil leaks at the timing chain cover.

21 Run the engine and check for oil leaks.

---

14 Driveplate - removal and installation

Refer to illustration 14.2

1 Raise the vehicle and support it securely on jackstands. Support the engine from above with a hoist or an engine support fixture, then refer to Chapter 7 and remove the transaxle.

2 Remove the bolts that secure the driveplate to the crankshaft flange. If difficulty is experienced in removing the bolts due to movement of the crankshaft, use a screwdriver or flywheel holding tool to keep the driveplate from turning **(see illustration)**. **Note:** *On some models, the driveplate will only bolt up in one orientation, but to be sure, make matching marks on the driveplate and*

the end of the crankshaft for alignment during installation.

4 Remove the driveplate from the crankshaft flange. **Caution:** *Wear gloves when handling the driveplate, as the teeth of the ring gear can be sharp.*

5 Check the starter ring gear for cracked or broken teeth. Lay the driveplate on a flat surface and use a straightedge to check for warpage. Clean and examine the metal between the large holes for evidence of cracks. **Note:** *If your driveplate has clip-on balancing weights (clipped onto the driveplate through one or more of the large holes), make sure they are attached firmly.*

6 Clean the mating surfaces of the driveplate and the crankshaft.

7 Position the driveplate against the crankshaft, matching the alignment marks made during removal. Before installing the bolts, clean their threads and apply a thread locking compound.

8 Wedge a screwdriver through the driveplate or into the ring gear teeth to keep the crankshaft from turning. Tighten the bolts to the torque listed in this Chapter's Specifications. Working in a criss-cross pattern, tighten the bolts in two or three steps.

9 The remainder of installation is the reverse of the removal procedure.

---

15 Rear main oil seal - replacement

1986 through 1990 models

Refer to illustrations 15.2, 15.3a, 15.3b, 15.4, 15.7a, 15.7b, 15.8a and 15.8b

1 Refer to Chapter 7 and remove the transaxle, then refer to Section 12 and remove the oil pan.

2 Unbolt the rear main bearing cap. Use the two main cap bolts as levers to rock the cap back and forth to loosen it for removal. **Note:** *It may help to tap the bolts on the side with a hammer to loosen the cap* **(see illustration).**

14.2 A flywheel holding tool can be used while removing the driveplate bolts, or an extension can be placed in one of the driveplate holes to stop the driveplate against the block (early driveplate shown, later models have 8 bolts)

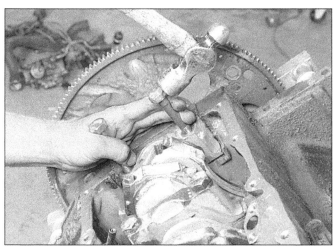

15.2 Give the bolts a few light taps with a hammer to loosen the rear main cap

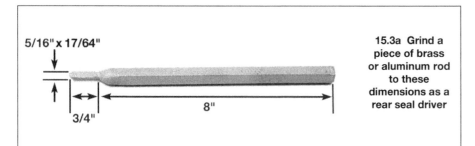

15.3a Grind a piece of brass or aluminum rod to these dimensions as a rear seal driver

15.3b Drive each end of the seal into the groove until it feels tightly packed

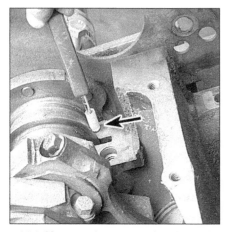

15.4 Measure the amount that the seal has been driven up into the groove - here a small screwdriver was used with a sleeve of masking tape (arrow) to serve as a depth indicator

15.7a The new seal material used in the cap can be settled into the groove with a hammer handle or by taping it in with a hammer and large-diameter socket

15.7b Trim the seal with a sharp razor blade flush with the bearing cap

3   Insert the special seal packing tool or equivalent (see illustration) against the seal. Drive the old seal into its groove until it is packed tight at each end (see illustration).
4   Measure the distance the seal was driven into the groove, then add 1/16-inch (see illustration). Cut two pieces, the calculated length, from the old seal taken from the bearing cap. Use the bearing cap as a guide when cutting.
5   Place a drop of sealant on each end of these seal pieces and then pack them into the upper groove to fill the gap made previously.
6   Trim the remaining material perfectly flush with the block. Be careful not to harm

the bearing surface. Push a thin piece of shim stock between the seal and the crankshaft, to protect the crankshaft, and trim the seal against the shim stock.
7   Install a new rope seal into the main bearing cap groove and push firmly all around using a hammer handle or special tool (see illustration). Make sure the seal is firmly seated, then trim the ends flush with the bearing cap mating surface (see illustration).
8   Apply a *thin* coat of RTV sealant to the bearing cap at the block mating surface (see illustration), install the cap and tighten the bolts finger tight. Soak the new side seals in clean engine oil for at least two minutes.

Insert the seals into the grooves on each side of the main bearing cap (see illustration). Make sure the side seals are seated at the bottom of the groove, then tighten the rear main bearing cap to the torque listed in this Chapter's Specifications. The side seals are slightly longer than the bearing cap; DO NOT

NEOPRENE COMPOSITION SEAL

APPLY SEALER
TO THESE TWO AREAS

15.8a Before installing the cap, coat the indicated area with a thin film of RTV sealant - do not get RTV on the bearing or seal

15.8b Soak the new side seals in clean engine oil, then install them in the grooves on either side of the rear main bearing cap (arrow) - install the seals after the bearing cap is installed

15.11 On 1991 through 1993 engines, the rear main seal can be pried out of the block

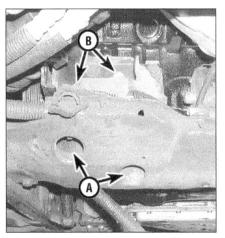

16.8a The right front engine mount is located between the front subframe and the engine - remove the two nuts (A) through the subframe and the two nuts (B) holding the mount to the engine bracket

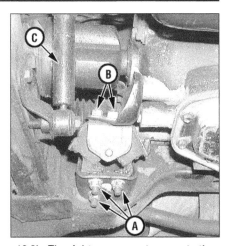

16.8b The right rear mount supports the transaxle - Remove the nuts at the subframe bracket (A), then the nuts (B) at the transaxle bracket - C is the vibration absorber; replace it if it shows signs of fluid leakage

trim the ends of the seals to fit. Install the oil pan and the remainder of the components. **Note:** *When new seals are installed, they may leak slightly until heat and oil swells the side seals sufficiently. Allow the engine to reach operating temperature before driving.*

1991 through 1993 models

Refer to illustration 15.11

9   These engines use a one-piece, neoprene rear seal that is more easily replaced than the older rope seal. Refer to Chapter 7 and remove the transaxle.
10   Refer to Section 14 and remove the driveplate.
11   Using a thin screwdriver or a seal removal tool, carefully remove the oil seal from the engine block **(see illustration)**. Be very careful not to damage the crankshaft surface while prying the seal out.
12   Clean the bore in the block and the seal contact surface on the crankshaft. Check the seal contact surface on the crankshaft for scratches and nicks that could damage the new seal and cause oil leaks - if the crankshaft is damaged, the only alternative is a new or different crankshaft. Inspect the seal bore for nicks or scratches. Carefully smooth it with a fine file if necessary, but don't nick the crankshaft in the process.
13   A special tool, available at some auto parts stores, is recommended to install the new seal. Lubricate the lips of the seal with clean engine oil. Slide the seal onto the mandrel until the dust lip bottoms squarely against the collar of the tool. **Note:** *If the tool isn't available, carefully work the seal over the crankshaft and tap it into place squarely and evenly with a hammer and large-diameter socket or length of pipe.*
14   Align the dowel pin on the tool with the dowel pin hole in the crankshaft and attach the tool to the crankshaft by hand-tightening the bolts.
15   Turn the tool handle until the collar bottoms against the case, seating the seal.
16   Loosen the tool handle, then remove the

bolts and the tool.
17   Check that the seal is installed evenly, and to the same depth as the original.
18   The remainder of installation is the reverse of removal.

---

16   Engine mounts - check and replacement

Warning: *A special engine support fixture should be used to support the engine during repair operations. These fixtures are available from equipment rental yards. Improper lifting methods or devices are hazardous and could result in severe injury or death. DO NOT place any part of your body under the engine/transaxle when it's supported only by a jack. Failure of the lifting device could result in serious injury or death.*

1   Engine mounts seldom require attention, but broken or deteriorated mounts should be replaced immediately or the added strain placed on the driveline components may cause damage or wear. The 3800 V6 engine has four main engine mounts, plus a torque strut, and on some models a vibration absorber between the bottom of the block and the frame at the timing chain end.

Check

2   During the check, the engine must be raised slightly to remove the weight from the mounts.
3   Raise the vehicle and support it securely on jackstands, then position a jack under the engine oil pan. Place a large block of wood between the jack head and the oil pan, then carefully raise the engine just enough to take the weight off the mounts. **Warning:** *DO NOT place any part of your body under the engine when it's supported only by a jack!*
4   Check the mounts to see if the rubber is cracked, hardened or separated from the

metal plates. Sometimes the rubber will split right down the center.
5   Check for relative movement between the mount plates and the engine or frame (use a large screwdriver or prybar to attempt to move the mounts). If movement is noted, lower the engine and tighten the mount fasteners.
6   Rubber preservative may be applied to the mounts to slow deterioration.

Replacement

Passenger's-side mounts

Refer to illustrations 16.8a and 16.8b

7   Disconnect the negative battery cable from the battery, then raise the vehicle and support it securely on jackstands (if not already done).
8   Raise the engine slightly with a jack or hoist. Install an engine support as described in the **Warning** above. Remove the fasteners and detach the mount from the frame bracket **(see illustrations)**.
9   Remove the mount-to-block bracket bolts/nuts and detach the mount.
10   Installation is the reverse of the removal procedure.

Driver's-side mounts

Refer to illustrations 16.14a, 6.14b and 16.14c

11   The 1986 through 1990 models have two driver's-side mounts, a front and a rear, while 1991 through 1993 models have a "cradle" attached to the left end of the transaxle and a single mount between this cradle and the subframe.
12   The left front mount on 1986 through 1990 models is attached to the transaxle near the engine/transaxle juncture, with the left-rear mount between the rear of the transaxle and the subframe.
13   Follow Steps 7 and 8 to support the engine.

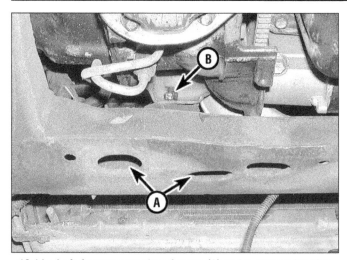

16.14a Left-front mount chassis nuts (A) and mount-to-bracket nut (B) - 1986 through 1990 models

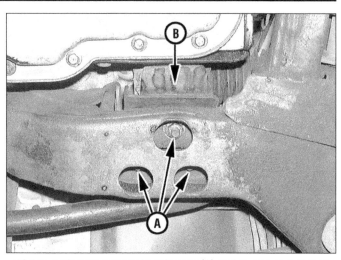

16.14b Left-rear mount chassis nuts (A) and mount-to-bracket nut (B) - 1986 through 1990 models

14   Remove the lower mounting nuts at the subframe and the nuts at the bracket to remove the mounts **(see illustrations)**.
15   Installation is the reverse of the removal procedure.

Engine torque strut

Refer to illustration 16.16
16   The engine torque strut mounts between the engine and the upper radiator support on the body **(see illustration)**. It can be replaced without jacking up the engine, by removing the through-bolts and the bolts attaching the strut to the mounts on the radiator support. **Note:** *When reinstalling this mount, you may have to use a prybar to rotate the engine toward or away from the radiator to align the through-bolts.*
17   1989 and earlier models are also equipped with a torque-axis strut located at the right-front of the engine, just ahead of the timing cover. This mounts to the engine and the chassis at the right side. It can be replaced without jacking up the engine, by removing the top and bottom torque-axis strut nuts and the bolts attaching the bracket to the timing cover.
17   Installation is the reverse of removal. Use thread-locking compound on the threads and be sure to tighten everything securely.

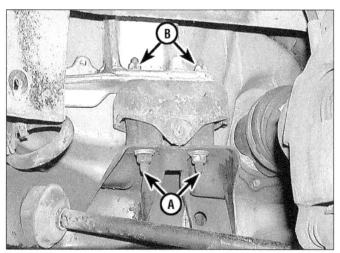

16.14c The 1991 through 1993 driver's side mount is secured by two nuts (A) to the subframe, and two nuts (B) at the "cradle"

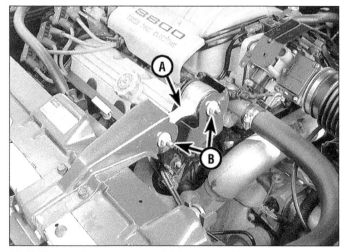

16.16 The engine torque strut (A) is mounted between the bracket at the radiator support and the bracket on the rear of the front cylinder head - remove the through-bolts (B) to replace the mount

Chapter 2 Part B
V8 engines

Contents

Specifications

General

Firing order	1-8-4-3-6-5-7-2
Displacement	
4.1L	250 cubic inches
4.5L	273 inches
4.9L	300 cubic inches

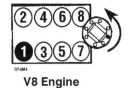

Cylinder location and distributor rotation

V8 Engine

Torque specifications

Ft-lbs (unless otherwise indicated)

Intake manifold bolts **(see illustration 7.15)**
    1986 through 1988
        Step 1 (bolts 1 through 4) 15
        Step 2 (bolts 5 through 16) 22
        Step 3 (all bolts, in sequence)... 22
        Step 4 (all bolts, in sequence)... 22
    1989 through 1993
        Step 1 (all bolts, in sequence)... 96 in-lbs
        Step 2 (all bolts, in sequence)... 144 in-lbs
        Step 3 (all bolts, in sequence)... 144 in-lbs
Exhaust manifold-to-cylinder head bolts.. 18
Power steering bracket at front exhaust 19
Camshaft sprocket bolt
    1986 through 1988.. 36
    1989 through 1993.. 31
Crankshaft pulley-to-hub bolts.. 18
Damper-to-crankshaft 70
Cylinder head bolts **(see illustration 9.7a)**
    Step 1 38
    Step 2 68
    Step 3 (bolts 1, 3 and 4) 90
Driveplate-to-crankshaft bolts
    1986 and 1987 35
    1988 through 1993 70
Oil pan bolts
    1986 through 1988... 132 in-lbs
    1989 through 1993... 168 in-lbs
Valve cover bolts .... 96 in-lbs
Rocker arm pivot-to-support bolts.. 22
Rocker arm support-to-head nuts (5) 36
Rocker arm support-to-head bolts (4) 7
Front cover bolts
    1986 through 1990 15
    1991 through 1993
        Lower 4 bolts 22
        Upper 4 bolts 40
Subframe to chassis bolts... 74

1   General information and engine identification

General information

This Part of Chapter 2 is devoted to in-vehicle repair procedures for the Cadillac V8 engines. All information concerning engine removal and installation and engine block and cylinder head overhaul can be found in Part C of this Chapter.

Since the repair procedures included in this Part are based on the assumption that the engine is still installed in the vehicle, if they are being used during a complete engine overhaul (with the engine already out of the vehicle and on a stand) many of the steps included here will not apply.

The specifications included in this Part of Chapter 2 apply only to the procedures found here. The specifications necessary for rebuilding the block and cylinder heads are included in Part C.

The Cadillac V8 engine in these models is constructed of an aluminum block with cast iron cylinder heads. Although there are three displacements of this engine (4.1L in 1986 and 1987, 4.5L from 1988 through 1990, and 4.9L from 1991 through 1993), the engines are virtually identical in all other respects. Because

the engine is largely aluminum, particular care must be taken to accommodate the characteristics of this metal. Bolts that thread into aluminum must not be over-tightened or the threads may be damaged. The cylinder head tightening procedure must be strictly followed to insure proper sealing of the head gasket. When cleaning the gasket surfaces of the aluminum, use soft wire brushes or dull scrapers. Blow out all debris from the bolt holes to avoid inaccurate torque readings.

There are some differences between the various years of this Cadillac engine, but overall, repair and maintenance procedures are nearly identical. Where differences occur, they will be noted.

2   Repair operations possible with the engine in the vehicle

Many major repair operations can be accomplished without removing the engine from the vehicle.

Clean the engine compartment and the exterior of the engine with some type of degreaser before any work is done. A clean engine will make the job easier and will help keep dirt out of the internal areas of the engine.

Depending on the components involved, it may be a good idea to remove the hood to improve access to the engine as repairs are performed (refer to Chapter 11 if necessary).

If oil or coolant leaks develop, indicating a need for gasket or seal replacement, the repairs can generally be made with the engine in the vehicle. The oil pan gasket, the cylinder head gaskets, intake and exhaust manifold gaskets, timing cover gaskets and the crankshaft oil seals are accessible with the engine in place.

Exterior engine components, such as the water pump, the starter motor, the alternator, the distributor and the fuel system components, as well as the intake and exhaust manifolds, can be removed for repair with the engine in place.

Since the cylinder heads can be removed without pulling the engine, valve component servicing can also be accomplished with the engine in the vehicle.

Replacement of, repairs to, or inspection of the timing chain and sprockets, and the oil pump are all possible with the engine in place. Camshaft replacement, however, requires the engine to be removed (see Chapter 2, Part C).

In extreme cases, caused by a lack of necessary equipment, repair or replacement of piston rings, pistons, connecting rods and

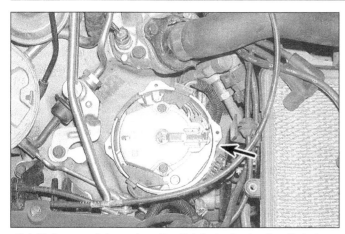

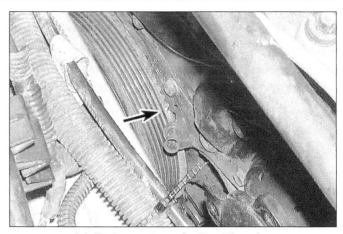

3.4 Make a mark (arrow) on the distributor body, directly below the number one plug wire terminal of the cap - align the rotor to this mark when finding TDC

3.6 Timing marks on Cadillac V8 engine

rod bearings is possible with the engine in the vehicle. However, this practice is not recommended because of the cleaning and preparation work that must be performed on the components involved.

3 Top Dead Center (TDC) for number one piston - locating

Refer to illustrations 3.4 and 3.6

1   Top Dead Center (TDC) is the highest point in the cylinder that each piston reaches as it travels up and down when the crankshaft turns. Each piston reaches TDC on the compression stroke and again on the exhaust stroke, but TDC generally refers to piston position on the compression stroke. The timing marks on the vibration damper installed on the front of the crankshaft are referenced to the number one piston at TDC on the compression stroke.

2   Positioning the piston(s) at TDC is an essential part of many procedures such as valve seal replacement, timing chain and sprocket replacement and distributor removal.

3   In order to bring any piston to TDC, the crankshaft must be turned using one of the methods outlined below. When looking at the front of the engine, normal crankshaft rotation is clockwise. **Warning:** *Before beginning this procedure, be sure to place the transmission in Neutral and unplug the wire connector at the distributor to disable the ignition system.*

a) *The preferred method is to turn the crankshaft with a large socket and breaker bar attached to the vibration damper bolt threaded into the front of the crankshaft.*

b) *A remote starter switch, which may save some time, can also be used. Attach the switch leads to the S (switch) and B (battery) terminals on the starter motor. Once the piston is close to TDC, use a socket and breaker bar as described in the previous paragraph.*

c) *If an assistant is available to turn the*

ignition switch to the Start position in short bursts, you can get the piston close to TDC without a remote starter switch. Use a socket and breaker bar as described in paragraph a) to complete the procedure.

4   Make a mark on the distributor housing directly below the number one spark plug wire terminal on the distributor cap **(see illustration)**. **Note:** *The terminal numbers may be marked on the spark plug wires near the distributor.*

5   Remove the distributor cap as described in Chapter 1.

6   Turn the crankshaft (see Step 3) until the line on the vibration damper is aligned with the zero mark on the timing indicator **(see illustration)**.

7   The rotor should now be pointing directly at the mark on the distributor housing **(see illustration 3.4)**. If it isn't, the piston may be at TDC on the exhaust stroke. To get the piston to TDC on the compression stroke, turn the crankshaft one complete turn (360-degrees) clockwise. The rotor should now be pointing at the mark.

8   When the rotor is pointing at the number one spark plug wire terminal in the distributor cap (which is indicated by the mark on the housing) and the ignition timing marks are aligned, the number one piston is at TDC on the compression stroke. **Warning:** *After set-*

ting TDC, remove the socket and breaker bar. If left in place on the crankshaft bolt when the engine is turned or started, it could cause injury to you or damage to the vehicle.

9   After the number one piston has been positioned at TDC on the compression stroke, TDC for any of the remaining cylinders can be located by turning the crankshaft 90-degrees at a time and following the firing order (refer to the Specifications).

4 Valve covers - removal and installation

Refer to illustrations 4.5, 4.6, 4.7, 4.8 and 4.11

1   If you are removing a rear valve cover and the fuel lines are in the way, refer to Chapter 4 and relieve the fuel system pressure. Disconnect the negative cable from the battery.

2   Remove the air cleaner assembly (see Chapter 4 if necessary).

3   Label and disconnect any emissions or vacuum lines and wires which are routed over the valve cover(s).

4   Refer to Chapter 10 and move the power steering pump reservoir when accessing the front valve cover.

5   Label and disconnect the spark plug wires and detach the spark plug wires and

4.5 Detach the spark plug wires and cover from the valve cover

4.6 Move aside the hoses and electrical harnesses above the rear valve cover on 1986 and 1987 models

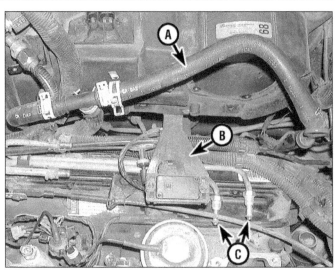

4.7 On 1988 and later engines, access to the rear valve cover is by removing the heater hose (A), the MAP sensor (B), and the fuel lines (C)

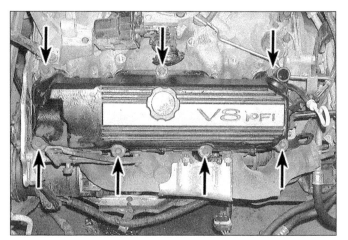

4.8 Valve cover bolt locations (arrows, front cover shown)

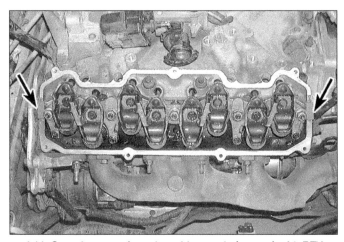

4.11 Coat the new triangular rubber seals (arrows) with RTV before installing the valve cover

cover from the valve cover **(see illustration)**.

6   Detach any remaining accessories as necessary. On 1986 and 1987 models, the rear covers can be removed by pulling aside the hoses and wires above the cover **(see illustration)**.

7   On later models, the rear cover is encumbered by several items. Refer to Chapter 6 and remove the MAP sensor, then refer to Chapter 4 and disconnect the fuel lines **(see illustration)**.

8   Remove the valve cover bolts **(see illustration)**. **Note:** *Keep track of the location of nuts/studs used as valve cover fasteners, as well as wiring tabs, hose clamps or other brackets retained by the valve cover fasteners.*

9   Remove the valve cover. **Caution:** *Do not pry on the sealing flange. To do so may damage the mating surface, causing oil leaks. Tap on the sides of the cover with a rubber mallet until it pops loose.*

10   Thoroughly clean the mating surfaces, removing all traces of old gasket material. Use acetone or lacquer thinner and a clean

rag to remove any traces of oil.

11   Remove the two triangular rubber plugs and clean their mounting points thoroughly. Apply RTV sealant to the mounting points and install new rubber triangles **(see illustration)**.

12   Install a new gasket in the cover and install it with cleaned bolts, tightening them to the torque listed in this Chapter's Specifications. Do not overtighten them, as the covers may be deformed.

13   Reinstall the remaining parts in the reverse order of removal.

14   Run the engine and check for leaks.

---

5   Rocker arms and pushrods - removal, inspection and installation

---

Removal

Refer to illustrations 5.2a and 5.2b

1   Refer to Section 4 and detach the valve

cover(s) from the cylinder head(s).

2   Remove each of the rocker arm support bolts and nuts, then remove the rocker arm assembly as a unit, loosening them evenly **(see illustration)**. Place them in order **(see illustration)**. *Note: The rocker arms can be unbolted and removed from the support for individual replacement if necessary. Be sure to mark each rocker arm pivot Left or Right to insure proper installation.*

3   Remove the pushrods and store them in order to make sure they don't get mixed up during installation **(see illustration 5.2b)**.

Inspection

Refer to illustration 5.6

4   Check each rocker arm for wear, cracks and other damage, especially where the pushrods and valve stems contact the rocker arm faces.

5   Make sure the hole at the pushrod end of each rocker arm is open.

6   Check each rocker arm pivot area for wear, cracks and galling. If the rocker arms

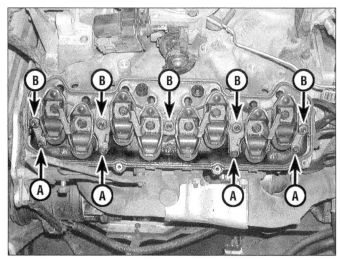

5.2a All of the rocker arms are bolted to a support, held to the head by four bolts (A) and five nuts (B) - remove the nuts/bolts and take the rockers/support out as an assembly

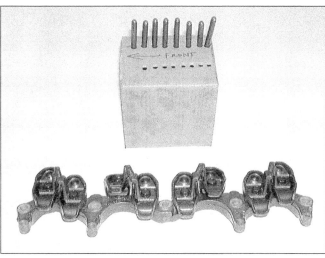

5.2b Keep the components in their original locations - if you must remove a rocker arm and fulcrum/bridge from the support, number the components so they can be returned to their original location

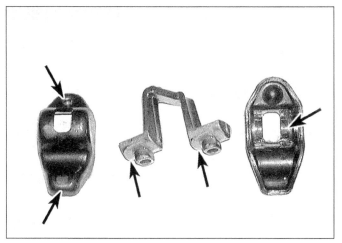

5.6 Inspect the pivots and wear surfaces of the rocker arms (arrows) - apply moly-based lubricant to these points before reassembly

6.1a Use a valve spring compressor to compress the spring far enough to remove the keepers with a small magnet

are worn or damaged, replace them with new ones and use new pivots as well **(see illustration)**.

7   Inspect the pushrods for cracks and excessive wear at the ends. Roll each pushrod across a piece of plate glass to see if it's bent (if it wobbles, it's bent).

Installation

8   Lubricate the lower end of each pushrod with clean engine oil or moly-base grease and install them in their original locations. Make sure each pushrod seats completely in the lifter.

9   Apply moly-base grease to the ends of the valve stems and the upper ends of the pushrods before positioning the rocker arm assembly over the studs on the head.

10   Bolt the rocker arm supports to the head, tightening the nuts first, then the bolts. Leave the rocker arm bolts loose until each

pushrod is seated under its rocker arm, then tighten the rocker arm bolts to the torque in this Chapter's Specifications.

11   The balance of the assembly is the reverse of the disassembly process. Start the engine and check for leaks and valvetrain noise.

---

6   Valve spring, retainer and seals - replacement

Refer to illustrations 6.1a, 6.1b, 6.1c and 6.1d

Refer to Chapter 2, Part A, for this procedure. Remove the valve covers, rocker arms and pushrods following the appropriate Sections in this Chapter. Refer to the **illustrations** in this Section for variances in V8 seals and valve assemblies.

6.1b Remove the valve guide seal (arrow) with pliers, being careful not to scratch the valve stem

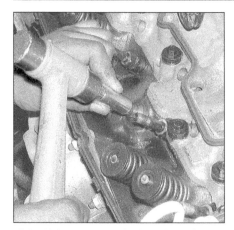

6.1c Valve seals can be installed with an appropriate-size deep socket - tap them down just until they seat on the valve guide

7 Intake manifold - removal and installation

Removal

Refer to illustrations 7.8 and 7.15

1   Relieve the fuel system pressure, then disconnect the cable from the negative terminal of the battery.

2   Drain the cooling system (see Chapter 1).

3   Remove the air cleaner assembly.

4   Detach all coolant hoses from the intake manifold.

5   Label and then disconnect any wiring and vacuum, emission and/or fuel lines which attach to the intake manifold. Detach the injector wires and remove the injectors and fuel rail (see Chapter 4).

6   Refer to Chapter 5 and remove the distributor, with the engine set to TDC (see Section 3). Mark the orientation of the distributor body and where the rotor is pointing.

7   Refer to Chapter 1 and remove the drivebelt.

8   Remove the brace between the strut towers **(see illustration)**.

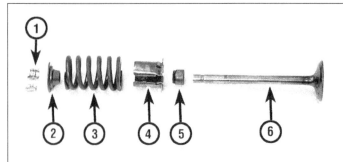

6.1d Relationship of Cadillac V8 valve components

1 *Keepers*
2 *Retainer*
3 *Valve spring*
4 *Valve spring damper*
5 *Intake valve seal*
6 *Intake valve*

7.8 Remove the bolts (A) and the brace between the strut towers - (B) is the cruise control actuator

9   Remove the coolant reservoir (see Chapter 3).

10   Refer to Chapter 5 and unbolt the alternator and set it aside.

11   Refer to Chapter 10 and unbolt the power steering pump (with belt tensioner attached) and set it aside. On 1986 and 1987 models, remove the belt idler pulley and bracket bolted to the right end of the intake manifold and block.

12   Disconnect the cruise control actuator and remove it **(see illustration 7.8)**.

13   Remove the EGR solenoid and its bracket (see Chapter 6).

14   Refer to Section 4 and remove the valve covers, then refer to Section 5 and remove the rocker arms and rocker arm supports as an assembly.

15   Remove the intake manifold bolts, keeping track of the location of each bolt **(see illustration)**. There are four lengths of intake

manifold bolts on early engines, three on later engines. Identify and mark each of the 16 intake manifold bolts as to exact location upon removal, to aid in later assembly.

16   When all of the bolts are removed, pry up the intake manifold at the ends. **Note:** *The intake manifold fits over a large dowel pin at each end of the block. When prying the manifold off, pry at both ends at the same time to pry the manifold straight up off the dowels* **(see illustration 7.15)**.

Installation

Refer to illustrations 7.18a, 7.18b and 7.18c

17   Clean the gasket surfaces of the intake manifold with a scraper followed by lacquer thinner or brake system cleaner. If the manifold was equipped with an EGR system, clean any carbon buildup from the EGR passages with carburetor spray cleaner, using a screwdriver, drill bit or lengths of coat-hanger wire to chip out hard carbon. Make sure the whole passage is clear.

18   Position new side gaskets and ridge seals, using RTV sealant in the corners to hold them in place **(see illustrations)**. Use plastic gasket retainer pins, if provided. Follow the gasket manufacturer's instructions and markings such as "UP", "TOP" or "FRONT". The end gaskets should be glued to the block, with the ends tucking into the pockets at the ends of the cylinder heads. Where the end gaskets and the side gaskets meet, apply a small dab of RTV sealant and a thin coat along the top surfaces that face the manifold.

19   Set the manifold into position until it starts onto the two dowels. Be sure all the bolt holes are lined up and start all the bolts by hand before tightening any of them (lightly oil the threads of the bolts before installation).

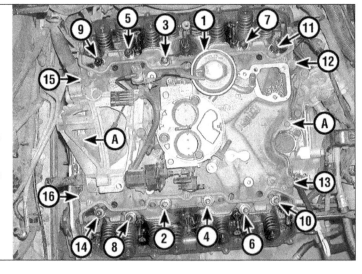

7.15 Intake manifold mounting bolts (arrows) - (A) indicates the two dowel pins that locate the manifold

7.18a At the four corners (arrows indicate two) where the heads and block come together, apply a dab of RTV sealant

7.18b The side gaskets usually come fitted with small plastic rivets (arrows) that locate the gasket precisely on the cylinder head

Note: *Locate each bolt/stud in its original location. Putting the wrong-length bolt in a hole could cause a leak or stripped threads.*

20   Tighten the bolts in stages, following the appropriate sequence **(see illustration 7.15)** until the specified torque is reached.

21   The remainder of installation is the reverse of removal. Be sure to refill the cooling system and change the engine oil and filter (see Chapter 1), as coolant often gets into the oil during manifold removal.

22   Run the engine and check for fluid and vacuum leaks. Adjust the ignition timing, if the distributor was disturbed (see Chapter 1). Road test the vehicle, checking for proper operation of engine and accessories.

8   Exhaust manifolds - removal and installation

Removal

1   Allow the engine to cool completely.

2   Place protective covers on the fenders.

3   Disconnect the negative cable from the battery.

4   Raise the vehicle and support it securely on jackstands.

5   Spray penetrating oil on the threads of fasteners holding the exhaust pipe(s) to the manifold(s) you wish to remove. Go on to the next steps, allowing the penetrating oil to soak in.

6   Remove the air cleaner and duct.

Front manifold

Refer to illustrations 8.7, 8.8 and 8.9

7   On front exhaust manifolds, remove the fasteners and take off the power steering reservoir bracket **(see illustration)**.

8   Under the front manifold, disconnect the nuts holding the crossover or crossunder pipe to the manifold (Deville models have a crossover pipe, Seville models a crossunder), and remove the brace from the exhaust stud to the motor mount **(see illustration)**. **Note:** *If more working room is needed, refer to Chapter 3 and remove the engine cooling fan.*

9   Detach the heat shield from the front manifold, then remove the bolts and studs **(see illustration)**. Keep track of the locations of any stud/nut fasteners.

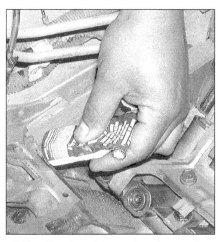

7.18c Glue the rubber end gaskets to the block with contact cement, then install the side gaskets and apply RTV in the corners where the side and end gaskets meet

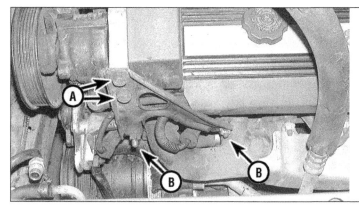

8.7 Remove the two bolts at the power steering reservoir (A), and the two nuts (B) to remove this bracket

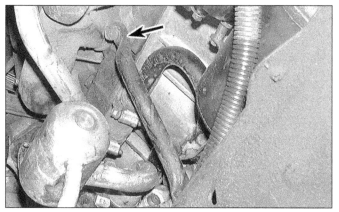

8.8 Remove the bolts and the brace (arrow) under the front manifold

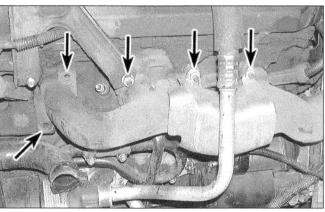

8.9 Remove the front manifold fasteners (arrows indicate five of the seven)

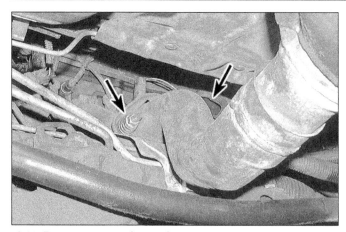

8.11 Remove the nuts (arrows) at the rear pipe-to-manifold joint

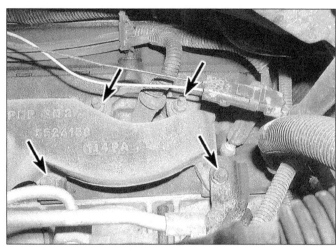

8.12a Remove the nuts (arrows indicate four of the seven) on the rear manifold - the lower right nut in this photo shows the top of a brace that connects to the right-rear powertrain mount; remove the lower bolt from the brace at the mount

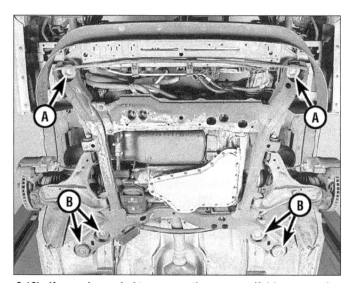

8.12b If room is needed to remove the rear manifold, support the subframe with a floor jack, loosen the front subframe bolts (A), remove the rear subframe bolts (B), and lower the subframe/engine just enough to withdraw the manifold - do not crawl underneath when the engine is supported only by the jack!

9.5 Remove the bolt (arrow) attaching the dipstick tube to the front cylinder head

10   Remove the manifold(s) from the engine compartment.

Rear manifold

Refer to illustrations 8.11, 8.12a and 8.12b

11   Remove the nuts (after soaking with penetrating oil) securing the rear exhaust pipe to the rear manifold **(see illustration)**. Disconnect the oxygen sensor wire, if applicable.

12   Remove the manifold retaining nuts and lower the manifold out and to the left (driver's side **(see illustration)**. **Note:** *On some cases, there may not be enough room to fully remove the rear manifold from the car. If so, support the subframe with a heavy-duty floor jack. Loosen the front subframe bolts and remove the rear subframe bolts **(see illustration)**. Lower the subframe/engine until the manifold can be withdrawn.* **Warning:** *Never place any part of your body under the vehicle or the engine/subframe while it is supported only by a jack!*

Installation

13   Thoroughly clean and inspect the sealing surfaces, removing all traces of old gasket material. Look for cracks and damaged threads. If the gasket was leaking, take the manifold to an automotive machine shop to inspect/correct for warpage.

14   If replacing the manifold, transfer all parts to the new unit. Using new gaskets (where applicable), install the manifold and start the bolts by hand. **Note:** *Apply a thin layer of graphite dry film lubricant to the exhaust manifold and cylinder head mating surfaces to prevent the exhaust manifold from cracking.*

15   Working from the center outward, tighten the bolts to the torque listed in this Chapter's Specifications.

16   Reinstall the remaining components in the reverse order of removal.

17   Run the engine and check for exhaust leaks.

9   Cylinder heads - removal and installation

Caution: *The engine must be completely cool when the heads are removed. Failure to allow the engine to cool off could result in head warpage.*

Removal

Refer to illustrations 9.5, 9.7a, 9.7b, 9.8, 9.9 and 9.11

1   Remove the intake manifold (see Section 7).

2   Remove the valve cover(s) as described in Section 4.

3   Remove the exhaust manifolds (see Section 8).

4   Unbolt the engine accessories and brackets as necessary and set them aside. See Chapter 5 for alternator removal. When removing the power steering pump and the drivebelt idler (see Chapter 10), leave the

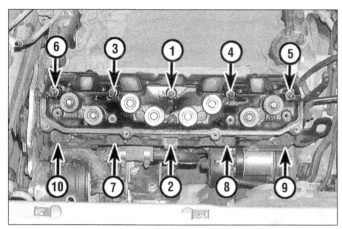

9.7a Cylinder head bolts, Cadillac V8 - numbers indicate
TIGHTENING sequence; loosen in the opposite sequence

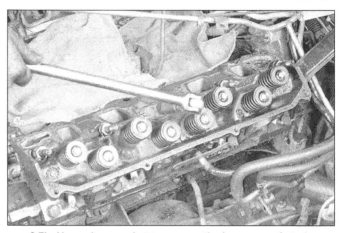

9.7b Use a deep socket to remove the inner row of studs

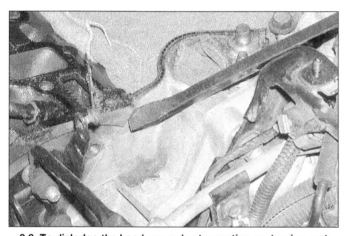

9.8 To dislodge the head, pry only at a casting protrusion, not
between the block and head

9.9 When the heads are off, keep a cylinder liner-holding tool in
place to keep the liners from moving and losing their bottom O-
ring seal

hoses attached and secure the unit in an
upright position. The refrigerant lines should
remain attached to the air conditioning compressor (see Chapter 3). **Warning:** *The air
conditioning system is under high pressure.
Do not loosen any hose or fitting until the
refrigerant has been discharged by an automotive air conditioning technician.*
5   Recheck to be sure nothing is still connected to the head(s) you intend to remove,
including ground wires bolted to the heads,
temperature senders or engine lifting brackets
(one on each cylinder head). If you're removing the front cylinder head, unbolt the oil dipstick tube from the head **(see illustration)**.
6   Using a new gasket, outline the cylinders and bolt pattern on a piece of cardboard. Be sure to indicate the front of the
engine for reference. Punch holes at the bolt
locations.
7   Loosen the head bolts in 1/4-turn increments until they can be removed by hand.
Work from bolt to bolt in a pattern that's the
reverse of the tightening sequence **(see illustration)**. Store the bolts in the cardboard
holder as they are removed; this will insure
the bolts are reinstalled in their original holes.
Note: *The top (inner) row of head bolts are
studs, which will require a deep socket for*

removal **(see illustration)**.
8   Lift the head(s) off the engine. If excessive resistance is felt, do not pry between the
head and block as damage to the mating surfaces may result. To dislodge the head, place
a block of wood against the end of it and
strike the wood with a hammer, or pry on a
casting protrusion **(see illustration)**.
9   Store the head(s) on their sides to prevent damage to the gasket surface. **Note:**
*Install a cylinder liner holder, available at auto
parts stores and automotive tool suppliers, to*

9.11 Scrape the block
surface, vacuum up any
debris , then clean the
surface with lacquer
thinner and a clean rag

the engine block to prevent the liners from
shifting in the event the engine is turned over
or disturbed **(see illustration)**. **Caution:** *If the
crankshaft is turned while the cylinder heads
are removed, the iron cylinder bore liners may
lose their sealing at the O-rings in the block,
causing water leaks later.*
10   Cylinder head disassembly and inspection procedures are covered in detail in
Chapter 2, Part C.
11   Use a gasket scraper to remove all
traces of carbon and old gasket material,

9.14 Install the new head gasket, locating it over the dowel pins (arrows) on the block

10.5 On 1991 and later models, remove the bolt in the center of the crankshaft pulley while using a chain or strap wrench to keep the pulley from turning - if using a chain wrench, protect the pulley's grooves with a leather strap or a piece of old drivebelt wrapped around the pulley

then clean the mating surfaces with lacquer thinner or acetone or brake system cleaner **(see illustration)**. If there's oil on the mating surfaces when the heads are installed, the gaskets may not seal properly and leaks may develop. When working on the block, cover the lifter valley with shop rags to keep out debris. Use a vacuum cleaner to remove any debris that falls into the cylinders. **Caution:** *The block is aluminum - use care not to gouge the surface when cleaning it, and do not disturb the liner-to-block sealing.*

12   Check the block and head mating surfaces for nicks, deep scratches and other damage. If damage is slight, it can be removed with a file. If the gasket seemed to be leaking, have the head(s) checked for cracks and warpage by an automotive machine shop. Surface damage or warpage can usually be corrected by machining. Cracks normally require replacement of the head.

Installation

Refer to illustration 9.14

13   Use a tap of the correct size to chase the threads in the head bolt holes. Mount each bolt in a vise and run a die down the threads to remove any corrosion and restore the threads. Dirt, corrosion, sealant and damaged threads will affect torque readings. This kind of fastener and thread cleaning is especially important on the aluminum block in the vehicles covered by this manual.

14   Most gaskets have "THIS SIDE UP" stamped into them, but follow the gasket manufacturer's instructions for sealant to use, if any. Most composition gaskets are installed without sealant. Position a new gasket over the dowel pins in the block **(see illustration)**.

15   Be sure the head bolt holes are clean. Apply a light coating of engine assembly lube on the threads and under the bolt heads. Head bolts that enter water jacket areas should be coated with a non-hardening sealant such as Permatex No. 2.

16   Install the bolts in their original locations and tighten them finger tight. Follow the recommended sequence **(see illustration 9.7a)** and tighten the bolts in three steps to the

torque values listed in this Chapter's Specifications.

17   The remaining installation steps are the reverse of removal.

18   Add coolant, change the engine oil and filter (see Chapter 1), then start the engine and check carefully for leaks.

10   Crankshaft front oil seal - replacement

Timing cover in place

Refer to illustrations 10.5, 10.6, 10.7 and 10.9

1   Remove the drivebelt (see Chapter 1).

2   Loosen the bolts on the right front wheel, raise the vehicle and support it securely on jackstands. Remove the right front wheel.

3   On some models, there isn't enough room between the engine and the right fenderwell for the damper or pulley to be removed. When this is the case, support the engine/subframe with a heavy-duty floor jack and loosen the rear subframe bolts. Remove the front subframe bolts **(see illustration 8.12b)**. Lower the engine/subframe several inches, until the crankshaft damper is clear of the fenderwell. **Warning:** *Do not place your body under the engine at any time that it is supported only by a jack!*

4   Remove the bolts from the crankshaft pulley on 1990 and earlier models (later models have only a damper without a pulley). Use a strap wrench or chain wrench over the pulley to hold it while the bolts are removed. If a chain wrench is used, place a rubber strap, a length of old leather belt or a length of old drivebelt between the pulley grooves and the chain to prevent damage.

5   On 1991 and later models, use the chain wrench as described above to hold the damper while you use a long breaker bar to remove the center bolt of the damper **(see illustration)**. **Caution:** *It will take lots of leverage and force to remove this bolt unless you have an air-powered impact wrench. Be certain that the engine is securely locked from turning (an assistant is helpful) while you use*

both hands and a long breaker bar to loosen the bolt. Make sure your hands are not going to hit anything when the bolt does break loose.

6   Attach a puller to the damper and draw it off the crankshaft. Be careful not to drop it as it comes free. On 1990 and earlier models that have threaded holes provided for a puller, a bolt-type puller (such as a steering wheel puller) can be used in these bolt holes. On 1991 and later models, a gear puller that grabs the hub must be used **(see illustration)**.

7   Carefully pull the seal out of the cover with a seal removal tool or pry it out with a large screwdriver **(see illustration)**. Be careful not to distort the cover or scratch the wall of the seal bore (cover the screwdriver tip with electrical tape to avoid nick-

10.6 Use a three-jaw puller to remove the damper from the crankshaft on 1991 and later models (the jaws must grasp the hub - not the outer circumference of the damper)

10.7 Use a screwdriver or seal removal tool (shown) to pull out the front seal, being careful not to nick the crankshaft surface

10.9 Use an appropriate-size socket to drive the new seal squarely into the front cover, to the same depth as the original seal

ing the crankshaft). It's a good idea to apply penetrating oil to the seal-to-cover joint and allow it to soak in before attempting to pull the seal out.

8   Clean the bore to remove any old seal material and corrosion. Position the new seal in the bore with the open end of the seal facing IN. A small amount of oil applied to the outer edge of the new seal will make installation easier.

9   Drive the seal squarely into the bore with a large socket and hammer until it's completely seated **(see illustration)**. Select a socket that's the same outside diameter as the seal (a section of pipe can be used if a socket isn't available). **Caution:** *If the crankshaft has a Woodruff key in place, the socket used must be big enough in diameter to clear the key while driving on the seal.* Proceed to Step 13.

Timing cover removed

Refer to illustrations 10.10 and 10.12

10   Use a punch or screwdriver and hammer to drive the seal out of the cover from the back side. Support the cover as close to the seal bore as possible **(see illustration)**. Be careful not to distort the cover or scratch the wall of the seal bore. If the engine has accumulated a lot of miles, apply penetrating oil to the seal-to-cover joint on each side and allow it to soak in before attempting to drive the seal out.

11   Clean the bore to remove any old seal material and corrosion. Support the cover on blocks of wood and position the new seal in the bore with the open end of the seal facing IN. A small amount of oil applied to the outer edge of the new seal will make installation easier.

12   Drive the seal into the bore with a wood block and hammer until it's completely seated **(see illustration)**. Proceed to the next Step.

Damper installation

13   Apply anti-seize compound to the snout

10.10 Support the front cover on wood blocks and drive the seal out from the backside with a hammer and punch - don't distort the seal bore

of the crankshaft and multi-purpose grease or clean engine oil to the lips of the seal in the front cover, then install the damper, making sure the keyway in the hub is aligned with the key on the crankshaft.

14   On 1990 and earlier models, a special installer tool (available at most auto parts stores) is recommended to press the damper onto the crankshaft. If this tool is not available, a threaded rod of the proper diameter and thread pitch, nut and large washer can be used to press the damper on. The damper hub is a press-fit on the crankshaft. When reinstalling it, use anti-seize lubricant on the crankshaft snout. When the hub has been pulled on far enough, it will bottom out on the crankshaft. Remove the installer and reinstall the crankshaft plug.

15   On 1991 and later models the damper retaining bolt can be used to pull the damper onto the crankshaft, although the tool described in the previous Step is recommended. When tightening the damper bolt, use the chain wrench and long breaker bar as described in Step 5.

10.12 Use a block of wood to squarely drive in the new seal

11   Timing chain - removal, inspection and installation

Removal

Refer to illustrations 11.7a, 11.7b, 11.9, 11.10 and 11.11

1   Disconnect the negative cable from the battery. See Chapter 5 and remove the battery.

2   Refer to Chapter 3 and remove the coolant recovery tank.

3   Drain the cooling system and disconnect the radiator hoses, heater hose (where applicable) and the small by-pass hose (where applicable).

4   Loosen the bolts on the water pump pulley, then remove the drivebelt (see Chapter 1). Remove the pulley.

5   Remove the crankshaft pulley and the vibration damper (see Section 10).

6   On 1986 and 1987 models, remove the idler pulley and bracket bolted to the left-front of the intake manifold and block.

7   Remove the nuts and bolts that attach the cover to the engine, then pull the cover

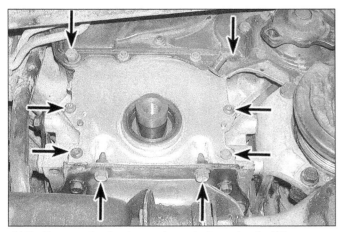

11.7a Remove the front cover lower bolts and pull off the cover - the front pan bolts must be loosened (arrows also indicate the two pan bolts that must be removed)

11.7b Remove the upper front cover bolts (arrows)

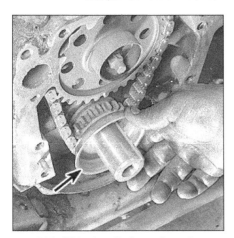

11.9 Slide the oil slinger from the crankshaft, noting which way it was installed

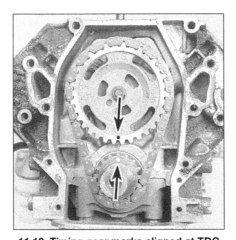

11.10 Timing gear marks aligned at TDC for number 1 cylinder

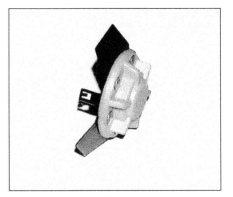

11.11 Remove the camshaft thrust bearing - it snaps out of the camshaft gear

free **(see illustrations)**. If it's stuck, knock it loose with a hammer and a wood block. Don't pry between the mating surfaces, or oil leaks may result. Keep track of the location of the studs. **Note:** *The water pump and the water pump inlet (on the back of the front cover) can remain bolted to the front cover during removal.*

8   Thoroughly clean all gasket mating surfaces (do not allow the old gasket material to fall into the oil pan), then wipe them with a cloth soaked in lacquer thinner. Clean sealant and gasket material from the exposed front edges of the oil pan.

9   Slide the oil slinger off of the end of the crankshaft **(see illustration)**.

10   Turn the crankshaft until the marks on the camshaft and crankshaft sprockets are perfectly aligned **(see illustration)**. DO NOT attempt to remove either sprocket or the chain until this is done. Also, do not turn the camshaft or the crankshaft after the sprockets are removed.

11   Remove the camshaft thrust bearing and the bolt securing the sprocket to the end of the camshaft **(see illustration)**.

12   Generally speaking, the camshaft

sprocket and the timing chain can be slipped off the crankshaft sprocket together. If resistance is encountered, it may be necessary to use two screwdrivers to carefully pry the sprocket off of the camshaft. If resistance is encountered on a crankshaft sprocket, a gear puller will be required.

13   If the crankshaft and the camshaft are not disturbed while the timing chain and sprockets are out of place, then installation can begin with Step 16. If the engine is completely dismantled, or if the crankshaft or camshaft are disturbed while the timing chain is off, then the engine must be positioned at TDC before the timing chain and sprockets are installed (see Section 3).

Installation

Refer to illustration 11.15

14   Align the hole in the camshaft sprocket with the dowel pin in the end of the camshaft, then slip the sprocket onto the end of the camshaft to align the timing marks **(see illustration 11.10)**. **Note:** *When the camshaft and crankshaft timing marks are aligned, the engine is positioned at TDC compression for cylinder number 1.*

15   Slip the crankshaft sprocket onto the end of the crankshaft (make sure the key and keyway are properly aligned), then turn the crankshaft until the timing mark on the sprocket is pointed straight up **(see illustration)**.

11.15 If a new timing chain set is installed, the new crankshaft sprocket may have to be tapped on with a large socket and hammer - make sure the keyway is aligned and that the socket doesn't hit the Woodruff key

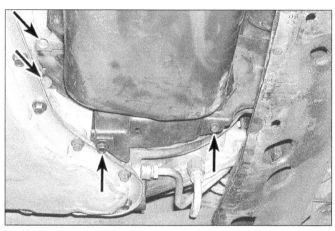

12.3a Remove the driveplate cover bolts (arrows) from below . . .

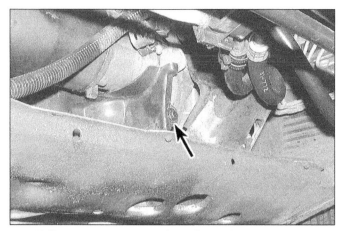

12.3b . . . and the one bolt (arrow) near the starter motor from the front

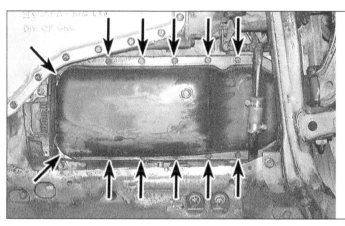

12.4 Remove the oil pan bolts (not all bolts are visible in this photo)

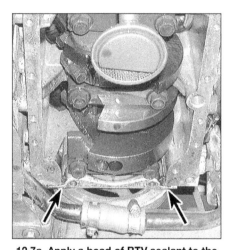

12.7a Apply a bead of RTV sealant to the joints (arrows) between the block and the front cover . . .

16   Now remove the camshaft sprocket and lay the timing chain over the sprocket. Slip the other end of the chain under the crankshaft sprocket (keep the timing marks aligned as this is done) and reinstall the camshaft sprocket.

17   Install the camshaft sprocket bolt, then recheck the alignment of the timing marks.

18   Tighten the camshaft sprocket bolt to the torque listed in this Chapter's Specifications, then install the camshaft thrust bearing. If the original bearing looks at all worn, install a new one.

19   Be sure to clean the mating surfaces of the front cover and cylinder block, and install a new gasket, using RTV sealant around the water passage holes. **Note:** *Be sure to reinstall the oil slinger on the crankshaft before installing the front cover.*

20   Apply a bead of RTV sealant to the front of the oil pan, especially where the oil pan meets the block.

21   Install the water pump and water pump inlet to the cover with new gaskets (if removed from cover).

22   Install the cover, aligning it over the two dowels in the block, and install all the bolts finger-tight.

23   Tighten all bolts to the torque values listed in this Chapter's Specifications and install all components in the reverse order of removal. Apply lubricant to the front hub seal

before installing the crankshaft damper (see Section 10).

12   Oil pan - removal and installation

Warning: *Do not place any part of your body below the engine when it is supported solely by a jack or hoist.*

Removal

Refer to illustrations 12.3a, 12.3b and 12.4

1   Disconnect the negative battery cable and drain the engine oil (see Chapter 1). Allow the engine to cool before removing the oil pan.

2   Raise the vehicle and support it securely on jackstands.

3   Remove the driveplate cover bolts and the cover **(see illustrations)**.

4   Remove the oil pan bolts **(see illustration)**. Note the different sizes and their locations.

5   Remove the oil pan by tilting it down at the rear and then working the front clear of the subframe. It may be necessary to use a rubber mallet to break the seal - don't pry between the pan and the block!

Installation

Refer to illustrations 12.7a and 12.7b

6   Before installing, thoroughly clean the

12.7b . . . and where the rear main cap meets the block (arrows)

gasket sealing surfaces on the engine block and on the oil pan. All sealant and gasket material must be removed.

7   Apply a bead of RTV sealant to the corners where the block meets the rear main cap and the front cover **(see illustrations)**.

13.2 The oil pump is bolted to the rear main bearing cap - remove the mounting bolts and nut (arrows)

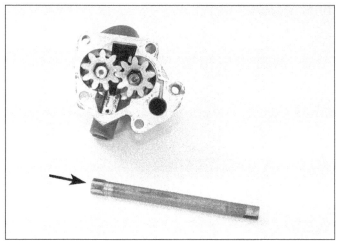

13.6 Lube the gears in the pump, install a new gasket on the pump body and install the hex end (arrow) of the driveshaft into the drive gear in the pump

8   Apply a continuous bead of RTV sealant to the pan, making sure the bead travels around the inside of the bolt holes. Install the pan while the sealant is still wet.
9   Lift the pan into position and install all bolts finger-tight. There is no specific order for tightening the bolts; however, it is a good policy to tighten the end bolts first, and all of the bolts should be started before any are tightened to Specifications.
10   The remainder of installation is the reverse of the removal procedure. Allow at least an hour after installing the pan for the sealant to cure, then fill the engine with the correct grade and quantity of oil (see Chapter 1), start the engine and check for leaks.

13   Oil pump - removal, inspection and installation

Removal

Refer to illustration 13.2
1   Remove the oil pan as described in Section 12.
2   Remove the bolts securing the oil pump assembly to the main bearing cap **(see illustration)**. Remove the oil pump with its pickup tube and screen as an assembly from the engine block. Once the pump is removed, the oil pump driveshaft can be withdrawn from the block.

Inspection

3   Remove the pump cover and fill the cavity with engine oil. If it slowly leaks out, this is an indication that the pressure relief valve is not closing properly.
4   Inspect the pump housing and cover for signs of wear or debris. Check the cover with a straightedge for flatness.
5   Because of the critical importance of the oil pump and the risk of engine damage if an overhauled pump doesn't perform like new,

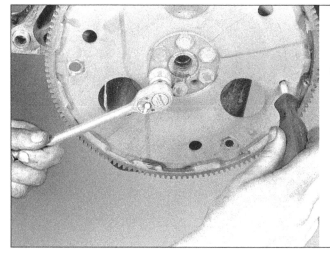

14.3 Mark the driveplate with reference to the end of the crankshaft before removal - hold the driveplate by jamming a large screwdriver through one of the holes and against the block, while removing the driveplate mounting bolts

we do not recommend overhauling the oil pump on these models. Always replace the pump with a new unit.

Installation

Refer to illustration 13.6
6   To install, slip the driveshaft extension into the oil pump driveshaft, with the hex into the pump **(see illustration)**. **Note:** *Pour clean engine oil into the pickup and turn the drive-shaft to prime the pump.*
7   Install new O-rings at each end of the oil pump's supply tube (at the pump and the block).
8   Move the pump assembly into position and align the extension with the lower end of the distributor. The distributor drives the oil pump, so it is essential that these two components mate properly. The driveshaft extension fits into the slot in the distributor drive gear.
9   Install the bolts and tighten them securely. Make sure the oil pump screen is parallel with the oil pan rails of the block. The screen must be in this position to fit into the oil pan properly.

10   Install the oil pan (see Section 12). Refill the crankcase with oil (see Chapter 1), run the engine and check for leaks.

14   Driveplate - removal and installation

Refer to illustration 14.3

Removal

1   Refer to Chapter 7 and remove the transaxle.
2   Use a scribe or a marking pen to draw a line from the driveplate to the end of the crankshaft for correct alignment during reinstallation.
3   Remove the bolts that secure the driveplate to the crankshaft rear flange. If difficulty is experienced in removing the bolts due to movement of the crankshaft, wedge a screwdriver to keep the driveplate from turning **(see illustration)**.
4   Remove the driveplate from the crankshaft flange, noting the location and installed direction of any spacer rings.

15.2 Use a seal removal tool or a screwdriver to pry out the old seal - tape the screwdriver tip to protect the crankshaft (typical application shown)

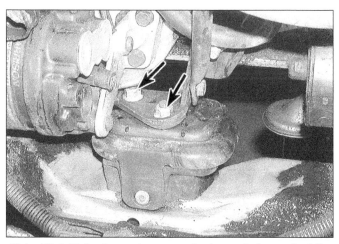

16.1a V8 right-front mount - remove the two nuts (arrows) at the engine bracket and two nuts below the subframe

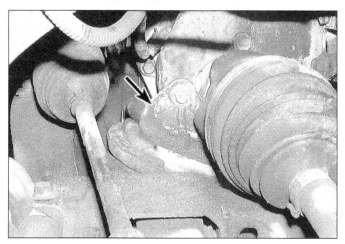

16.1b The right-rear mount is attached at the top with nuts (nuts are hidden here, but arrow shows where) . . .

16.1c . . . and below with two nuts (arrows) at the subframe

Caution: *Wear gloves or protect your hands with rags; the ring gear teeth can be sharp.*

5   Clean any grease or oil from the driveplate. Check for cracked or broken starter ring gear teeth. Lay the driveplate on a flat surface and use a straightedge to check for warpage.

6   Clean the mating surfaces of the driveplate and the crankshaft.

Installation

7   Position the driveplate against the crankshaft, matching the alignment marks made during removal. Before installing the bolts, apply a non-hardening thread-locking compound to the threads.

8   Wedge a screwdriver through the driveplate or into the ring gear teeth to keep the crankshaft from turning. Tighten the bolts to the torque listed in this Chapter's Specifications in two or three steps, working in a criss-cross pattern.

9   The remainder of installation is the reverse of the removal procedure.

15   Rear main oil seal - replacement

Refer to illustration 15.2

1   Refer to Section 14 and remove the driveplate.

2   Using a thin screwdriver or a seal removal tool, carefully remove the oil seal from the engine block **(see illustration)**. Be very careful not to damage the crankshaft surface while prying the seal out.

3   Clean the bore in the block and the seal contact surface on the crankshaft. Check the seal contact surface on the crankshaft for scratches and nicks that could damage the new seal and cause oil leaks - if the crankshaft is damaged, the only alternative is a new or different crankshaft. Inspect the seal bore for nicks or scratches. Carefully smooth it with a fine file if necessary, but don't nick the crankshaft in the process.

4   A special tool, available at most auto parts stores, is recommended to install the new seal. Lubricate the lips of the seal with clean engine oil. Slide the seal onto the mandrel until the dust lip bottoms squarely

against the collar of the tool. **Note:** *If the tool isn't available, carefully work the seal over the crankshaft and tap it into place squarely and evenly with a hammer and large-diameter socket or length of pipe.*

5   Align the dowel pin on the tool with the dowel pin hole in the crankshaft and attach the tool to the crankshaft by hand-tightening the bolts.

6   Turn the tool handle until the collar bottoms against the case, seating the seal.

7   Loosen the tool handle, then remove the bolts and the tool.

8   Check that the seal is installed evenly, and to the same depth as the original.

9   The remainder of installation is the reverse of removal.

16   Engine mounts - check and replacement

Refer to illustrations 16.1a, 16.1b, 16.1c, 16.1d, 16.1e and 16.1f

Engine mounts seldom require attention, but broken or deteriorated mounts

16.1d On some models, the transaxle end of the powertrain is supported by two mounts, arrow indicates the left-front

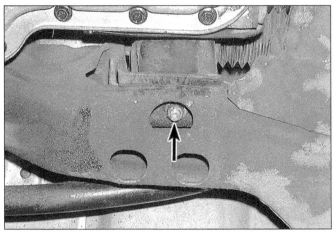

16.1e The left-rear is mounted to the engine bracket and the subframe (arrow)

should be replaced immediately or the added strain placed on the driveline components may cause damage. The procedures to check and replace V8 engine mounts are essentially the same as for the V6 engine, as covered in Part A of this Chapter, but refer to these **illustrations** for differences in the V8 engine mounts.

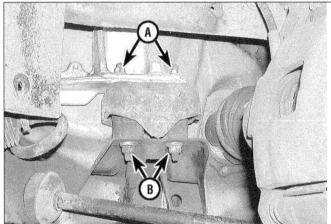

16.1f Instead of two left-side mounts, some models have an aluminum "cradle" attached to the transaxle, which is isolated from the chassis by a single mount - remove the upper nuts (A) at the cradle, then the lower nuts (B) at the subframe

Chapter 2 Part C
General engine overhaul procedures

Contents

Specifications

V6 engine, Buick and Oldsmobile

See Chapter 2, Part A, for additional information

General
Displacement	
3.8L	231 cubic inches
Oil pressure at 1850 rpm	
1986 to 1990	40 psi
1991 to 1993	60 psi
Compression pressure	
Minimum	100 psi
Allowable variation between cylinders	20%
Cylinder head warpage limit	0.003 inch per 6 inches

Camshaft and lifters
Lifter to lifter-bore clearance	0.0008 to 0.0025 inch
Camshaft lobe lift	
1986 (VIN code 3)	
Intake	0.237 inch
Exhaust	0.247 inch

V6 engine, Buick and Oldsmobile (continued)

Camshaft and lifters (continued)

Camshaft lobe lift
    1987, intake and exhaust .... 0.256 inch
    1988, intake and exhaust .... 0.245 inch
    1989 through 1993
        Intake .... 0.250 inch
        Exhaust .... 0.255 inch
Camshaft to bearings clearance
    New .... 0.0005 to 0.0035 inch
    Used, limit .... 0.004 inch

Balance shaft

    Endplay .... 0.008 inch
    Radial play, front .... 0.0 to 0.028 inch
    Radial play, rear .... 0.0127 to 0.119 inch

Valves

Valve stem diameter
    Intake .... 0.3412 to 0.3401 inch
    Exhaust .... 0.3412 to 0.3405 inch
Valve margin width (limit) .... 0.025 inch
Valve spring installed height .... 1.690 to 1.720 inches
Valve stem-to-guide clearance
    Intake .... 0.0015 to 0.0035 inch
    Exhaust .... 0.0015 to 0.0032 inch

Engine block

Cylinder bore diameter .... 3.800 inches
Cylinder bore taper limit 0.0005
Cylinder bore out-of-round limit .... 0.0004 inch

Pistons and rings

Piston-to-bore clearance (measured at top of skirt)
    1986 through 1990 .... 0.0007 to 0.0027 inch
    1991 through 1993 .... 0.0004 to 0.0022 inch
Piston ring end gap
    Compression rings .... 0.010 to 0.0025 inch
    Oil rings .... 0.015 to 0.055 inch

Crankshaft, connecting rods and bearings

Connecting rod side clearance (total - both rods) .... 0.003 to 0.015 inch
Connecting rod bearing oil clearance
    1986 through 1991 .... 0.0005 to 0.0026 inch
    1992 and 1993 .... 0.0008 to 0.0022 inch
Rod and main journal taper/out-of-round limit .... 0.0003 inch
Crankshaft endplay .... 0.003 to 0.0011 inch
Main bearing journal diameter .... 2.4988 to 2.4998 inches
Main bearing oil clearance .... 0.0008 to 0.0022 inch

Torque specifications*

                                                        Ft-lbs (unless otherwise indicated)

Main bearing cap bolts
    1986 and 1987 100
    1988 through 1990 90
    1991 through 1993
        Step 1 26
        Step 2 .... Tighten an additional 50-degrees rotation
Connecting rod cap nuts
    1986 through 1990 40
    1991 through 1993
        Step 1 20
        Step 2 .... Tighten an additional 50-degrees rotation
Balance shaft bearing retainer bolts 26
Balance shaft gear bolt
    Step 1 .... 168 in-lbs
    Step 2 .... Tighten an additional 35-degrees rotation

***Note:** Refer to Part A for additional torque specifications.*

Cadillac V8 engines

See Chapter 2, Part B, for additional information

General

Displacement	
4.1L	250 cubic inches
4.5L	273 inches
4.9L	300 cubic inches
Oil pressure	
Idle..	10 psi
2000 rpm	53 psi
Compression pressure	
Standard..	140 to165 psi
Allowable variation between cylinders	20%
Cylinder head warpage limit	0.003 inch per 6 inches
Bore diameter	
4.1L	3.465 inches
4.5L	3.465 inches
4.9L	3.465 inches
Oil pressure (minimum)...	30 psi @ 2600 rpm
Compression pressure	
Allowable variation between cylinders	20%
Minimum..	140 psi
Cylinder head warpage limit	0.006 inch

Camshaft and lifters

Lifter-to-lifter carrier clearance	0.007 to 0.0027 inch
Camshaft lobe lift	
Intake	0.240 inch
Exhaust	0.247 inch
Camshaft-to-bearings clearance	
New	0.0018 to 0.0037 inch
Used, limit	0.004 inch

Valves

Valve stem diameter	
Intake	0.3420 to 0.3413 inch
Exhaust	0.3401 to 0.3408 inch
Valve margin width (minimum)..	0.025 inch
Valve stem-to-guide clearance (new)	
Intake	0.001 to 0.003 inch
Exhaust	0.002 to 0.004 inch
Valve stem-to-guide clearance limit	0.005 inch

Engine block

Cylinder bore diameter	
4.1L	3.465 inches
4.5L	3.622 inches
4.9L	3.623 inches
Piston-to-bore clearance*	0.0010 to 0.0018 inch

***Note:** Measure pistons at top of skirt, perpendicular to pin.*

Pistons and rings

Piston ring end gap	
4.1L and 4.5L	
Compression rings..	0.015 to 0.024 inch
Oil rings..	0.010 to 0.050 inch
4.9L	
Compression rings..	0.012 to 0.022 inch
Oil rings..	0.0004 to 0.020 inch
Compression ring side clearance	0.0016 to 0.0037 inch
Oil ring side clearance	None (side sealing)

Crankshaft, connecting rods and main bearings

Connecting rod endplay (side clearance)	0.008 to 0.020 inch
Crankshaft endplay	
1986 and 1987	
New limits	0.001 to 0.007 inch

Cadillac V8 engines (continued)

Crankshaft, connecting rods and main bearings (continued)

Crankshaft endplay	
1989 through 1993	
New limits	0.001 to 0.008 inch
Worn limits (all)	0.015 inch
Main bearing journal diameter	2.637 to 2.638 inches
Connecting rod journal diameter	
1986 through 1990	1.929 inches
1991 through 1993	1.927 to 1.928 inches
Connecting rod bearing oil clearance	
Production	0.0005 to 0.0028 inch
Wear limit	0.0035 inch
Connecting rod bearing journal out-of-round limit	0.0003 inch
Main bearing oil clearance	
No. 1	0.0008 to 0.0031 inch
Nos. 2, 3, 4 and 5	0.0016 to 0.0039 inch

Torque specifications*

	Ft-lbs (unless otherwise specified)
Main bearing cap bolts	85
Connecting rod bolts	22

*Note: *Refer to Part B for additional torque specifications.*

---

1   General information - engine overhaul

Included in this portion of Chapter 2 are the general overhaul procedures for the cylinder head and internal engine components.

The information ranges from advice concerning preparation for an overhaul and the purchase of replacement parts to detailed, step-by-step procedures covering removal and installation of internal engine components and the inspection of parts.

The following Sections have been written based on the assumption that the engine has been removed from the vehicle. For information concerning in-vehicle engine repair, as well as removal and installation of the external components necessary for the overhaul, see Chapter 2A (Buick and Oldsmobile V6 models) or 2B (Cadillac V8 models).

The Specifications included in this Part are only those necessary for the inspection and overhaul procedures which follow. Refer to Chapter 2, Part A or Part B for additional Specifications.

It's not always easy to determine when, or if, an engine should be completely overhauled, as a number of factors must be considered.

High mileage is not necessarily an indication that an overhaul is needed, while low mileage doesn't preclude the need for an overhaul. Frequency of servicing is probably the most important consideration. An engine that's had regular and frequent oil and filter changes, as well as other required maintenance, will most likely give many thousands of miles of reliable service. Conversely, a neglected engine may require an overhaul very early in its life.

Excessive oil consumption is an indication that piston rings, valve seals and/or valve guides are in need of attention. Make sure that oil leaks aren't responsible before deciding that the rings and/or guides are bad. Perform a cylinder compression check to determine the extent of the work required (see Section 3). Also check the vacuum readings under various conditions (see Section 4).

Loss of power, rough running, knocking or metallic engine noises, excessive valve train noise and high fuel consumption rates may also point to the need for an overhaul, especially if they're all present at the same time. If a complete tune-up doesn't remedy the situation, major mechanical work is the only solution.

An engine overhaul involves restoring the internal parts to the specifications of a new engine. During an overhaul, the piston rings are replaced and the cylinder walls are reconditioned (re-bored and/or honed). If a re-bore is done by an automotive machine shop, new oversize pistons will also be installed. The main bearings, connecting rod bearings and camshaft bearings are generally replaced with new ones and, if necessary, the crankshaft may be reground to restore the journals. Generally, the valves are serviced as well, since they're usually in less-than-perfect condition at this point. While the engine is being overhauled, other components, such as the distributor, starter and alternator, can be rebuilt as well. The end result should be a like new engine that will give many trouble free miles. **Note:** *Critical cooling system components such as the hoses, drivebelts, thermostat and water pump should be replaced with new parts when an engine is overhauled. The radiator should be checked carefully to ensure that it isn't clogged or leaking (see Chapter 3). If you purchase a rebuilt engine or*

short block, some rebuilders will not warranty their engines unless the radiator has been professionally flushed. Also, we don't recommend overhauling the oil pump - always install a new one when an engine is rebuilt.

Before beginning the engine overhaul, read through the entire procedure to familiarize yourself with the scope and requirements of the job. Overhauling an engine isn't difficult, but it is time-consuming. Plan on the vehicle being tied up for a minimum of two weeks, especially if parts must be taken to an automotive machine shop for repair or reconditioning. Check on availability of parts and make sure that any necessary special tools and equipment are obtained in advance. Most work can be done with typical hand tools, although a number of precision measuring tools are required for inspecting parts to determine if they must be replaced. Often an automotive machine shop will handle the inspection of parts and offer advice concerning reconditioning and replacement. **Note:** *Always wait until the engine has been completely disassembled and all components, especially the engine block, have been inspected before deciding what service and repair operations must be performed by an automotive machine shop. Since the block's condition will be the major factor to consider when determining whether to overhaul the original engine or buy a rebuilt one, never purchase parts or have machine work done on other components until the block has been thoroughly inspected.* As a general rule, time is the primary cost of an overhaul, so it doesn't pay to install worn or substandard parts.

As a final note, to ensure maximum life and minimum trouble from a rebuilt engine, everything must be assembled with care in a spotlessly-clean environment.

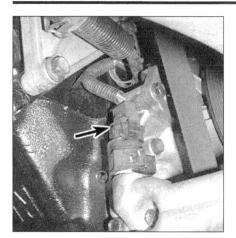

2.2a Oil pressure sending unit location - V6 engines, seen through left fenderwell

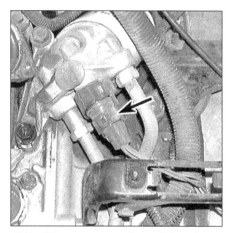

2.2b Oil pressure sending unit location - Cadillac V8 - above the oil filter at the right rear of the block

3.6 Use a compression gauge with a threaded fitting for the spark plug hole; don't use the type with a conical rubber tip - it relies on hand pressure to create a seal between the gauge and the combustion chamber, and usually gives an inaccurate reading

2   Oil pressure check

Refer to illustrations 2.2a and 2.2b

1   Low engine oil pressure can be a sign of an engine in need of rebuilding. A "low oil pressure" indicator (often called an "idiot light") is not a test of the oiling system. Such indicators only come on when the oil pressure is dangerously low. Even an original pressure gauge in the instrument panel is only a relative indication, although it's much better for driver information than a warning light. An accurate test can only be performed with a mechanical (not electrical) oil pressure gauge. When used in conjunction with an accurate tachometer, the engine's oil pressure performance can be compared to the manufacturer's Specifications for that year and model.

2   Locate the oil pressure sending unit **(see illustrations)**.

3   Remove the oil pressure sending unit and install a fitting that will allow you to directly connect your hand-held, mechanical oil pressure gauge. Use Teflon tape or sealant on the threads of the adapter and the fitting on the end of your gauge's hose.

4   Connect an accurate tachometer to the engine, according to the tachometer manufacturer's instructions.

5   Check the oil pressure with the engine running (full operating temperature) at the specified engine speed, and compare it to this Chapter's Specifications. If it's extremely low, the bearings and/or oil pump are probably worn out.

3   Cylinder compression check

Refer to illustration 3.6

1   A compression check will tell you what mechanical condition the upper end of your engine (pistons, rings, valves, head gaskets) is in. Specifically, it can tell you if the compression is down due to leakage caused by worn piston rings, defective valves and seats

or a blown head gasket. **Note:** *The engine must be at normal operating temperature and the battery must be fully charged for this check.*

2   Begin by cleaning the area around the spark plugs before you remove them (compressed air should be used, if available). The idea is to prevent dirt from getting into the cylinders as the compression check is being done.

3   Remove all of the spark plugs from the engine (see Chapter 1).

4   Block the throttle wide open.

5   Remove the IGN fuse from the fuse block to disable the ignition system. The fuel pump circuit should also be disabled (see Chapter 4).

6   Install the compression gauge in the spark plug hole **(see illustration)** .

7   Crank the engine over at least seven compression strokes and watch the gauge. The compression should build up quickly in a healthy engine. Low compression on the first stroke, followed by gradually increasing pressure on successive strokes, indicates worn piston rings. A low compression reading on the first stroke, which doesn't build up during successive strokes, indicates leaking valves or a blown head gasket (a cracked head could also be the cause). Deposits on the undersides of the valve heads can also cause low compression. Record the highest gauge reading obtained.

8   Repeat the procedure for the remaining cylinders and compare the results to this Chapter's Specifications.

9   Add some engine oil (about three squirts from a plunger-type oil can) to each cylinder, through the spark plug hole, and repeat the test on each cylinder.

10   If the compression increases after the oil is added, the piston rings are definitely worn. If the compression doesn't increase significantly, the leakage is occurring at the valves or head gasket. Leakage past the valves may be caused by burned valve seats and/or faces or warped, cracked or bent valves.

11   If two adjacent cylinders have equally

low compression, there's a strong possibility that the head gasket between them is blown. The appearance of coolant in the combustion chambers or the crankcase would verify this condition.

12   If one cylinder is slightly lower than the others, and the engine has a slightly rough idle, a worn lobe on the camshaft could be the cause.

13   If the compression is unusually high, the combustion chambers are probably coated with carbon deposits. If that's the case, the cylinder head(s) should be removed and decarbonized.

14   If compression is way down or varies greatly between cylinders, it would be a good idea to have a leak-down test performed by an automotive repair shop. This test will pinpoint exactly where the leakage is occurring and how severe it is.

4   Vacuum gauge diagnostic check

Refer to illustrations 4.4 and 4.6

A vacuum gauge provides valuable information about what is going on in the engine at a low cost. You can check for worn rings or cylinder walls, leaking head or intake manifold gaskets, incorrect fuel mixtures, restricted exhaust, stuck or burned valves, weak valve springs, improper ignition or valve timing and ignition problems.

Unfortunately, vacuum gauge readings are easy to misinterpret, so they should be used in conjunction with other tests to confirm the diagnosis.

Both the gauge readings and the rate of needle movement are important for accurate interpretation. Most gauges measure vacuum in inches of mercury (in-Hg). As vacuum increases (or atmospheric pressure decreases), the reading will increase. Also, for

4.4 Use a vacuum gauge connected to manifold vacuum (full vacuum available at idle) to test an engine's condition

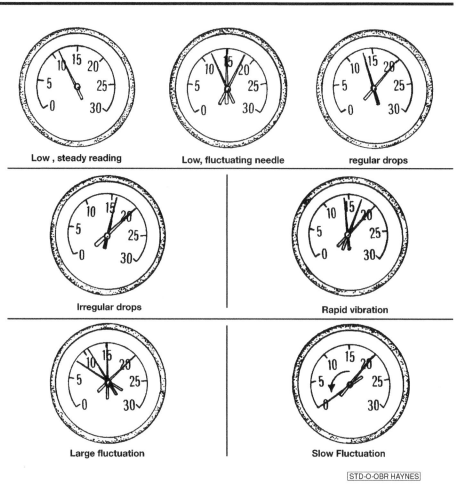

Low , steady reading   Low, fluctuating needle   regular drops

Irregular drops   Rapid vibration

Large fluctuation   Slow Fluctuation

STD-O-OBR HAYNES

4.6 Typical vacuum gauge readings

every 1,000-foot increase in elevation above sea level, the gauge readings will decrease about one inch of mercury.

Connect the vacuum gauge directly to intake manifold vacuum **(see illustration)**. Be sure no hoses are left disconnected during the test or false readings will result.

Before you begin the test, allow the engine to warm up completely. Block the wheels and set the parking brake. With the transaxle in Park, start the engine and allow it to run at normal idle speed. **Warning:** *Carefully inspect the fan blades for cracks or damage before starting the engine. Keep your hands and the vacuum tester clear of the fan and do not stand in front of the vehicle or in line with the fan when the engine is running.*

Read the vacuum gauge; an average, healthy engine should normally produce about 17 to 22 inches of vacuum with a fairly steady needle **(see illustration)**. Refer to the following vacuum gauge readings and what they indicate about the engine's condition:

1   A low, steady reading usually indicates a leaking gasket between the intake manifold and throttle body, a leaky vacuum hose, late ignition timing or incorrect camshaft timing. Eliminate all other possible causes, utilizing the tests provided in this Chapter before you remove the timing chain cover to check the timing marks.

2   If the reading is three to eight inches below normal and it fluctuates at that low reading, suspect an intake manifold gasket leak at an intake port.

3   If the needle has regular drops of about two to four inches at a steady rate, the valves are probably leaking. Perform a compression or leak-down test to confirm this.

4   An irregular drop or down-flick of the needle can be caused by a sticking valve or an ignition misfire. Perform a compression or leak-down test and read the spark plugs.

5   A rapid vibration of about four inches-Hg vibration at idle combined with exhaust smoke indicates worn valve guides. Perform a leak-down test to confirm this. If the rapid vibration occurs with an increase in engine

speed, check for a leaking intake manifold gasket or head gasket, weak valve springs, burned valves or ignition misfire.

6   A slight fluctuation, say one inch up and down, may mean ignition problems. Check all the usual tune-up items and, if necessary, run the engine on an ignition analyzer.

7   If there is a large fluctuation, perform a compression or leak-down test to look for a weak or dead cylinder or a blown head gasket.

8   If the needle moves slowly through a wide range, check for a clogged PCV system, incorrect idle fuel mixture, throttle body or intake manifold gasket leaks.

9   Check for a slow return after revving the engine by quickly snapping the throttle open until the engine reaches about 2,500 rpm and let it shut. Normally the reading should drop to near zero, rise above normal idle reading (about 5 in-Hg over) and then return to the previous idle reading. If the vacuum returns slowly and doesn't peak when the throttle is snapped shut, the rings may be worn. If there is a long delay, look for a restricted exhaust system (often the muffler or catalytic converter). An easy way to check this is to temporarily disconnect the exhaust ahead of the suspected part and re-test.

5   Engine rebuilding alternatives

The do-it-yourselfer is faced with a number of options when performing an engine overhaul. The decision to replace the engine block, piston/connecting rod assemblies and crankshaft depends on a number of factors, with the number one consideration being the condition of the block. Other considerations are cost, access to machine shop facilities, parts availability, time required to complete the project and the extent of prior mechanical experience on the part of the do-it-yourselfer.

Some of the rebuilding alternatives include:

Individual parts - If the inspection procedures reveal that the engine block and most engine components are in reusable condition, purchasing individual parts may be the most economical alternative. The block, crankshaft and piston/connecting rod assemblies should all be inspected carefully. Even if the block shows little wear, the cylinder bores should be surface-honed.

Crankshaft kit - This rebuild package consists of a reground crankshaft and a matched set of pistons and connecting rods.

The pistons will already be installed on the connecting rods. Piston rings and the necessary bearings will be included in the kit. These kits are commonly available for standard cylinder bores, as well as for engine blocks that have been bored to a regular oversize.

Short block - A short block consists of an engine block with renewed crankshaft and piston/connecting rod assemblies already installed. All new bearings are incorporated and all clearances will be correct. The existing cylinder head(s), camshaft, valve train components and external parts can be bolted to the short block with little or no machine shop work necessary.

Long block - A long block consists of a short block plus an oil pump, oil pan, cylinder heads, valve covers, camshaft and valve train components, timing sprockets, timing chain and timing cover. All components are installed with new bearings, seals and gaskets incorporated throughout. The installation of manifolds and external parts is all that is necessary.

Used engine assembly - While overhaul provides the best assurance of a like-new engine, used engines available from wrecking yards and importers are often a very simple and economical solution. Many used engines come with warranties, but always give any engine a thorough diagnostic check-out before purchase. Check compression, vacuum and also for signs of oil leakage. If possible, have the seller run the engine, ether in the vehicle or on a test stand so you can be sure it runs smoothly with no knocking or other noises.

Give careful thought to which alternative is best for you and discuss the situation with local automotive machine shops, auto parts dealers or parts store countermen before ordering or purchasing replacement parts.

6 Engine removal - methods and precautions

If you've decided that an engine must be removed for overhaul or major repair work, several preliminary steps should be taken.

Locating a suitable place to work is extremely important. Adequate work space, along with storage space for the vehicle, will be needed. If a shop or garage isn't available, at the very least a flat, level, clean work surface made of concrete or asphalt is required. Cleaning the engine compartment and engine before beginning the removal procedure will help keep tools clean and organized.

An engine hoist or A-frame will also be necessary. Make sure the equipment is rated in excess of the combined weight of the engine and transaxle. Safety is of primary importance, considering the potential hazards involved in lifting the engine out of the vehicle.

If the engine is being removed by a novice, a helper should be available. Advice and aid from someone more experienced would also be helpful. There are many instances when one person cannot simultaneously perform all of the operations required when lifting the engine out of the vehicle.

Plan the operation ahead of time. Arrange for or obtain all of the tools and equipment you'll need prior to beginning the job. Some of the equipment necessary to perform engine removal and installation safely and with relative ease are (in addition to an engine hoist) a heavy-duty floor jack, complete sets of wrenches and sockets as described in the front of this manual, wooden blocks and plenty of rags and cleaning solvent for mopping up spilled oil, coolant and gasoline. If the hoist must be rented, make sure that you arrange for it in advance and perform all of the operations possible without it beforehand. This will save you money and time.

Plan for the vehicle to be out of use for quite a while. A machine shop will be required to perform some of the work that the do-it-yourselfer can't accomplish without special equipment. These shops often have a busy schedule, so it would be a good idea to consult them before removing the engine in order to accurately estimate the amount of time required to rebuild or repair components that may need work.

Always be extremely careful when removing and installing the engine. Serious injury can result from careless actions. Plan ahead, take your time and a job of this nature, although major, can be accomplished successfully.

7 Engine - removal and installation

Warning 1: *Gasoline is extremely flammable, so take extra precautions when you work on any part of the fuel system. Don't smoke or allow open flames or bare light bulbs near the work area, and don't work in a garage where a natural gas-type appliance (such as a water heater or clothes dryer) with a pilot light is present. Since gasoline is carcinogenic, wear latex gloves when there's a possibility of being exposed to fuel, and, if you spill any fuel on your skin, rinse it off immediately with soap and water. Mop up any spills immediately and do not store fuel-soaked rags where they could ignite. The fuel system on fuel-injected models is under constant pressure, so, if any fuel lines are to be disconnected, the fuel pressure in the system must be relieved first (see Chapter 4 for more information). When you perform any kind of work on the fuel system, wear safety glasses and have a Class B type fire extinguisher on hand.*
Warning 2: *The air conditioning system is under high pressure! Have a dealer service department or service station discharge the system before disconnecting any air conditioning system hoses or fittings.*

Warning 3: *Some models covered by this manual are equipped with Supplemental Restraint Systems (SRS), more commonly known as airbags. Always disable the airbag system before working in the vicinity of any airbag system components to avoid the possibility of accidental deployment of the airbag(s), which could cause personal injury (see Chapter 12).*

Removal

Refer to illustrations 7.5, 7.15, 7.19a, 7.19b, 7.19c and 7.24

1   Relieve the fuel system pressure (see Chapter 4). Disconnect the negative cable from the battery.
2   Cover the fenders and cowl and remove the hood (see Chapter 11). Special pads are available to protect the fenders, but an old bedspread or blanket will also work.
3   Remove the air cleaner assembly.
4   Drain the cooling system (see Chapter 1). **Warning:** *Wait until the engine is completely cool before draining the coolant.*
5   Label the vacuum lines, emissions system hoses, electrical connectors, ground straps and fuel lines, to ensure correct reinstallation, then detach them. Pieces of masking tape with numbers or letters written on them work well **(see illustration)**. If there's any possibility of confusion, make a sketch of the engine compartment and clearly label the lines, hoses and wires.
6   Label and detach all coolant hoses from the engine.
7   Remove the cooling fan, shroud and radiator (see Chapter 3).
8   Remove the drivebelt (see Chapter 1).
9   Disconnect the fuel lines running from the engine to the chassis (see Chapter 4). Plug or cap all open fittings/lines.
10   Disconnect the accelerator linkage (and TV linkage/speed control cable, if equipped) from the engine (see Chapter 4 and Chapter 7).
11   Unbolt the power steering pump (see Chapter 10). Leave the lines/hoses attached and make sure the pump is kept in an upright position in the engine compartment (use wire or rope to restrain it out of the way).

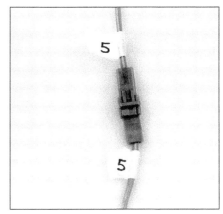

7.5 Label each wire and hose before unplugging the connector

7.15 Remove the four nuts (arrows, 1987 model shown, there are bolts on later models) and the strut tower brace

7.19a Engine lifting eyes (arrows) - V6 models

12   Unbolt the air-conditioning compressor (see Chapter 3) and set it aside. Do not disconnect the hoses.
13   Drain the engine oil (see Chapter 1) and remove the filter.
14   Remove the starter motor and alternator (see Chapter 5).
15   On models with a brace between the two front strut towers, remove the four

7.19b Engine lifting eye (arrow) - V8 models, right side

7.19c Engine lifting eye (arrow) - V8 models, left side

nuts/bolts and the brace **(see illustration)**.
16   Unbolt the exhaust pipe from below the engine (see Chapter 2A or 2B). **Note:** *It's possible to remove the engine with both exhaust manifolds in place.*
17   Refer to Chapter 7 and remove the torque converter-to-driveplate fasteners. **Note:** *Mark the relationship of the converter to the driveplate for later assembly.*
18   Support the transaxle with a jack. Position a wood block between them to prevent damage to the transaxle. Special transaxle jacks with safety chains are available - use one if possible.
19   Attach an engine sling or a length of chain to the lifting brackets on the engine **(see illustrations)**.
20   Roll the hoist into position and connect the sling to it. Take up the slack in the sling or chain, but don't lift the engine. You may have to experiment with the location of the hook from the hoist with the lifting sling to get the engine to lift out level or at the angle you want. **Warning:** *DO NOT place any part of your body under the engine when it's supported only by a hoist or other lifting device.*
21   Remove the transaxle-to-engine block bolts.
22   Remove the torque strut on V6 models,

and the right-side engine mount-to-frame bolts on all models (see Chapter 2A or 2B).
23   Recheck to be sure nothing is still connecting the engine to the transaxle or vehicle. Disconnect anything still remaining.
24   Raise the engine slightly. Carefully work it forward to separate it from the transaxle. Be sure the torque converter stays in the transaxle (clamp a pair of vise-grips to the housing to keep the converter from sliding out). Slowly raise the engine out of the engine compartment **(see illustration)**. Check carefully to make sure nothing is hanging up.
25   Remove the driveplate and mount the engine on an engine stand.

Installation

26   Check the engine and transaxle mounts. If they're worn or damaged, replace them.
27   Raise the transaxle slightly to accept the angle of the engine when it is lowered into place.
28   Carefully lower the engine into the engine compartment - make sure the engine mounts line up.
29   Install the transaxle-to-engine bolts and tighten them securely. **Caution:** *DO NOT use the bolts to force the transaxle and engine together!*

7.24 With the accessories tied out of the way and the engine mount bolts removed, tilt, twist and raise the engine with the hoist until it clears the engine compartment

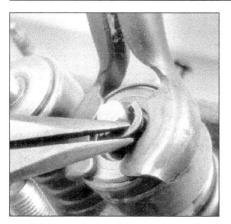

9.2 Use a valve spring compressor to compress the valve spring, then remove the keepers from the valve stem with a magnet or needle-nose pliers

9.3 A small plastic bag, with an appropriate label, can be used to store the valve train components so they can be kept together and reinstalled in their original locations

9.4 If the valve won't pull through the guide, deburr the edge of the stem end and the area around the top of the keeper groove with a file or whetstone

30   Install the driveplate-to-torque converter bolts following the procedure outlined in Chapter 7.

31   Reinstall the remaining components in the reverse order of removal.

32   Add coolant, oil, power steering and transaxle fluid as needed.

33   Run the engine and check for leaks and proper operation of all accessories, then install the hood and test drive the vehicle.

34   Have the air conditioning system recharged and leak tested, if it had been disconnected.

8 Engine overhaul - disassembly sequence

1   It's much easier to disassemble and work on the engine if it's mounted on a portable engine stand. A stand can often be rented quite cheaply from an equipment rental yard. Before it's mounted on a stand, the flywheel/driveplate should be removed from the engine.

2   If a stand isn't available, it's possible to disassemble the engine with it blocked up on the floor. Be extra careful not to tip or drop the engine when working without a stand.

3   If you're going to obtain a rebuilt engine, all external components must come off first, to be transferred to the replacement engine, just as they will if you're doing a complete engine overhaul yourself. These include:

Alternator and brackets
Power steering pump and brackets
Emissions control components
Coilpack, spark plug wires and spark
  plugs
Thermostat and housing cover
Water pump bypass pipe/hose
Fuel injection components
Intake/exhaust manifolds
Oil filter
Engine mounts
Driveplate

Note: *When removing the external components from the engine, pay close attention to details that may be helpful or important during installation. Note the installed position of gaskets, seals, spacers, pins, brackets, washers, bolts, wiring and other small items.*

4   If you're obtaining a short block, which consists of the engine block, crankshaft, pistons and connecting rods all assembled, then the cylinder head(s), oil pan and oil pump will have to be removed as well. See *Engine rebuilding alternatives* for additional information regarding the different possibilities to be considered.

5   If you're planning a complete overhaul, the engine should be disassembled and the internal components removed in the following general order:

Intake and exhaust manifolds
Valve cover(s)
Rocker arm assembly and pushrods
Cylinder heads
Timing chain cover and bolts
Timing chain and sprockets
Lifters and camshaft
Balance shaft (1988 and later V6's)
Oil pan
Oil pick-up tube
Oil pump
Piston/connecting rod assemblies
Rear main oil seal retainer
Crankshaft and main bearings

6   Before beginning the disassembly and overhaul procedures, make sure the following items are available. Also, refer to *Engine overhaul - reassembly sequence* for a list of tools and materials needed for engine reassembly.

Common hand tools
Small cardboard boxes or plastic bags for
  storing parts
Gasket scraper
Ridge reamer
Vibration damper puller
Micrometers
Telescoping gauges
Dial indicator set

Valve spring compressor
Cylinder surfacing hone
Piston ring groove cleaning tool
Electric drill motor
Tap and die set
Wire brushes
Oil gallery brushes
Cleaning solvent
Slide-hammer (V6's with balance shaft)

9 Cylinder head - disassembly

Refer to illustrations 9.2, 9.3 and 9.4
Note: *New and rebuilt cylinder heads are commonly available for most engines at dealerships and auto parts stores. Due to the fact that some specialized tools are necessary for the disassembly and inspection procedures, and some replacement parts may not be readily available, it may be more practical and economical for the home mechanic to purchase replacement head(s) rather than taking the time to disassemble, inspect and recondition the original(s).*

1   Cylinder head disassembly involves removal of the intake and exhaust valves and related components. It's assumed that the rocker arms have already been removed (see Chapter 2, Part A or B as needed).

2   Compress the spring on the first valve with a spring compressor and remove the keepers **(see illustration)**. **Note:** *If the keepers are stuck in the retainer, remove the compressor, place a socket over the retainer and strike it with a hammer to break the bond between the components.* Carefully release the valve spring compressor and remove the retainer, the spring and the spring seat (if used).

3   After the valves are removed, label and store them, along with their related components, so they can be kept separate and reinstalled in the same valve guides they are removed from **(see illustration)**.

4   Pull the valve out of the head, then remove the oil seal from the guide. If the valve binds in the guide (won't pull through), push it back into the head and deburr the

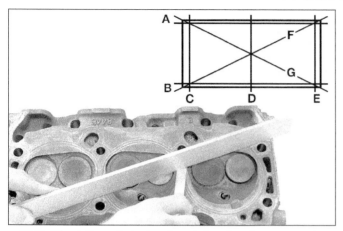

10.11 Check the cylinder head gasket surface for warpage by trying to slip a feeler gauge under the straightedge (see this Chapter's Specifications for the maximum warpage allowed and use a feeler gauge of that thickness) - check straight across and diagonally

10.13 A dial indicator can be used to determine the valve stem-to-guide clearance (move the valve head toward the indicator and back)

area around the keeper groove with a fine file or whetstone **(see illustration)**.

5   Repeat the procedure for the remaining valves. Remember to keep all the parts for each valve together so they can be reinstalled in the same locations.

6   Once the valves and related components have been removed and stored in an organized manner, the head should be thoroughly cleaned and inspected. If a complete engine overhaul is being done, finish the engine disassembly procedures before beginning the cylinder head cleaning and inspection process.

10 Cylinder head - cleaning and inspection

Cleaning

1   Thorough cleaning of the cylinder head(s) and related valvetrain components, followed by a detailed inspection, will enable you to decide how much valve service work must be done during the engine overhaul. **Note:** *If the engine was severely overheated, the cylinder head is probably warped (see Step 11)*.

2   Scrape all traces of old gasket material and sealing compound off the head gasket, intake manifold and exhaust manifold sealing surfaces. Be very careful not to gouge the cylinder head. Special gasket-removal solvents that soften gaskets and make removal much easier are available at auto parts stores.

3   Remove all built up scale from the coolant passages.

4   Run a stiff wire brush through the various holes to remove deposits that may have formed in them.

5   Run an appropriate-size tap into each of the threaded holes to remove corrosion and thread sealant that may be present. If compressed air is available, use it to clear the

holes of debris produced by this operation. **Warning:** *Wear eye protection when using compressed air!*

6   Clean the cylinder head with solvent and dry it thoroughly. Compressed air will speed the drying process and ensure that all holes and recessed areas are clean. **Note:** *Decarbonizing chemicals are available and may prove very useful when cleaning cylinder heads and valvetrain components. These chemicals are very caustic and should be used with caution. Wear rubber gloves, goggles, and be sure to follow the instructions on the container.*

7   Clean the rocker arms with solvent and dry them thoroughly (don't mix them up during the cleaning process). Compressed air will speed the drying process and can be used to clean out the oil passages.

8   Clean all the valve springs, spring seats, keepers and retainers with solvent and dry them thoroughly. Work on the components from one valve at a time to avoid mixing up the parts.

9   Scrape off any heavy deposits that may have formed on the valves, then use a motorized wire brush to remove deposits from the valve heads and stems. Again, make sure the valves don't get mixed up.

Inspection

Note: *Be sure to perform all of the following inspection procedures before concluding that machine shop work is required. Make a list of the items that need attention. The inspection procedures for the rocker arms can be found in Part A and Part B.*

Cylinder head

Refer to illustrations 10.11 and 10.13

10   Inspect the head very carefully for cracks, evidence of coolant leakage and other damage. If cracks are found, check with an automotive machine shop concerning repair. If repair isn't possible, a new cylinder head should be obtained.

11   Using a straightedge and feeler gauge, check the head-gasket mating surface for warpage **(see illustration)**. If the warpage exceeds the specified limit, it can be resurfaced at an automotive machine shop.

12   Examine the valve seats in each of the combustion chambers. If they're pitted, cracked or burned, the head will require valve service that's beyond the scope of the home mechanic.

13   Check the valve stem-to-guide clearance with a small hole gauge and micrometer. Also check the valve stem deflection crosswise (parallel to the rocker arm) with a dial indicator attached securely to the head **(see illustration)**. The valve must be in the guide and approximately 1/16-inch off the seat. The total valve stem movement indicated by the gauge needle must be noted and divided in half to obtain the stem-to-guide clearance specification. If it exceeds the stem-to-guide clearance limit listed in this Chapter's specifications, the valve guides should be replaced. After this is done, if there's still some doubt regarding the condition of the valve guides, they should be checked by an automotive machine shop (the cost should be minimal).

Valves

Refer to illustrations 10.14 and 10.15

14   Carefully inspect each valve face for uneven wear, deformation, cracks, pits and burned areas. Check the valve stem for scuffing and galling and the neck for cracks. Rotate the valve and check for any obvious indication that it's bent. Look for pits and excessive wear on the end of the stem. The presence of any of these conditions indicates the need for valve service by an automotive machine shop **(see illustration)**.

15   Measure the margin width on each valve **(see illustration)**. Any valve with a margin narrower than that listed in this Chapter's Specifications will have to be replaced with a new one.

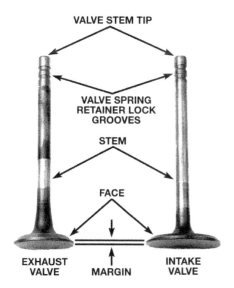

10.14 Using a micrometer or vernier caliper, check for valve wear at the points shown here

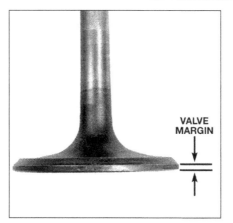

10.15 The margin width on the valve must be as specified (if no margin exists, the valve cannot be reused)

10.16 Check each valve spring for squareness

Valve components

Refer to illustration 10.16

16   Check each valve spring for wear (on the ends) and pits. Stand each spring on a flat surface and check it for squareness **(see illustration)**. If any of the springs are distorted or sagged, replace all of them with new parts.

17   The tension of all springs should be checked with a special fixture before deciding if they are suitable for use (an automotive machine shop will perform this service for you). If in doubt, replace all of the valve springs during an overhaul.

18   Check the spring retainers and keepers for obvious wear and cracks. Any questionable parts should be replaced with new ones, as extensive damage will occur if they fail during engine operation.

19   Any damaged or excessively worn parts must be replaced with new ones.

20   If the inspection process indicates that the valve components are in generally poor condition and worn beyond the limits specified, which is usually the case in an engine that's being overhauled, reassemble the valves in the cylinder head and refer to Section 11 for valve servicing recommendations.

11 Valves - servicing

1   Because of the complex nature of the job and the special tools and equipment needed, servicing of the valves, the valve seats and the valve guides, commonly known as a valve job, should be done by a professional.

2   The home mechanic can remove and disassemble the head, do the initial cleaning and inspection, then reassemble and deliver it to a dealer service department or an automo-

tive machine shop for the actual service work. Doing the inspection will enable you to see what condition the head and valvetrain components are in and will ensure that you know what work and new parts are required when dealing with an automotive machine shop.

3   The dealer service department, or automotive machine shop, will remove the valves and springs, recondition or replace the valves and valve seats, recondition the valve guides, check and replace the valve springs, spring retainers and keepers (as necessary), replace the valve seals with new ones, reassemble the valve components and make sure the installed spring height is correct. The cylinder head gasket surface should also be resurfaced if it's warped. If one of the heads is warped, have both of them resurfaced. **Note:** *On V6 cylinder heads, there is a minimum height, indicated by cast pads on the block side of the head. If the heads have been surfaced before and the machining gets to the pads, the factory recommends replacing the head. Your machine shop will be able to tell you if the heads are serviceable.*

4   After the valve job has been performed by a professional, the head will be in like new condition. When the head is returned, be sure to clean it again before installation on the engine to remove any metal particles and abrasive grit that may still be present from the valve service or head-resurfacing operations. Use compressed air, if available, to blow out all the oil holes and passages. Keep the completed heads in plastic bags to keep them clean until you assemble the engine.

12 Cylinder head - reassembly

1   Regardless of whether or not the heads were sent to an automotive repair shop for valve servicing, make sure they are clean before beginning reassembly.

2   If the heads were sent out for valve servicing, the valves and related components will already be in place. Begin the reassembly procedure with Step 7.

3   Install new seals on each of the valve

guides. The valves should all be in place, and lubricated with clean engine oil. Slip the seal carefully over the stem of the valve until the seal aligns squarely with the guide. Using a hammer and a deep socket, gently tap each seal into place until it is completely seated on the guide. Do not twist or cock the seals during installation or they will not seal properly on the valve stems. Refer to Chapter 2A or 2B for valve seal installation. **Note:** *On many engines, the exhaust and intake seal are slightly different. Replacement seals may be colored differently than the originals, so follow the seal manufacturer's instructions for identification and installation.*

4   Drop the spring seat or shims (if used) over the valve guide and set the valve springs, shield and retainer in place.

5   Compress the springs with a valve spring compressor, position the keepers in the upper groove, then slowly release the compressor and make sure the keepers seat properly. Apply a small dab of grease to each keeper to hold it in place if necessary.

6   Repeat the procedure for the remaining valves. Be sure to return the components to their original locations - do not mix them up!

7   Apply moly-base grease to the rocker arm faces and the pivots, then install the rocker arms. On early V-6 engines, assemble the rocker arms to the shafts, with moly-based lube on the shafts and the rocker arms properly oriented (in their original location on their original shaft). Do not tighten the rocker shaft/arm bolts completely at this time. When the heads are installed later, the pushrods must be installed before the rocker arms are fully tightened.

13 Balance shaft - inspection and removal (V6 engine)

Inspection

1   1988 and later V6 engines have a balance shaft, which is geared to the camshaft, installed in the lifter valley area of the block.

2   While the camshaft and its balance-

13.5 Remove the retaining bolts and the balance shaft thrust plate

13.6 A slide hammer must be threaded into the front of the balance shaft to pull the balance shaft front bearing out of the block

shaft drive gear are still in place, set up a dial indicator against the top teeth of the balance shaft driven gear (on the balance shaft) and rock the balance shaft by hand. Compare this backlash to this Chapter's Specifications. The gears will have to be replaced if the backlash exceeds the Specification.

3   Remove the two gears and mount a dial indicator on the front of the block with the tip on the nose of the balance shaft and zero it. Reaching inside the lifter valley, grab the balance shaft and move it forward and back in the block while watching the dial indicator. Compare this endplay measurement with this Chapter's Specifications. If it's incorrect, a new balance shaft thrust plate will have to be installed.

4   Mount the dial indicator on the block with the tip on the front of the balance shaft (inside the lifter valley). Pull the shaft up and down and compare the radial play with this Chapter's Specifications. Repeat the procedure for radial play at the rear of the balance shaft. If the play exceeds Specifications, the balance shaft and its bearings will have to be replaced. The bearings are not available sep-

arately. The rear, sleeve-type bearing can be replaced by your automotive machine shop just like a camshaft bearing.

Removal

Refer to illustrations 13.5 and 13.6

5   Remove the balance shaft drive gear from the camshaft (see Chapter 2, Part A), then remove the bolts and the balance shaft thrust plate **(see illustration)**.

6   The balance shaft rides in a sleeve-type bearing at the back of the block (like a camshaft bearing, and a ball-bearing assembly pressed into the front of the block. To remove the balance shaft, screw a slide-hammer into the front of the balance shaft and slap it until the shaft and bearing are free **(see illustration)**. Once the bearing is free of the block, withdraw the balance shaft straight out by hand, being careful to not cock it against its rear bearing.

7   Inspect the balance shaft drive gear (mounts on the camshaft) and the balance shaft driven gear (mounts on balance shaft) for wear, nicks and burrs.

14 Camshaft and lifters - removal

Lifters

Refer to illustrations 14.1, 14.2a, 14.2b and 14.2c

1   The lifters must be removed before removing the camshaft, and there are several ways to extract the lifters from the bores. A special tool designed to grip and remove lifters is manufactured by many tool companies and is widely available, but it may not be required in every case. On engines without a lot of varnish buildup, the lifters can often be removed with a small magnet or even with your fingers **(see illustration)**. A machinist's scribe with a bent end can be used to pull the lifters out by positioning the point under the retainer ring in the top of each lifter. **Caution:** *Do not use pliers to remove the lifters unless you intend to replace them with new ones (along with the camshaft). The pliers may damage the precision machined and hardened lifters.*

2   Most V6 engines, and all 1988 and later

14.1 Most lifters can be removed with a mechanic's magnet (Cadillac V8 shown); otherwise use pliers but replace the lifters with new ones

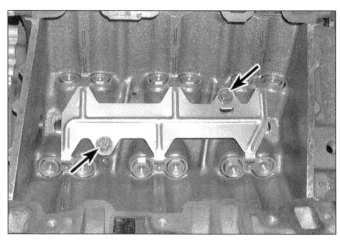

14.2a Non-balance-shaft V6 engines use a single, metal guide retainer - remove the bolts (arrows) and the retainer

14.2b Remove the lifter guides from each pair of lifters

14.2c On 1993 V6 engines, there are two long lifter guides without a metal retainer plate - remove the bolts (arrows) and remove the two guides, marking them Left and Right

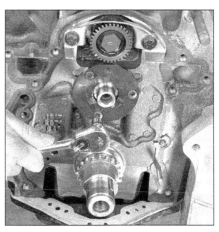

14.6 Remove the retaining bolts and the camshaft thrust plate (1988 and later V6 engines only)

Cadillac V8 engines, are equipped with roller-tipped lifters. These lifters must be kept with their rollers oriented to the camshaft lobes, so there are lifter guides over each pair of lifters, and a lifter guide retainer to hold the guides in place in the block. Remove the guide retainer and the guides **(see illustrations)**.

3   Before removing the lifters, arrange to store them in a clearly labeled box to ensure that they are reinstalled in their original locations. Remove the lifters and store them where they will not get dirty. Do not attempt to withdraw the camshaft with the lifters in place.

Camshaft

Refer to illustrations 14.6 and 14.7

4   Refer to Chapter 2, Part A (V6 engines) or Part B (V8 engines) and remove the timing chain and sprockets.

5   On 1988 and later V6 engines (equipped with balance shaft), the camshaft is retained by a plate at the front of the block. Early V6 engines, and all Cadillac V8 engines, have camshaft endplay controlled by a thrust bearing at the front of the camshaft. See Chapter 2, Part A or B for replacement of the thrust bearing.

6   On balance-shaft V6 engines, remove the camshaft thrust plate **(see illustration)**.

7   To remove the camshaft from the front of the block without cocking it and causing damage to the bearings in the block, thread a long bolt into the front of the camshaft to act as a handle during removal **(see illustration)**. Draw the camshaft straight out.

8   For camshaft and lifter inspection, refer to Section 23.

15 Piston/connecting rod assembly - removal

Refer to illustrations 15.1, 15.3 and 15.4

Note: *Prior to removing the piston/connecting rod assemblies, remove the cylinder heads, the oil pan and the oil pump/front cover by referring to the appropriate Sections in Chapter 2A or 2B.*

1   Use your fingernail to feel if a ridge has formed at the upper limit of ring travel (about 1/4-inch down from the top of each cylinder).

If carbon deposits or cylinder wear have produced ridges, they must be completely removed with a special tool **(see illustration)**. Follow the manufacturer's instructions provided with the tool. Failure to remove the ridges before attempting to remove the piston/connecting rod assemblies may result in piston breakage.

2   After the cylinder ridges have been removed, turn the engine upside-down so the crankshaft is facing up.

3   Before the connecting rods are removed, check the side clearance with feeler gauges. Slide them between the first connecting rod and the crankshaft throw until the play is removed **(see illustration)** . The endplay is equal to the thickness of the feeler gauge(s). If the endplay exceeds the service limit, new connecting rods will be required. If new rods (or a new crankshaft) are installed, the endplay may fall under the specified minimum (if it does, the rods will have to be machined to restore it - consult an automotive machine shop for advice if necessary). Repeat the procedure for the remaining connecting rods.

4   Check the connecting rods and caps for

14.7 Thread a long bolt into the camshaft to aid in withdrawing the camshaft, being careful not to nick the camshaft bearings in the block with the lobes or journals

15.1 A ridge reamer is required to remove the ridge from the top of each cylinder - do this before removing the pistons!

15.3 Check the connecting rod side clearance with a feeler gauge as shown

15.4 A liner-holding tool must be used on Cadillac V8 engines to keep the liners from moving and losing their bottom seal

16.1 Checking the crankshaft endplay with a dial indicator

identification marks. If they aren't plainly marked, use a small center punch to make the appropriate number of indentations on each rod and cap (1, 2, 3, etc., depending on the engine type and cylinder they're associated with). **Note:** *On Cadillac V8 engines, if the original cylinder liners are to be reused, install a cylinder liner holder, available at some auto parts stores, to the engine block to prevent the liners from shifting when the pistons are withdrawn* **(see illustration)**. *Also, mark the position of the liners relative to the block with indelible ink to ensure they do not shift.*

5   Loosen each of the connecting rod cap bolts 1/2-turn at a time until they can be removed by hand. Remove the number one connecting rod cap and bearing insert. Don't drop the bearing insert out of the cap.

6   Remove the bearing insert and push the connecting rod/piston assembly out through the top of the engine. Use a wooden hammer handle to push on the upper bearing surface in the connecting rod. If resistance is felt, double-check to make sure that all of the ridge was removed from the cylinder.

7   Repeat the procedure for the remaining cylinders.

8   After removal, reassemble the connecting rod caps and bearing inserts in their respective connecting rods and install the cap bolts finger tight. Leaving the old bearing inserts in place until reassembly will help prevent the connecting rod bearing surfaces from being accidentally nicked or gouged.

9   Don't separate the pistons from the connecting rods (see Section 20 for additional information).

16 Crankshaft - removal

Refer to illustrations 16.1 and 16.3

Note: *The crankshaft can be removed only after the engine has been removed from the vehicle. It's assumed that the driveplate, vibration damper, timing chain, oil pan, oil pump and piston/connecting rod assemblies have already been removed. If your engine is equipped with a one-piece rear main oil seal,*

the seal must separated from the block before proceeding with crankshaft removal (see Section 27).

1   Before the crankshaft is removed, check the endplay. Mount a dial indicator with the stem in line with the crankshaft and touching one of the crank throws or the nose of the crankshaft **(see illustration)** .

2   Push the crankshaft all the way to the rear and zero the dial indicator. Next, pry the crankshaft to the front as far as possible and check the reading on the dial indicator. The distance that it moves is the endplay. If it's greater than specified, check the crankshaft thrust surfaces for wear. If no wear is evident, new main bearings should correct the endplay.

3   If a dial indicator isn't available, feeler gauges can be used. Gently pry or push the crankshaft all the way to the front of the engine. Slip feeler gauges between the crankshaft and the front face of the thrust main bearing to determine the clearance **(see illustration)**.

4   Check the main bearing caps to see if they're marked to indicate their locations. They should be numbered consecutively from the front of the engine to the rear, usually also marked with a cast-in arrow, which points to the front of the engine. If they aren't, mark them with number stamping dies or a center punch. Loosen the main bearing cap bolts 1/4-turn at a time each, until they can be removed by hand. Note if any stud bolts are used and make sure they're returned to their original locations when the crankshaft is reinstalled.

5   Gently tap the caps with a soft-face hammer, then separate them from the engine block. If necessary, use the bolts as levers to remove the caps. Try not to drop the bearing inserts if they come out with the caps.

6   Carefully lift the crankshaft out of the engine. It may be a good idea to have an assistant available, since the crankshaft is quite heavy. With the bearing inserts in place in the engine block and main bearing caps, return the caps to their respective locations on the engine block and tighten the bolts finger tight.

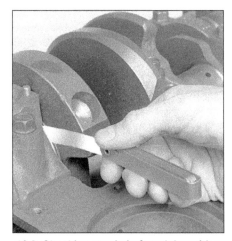

16.3 Checking crankshaft endplay with a feeler gauge

17 Engine block - cleaning

Refer to illustrations 17.1a, 17.1b, 17.9 and 17.11

1   Remove the core plugs from the engine block. To do this, knock one side of the plugs into the block with a hammer and punch, then grasp them with large pliers and pull them back through the holes **(see illustrations)**. **Caution:** *The core plugs (also known as freeze or soft plugs) may be difficult or impossible to retrieve if they're driven into the block coolant passages.*

2   Using a gasket scraper, remove all traces of gasket material from the engine block. Be very careful not to nick or gouge the gasket sealing surfaces.

3   Remove the main bearing caps and separate the bearing inserts from the caps and the engine block. Tag the bearings, indicating which cylinder they were removed from and whether they were in the cap or the block, then set them aside.

4   Remove all of the threaded oil gallery plugs from the block. The plugs are usually very tight - they may have to be drilled out and the holes retapped. Use new plugs when

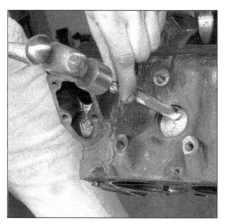

17.1a Use a hammer and large punch to knock the core plugs sideways in their bores, then pull them out with pliers

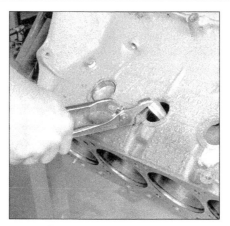

17.1b Pull the core plugs from the block with pliers

17.9 All bolt holes in the block - particularly the main bearing cap and head bolt holes - should be cleaned and restored with a tap (be sure to remove debris from the holes after this is done)

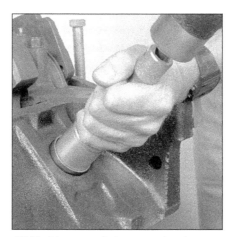

17.11 A large socket on an extension can be used to drive the new core plugs into the bores

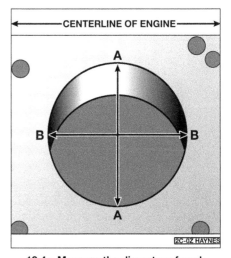

18.4a Measure the diameter of each cylinder at a right angle to the engine centerline (A), and parallel to the engine centerline (B) - out-of-round is the difference between A and B; taper is the difference between A and B at the top of the cylinder and A and B at the bottom of the cylinder

the engine is reassembled.

5   **Caution:** *On Cadillac V8 engines, it is necessary to remove the cylinder liners in order to properly clean the block. Carefully mark the alignment of each liner to the cylinder block if they are to be reused. However, this particular engine is a difficult one for the home mechanic to rebuild, because of the cylinder liners and other rebuilding problems. It is suggested that when rebuilding this engine, have the short-block cleaned and rebuilt at a professional engine shop. They will install new pistons with new, matching liners, carefully fit new O-rings to the bottom of the liners and set the correct liner height in the block.*

6   If the engine is extremely dirty it should be taken to an automotive machine shop to be steam cleaned or hot-tanked.

7   After the block is returned, clean all oil holes and oil galleries one more time. Brushes specifically designed for this purpose are available at most auto parts stores. Flush the passages with warm water until the water runs clear, dry the block thoroughly and wipe all machined surfaces with a light, rust preventive oil. If you have access to compressed air, use it to speed the drying process and to blow out all the oil holes and galleries. **Warning:** *Wear eye protection when using compressed air!*

8   If the block isn't extremely dirty or sludged up, you can do an adequate cleaning job with hot soapy water and a stiff brush. Take plenty of time and do a thorough job. Regardless of the cleaning method used, be sure to clean all oil holes and galleries very thoroughly, dry the block completely and coat all machined surfaces with light oil.

9   The threaded holes in the block must be clean to ensure accurate torque readings during reassembly. Run the proper size tap into each of the holes to remove rust, corrosion, thread sealant or sludge and restore damaged threads **(see illustration)**. If possible, use compressed air to clear the holes of debris produced by this operation. Now is a good time to clean the threads on the head

bolts and the main bearing cap bolts as well. **Caution:** *On the Cadillac V8 engine with it's aluminum block, it is important to clean all of the block's threaded holes with taps to remove dirt or corrosion.*

10   Reinstall the main bearing caps and tighten the bolts finger tight.

11   After coating the sealing surfaces of the new core plugs with Permatex no. 2 sealant, install them in the engine block **(see illustration)** . Make sure they're driven in straight and seated properly or leakage could result. Special tools are available for this purpose, but a large socket, with an outside diameter that will just slip into the core plug, a 1/2-inch drive extension and a hammer will work just as well. **Warning:** *On Cadillac V8 engines, tap the plugs into the block slowly and gently to avoid cracking the aluminum block.*

12   Apply non-hardening sealant (such as Permatex no. 2 or Teflon pipe sealant) to the new oil gallery plugs and thread them into the holes in the block. Make sure they're tightened securely.

13   If the engine isn't going to be reassembled right away, cover it with a large plastic trash bag to keep it clean.

18 Engine block - inspection

Refer to illustrations 18.4a, 18.4b, 18.4c and 18.8

1   Before the block is inspected, it should be cleaned as described in Section 17.

2   Visually check the block for cracks, rust and corrosion. Look for stripped threads in the threaded holes. It's also a good idea to have the block checked for hidden cracks by an automotive machine shop that has the special equipment to do this type of work. If defects are found, have the block repaired, if possible, or replaced.

3   Check the cylinder bores for scuffing and scoring.

4   Check the cylinders for taper and out-of-round conditions as follows **(see illustrations)**:

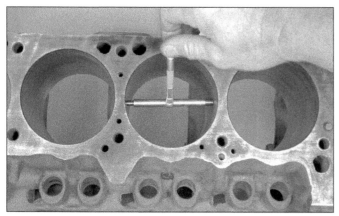

18.4b The ability to "feel" when the telescoping gauge is at the correct point will be developed over time, so work slowly and repeat the check until you're satisfied the bore measurement is accurate

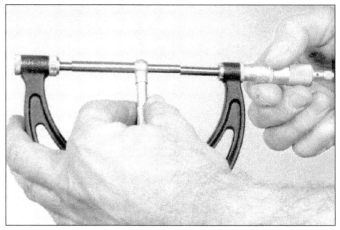

18.4c The gauge is then measured with a micrometer to determine the bore size

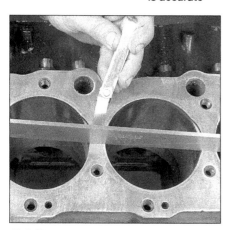

18.8 When the top of the block is cleaned, place a straightedge across the deck and try to slip a feeler gauge under it - if the specified gauge fits under the straightedge, the block must be surfaced at a machine shop

a) *Measure the diameter of each cylinder at the top (just under the ridge area), center and bottom of the cylinder bore, parallel to the crankshaft axis.*

b) *Next measure each cylinder's diameter at the same three locations perpendicular to the crankshaft axis.*

c) *The taper of the cylinder is the difference between the bore diameter at the top of the cylinder and the diameter at the bottom. The out-of-round specification of the cylinder bore is the difference between the parallel and perpendicular readings.*

d) *Compare your results to those listed in this Chapter's Specifications.*

5   Repeat the procedure for the remaining pistons and cylinders.

6   If the cylinder walls are badly scuffed or scored, or if they're out-of-round or tapered beyond the limits given in this Chapter's Specifications, have the engine block rebored and honed at an automotive machine shop. If a rebore is done, oversize pistons

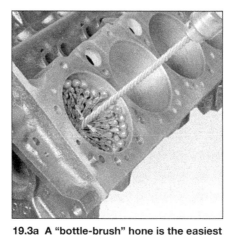

19.3a A "bottle-brush" hone is the easiest type of hone to use

and rings will be required.

7   If the cylinders are in reasonably good condition and not worn to the outside of the limits, and if the piston-to-cylinder clearances can be maintained properly, then they don't have to be rebored. Honing is all that's necessary (see Section 19).

8   Using a precision straightedge and feeler gauge, check the block deck (the surface that mates with the cylinder head) for distortion **(see illustration)**.

19 Cylinder honing

Refer to illustrations 19.3a and 19.3b

1   Prior to engine reassembly, the cylinder bores must be honed so the new piston rings will seat correctly and provide the best possible combustion chamber seal (except for Cadillac V8 engines, see Step 4 of Section 15). **Note:** *If you don't have the tools or don't want to tackle the honing operation, most automotive machine shops will do it for a reasonable fee.*

2   Before honing the cylinders, install the main bearing caps and tighten the bolts to the specified torque.

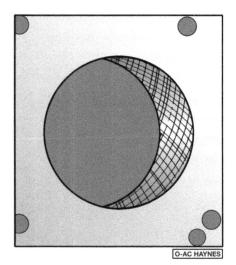

19.3b The cylinder hone should leave a smooth, crosshatch pattern with the lines intersecting at approximately a 45 to 60-degree angle

3   Two types of cylinder hones are commonly available - the flex hone, or "bottle brush", type and the more traditional surfacing hone with spring-loaded stones. Both will do the job, but for the less experienced mechanic the "bottle brush" hone will probably be easier to use. You'll also need some kerosene or honing oil, rags and an electric drill motor. Proceed as follows:

a) *Mount the hone in the drill motor, compress the stones and slip it into the first cylinder* **(see illustration)**. *Be sure to wear safety goggles or a face shield!*

b) *Lubricate the cylinder with plenty of honing oil, turn on the drill and move the hone up and down in the cylinder at a pace that will produce a fine crosshatch pattern on the cylinder walls. Ideally, the crosshatch lines should intersect at approximately a 60-degree angle* **(see illustration)**. *Be sure to use plenty of lubricant and don't take off any more*

20.4a The piston ring grooves can be cleaned with a special tool, as shown here . . .

20.4b . . . or a section of broken ring

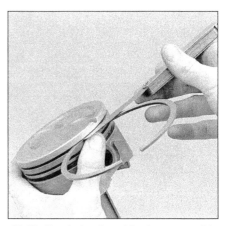

20.10 Check the ring side clearance with a feeler gauge at several points around the groove

material than is absolutely necessary to produce the desired finish. **Note:** *Piston ring manufacturers may specify a smaller crosshatch angle than the traditional 60-degrees - read and follow any instructions included with the new rings.*

c) *Don't withdraw the hone from the cylinder while it's running. Instead, shut off the drill and continue moving the hone up- and down in the cylinder until it comes to a complete stop, then compress the stones and withdraw the hone. If you're using a "bottle brush" type hone, stop the drill motor, then turn the chuck in the normal direction of rotation while withdrawing the hone from the cylinder.*

d) *Wipe the oil out of the cylinder and repeat the procedure for the remaining cylinders.*

4   After the honing job is complete, chamfer the top edges of the cylinder bores with a small file so the rings won't catch when the pistons are installed. Be very careful not to nick the cylinder walls with the end of the file.

5   The entire engine block must be washed again very thoroughly with warm, soapy water to remove all traces of the abrasive grit produced during the honing operation. **Note:** *The bores can be considered clean when a lint-free white cloth - dampened with clean engine - oil is used to wipe them and doesn't pick up any more honing residue, which will show up as gray areas on the cloth. Be sure to run a brush through all oil holes and galleries and flush them with running water.*

6   After rinsing, dry the block and apply a coat of light rust preventive oil to all machined surfaces. Wrap the block in a plastic trash bag to keep it clean and set it aside until reassembly.

20 Piston/connecting rod assembly - inspection

Refer to illustrations 20.4a, 20.4b, 20.10 and 20.11

1   Before the inspection process can be

carried out, the piston/connecting rod assemblies must be cleaned and the original piston rings removed from the pistons. **Note:** *Always use new piston rings when the engine is reassembled.*

2   Using a piston ring installation tool, carefully remove the rings from the pistons. Be careful not to nick or gouge the pistons in the process.

3   Scrape all traces of carbon from the top of the piston. A hand-held wire brush or a piece of fine emery cloth can be used once the majority of the deposits have been scraped away. Do not, under any circumstances, use a wire brush mounted in a drill motor to remove deposits from the pistons. The piston material is soft and may be eroded away by the wire brush.

4   Use a piston ring groove-cleaning tool to remove carbon deposits from the ring grooves. If a tool isn't available, a piece broken off the old ring will do the job. Be very careful to remove only the carbon deposits - don't remove any metal and do not nick or scratch the sides of the ring grooves **(see illustrations)**.

5   Once the deposits have been removed, clean the piston/rod assemblies with solvent and dry them with compressed air (if available). Make sure the oil return holes in the back sides of the ring grooves are clear.

6   If the pistons and cylinder walls aren't damaged or worn excessively, and if the engine block is not rebored, new pistons won't be necessary. Normal piston wear appears as even vertical wear on the piston thrust surfaces and slight looseness of the top ring in its groove. New piston rings, however, should always be used when an engine is rebuilt.

7   Carefully inspect each piston for cracks around the skirt, at the pin bosses and at the ring lands.

8   Look for scoring and scuffing on the thrust faces of the skirt, holes in the piston crown and burned areas at the edge of the crown. If the skirt is scored or scuffed, the engine may have been suffering from overheating and/or abnormal combustion, which caused excessively high operating tempera-

tures. The cooling and lubrication systems should be checked thoroughly. A hole in the piston crown is an indication that abnormal combustion (preignition) was occurring. Burned areas at the edge of the piston crown are usually evidence of spark knock (detonation). If any of the above problems exist, the causes must be corrected or the damage will occur again. The causes may include intake air leaks, incorrect fuel/air mixture, incorrect ignition timing and EGR system malfunctions.

9   Corrosion of the piston, in the form of small pits, indicates that coolant is leaking into the combustion chamber and/or the crankcase. Again, the cause must be corrected or the problem may persist in the rebuilt engine.

10   Measure the piston ring side clearance by laying a new piston ring in each ring groove and slipping a feeler gauge in beside it **(see illustration)**. Check the clearance at three or four locations around each groove. Be sure to use the correct ring for each groove - they are different. If the side clearance is greater than specified, new pistons will have to be used.

11   Check the piston-to-bore clearance by measuring the bore (see Section 18) and the

20.11 Measure the piston diameter at a 90-degree angle to the piston pin and in line with it

21.1 The oil holes should be chamfered so sharp edges don't gouge or scratch the new bearings

21.2 Use a wire or stiff plastic bristle brush to clean the oil passages in the crankshaft

piston diameter. Make sure the pistons and bores are correctly matched. Measure the piston across the skirt, at a 90-degree angle to and in line with the piston pin **(see illustration)** and 3/16-inch below the cross slot. Subtract the piston diameter from the bore diameter to obtain the clearance. If it's greater than specified, the block will have to be rebored and new pistons and rings installed. **Note:** *On Cadillac V8 engines, if the clearances exceed the specifications, then the cylinder liner must be replaced by an automotive machine shop as the installation requires a special tool to set the cylinder liner height. It is best to purchase new pistons with matched liners.*

12   Check the piston-to-rod clearance by twisting the piston and rod in opposite directions. Any noticeable play indicates excessive wear, which must be corrected. The piston/connecting rod assemblies should be taken to an automotive machine shop to have the pistons and rods resized and new pins installed.

13   If the pistons must be removed from the connecting rods for any reason, they should be taken to an automotive machine shop.

While they are there have the connecting rods checked for bend and twist, since automotive machine shops have special equipment for this purpose. **Note:** *Unless new pistons and/or connecting rods must be installed, do not disassemble the pistons from the connecting rods.*

14   Check the connecting rods for cracks and other damage. Temporarily remove the rod caps, lift out the old bearing inserts, wipe the rod and cap bearing surfaces clean and inspect them for nicks, gouges and scratches. After checking the rods, replace the old bearings, slip the caps into place and tighten the nuts finger tight. **Note:** *If the engine is being rebuilt because of a connecting rod knock, be sure to install new or rebuilt rods.*

21   Crankshaft - inspection

Refer to illustrations 21.1, 21.2, 21.5 and 21.7

1   Remove all burrs from the crankshaft oil holes with a stone, file or scraper **(see illustration)**.

2   Clean the crankshaft with solvent and

dry it with compressed air (if available). Be sure to clean the oil holes with a stiff brush **(see illustration)** and flush them with solvent.

3   Check the main and connecting rod bearing journals for uneven wear, scoring, pits and cracks.

4   Check the rest of the crankshaft for cracks and other damage. It should be magnafluxed to reveal hidden cracks - an automotive machine shop will handle the procedure.

5   Using a micrometer, measure the diameter of the main and connecting rod journals and compare the results to this Chapter's Specifications **(see illustration)**. By measuring the diameter at a number of points around each journal's circumference, you'll be able to determine whether or not the journal is out-of-round. Take the measurement at each end of the journal, near the crank throws, to determine if the journal is tapered.

6   If the crankshaft journals are damaged, tapered, out-of-round or worn beyond the limits given in the Specifications, have the crankshaft reground by an automotive machine shop. Be sure to use the correct size bearing inserts if the crankshaft is reconditioned.

7   Check the oil seal journals at each end of the crankshaft for wear and damage. If the seal has worn a groove in the journal, or if it's nicked or scratched **(see illustration)**, the new seal may leak when the engine is reassembled. In some cases, an automotive machine shop may be able to repair the journal by pressing on a thin sleeve. If repair isn't feasible, a new or different crankshaft should be installed.

8   Refer to Section 22 and examine the main and rod bearing inserts.

22   Main and connecting rod bearings - inspection and selection

Refer to illustration 22.1

1   Even though the main and connecting rod bearings should be replaced with new

21.5 Measure the diameter of each crankshaft journal at several points to detect taper and out-of-round conditions

21.7 If the seals have worn grooves in the crankshaft journals, or if the seal contact surfaces are nicked or scratched, the new seals will leak

ones during the engine overhaul, the old bearings should be retained for close examination, as they may reveal valuable information about the condition of the engine **(see illustration)**.

2   Bearing failure occurs because of lack of lubrication, the presence of dirt or other foreign particles, overloading the engine and corrosion. Regardless of the cause of bearing failure, it must be corrected before the engine is reassembled to prevent it from happening again.

3   When examining the bearings, remove them from the engine block, the main bearing caps, the connecting rods and the rod caps and lay them out on a clean surface in the same general position as their location in the engine. This will enable you to match any bearing problems with the corresponding crankshaft journal.

4   Dirt and other foreign particles get into the engine in a variety of ways. It may be left in the engine during assembly, or it may pass through filters or the PCV system. It may get into the oil, and from there into the bearings. Metal chips from machining operations and normal engine wear are often present. Abrasives are sometimes left in engine components after reconditioning, especially when parts are not thoroughly cleaned using the proper cleaning methods. Whatever the source, these foreign objects often end up embedded in the soft bearing material and are easily recognized. Large particles will not embed in the bearing and will score or gouge the bearing and journal. The best prevention for this cause of bearing failure is to clean all parts thoroughly and keep everything spotlessly clean during engine assembly. Frequent and regular engine oil and filter changes are also recommended.

5   Lack of lubrication (or lubrication breakdown) has a number of interrelated causes. Excessive heat (which thins the oil), overloading (which squeezes the oil from the bearing face) and oil leakage or throw off (from excessive bearing clearances, worn oil pump or high engine speeds) all contribute to lubrication breakdown. Blocked oil passages, which usually are the result of misaligned oil holes in a bearing shell, will also oil starve a bearing and destroy it. When lack of lubrication is the cause of bearing failure, the bearing material is wiped or extruded from the steel backing of the bearing. Temperatures may increase to the point where the steel backing turns blue from overheating.

6   Driving habits can have a definite effect on bearing life. Driving at low speed in too high a gear (lugging the engine) puts very high loads on bearings, which tends to squeeze out the oil film. These loads cause the bearings to flex, which produces fine cracks in the bearing face (fatigue failure). Eventually the bearing material will loosen in pieces and tear away from the steel backing. Short-trip driving leads to corrosion of bearings because insufficient engine heat is produced to drive off the condensed water and corrosive gases. These products collect in

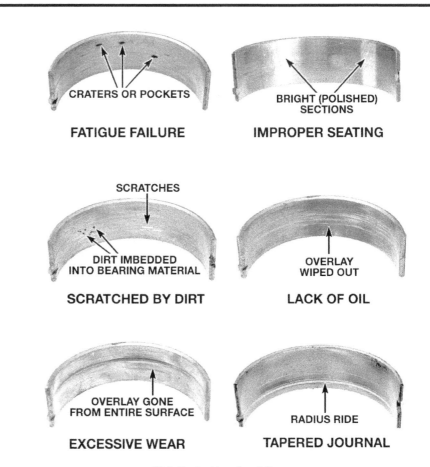

CRATERS OR POCKETS

FATIGUE FAILURE

BRIGHT (POLISHED) SECTIONS

IMPROPER SEATING

SCRATCHES

DIRT IMBEDDED INTO BEARING MATERIAL

SCRATCHED BY DIRT

OVERLAY WIPED OUT

LACK OF OIL

OVERLAY GONE FROM ENTIRE SURFACE

EXCESSIVE WEAR

RADIUS RIDE

TAPERED JOURNAL

22.1 Typical bearing failures

the engine oil, forming acid and sludge. As the oil is carried to the engine bearings, the acid attacks and corrodes the bearing material.

7   Incorrect bearing installation during engine assembly will lead to bearing failure as well. Tight fitting bearings leave insufficient bearing oil clearance and will result in oil starvation. Dirt or foreign particles trapped behind a bearing insert result in high spots on the bearing which lead to failure.

Selection

8   If the original bearings are worn or damaged, or if the oil clearances are incorrect, new bearings will have to be purchased. It is rare during a thorough rebuild of an engine with many miles on it that new replacement bearings would not be employed. However, if the crankshaft has been reground, new **undersize** bearings must be installed.

9   The automotive machine shop that reconditions the crankshaft will provide or help you select the correct size bearings. Depending on how much material has to be ground from the crankshaft to restore it, different undersize bearings are required. Crankshafts are normally ground in increments of 0.010-inch. Sometimes the amount of material removed will differ between the main and rod journals, especially if a rod jour-

nal was damaged. Markings on most reground crankshafts indicate how much was machined, such as "10-10", meaning that 0.010-inch was removed from both the main and rod journals. Such a crankshaft would require 0.010-inch undersize bearings, a common replacement size.

10   Regardless of how the bearing sizes are determined, use the oil clearance, measured with Plastigage, as a final guide to ensure the bearings are the right size. If you have any questions or are unsure which bearings to use, get help from your machine shop or dealer parts/service department.

23 Camshaft, lifters and bearings - inspection

Camshaft and bearings

Refer to illustrations 23.2 and 23.4

1   After the camshaft has been removed from the engine, cleaned with solvent and dried, inspect the bearing journals for uneven wear, pitting and evidence of seizure. If the journals are damaged, the bearing inserts in the block are probably damaged as well. Both the camshaft and bearings will have to be replaced.

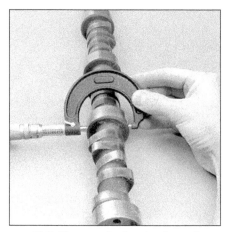

23.2 Measure the bearing journal diameters on the camshaft, then the bearing inside diameter in the block - compare the difference (the clearance) to this Chapter's Specifications

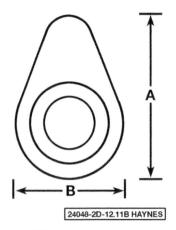

24048-2D-12.11B HAYNES

23.4 Measure the camshaft lobe maximum diameter (A) and the minimum (B) - subtract B from A to calculate the lobe lift

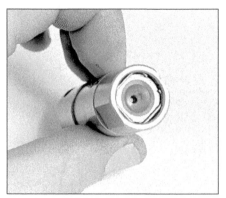

23.8a Check the pushrod seat in the top of each lifter for wear

2   Measure the bearing journals with a micrometer to determine if they are excessively worn or out-of-round **(see illustration)**.
3   Check the camshaft lobes for heat discoloration, score marks, chipped areas, pitting and uneven wear. If the lobes are in good condition and if the lobe lift measurements are as specified, the camshaft can be reused.
4   Check the lobes for wear by measuring all the intake lobe heights with a micrometer. The shortest lobe should be no more than 0.005-inch shorter than the tallest. Make the same checks for the exhaust lobes. Determine the lobe heights by measuring the tallest dimension to the shortest dimension for each lobe, then comparing the result to this Chapter's Specifications **(see illustration)**.
5   Check the condition and inside diameter of the camshaft bearings. Look for wear or damage.
6   The inside diameter of each bearing can be determined with an inside micrometer or a small-hole gauge and an outside micrometer. Subtract the camshaft journal diameter from the corresponding bearing inside diameter to obtain the bearing oil clearance. If it's excessive, new bearings will be required. Whenever an engine has accumulated enough miles to warrant a complete rebuild, it is common to replace the camshaft and bearings as a general rule, regardless of their original condition. The camshaft bearings are not expensive, but they can only be replaced during a complete rebuild, when the block is bare. Specialized equipment is needed to install camshaft bearings, but the same machine shop that hot-tanks your block can install the new camshaft bearings (and the rear balance shaft bearing on V6 engines).

Lifters

Refer to illustrations 23.8a, 23.8b and 23.8c
Caution: *Do not unbolt and remove the lifter carrier assembly on the Cadillac V8 engine. This will destroy the alignment of the lifter*

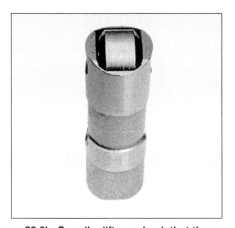

23.8b On roller lifters, check that the roller is free of pits or wear and rolls freely

bores. Also, don't substitute an incorrect lifter on the Cadillac V8 engine. These special lifters are designed for the valve system on the Cadillac V8 only, and since other types fit the lifter bore, incorrect lifters can easily be installed.
7   Clean the lifters with solvent and dry them thoroughly without mixing them up.
8   Check each lifter wall, pushrod seat and foot for scuffing, score marks and uneven wear **(see illustrations)**. Each lifter foot (the surface that rides on the cam lobe) must be slightly convex, although this can be difficult to determine by eye. If the base of the lifter is concave, the lifters and camshaft must be replaced **(see illustration)**. If the lifter walls are damaged or worn (which is not very likely), inspect the lifter bores in the engine block as well. If the pushrod seats are worn, check the pushrod ends.
9   If new lifters are being installed, a new camshaft must also be installed. If a new camshaft is installed, use new lifters as well. It is possible to install new roller lifters on a used camshaft, but the factory doesn't recommend it. Never install used lifters unless the original camshaft is used and the lifters can be installed in their original locations.

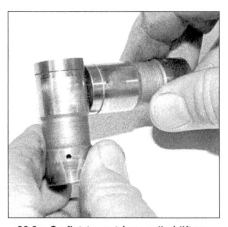

23.8c On flat-tappet (non-roller) lifters, check the foot of each lifter for curvature against the flat side of another lifter - if it appears concave, it's worn and a new camshaft and lifters will be required

Caution: *On Cadillac V8 engines, do not use a reground or "rebuilt" camshaft. Use only a brand-new factory camshaft to ensure camshaft longevity.* **Note:** *Some engines may be factory-fitted with oversized lifters in some bores. The lifters and the block near their lifter bore is usually marked with an "O", which indicates a .010-inch-oversize lifter is required there.*

24 Engine overhaul - reassembly sequence

1   Before beginning engine reassembly, make sure you have all the necessary new parts, gaskets and seals as well as the following items on hand:

Common hand tools
A 1/2-inch drive torque wrench
Piston ring installation tool
Piston ring compressor
Vibration damper installation tool
Short lengths of rubber or plastic hose to fit over connecting rod bolts
Plastigage
Feeler gauges
A fine-tooth file

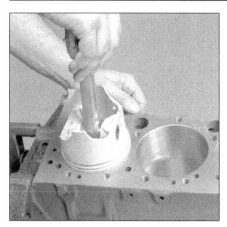

25.3 When checking piston ring end gap, the ring must be square in the cylinder bore (this is done by pushing the ring down with the top of a piston as shown)

New engine oil
Engine assembly lube or moly-base grease
Gasket sealant
Thread-locking compound

2   In order to save time and avoid problems, engine reassembly must be done in the following general order:

New camshaft bearings (must be installed by an automotive machine shop)
Piston rings
Crankshaft and main bearings
Piston/connecting rod assemblies
Camshaft and lifters
Balance shaft (1988 and later V6 engines)
Timing chain and sprockets
Engine front cover
Oil pump (V8 engines)
Oil pan
Cylinder heads, pushrods and rocker arms
Intake and exhaust manifolds
Valve covers
Engine rear plate
Driveplate

25   Piston rings - installation

Refer to illustrations 25.3, 25.4, 25.9a, 25.9b and 25.12

1   Before installing the new piston rings, the ring end gaps must be checked. It's assumed that the piston ring side clearance has been checked and verified correct (see Section 19).
2   Lay out the piston/connecting rod assemblies and the new ring sets so the ring sets will be matched with the same piston and cylinder during the end-gap measurement and engine assembly.
3   Insert the top (number one) ring into the first cylinder and square it up with the cylinder walls by pushing it in with the top of the piston **(see illustration)**. The ring should be near the bottom of the cylinder, at the lower limit of ring travel.

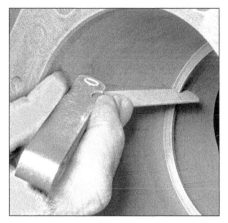

25.4 With the ring square in the cylinder, measure the end gap with a feeler gauge

25.9b DO NOT use a piston ring installation tool when installing the oil ring side rails

4   To measure the end gap, slip feeler gauges between the ends of the ring until a gauge equal to the gap width is found **(see illustration)**. The feeler gauge should slide between the ring ends with a slight amount of drag. Compare the measurement to this Chapter's Specifications. If the gap is larger or smaller than specified, double-check to make sure you have the correct rings before proceeding.
5   If the gap is too small, try another set of rings - DO NOT file the ends to increase the clearance.
6   Excess end gap isn't critical unless it's greater than the service limit listed in this Chapter's Specifications. Again, double-check to make sure you have the correct rings for your engine.
7   Repeat the procedure for each ring that will be installed in the first cylinder and for each ring in the remaining cylinders. Remember to keep rings, pistons and cylinders matched up.
8   Once the ring end gaps have been checked/corrected, the rings can be installed on the pistons.
9   The oil control ring (lowest one on the piston) is usually installed first. It's composed of three separate components. Slip the

25.9a Install the spacer/expander in the oil control ring groove

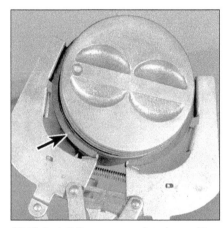

25.12 Install the compression rings with a ring expander - the mark (arrow) must face up

spacer/expander into the groove **(see illustration)**. If an anti-rotation tang is used, make sure it's inserted into the drilled hole in the ring groove. Next, install the lower side rail. Do not use a piston-ring installation tool on the oil ring side rails, as they may be damaged. Instead, place one end of the side rail into the groove between the spacer/expander and the ring land, hold it firmly in place and slide a finger around the piston while pushing the rail into the groove **(see illustration)**. Next, install the upper side rail in the same manner.
10   After the three oil ring components have been installed, check to make sure that both the upper and lower side rails can be turned smoothly in the ring groove.
11   The number two (middle) ring is installed next. It's usually stamped with a mark that must face up, toward the top of the piston. **Note:** *Always follow the instructions printed on the ring package or box - different manufacturers may require different approaches.* Do not mix up the top and middle rings, as they have different cross-sections.
12   Use a piston-ring installation tool and make sure the ring's identification mark is facing the top of the piston, then slip the ring into the middle groove on the piston **(see**

illustration). Don't expand the ring any more than necessary to slide it over the piston.

13 Install the number one (top) ring in the same manner. Make sure the mark is facing up. Be careful not to confuse the number one and number two rings.

14 Repeat the procedure for the remaining pistons and rings.

26 Crankshaft - installation and main bearing oil clearance check

Refer to illustrations 26.11 and 26.15

1 Crankshaft installation is the first step in engine reassembly. It's assumed at this point that the engine block and crankshaft have been cleaned, inspected and repaired or reconditioned.

2 Position the engine with the bottom facing up.

3 Remove the main bearing cap bolts and lift out the caps. Lay them out in the proper order to ensure correct installation. **Note:** *On Cadillac V8 engines, each bearing cap has a cast number (1/8, 2/8, 3/8, 4/8 and 5/8) on the bottom starting from the front of the engine. Do not mismatch them, as they are manufactured specifically for that particular journal.*

4 If they're still in place, remove the original bearing inserts from the block and the main bearing caps. Wipe the bearing surfaces of the block and caps with a clean, lint-free cloth. They must be kept spotlessly clean.

Main bearing oil clearance check

5 Clean the back sides of the new main bearing inserts and lay one in each main bearing saddle in the block. If one of the bearing inserts from each set has a large groove in it, make sure the grooved insert is installed in the block. Lay the other bearing from each set in the corresponding main bearing cap. Make sure the tab on the bearing insert fits into the recess in the block or cap. **Caution:** *The oil holes in the block must line up with the oil holes in the bearing insert.* Do not hammer the bearing into place and don't nick or gouge the bearing faces. No lubrication should be used at this time.

6 The flanged thrust bearing must be installed in the proper cap and saddle.

7 Clean the faces of the bearings in the block and the crankshaft main bearing journals with a clean, lint-free cloth.

8 Check or clean the oil holes in the crankshaft, as any dirt here can go only one way - straight through the new bearings.

9 Once you're certain the crankshaft is clean, carefully lay it in position in the main bearings.

10 Before the crankshaft can be permanently installed, the main bearing oil clearance must be checked.

11 Cut several pieces of the appropriate-size Plastigage (they must be slightly shorter than the width of the main bearings) and place one piece on each crankshaft main

26.11 Lay the Plastigage strips (arrow) on the main bearing journals, parallel to the crankshaft centerline

bearing journal, parallel with the journal axis **(see illustration)**.

12 Clean the faces of the bearings in the caps and install the caps in their respective positions (don't mix them up) with the arrows pointing toward the front of the engine. Don't disturb the Plastigage.

13 Starting with the center main and working out toward the ends, tighten the main bearing cap bolts, in three steps, to the specified torque. Don't rotate the crankshaft at any time during this operation.

14 Remove the bolts and carefully lift off the main bearing caps. Keep them in order. Don't disturb the Plastigage or rotate the crankshaft. If any of the main bearing caps are difficult to remove, tap them gently from side to side with a soft-face hammer to loosen them.

15 Compare the width of the crushed Plastigage on each journal to the scale printed on the Plastigage envelope to obtain the main bearing oil clearance **(see illustration)**. Check the Specifications to make sure it's correct.

16 If the clearance is not as specified, the bearing inserts may be the wrong size (which means different ones will be required). Before deciding that different inserts are needed, make sure that no dirt or oil was between the bearing inserts and the caps or block when the clearance was measured. If the Plastigage was wider at one end than the other, the journal may be tapered (refer to Section 21).

17 Carefully scrape all traces of the Plastigage material off the main bearing journals and/or the bearing faces. Use your fingernail or the edge of a credit card - don't nick or scratch the bearing faces.

Final crankshaft installation

18 Carefully lift the crankshaft out of the engine.

19 Clean the bearing faces in the block, then apply a thin, uniform layer of moly-base grease or engine assembly lube to each of the bearing surfaces. Be sure to coat the thrust faces as well as the journal face of the thrust bearing.

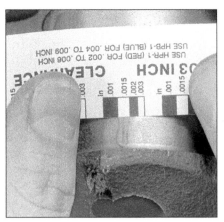

26.15 Measure the width of the crushed Plastigage to determine the main bearing oil clearance (be sure to use the correct scale - standard and metric ones are included)

20 Make sure the crankshaft journals are clean, then lay the crankshaft back in place in the block. On engines with a two-piece rear seal, the upper half of the rear main seal should be in place in the block (see Section 27).

21 Clean the faces of the bearings in the caps, then apply lubricant to them.

22 Install the caps in their respective positions with the arrows pointing toward the front of the engine. The lower half of the rear main seal should be installed in the cap (see Section 27).

23 Install the bolts.

24 Tighten all except the thrust bearing cap bolts to the specified torque (work from the center out and approach the final torque in three steps).

25 Tighten the thrust bearing cap bolts to 10 to 12 ft-lbs.

26 Tap the ends of the crankshaft forward and backward with a lead or brass hammer to line up the main bearing and crankshaft thrust surfaces.

27 Retighten all main bearing cap bolts to the specified torque, starting with the center main and working out toward the ends.

28 Rotate the crankshaft a number of times by hand to check for any obvious binding.

29 The final step is to check the crankshaft endplay with a feeler gauge or a dial indicator as described in Section 16. The endplay should be correct if the crankshaft thrust faces aren't worn or damaged and new bearings have been installed.

27 Rear main oil seal installation

Refer to illustration 27.7

1 In Parts A and B of this Chapter, the procedure is given for replacement of the rear main seal while the engine is in the car, and the crankshaft still in the engine. The procedure is somewhat simpler when the engine is being overhauled, but refer to the illustrations and procedures in those Parts first.

27.7 Force RTV into the Cadillac V8 rear main cap grooves (arrow), using a sealant tube fitted with a washer

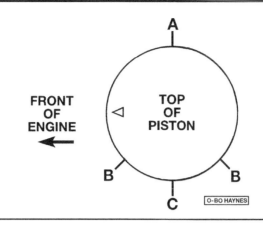

28.5 Ring end gap positions

A *Top compression ring gap, oil ring spacer gap*
B *Oil ring side rail gaps*
C *Second compression ring gap*

2   On engines with neoprene seals, install the upper half in the groove in the block. Most neoprene seals come with a wax coating to ease installation, don't remove this coating. Seat the seal in the block by pressing the ends flush with the block, pushing evenly on both ends of the seal at the same time. **Caution:** *Do not use any kind of silicone lubricant on the seal lips or oil leakage could develop.*
3   On engines with fabric rear seals (early V6's), apply some Loctite 414 (or equivalent) to the seal's groove, install one-half of the new seal into the block, and work it into the channel with a large-diameter socket or a wooden hammer handle. Trim the ends with a sharp razor or utility knife, leaving each end sticking up 1/16-inch above the surface of the block. **Note:** *When trimming the seal ends, use a metal backing to cut against, so the seal end is cut clean and square.*
4   With either type of seal, proceed with the crankshaft installation and main bearing selection (see Section 26) after Steps 1 through 3. The following steps should be performed after the Plastigage is removed, the bearings are lubed and the caps are being installed for the last time.
5   Install the lower seal half (neoprene) into the main cap until it is flush with the cap at each end. On fabric seals, install and trim the lower seal to the cap as in Step 3 above. On V6 engines, apply a drop of RTV sealant to the ends of the seals and the rear of the cap (see Chapter 2A).
6   On some V6 models, there are two side seals between the cap and the block. Install and tighten the cap bolts to Specifications. Soak the side seals in light engine oil for five minutes, then push them into the grooves between the main cap and the block (see Chapter 2A).
7   On Cadillac V8 engines, the side seals on the rear main are not separate neoprene seals, but RTV sealant. Clean any old sealant from the cap and block, then install the cap with new main seals in place. Fit a small washer over the tip on your RTV tube, put the

tip into the cap-to-block groove and force RTV in **(see illustration).** Both grooves should be filled completely.
8   On all engines, after the rear cap is tightened to the specified torque, rotate the crankshaft to make sure everything turns properly.

28 Piston/connecting rod assembly - installation and rod bearing oil clearance check

Refer to illustrations 28.5, 28.10, 28.12 and 28.16
1   Before installing the piston/connecting rod assemblies, the cylinder walls must be perfectly clean, the top edge of each cylinder must be chamfered, and the crankshaft must be in place.
2   Remove the cap from the end of the number one connecting rod (refer to the marks made during removal). Remove the original bearing inserts and wipe the bearing surfaces of the connecting rod and cap with a clean, lint-free cloth. They must be kept spotlessly clean.

Connecting rod bearing oil clearance check

3   Clean the back side of the new upper bearing insert, then lay it in place in the connecting rod. Make sure the tab on the bearing fits into the recess in the rod. Don't hammer the bearing insert into place and be very careful not to nick or gouge the bearing face. Don't lubricate the bearing at this time.
4   Clean the back side of the other bearing insert and install it in the rod cap. Again, make sure the tab on the bearing fits into the recess in the cap, and don't apply any lubricant. It's critically important that the mating surfaces of the bearing and connecting rod are perfectly clean and oil-free when they're assembled.
5   Stagger the piston rings around the piston as shown in the accompanying illustration **(see illustration).** This will prevent blowby, loss of compression and excessive oil consumption.
6   Lubricate the piston and rings with clean engine oil and attach a piston ring compres-

sor to the piston. Leave the skirt protruding about 1/4-inch to guide the piston into the cylinder. The rings must be compressed until they're flush with the piston.
7   Rotate the crankshaft until the number one connecting rod journal is at BDC (bottom dead center) and apply a coat of engine oil to the cylinder walls. **Caution:** *During this procedure on Cadillac V8 engines, the crankshaft must never be turned unless either the cylinder heads are in place or a suitable tool is bolted to the head surface of the block to hold the cylinder liners from moving (see Part B of this Chapter). If the engine is turned without the plate tightened in place, the cylinder liners can move enough to break the seal on the O-rings sealing the bottom of the liners to the block, causing coolant leaks later on. It is one of the reasons that we recommend having this engine rebuilt professionally.*
8   With the mark or notch on top of the piston facing the front of the engine, gently insert the piston/connecting rod assembly into the number one cylinder bore and rest the bottom edge of the ring compressor on the engine block.
9   Tap the top edge of the ring compressor to make sure it's contacting the block around its entire circumference.
10   Gently tap on the top of the piston with the end of a wooden hammer handle while guiding the end of the connecting rod into

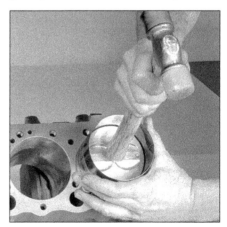

28.10 Drive the piston gently into the cylinder bore with the end of a wooden or plastic hammer handle

28.12 Lay the Plastigage strips on each rod bearing journal, parallel to the crankshaft centerline

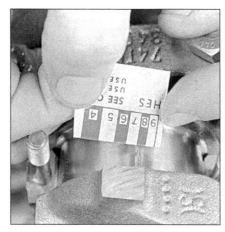

28.16 Measure the width of the crushed Plastigage to determine the rod bearing oil clearance (be sure to use the correct scale - standard and metric ones are included)

place on the crankshaft journal **(see illustration)**. The piston rings may try to pop out of the ring compressor just before entering the cylinder bore, so keep some downward pressure on the ring compressor. Work slowly, and if any resistance is felt as the piston enters the cylinder, stop immediately. Find out what's hanging up and fix it before proceeding. Do not, for any reason, force the piston into the cylinder - you might break a ring and/or the piston.

11   Once the piston/connecting rod assembly is installed, the connecting rod bearing oil clearance must be checked before the rod cap is permanently bolted in place.

12   Cut a piece of the appropriate-size Plastigage slightly shorter than the width of the connecting rod bearing and lay it in place on the number one connecting rod journal, parallel with the journal axis **(see illustration)**.

13   Clean the connecting rod cap bearing face and install the rod cap. Make sure the mating mark on the cap is on the same side as the mark on the connecting rod.

14   Install the bolts and tighten them to the torque listed in this Chapter's Specifications, working up to it in three steps. **Note:** *Use a thin-wall socket to avoid erroneous torque readings that can result if the socket is wedged between the rod cap and nut. If the socket tends to wedge itself between the nut and the cap, lift up on it slightly until it no longer contacts the cap. Do not rotate the crankshaft at any time during this operation.*

15   Remove the bolts and detach the rod cap, being very careful not to disturb the Plastigage.

16   Compare the width of the crushed Plastigage to the scale printed on the Plastigage envelope to obtain the oil clearance. Compare it to the Specifications to make sure the clearance is correct **(see illustration)**.

17   If the clearance is not as specified, the bearing inserts may be the wrong size (which means different ones will be required). Before deciding that different inserts are needed, make sure that no dirt or oil was between the bearing inserts and the connecting rod or cap

when the clearance was measured. Also, recheck the journal diameter. If the Plastigage was wider at one end than the other, the journal may be tapered (refer to Section 21).

Final connecting rod installation

18   Carefully scrape all traces of the Plastigage material off the rod journal and/or bearing face. Be very careful not to scratch the bearing - use your fingernail or the edge of a credit card.

19   Make sure the bearing faces are perfectly clean, then apply a uniform layer of clean moly-base grease or engine assembly lube to both of them. You'll have to push the piston into the cylinder to expose the face of the bearing insert in the connecting rod - be sure to slip the protective hoses over the rod bolts first.

20   Slide the connecting rod back into place on the journal, install the rod cap and tighten the bolts to the specified torque. Again, work up to the torque in three steps.

21   Repeat the entire procedure for the remaining pistons/connecting rods.

22   The important points to remember are:

a) *Keep the back sides of the bearing inserts and the insides of the connecting rods and caps perfectly clean when assembling them.*

b) *Make sure you have the correct piston/rod assembly for each cylinder.*

c) *The notch or mark on the piston must face the front of the engine.*

d) *Lubricate the cylinder walls with clean oil.*

e) *Lubricate the bearing faces when installing the rod caps after the oil clearance has been checked.*

23   After all the piston/connecting rod assemblies have been properly installed, rotate the crankshaft a number of times by hand to check for any obvious binding.

24   As a final step, the connecting rod endplay must be checked. Refer to Section 15 for this procedure.

25   Compare the measured endplay to the Specifications to make sure it's correct. If it was correct before disassembly and the original crankshaft and rods were reinstalled, it should still be right. If new rods or a new crankshaft were installed, the endplay may be inadequate. If so, the rods will have to be removed and taken to an automotive machine shop for resizing.

29   Camshaft and lifters - installation

Refer to illustrations 29.2 and 29.3

1   If new lifters are being installed, a new camshaft must also be installed. If the camshaft is replaced, then install new lifters as well. Never install used lifters unless the original camshaft is used and the lifters can be installed in their original locations!

2   Prior to installing the camshaft, coat each of the lobes and journals with camshaft installation lubricant **(see illustration)**.

3   Slide the camshaft into the engine block, again taking extra care so you don't damage the bearings. Do not push the camshaft too hard; there is a plug at the back of the block that could be loosened, causing an oil leak **(see illustration)**. On balance-shaft V6 engines, install the camshaft thrust plate and bolts (see Section 14).

4   Soak new lifters in oil to remove trapped

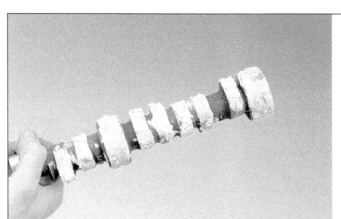

29.2 Coat the journals and lobes of the camshaft with camshaft installation lube before installation

29.3 The back of the engine has a camshaft plug (arrow, Cadillac V8 shown) - do not push the camshaft in hard enough to disturb this plug, or an oil leak will result

30.1 Drive the front balance shaft bearing in (it's attached to the balance shaft) just until the retainer plate can be installed

air, and lube the bottom of the lifters with assembly lube and install them into the engine block.

5   Section 14 covers the removal of the lifter guides and lifter guide retainer for roller-lifter engines. Lubricate the lifters with engine oil and use the illustrations in Section 14 to install the lifter guides and lifter guide retainer. Unless the lifter guide retainer is in place, the engine must remain upright on the engine stand until the cylinder heads, pushrods and rocker assemblies are installed, or the lifters could fall out of the engine.

30 Balance shaft (V6 engine) - installation

Refer to illustrations 30.1, 30.4 and 30.5

1   Lubricate the front and rear journals of the balance shaft with new engine oil and insert the balance shaft carefully into the block. When the front bearing approaches the insert in the front of the block, use a ham-

mer and an appropriate-size socket to drive the front bearing into its insert. Drive it in just enough to allow installation of the balance shaft bearing retainer **(see illustration)**. Tighten the balance shaft retainer bolts to the torque listed in this Chapter's Specifications.

2   Install the balance shaft driven gear and its bolt to the balance shaft.

3   Turn the camshaft so that with the camshaft sprocket temporarily installed, the timing mark points straight down.

4   With the camshaft sprocket removed, turn the balance shaft so that the timing mark on the driven gear points straight down **(see illustration)**.

5   Install the camshaft gear (that drives the balance shaft) onto the camshaft, aligning it with the keyway and align the marks on the balance shaft gear and camshaft gear by turning the balance shaft, not the camshaft **(see illustration)**.

6   After the camshaft sprocket, crankshaft sprocket and timing chain have been installed (see Chapter 2, Part A), tighten the balance shaft driven gear bolt to the torque listed in this Chapter's Specifications.

31 Initial start-up and break-in after overhaul

Warning: *Have a fire extinguisher handy when starting the engine for the first time.*

1   Once the engine has been installed in the vehicle, double-check the engine oil and coolant levels.

2   With the spark plugs out of the engine and the ignition system and fuel system disabled (see Section 3 and Chapter 4), crank the engine until oil pressure registers on the gauge or the light goes out.

3   Install the spark plugs, connect the plug wires and restore the ignition system functions (see Section 3).

4   Start the engine. It may take a few moments for the fuel system to build up pressure, but the engine should start without a great deal of effort.

5   After the engine starts, allow it to warm up to normal operating temperature. While the engine is warming up, make a thorough check for fuel, oil and coolant leaks. When all new bearings, rings and camshaft(s) are

30.4 The balance shaft gear mark (arrow) should point straight down

30.5 Align the marks (arrows) on both balance shaft gears as shown

installed, the engine should run for 15 minutes at normal operating temperature on the first start-up to help break in the new components.

6   Shut the engine off and recheck the engine oil and coolant levels.

7   Drive the vehicle to an area with no traffic, accelerate from 30 to 50 mph, then allow the vehicle to slow to 30 mph with the throttle closed. Repeat the procedure 10 or 12 times. This will load the piston rings and cause them to seat properly against the cylinder walls. Check again for oil and coolant leaks.

8   Drive the vehicle gently for the first 500 miles (no sustained high speeds) and keep a constant check on the oil level. It is not unusual for an engine to use oil during the break-in period.

9   At approximately 500 to 600 miles, change the oil and filter.

10   For the next few hundred miles, drive the vehicle normally. Do not pamper it or abuse it.

11   After 2000 miles, change the oil and filter again and consider the engine broken in.

Chapter 3
Cooling, heating and air conditioning systems

Contents

Specifications

General

Coolant capacity	See Chapter 1
Drivebelt tension	See Chapter 1
Radiator pressure cap rating	12 to 16 psi
Thermostat rating	
Begins to open	192 to 199-degrees F
Fully open	219-degrees F

Torque specifications

	Ft lbs (unless otherwise indicated)
Thermostat housing cover bolts	
V8 engine	20
V6 engine	120 in-lbs
Torque strut to radiator support bolts	18
Water pump pulley bolts	
V8 engine	22
V6 engine	115 in-lbs
Water pump fasteners	
V8 engine	
Torx-head bolts	30
Lower hex-head screws	62 in-lbs
Upper hex-head screws	30
Stud nuts	62 in-lbs
V6 engine	
Lower bolts	29
Upper bolts	97 in-lbs
Engine oil cooler/filter adapter (V8 engine)	
Cooler-to-block mounting bolts	15
Oil lines to cooler	156 in-lbs

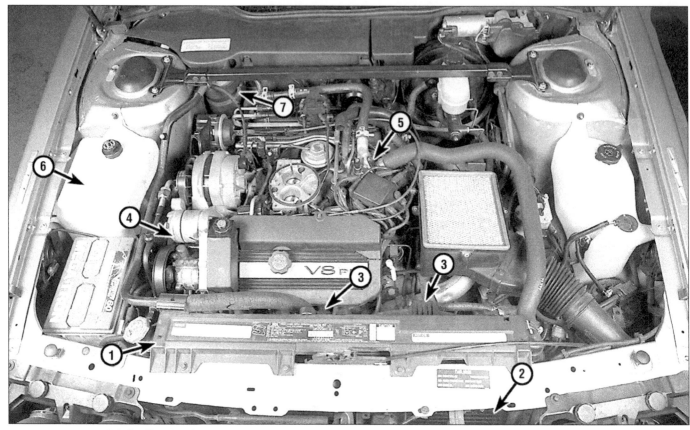

1.1 Cooling, heating and air conditioning components (1991 Cadillac shown, other models similar)

1	Radiator	4	Water pump (below alternator)	6	Coolant recovery tank
2	Condenser	5	Thermostat	7	Air conditioning accumulator
3	Radiator and condenser fans				

1   General information

Engine cooling system

Refer to illustrations 1.1 and 1.2

All vehicles covered by this manual employ a pressurized engine cooling system with thermostatically-controlled coolant circulation **(see illustration)**. An impeller-type water pump mounted on the engine block pumps coolant through the engine. The coolant flows around each cylinder and toward the rear of the engine. Cast-in coolant passages direct coolant around the intake and exhaust ports, near the spark plug areas and in close proximity to the exhaust valve guides.

A wax-pellet type thermostat controls engine coolant temperature. During warm up, the closed thermostat prevents coolant from circulating through the radiator. As the engine nears normal operating temperature, the thermostat opens and allows hot coolant to travel through the radiator, where it's cooled before returning to the engine **(see illustration)**.

The cooling system is sealed by a pressure-type radiator cap, which raises the boiling point of the coolant and increases the cooling efficiency of the radiator. If the sys-

tem pressure exceeds the cap pressure relief value, the excess pressure in the system forces the spring-loaded valve inside the cap off its seat and allows the coolant to escape through the overflow tube into a coolant reservoir. When the system cools, the excess coolant is automatically drawn from the reservoir back into the radiator.

The coolant reservoir serves as both the point at which fresh coolant is added to the cooling system to maintain the proper fluid level and as a holding tank for overheated coolant. This type of cooling system is known as a closed design because coolant that escapes past the pressure cap is saved and reused.

Heating system

The heating system consists of a blower fan, a heater core located in the heater box, the hoses connecting the heater core to the engine cooling system and the heater/air conditioning control head on the dashboard. Hot engine coolant is circulated through the heater core. When the heater mode is activated, a flap door opens to expose the heater box to the passenger compartment. A fan switch on the control head activates the blower motor, which forces air through the core, heating the air.

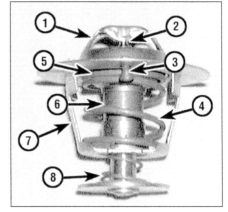

1.2 Typical thermostat

1	Flange	5	Valve seat
2	Piston	6	Valve
3	Jiggle valve	7	Frame
4	Main coil	8	Secondary
	spring		coil spring

Air conditioning system

The air conditioning system consists of a condenser mounted in front of the radiator, an evaporator mounted in the blower housing, a compressor mounted on the engine, an accumulator which filters moisture out of the

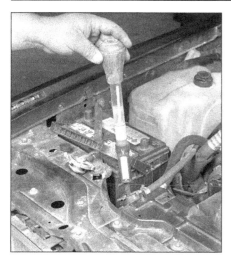

2.4 An inexpensive hydrometer can be used to test the condition of your coolant

3.10a To replace the V6 engine thermostat, remove the hose, then the bolt (arrow) holding the cover, pull off the cover and remove the thermostat from the housing

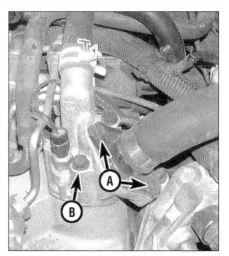

3.10b V8 engine thermostat cover bolts (A) - if there is evidence of coolant stains below the thermostat housing (B), unbolt it and install a new gasket between the housing and the intake manifold

refrigerant and the plumbing connecting all of the above components.

A blower fan forces the warmer air of the passenger compartment through the evaporator core (sort of a radiator-in-reverse), transferring the heat from the air to the refrigerant. The liquid refrigerant boils off into low pressure vapor, taking the heat with it when it leaves the evaporator.

2   Antifreeze - general information

Refer to illustration 2.4

Warning: *Do not allow antifreeze to come in contact with your skin or painted surfaces of the vehicle. Rinse off spills immediately with plenty of water. Antifreeze is highly toxic if ingested. Never leave antifreeze lying around in an open container or in puddles on the floor; children and pets are attracted by its sweet smell and may drink it. Check with local authorities about disposing of used antifreeze. Many communities have collection centers that will see that antifreeze is disposed of safely. Never dump used anti-freeze on the ground or into drains.*
Note: *Non-Toxic coolant is available at local auto parts stores. Although the coolant is non-toxic when fresh, proper disposal is still required.*

The cooling system should be filled with a water/ethylene glycol based antifreeze solution, which will prevent freezing down to at least -20-degrees F, or lower if local climate requires it. It also provides protection against corrosion and increases the coolant boiling point. Cadillac models have an aluminum-block V8 engine, which requires a coolant made specifically for aluminum engines.

The cooling system should be drained, flushed and refilled at the specified intervals (see Chapter 1). Old or contaminated antifreeze solutions are likely to cause damage and encourage the formation of rust and

scale in the system. Use distilled water with the antifreeze.

Before adding antifreeze, check all hose connections, because antifreeze tends to leak through very minute openings. Engines don't normally consume coolant, so if the level goes down, find the cause and correct it.

The exact mixture of antifreeze-to-water that you should use depends on the relative weather conditions. The mixture should contain at least 50-percent antifreeze, but should never contain more than 70-percent antifreeze. Consult the mixture ratio chart on the antifreeze container before adding coolant. Hydrometers are available at most auto parts stores to test the coolant **(see illustration)**. Use antifreeze that meets the vehicle manufacturer's specifications.

3   Thermostat - check and replacement

Warning: *Do not remove the radiator cap, drain the coolant or replace the thermostat until the engine has cooled completely.*

Check

1   Before assuming the thermostat is to blame for a cooling system problem, check the coolant level, drivebelt tension (see Chapter 1) and temperature gauge operation.
2   If the engine seems to be taking a long time to warm up (based on heater output or temperature gauge operation), the thermostat is probably stuck open. Replace the thermostat with a new one.
3   If the engine runs hot, use your hand to check the temperature of the upper radiator hose. If the hose isn't hot, but the engine is, the thermostat is probably stuck closed, preventing the coolant inside the engine from

escaping to the radiator. Replace the thermostat. **Caution:** *Don't drive the vehicle without a thermostat. The computer may stay in open loop and emissions and fuel economy will suffer.*
4   If the upper radiator hose is hot, it means that the coolant is flowing and the thermostat is open. Consult the *Troubleshooting* section at the front of this manual for cooling system diagnosis.

Replacement

Refer to illustrations 3.10a, 3.10b, 3.13 and 3.14

5   Disconnect the negative battery cable from the battery.
6   Drain the cooling system (see Chapter 1). If the coolant is relatively new or in good condition (see Chapter 1), save it and reuse it. Read the **Warning** in Section 2.
7   Follow the upper radiator hose to the engine to locate the thermostat housing cover.
8   Loosen the hose clamp, then detach the hose from the fitting. If it's stuck, grasp it near the end with a pair of adjustable pliers and twist it to break the seal, then pull it off. If the hose is old or deteriorated, cut it off and install a new one.
9   If the outer surface of the large fitting that mates with the hose is deteriorated (corroded, pitted, etc.) it may be damaged further by hose removal. If it is, the thermostat housing cover will have to be replaced.
10   Remove the thermostat cover bolts **(see illustrations)** and detach the housing cover. If the cover is stuck, tap it with a soft-face hammer to jar it loose. Be prepared for some coolant to spill as the gasket seal is broken.
11   Note how it's installed, then remove the thermostat.
12   Remove all traces of old gasket material and/or sealant from the housing and cover.
13   Install a new rubber gasket over the

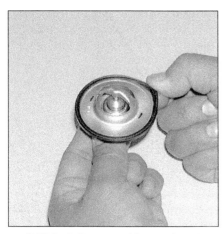

3.13 Install a new rubber seal over the thermostat

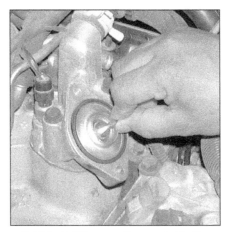

3.14 Install the new thermostat in the housing with the spring towards the engine (V8 engine shown)

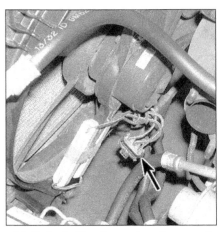

4.1 To test either fan motor, disconnect the electrical connector (arrow) and use jumper wires to connect the fan directly to the battery and ground - if the fan still doesn't work, replace the motor

thermostat **(see illustration)**.
14   Install the new thermostat in the housing without using sealant. Make sure the spring end is directed into the engine **(see illustration)**. On V6 models, replace both O-rings on the thermostat cover. Apply clean anti-freeze to the O-rings for lubrication before installation.
15   Install the housing cover and bolts. Tighten the bolts to the torque listed in this Chapter's Specifications.
16   Reattach the hose and tighten the hose clamp securely. Install all components that were removed for access.
17   Refill the cooling system (see Chapter 1).
18   Start the engine and allow it to reach normal operating temperature, then check for leaks and proper thermostat operation (as described in Steps 2 through 4).

---

4   Engine cooling fans - check and replacement

Warning: *To avoid possible injury or damage, DO NOT operate the engine with a damaged fan. Do not attempt to repair fan blades -* replace a damaged fan with a new one.
Note: *All models have two fans. The following procedures apply to both.*

Check

Refer to illustrations 4.1 and 4.3

1   To test a fan motor, disconnect the electrical connector at the motor **(see illustration)** and use fused jumper wires to connect the fan directly to the battery and to ground. If the fan still doesn't work, replace the motor.
2   If the motor tests OK, check the fuses and relays, located in the engine compartment fuse/relay box. Some models have two fan relays, others have three; look for symbol of a fan on top of the relay, or identify them from the chart on the inside of the relay box cover. See Chapter 12 for relay testing. Also check the wiring which connects the components (refer to the wiring diagrams for the fan circuit in Chapter 12).
3   The fans are controlled by the ECM, which takes input from several sources to determine when the fans should operate. The main input is from the engine's coolant temperature sensor. See Chapter 6 for location and testing of the coolant temperature sensor. The body control module also provides

input, to tell the ECM when the air conditioning is on. On 1986 and 1987 Deville models and 1986 through 1989 Seville models, fan speed is controlled by a fan control module, mounted in front of the radiator **(see illustration)**. **Warning:** *Part of the fan control module is a heat sink, which may be hot after fan operation. Allow it to cool before removing or working around the module.*
4   All models equipped with Electronic Climate Control can display trouble codes, which can be used to diagnose engine and body components. See Chapter 6 for entering the diagnostic mode, and how to access the Body Control Module (BCM) codes, which include those that cover heating, cooling and air conditioning. A problem with the engine cooling fan circuit will display a code. A problem with the CTS will display a code during the ECM diagnostics.

Replacement

Refer to illustrations 4.8a, 4.8b, 4.9a, 4.9b, 4.11 and 4.12

5   Disconnect the negative battery cable

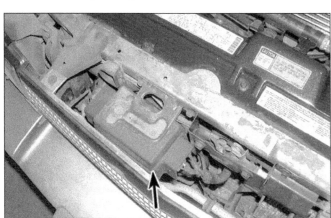

4.3 The cooling fans on some models are controlled by the Fan Control Module (arrow) mounted on front of the radiator

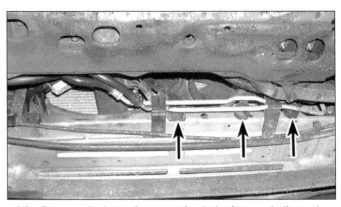

4.8a Remove the lower fan mounting bolts (arrows indicate the two main fan lower bolts and one of the two lower condenser fan bolts, seen from below) - later model shown, with two fans on engine side of radiator

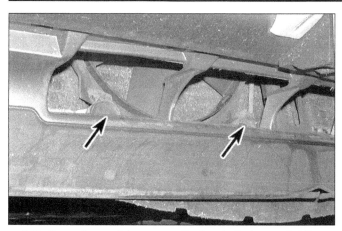

4.8b On earlier models, remove the two bolts (arrows) at the bottom of the engine-side fan

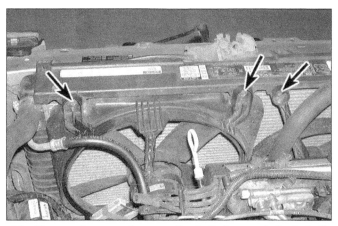

4.9a Remove the upper fan mounting bolts (arrows indicate both of the main fan bolts and one of the two condenser fan bolts at right)

4.9b The single upper fan bolt (arrow) on front-of-radiator fans is accessible, the two lower bolts only after the grille and fan module are removed

4.11 To remove the fan, unscrew the left-hand-threaded nut in the center (arrow), then pull the fan blade from the motor shaft

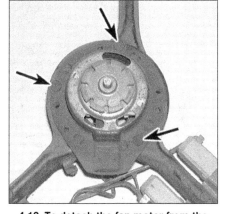

4.12 To detach the fan motor from the shroud, remove these bolts (arrows)

from the battery.

6   Set the parking brake and block the rear wheels to prevent the vehicle from rolling. Raise the front of the vehicle and support it securely with jackstands. Remove the lower splash pan, if equipped, from under the radiator.

7   Refer to Step 1 and disconnect the fan electrical connector. Early models have one fan in front of the radiator and one on the engine side, while later models have both fans on the engine side.

8   Remove the fan(s) lower mounting bolts **(see illustrations)**.

9   Unbolt the fan(s) from the radiator at the top **(see illustrations)**. **Note:** *When removing the front fan on early models, refer to Chapter 11 for grille removal, then unbolt the fan control module* **(see illustration 4.3)** *to access the two lower fan bolts.*

10   Carefully lift the fan out of the engine compartment.

11   To detach the fan from the motor, remove the motor shaft nut **(see illustration)**. **Note:** *The nut has a **left-hand** thread.*

12   To detach the fan motor from the shroud, remove the mounting bolts **(see illustration)**.

13   Installation is the reverse of removal.

---

5   Radiator and coolant reservoir-removal and installation

Warning: *Wait until the engine is completely cool before beginning this procedure.*

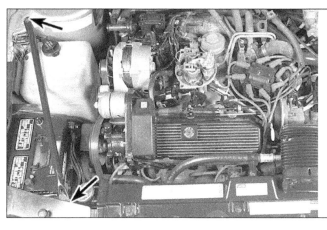

Coolant reservoir

Refer to illustrations 5.2a and 5.2b

1   Remove the overflow hose (that connects to the reservoir) from the top of the radiator and pinch it shut to avoid spilling coolant.

2   Remove the mounting bolts and pull the coolant reservoir straight up and out of the vehicle **(see illustrations)**.

3   Installation is the reverse of removal. To better see the coolant level in the tank, clean

5.2a On some models, remove the two bolts (arrows and the shock tower-to-radiator support brace

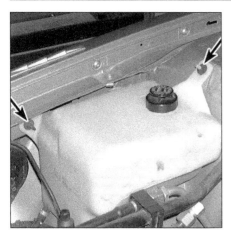

5.2b Remove the nuts (arrows) and lift the coolant reservoir straight up out of its bracket - location of the mounting nuts varies with model

5.6a Disconnect the upper and lower radiator hoses (arrow) indicates upper hose - mark the end of the hose and the radiator connection for later assembly

5.6b Disconnect the cooler lines from the radiator (arrow indicates lower of two fittings for the engine oil cooler on V8 models, attached to the left tank) - disconnect the two transmission cooler lines at the right side of the radiator

5.7 Remove the four bolts (arrows) and lift the radiator top panel off to allow radiator removal

the inside of the tank of any rust deposits or scale with hot, soapy water. Use a wad of cloth taped to the end of a bent coat hanger to scrub hard-to-reach areas inside the tank, then rinse with clear water.

Radiator

Refer to illustrations 5.6a. 5.6b and 5.7

4   Disconnect the negative battery cable and drain the radiator, referring to Chapter 1 if necessary. See the coolant **Warning** in Section 2.
5   Refer to Section 4 and remove the engine-side fan(s) and shroud(s). On early models, it's not necessary to remove the front-of-radiator fan.
6   Disconnect the radiator upper and lower hoses and the automatic transmission fluid cooling lines **(see illustration)**. Have a pan ready to catch drips of transmission fluid that may come out. On V8 models, also disconnect the engine oil cooler lines at the left side of the radiator **(see illustration)**.
7   Remove the upper metal panel at the top of the radiator **(see illustration)**.
8   Lift the radiator straight up and out of the engine compartment. Be careful not to drip coolant on any body paint; immediately wash it off with clear water as the antifreeze

solution can damage the finish.
9   With the radiator removed, it can be inspected for leaks or damage. Green stains (on copper radiators, or white stains (on aluminum radiators) indicate leakage. If in need of repairs, have a professional radiator shop or dealer perform the work, as special equipment and techniques are required.
10   Bugs and dirt can be cleaned from the radiator by using compressed air and a soft brush. Do not bend the cooling fins as this is done.
11   Inspect the rubber mounting pads which the radiator sits on and replace as necessary. Most models will have two rubber-lined insulators at the top and two at the bottom, in the radiator cradle.
12   Lift the radiator into position, making sure it is seated in the mounting pads.
13   Install the radiator cover, fan/shroud, hoses and lines in the reverse order of removal.
14   Connect the negative battery cable and fill the radiator as described in Chapter 1.
15   Start the engine and check for leaks. Allow the engine to reach normal operating temperature (upper radiator hose hot) and add coolant until the level reaches the bottom of the filler neck.

6   Water pump - check and replacement

Check

Refer to illustration 6.3

1   A failure in the water pump can cause serious engine damage due to overheating.
2   There are two ways to check the operation of the water pump while it's installed on the engine. If the pump is defective, it should be replaced with a new or rebuilt unit.
3   Water pumps are equipped with weep (or vent) holes **(see illustration)**. If a failure occurs in the pump seal, coolant will leak from the hole. With the timing belt cover removed, you'll need a flashlight and small mirror to find the hole on the water pump from underneath to check for leaks.
4   If the water pump shaft bearings fail, there may be a howling sound at the pump while it's running. Shaft wear can be felt with the drivebelt removed if the water pump pul-

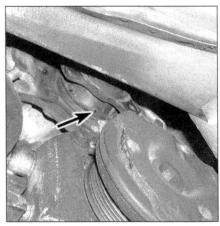

6.3 The weep hole (arrow) is on the underside of the pump - you'll need a flashlight and small mirror to inspect it

6.9 Loosen the water pump pulley bolts slightly with the drivebelt in place - screwdriver helps keep pulley from turning

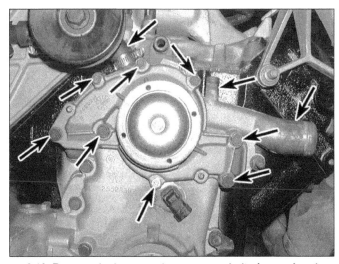

6.10 Remove the hoses and water pump bolts (arrows) and detach the water pump from the engine (V6 models)

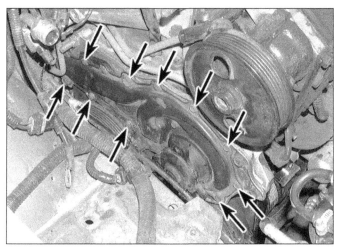

6.11a Water pump mounting bolts (arrows indicate most of the bolts) - V8 engine

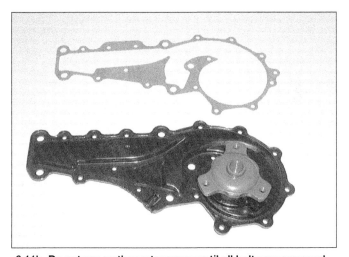

6.11b Do not pry on the water pump until all bolts are removed - new pump here illustrates all the bolt locations

ley is rocked up and down (with the engine off). Don't mistake drivebelt slippage, which causes a squealing sound, for water pump bearing failure.

5   Even a pump that exhibits no outward signs of a problem, such as noise or leakage, can still be due for replacement. Removal for close examination is the only sure way to tell. Sometimes the fins on the back of the impeller can corrode to the point that cooling efficiency is hampered.

Replacement

Refer to illustrations 6.9, 6.10, 6.11a, 6.11b, 6.12 and 6.13

Note: *It is not economical or practical to overhaul a water pump. If failure occurs, a new or rebuilt unit should be purchased to replace the faulty water pump.*

6   Disconnect the negative battery cable.
7   Refer to Chapter 1 to drain the radiator.
8   Refer to Section 5 and remove the coolant reservoir.
9   Loosen the bolts on the water pump

pulley while the drivebelt is still tight, then refer to Chapter 1 and remove the drivebelt **(see illustration)**.
10   On V6 models, disconnect the lower radiator hose, the heater and bypass hoses **(see illustration)**.
11   Remove the bolts that secure the water pump to the front cover of the engine. The bolts are different sizes, so be sure to note their locations and which ones are studs **(see illustrations)**. Pull the water pump away from the engine. If the pump is stuck to the engine, tap it gently with a soft-face hammer.
12   On V8 models, the water pump inlet is actually bolted to the backside of the timing chain cover (the pump is bolted to the front side). Remove the hose from the inlet and examine the inlet pipe for corrosion **(see illustration)**. If there are leaks evident on the inlet-to-timing-cover gasket, refer to Chapter 2, Part B for timing cover removal, since the inlet can't be removed in-vehicle.
13   Clean the gasket surfaces of the front cover completely using a gasket scraper or

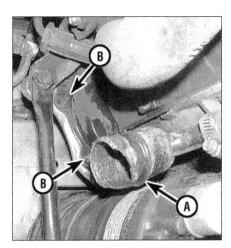

6.12 On V8 models, the inlet is on the back of the timing cover; inspect the neck (A) (this one needs replacing) - (B) indicates two of the mounting bolts, but timing cover removal is necessary to replace the pump inlet

6.13 Remove all traces of old gasket materials with a scraper

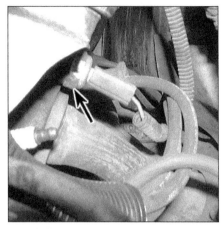

7.3 On V8 models, the engine metal temperature switch (arrow) is located on the left end of the rear cylinder head, on the exhaust side

putty knife, then follow with lacquer thinner **(see illustration)**.

14   Use a thin coat of gasket sealant on the new gasket and install the new water pump. Place the pump into position on the front cover and secure it loosely with the bolts. **Note:** *If the engine front cover had been removed to replace the water pump inlet on a V8 engine, assemble the inlet with a new gasket to the back of the front cover, then the water pump to the front of the cover. The bolts to the inlet can't be accessed once the front cover is installed on the engine.*

15   Tighten all water pump bolts to the torque listed in this Chapter's Specifications. Use sealant on those bolts immediately next to the water passages in the block.

16   Install the engine components in the reverse order of removal, tightening the fasteners securely.

17   Refill and bleed the cooling system and check the drivebelt tension (see Chapter 1). Run the engine and check for leaks.

---

7   Coolant temperature gauge sending unit - check and replacement

---

Warning: *Wait until the engine is completely cool before beginning this procedure.*

Check

Refer to illustration 7.3

1   The coolant temperature indicator system consists of two warning lights on the instrument panel, or on models with digital instrument cluster, a warning light and a temperature gauge. The temperature gauge or warning light is supplied information by the main coolant temperature sensor (CTS), which routes coolant temperature input to the ECM and BCM, then to the instrument panel. Early V6 models and all V8 models also have a engine metal temperature switch which is set to come on at 302 degrees F. When this switch closes, it turns on a red light (Eng

Temp STOP) on the instrument panel and sets off a warning chime to indicate that serious engine overheating has occurred.

2   To test the coolant temperature sensor (CTS), refer to Chapter 6.

3   To test the engine metal temperature switch circuit on Cadillac models **(see illustration)**, disconnect the electrical connector and temporarily attach a ground wire to the connector. Turn the ignition key On and the warning light and chime should go on. If they don't, look for a problem in the circuit. If the warning light comes on intermittently or all the time (with no jumper wire and the connector in place), and you know the engine is not overheating, look for a short to ground somewhere in the circuit from the switch to the chime module and instrument panel. Consult the wiring diagrams in Chapter 12.

Replacement

6   If the sending unit must be replaced, simply unscrew it from the engine and quickly install the replacement. Use a conductive sealant on the threads (not Teflon tape). Make sure the engine is cool before removing the defective sending unit. When replacing the CTS, there will be some coolant loss as

the unit is removed, so be prepared to catch it. Check the coolant level after the replacement part has been installed. **Note:** *The engine metal temperature switch does not have coolant behind it.*

---

8   Engine oil cooler - replacement

Refer to illustration 8.3

1   On V8 models, extra engine cooling is provided by an oil cooler mounted on the top-left end of the engine. The cooler housing doubles as the oil filter adapter, and has two metal oil lines connecting the cooler to the tank on the left side of the radiator. Engine oil is circulated to a finned cooler inside the radiator tank, where the oil temperature is lowered by the radiator coolant.

2   Remove the oil filter, wrapping it with a rag to catch any oil that may spill on the engine. **Caution:** *The engine should be cool before beginning this procedure.*

3   Disconnect the oil pressure sending unit electrical connector **(see illustration)**.

4   Use a flare-nut wrench to disconnect the two oil cooler lines. Use care not to round off the fitting nuts.

5   Remove the two mounting bolts and the oil cooler body.

6   Scrape the oil cooler and block mounting surfaces of any old gasket material or sealant, then clean with lacquer thinner.

7   Install the cooler body to the block with a new gasket. If a new oil cooler is being installed, transfer the oil pressure switch from the old cooler to the new one, using conductive sealant on the threads.

8   Install new O-rings on each line fitting and start them carefully by hand into the cooler body. Make sure the threads on the cooler and fitting are clean, and do not tighten the line fittings until you are sure the fitting isn't cross-threaded. Tighten the fittings to the torque listed in this Chapter's Specifications. Do not overtighten!

9   Install a new oil filter, then run the engine, checking for leaks. Turn off the engine for five minutes and check the oil level, adding oil if necessary.

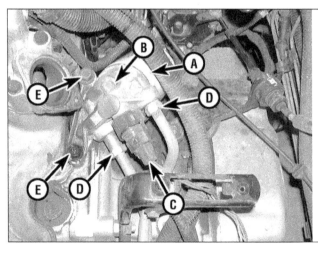

8.3 Components of the V8 engine oil cooler

A   Oil filter
B   Oil cooler/filter adapter
C   Oil pressure switch
D   Oil cooler line fittings
E   Oil cooler mounting bolts

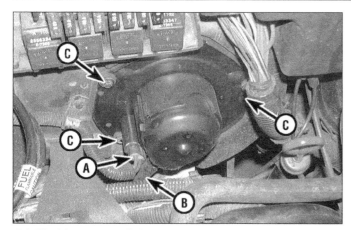

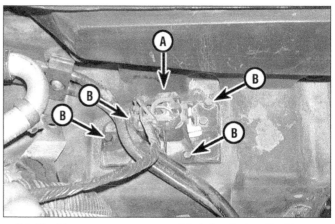

9.3 The blower motor is located on the front of the firewall, (A) is the electrical connector, (B) the cooling hose, and (C) are the mounting screws

9.4 The blower motor power module is located on the blower housing - (A) is the electrical connector, (B) are the mounting screws

9 Blower motor circuit - check

Refer to illustrations 9.3, 9.4 and 9.6

1   All vehicles covered by this manual are equipped with an Electronic Climate Control (ECC) system. In this system, the driver can select blower motor operation manually, or set a desired temperature setting and allow the system to continuously select the blower speed to achieve that comfort level. In the automatic mode, the climate controls signal the Body Computer Module (BCM), which sends varying low-voltage signals (to correspond with blower speeds) to the HVAC Programmer, which sends varying voltages to the Power Module. The Power Module performs the function of a multi-stage blower motor resistor in a conventional blower circuit, and boosts the blower speed input signal to deliver higher-voltage input to the blower motor.

2   ECC systems can be diagnosed with a self-check program that displays trouble codes on the digital display. Refer to Chapter 6 for the code lists and procedure to enter the self-diagnostic mode.

3   If the blower motor doesn't operate at any speed, disconnect the two-pin connector at the blower motor **(see illustration)**. Connect a voltmeter across the two pins on the harness side of the connector. With the engine running, select the High blower speed and check the voltmeter. If the voltage is greater than 4 volts, the blower motor is faulty. If less than 4 volts, check the voltage at the positive wire (the ground wire is always the black one) on the connector and the other probe/clip of the meter on a known chassis ground. If the voltage is now greater than 4 volts, the problem is in the circuit between the motor and ground and/or between the power module and ground. Refer to the wiring diagrams at the end of Chapter 12 to trace the wires.

4   If the voltage in the previous test was less than 4 volts, check for a short to ground in the wire from the positive side of the blower motor to the power module, and from

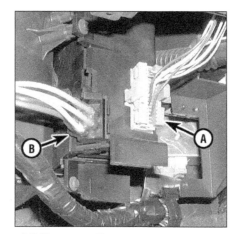

9.6 To access the programmer, remove the insulator panel under the right side of the dash (see Chapter 11) - (A) is the programmer electrical connector, (B) is the vacuum connector

that terminal of the power module back to the programmer (labeled blower speed feedback input). Backprobe the red/black wire terminal on the power module **(see illustration)**. If there isn't battery voltage present, check for an open in that circuit.

5   Reconnect the two-pin connector to the blower motor, check for voltage at the power module at the blower control input. If the voltage is 5 volts or more, remove the four-pin electrical connector from the power module. Using a fused (20 amp) jumper wire, connect the A pin of that connector to battery voltage for two minutes. If the motor now runs the entire time and the fuse doesn't blow, replace the power module. If the fuse blew, replace the power module and the blower motor.

6   If you measured less than 5 volts in Step 5, connect a voltmeter to ground and pin P on the programmer electrical connector **(see illustration)**. If there is at least 5 volts present, check for a short or open in the circuit from Pin R to the power module. **Note:** *The terminal lettering should be indicated on the connector.*

10.3 The fan can be removed after removing the nut (arrow) from the blower motor shaft

7   If the voltage is Step 5 was less than 5, check the voltage at the programmer pin N (the signal from the BCM). If the voltage is greater than 1.5 volts, the programmer should be replaced (check the cleanliness of the contacts there first). If the voltage is less than 1.5 volts, there is either a short in the circuit from the BCM to the programmer, a bad connector at the BCM, or a defective BCM.

10 Blower motor and power module - removal and installation

Refer to illustration 10.3

1   Disconnect the battery negative cable.

2   Disconnect the wiring connector to the blower motor and remove the cooling hose. Remove the mounting screws and pull the blower assembly out of the housing **(see illustration 9.3)**. On Cadillac models, refer to Chapter 2, Part B and remove the lateral brace between the shock towers.

3   The blower motor can be removed as an assembly, but there is a strong chance of bending the fan. Angle the blower motor out enough to reach the nut on the fan, force the

11.2a On Cadillac Deville and Fleetwood models, remove the screw (arrow) and the trim panel at each end of the instrument panel (right side shown, left side similar)

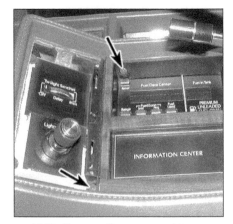

11.2b Remove the upper and lower screws (arrows) on the left side . . .

11.2c . . . and the three screws (arrows) on the right side to remove the central cluster panel

11.3 Remove the two screws (arrows) securing the climate control panel to the dash

11.4 Gently pry up the clip to disconnect the electrical connector on the back of the climate control panel

11.8 On Eldorado and Seville models, remove the four screws and the climate control panel - the Driver Information Center is integral with the climate control panel, they are disconnected as a unit

fan off the motor shaft, then remove the motor and fan separately **(see illustration)**. The fan can be used again on the new blower motor.

4   Installation is the reverse of removal. Check for proper operation.

5   If the power module failed the tests in Section 9, it must be replaced **(see illustration 9.4)**. On Buick models, there is direct access to the power module on top of the blower housing. On Cadillac and Oldsmobile models, twist the two plastic wingnuts and remove the cover over the engine compartment fuse/relay panel on the cowl. Remove two bolts and move the fuse/relay panel out of the way for access to the power module mounting screws. **Warning:** *The bottom of the power module is a heat sink. Allow the unit to cool thoroughly before handling it.*

---

11   Heater and air conditioning control assembly - removal and installation

---

Warning: *On models equipped with a Supplemental Inflatable Restraint (SIR) system*

(more commonly known as airbags), always disable the airbag system before working in the vicinity of any airbag system component to avoid the possibility of accidental deployment of the airbag, which could cause personal injury (see Chapter 12).

Toronado, Riviera, Deville and Fleetwood models

Refer to illustrations 11.2a, 11.2b, 11.2c, 11.3 and 11.4

1   The climate control panel is located above and to the right of the steering wheel. Before removing the controls, disconnect the negative battery cable.

2   The instrument cluster panel must be removed first to access the climate control panel. On Toronado and Riviera models, the cluster panel is one-piece, and comes out with the removal of four screws. On Cadillac Deville and Fleetwood models, the cluster panel is a three-piece assembly. Remove the left and right-hand end panels first, then the central panel **(see illustrations)**.

3   With the cluster panel out, remove the screws on the climate control panel **(see**

illustration).

4   Pull the climate control panel out enough to disconnect the electrical connector behind it **(see illustration)**.

5   Installation is the reverse of the removal procedure.

Eldorado and Seville models

Refer to illustration 11.8

6   The climate control panel on these models is located in the central lower dash panel. Before removing the controls, disconnect the negative battery cable.

7   Remove the top screws and the trim bezel that surrounds the heater/air conditioning control assembly, Driver Information Center, and the center dash vent.

8   Remove the control assembly retaining screws **(see illustration)**.

9   Pull the heater/air conditioning control assembly from the instrument panel enough to disconnect the electrical connector.

10   To install the control assembly, reverse the removal procedure.

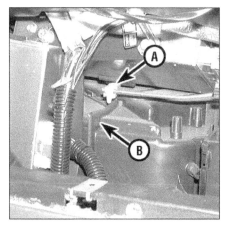

12.4 Loosen the two heater hose clamps and disconnect the heater hoses (arrows) from the heater core inlet and outlet pipes at the firewall

12.6 Disconnect the blend-air door link rod (A) - (B) is the upper heater core cover mounting screw

Note: *Make sure to transfer any weather-stripping/sealing material from the old core or cover to the new one.*

12 Heater core - replacement

Refer to illustrations 12.4, 12.6, 12.7 and 12.9
Warning: *On models equipped with a Supplemental Inflatable Restraint (SIR) system (more commonly known as airbags), always disable the airbag system before working in the vicinity of any airbag system component to avoid the possibility of accidental deployment of the airbag, which could cause personal injury (see Chapter 12).*
Note: *On Riviera and Toronado models, it may be necessary to remove the instrument panel, console, ECM and BCM to access the heater core cover. If the procedure described below does not allow sufficient access to remove the heater core cover, removal of the complete instrument panel may be necessary. It is suggested that the vehicle be taken to a dealer service department or other properly equipped repair facility for instrument panel removal.*
1   The heater core is contained within the heater/air conditioning assembly. The assembly is divided by the firewall, with the blower motor and evaporator located in a housing in the engine compartment, and the heater core located in a housing in the passenger compartment, under the instrument panel. If only the heater core is being removed, there is no

need to evacuate the refrigerant from the air conditioning system or remove the blower motor/evaporator housing.
2   Disconnect the negative battery cable.
3   Drain the cooling system (see Chapter 1).
4   Disconnect the heater hoses at the firewall **(see illustration)**.
5   Remove the glove box and the sound insulator panel below the right side of the dash (see Chapter 11).
6   Disconnect the rod from the programmer to the blend-air door by squeezing the plastic connector and pulling it out of the arm **(see illustration)**.
7   Disconnect the vacuum and electrical connectors from the programmer, just to the right of the heater core housing **(see illustration)**.
8   Remove the two heater core cover mounting screws **(see illustrations 12.6 and 12.7)** and pull the heater core cover (with programmer still attached) away from the heater housing.
9   Remove the two core mounting screws and carefully lower the heater/air conditioning assembly from under the dash **(see illustration)**. **Caution:** *Have some rags or old towels on the floor to protect the carpet from any coolant that may spill.*
10   Installation is the reverse of removal.

13   Air conditioning and heating system - check and maintenance

Air conditioning system

Refer to illustration 13.1
Warning: *The air conditioning system is under high pressure. Do not loosen any hose fittings or remove any components until after the system has been discharged. Air conditioning refrigerant should be properly discharged into an EPA-approved recovery/recycling unit at a dealer service department or an automotive air conditioning repair facility. Always wear eye protection when disconnecting air conditioning system fittings.*
Caution: *When replacing entire components, additional refrigerant oil should be added equal to the amount that is removed with the component being replaced. Be sure to read the can before adding any oil to the system, to make sure it is compatible with the R-12 system.*
1   The following maintenance checks should be performed on a regular basis to ensure that the air conditioning continues to operate at peak efficiency.

a) *Inspect the condition of the compressor drivebelt. If it is worn or deteriorated, replace it (see Chapter 1).*
b) *Check the drivebelt tension and, if necessary, adjust it (see Chapter 1).*
c) *Inspect the system hoses. Look for cracks, bubbles, hardening and deterioration. Inspect the hoses and all fittings for oil bubbles or seepage. If there is any evidence of wear, damage or leakage, replace the hose(s).*
d) *Inspect the condenser fins for leaves, bugs and any other foreign material that may have embedded itself in the fins. Use a "fin comb" or compressed air to remove debris from the condenser.*
e) *Make sure the system has the correct refrigerant charge.*

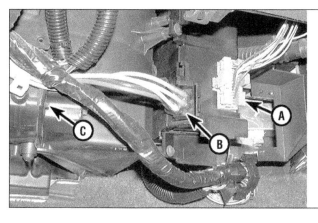

12.7 Disconnect the electrical connector (A) and the vacuum connector (B) from the programmer - (C) is the lower heater core cover mounting screw

12.9 Remove the screws (arrows) holding the heater core to the housing

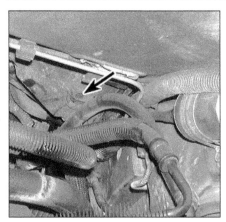

13.1 Check that the evaporator housing drain tube (arrow) at the firewall is clear of any blockage

13.9 Insert a thermometer in the center duct while operating the air conditioning system - the output air should be 35 to 40 degrees F less than the ambient temperature, depending on humidity (but no lower than 40-degrees F)

f) *Check the evaporator housing drain tube* **(see illustration)**. *Insert a piece of wire into the opening to check for blockage.*

2   It's a good idea to operate the system for about ten minutes at least once a month. This is particularly important during the winter months because long term non-use can cause hardening, and subsequent failure, of the seals. Note that using the Defrost function operates the compressor.

3   If the air conditioning system is not working properly, first make sure the compressor clutch is operating (see Section 14).

4   Because of the complexity of the air conditioning system and the special equipment necessary to service it, in-depth troubleshooting and repairs are not included in this manual. However, simple checks and component replacement procedures are provided in this Chapter. For more complete information on the air conditioning system, refer to the *Haynes Automotive Heating and Air Conditioning Manual*. However, simple component replacement procedures are provided in this Chapter.

5   The most common cause of poor cooling is simply a low system refrigerant charge. If a noticeable drop in system cooling ability occurs, one of the following quick checks will help you determine whether the refrigerant level is low. Should the system lose its cooling ability, the following procedure will help you pinpoint the cause.

Check

Refer to illustration 13.9

6   Warm the engine up to normal operating temperature.

7   Place the air conditioning temperature selector at the coldest setting and put the blower at the highest setting. Open the doors (to make sure the air conditioning system doesn't cycle off as soon as it cools the passenger compartment).

8   After the system reaches operating temperature, feel the two pipes connected to the evaporator at the firewall.

9   The pipe (thinner tubing) leading from the condenser outlet to the evaporator should be

cold, and the evaporator outlet line (the thicker tubing that leads back to the compressor) should be slightly colder (3 to 10 degrees F). If the evaporator outlet is considerably warmer than the inlet, the system needs a charge. Insert a thermometer in the center air distribution duct **(see illustration)**, while operating the air conditioning system - the temperature of the output air should be 35 to 40 degrees F below the ambient air temperature (down to approximately 40 degrees F). If the ambient (outside) air temperature is very high, say 110 degrees F, the duct air temperature may be as high as 60 degrees F, but generally the air conditioning is 35 to 40 degrees F cooler than the ambient air.

10   If the air isn't as cold as it used to be, the system probably needs a charge. Further inspection or testing of the system is beyond the scope of the home mechanic and should be left to a professional. Some on-board diagnostics capability on your vehicle can help point to areas for testing or repair (see subhead below).

Adding refrigerant

11   The vehicles covered by this manual use R-12 refrigerant. Make sure any refrigerant, refrigerant oil or replacement component your purchase is designated as compatible with R-12 systems.

12   Because of federal regulations imposed by the Environmental Protection Agency, 14-ounce cans of R-12 refrigerant are not available to the home mechanic. R-12 is no longer produced in the US, but remaining supplies of R-12 are still available to professional air-conditioning service shops. R-12 has been superceded by the environmentally-friendly R-134a refrigerant on newer vehicles, and R-134a is available for home use. If your R-12 system needs charging, you must take it to a dealer service department or air-conditioning shop. Check with your dealer service department for the availability of kits to convert your system to the newer R-134a refrigerant.

Heating systems

13   If the carpet under the heater core is damp, or if antifreeze vapor or steam is coming through the vents, the heater core is leaking. Remove it (see Section 12) and install a new unit (most radiator shops will not repair a leaking heater core).

14   If the air coming out of the heater vents isn't hot, the problem could stem from any of the following causes:

a) *The thermostat is stuck open, preventing the engine coolant from warming up enough to carry heat to the heater core. Replace the thermostat (see Section 3).*

b) *There is a blockage in the system, preventing the flow of coolant through the heater core. Feel both heater hoses at the firewall. They should be hot. If one of them is cold, there is an obstruction in one of the hoses or in the heater core, or the heater control valve is shut. Detach the hoses and back flush the heater core with a water hose. If the heater core is clear but circulation is impeded, remove the two hoses and flush them out with a water hose.*

c) *If flushing fails to remove the blockage from the heater core, the core must be replaced (see Section 12).*

On-board diagnostics

15   A comprehensive system of self-diagnostics is used on these models, with the LED portion of the climate control panel allowing display of trouble codes.

16   Refer to Chapter 6 for the procedures to enter the self-diagnostics program, how to access the system that deals with heating and air conditioning, and for a listing of trouble codes.

17   After any repair work, repeat the self-diagnostics procedure to check your work.

Programmer vacuum check

18   Put the controls through all of their modes with the vehicle running. If necessary, have someone else operate the controls while you look under the dashboard to observe the action of the various vacuum motors controlling airflow and temperature modes. If there is any position of the controls that gives air from the heater ducts *and* the air conditioning ducts, this indicates a vacuum leak.

19   Vacuum is supplied to the heater/air conditioning control assembly by tubing attached to the intake manifold. Inspect the supply line, from the intake manifold to the programmer, for damage or leaks **(see illustration 12.7)**. The programmer is the black box mounted to the right side of the heater/air conditioning housing under the dashboard. To inspect it, remove the mounting screws and snap loose the rod connecting it to the blend-air door.

20   The vacuum lines run from the programmer to each of the mode actuators. Apply 8 inches of vacuum to each of the ports on the harness connector and observe the functions

13.26 Using a can of automotive disinfectant and a long hose, spray through the heater core cover as shown to access the interior side of the evaporator core, and through the power module hole to spray the front side

14.3 On most models, the pressure switch (arrow) is located on the accumulator - on other models the switch is located in the refrigerant line leading from the evaporator to the compressor

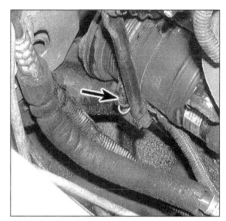

14.7 Disconnect the electrical connector (arrow) at the clutch (always at the front of the compressor) to check the power and ground circuits

of that vacuum circuit.

21   If any of the circuits don't hold vacuum, follow the lines to their destination, isolate the source of the leak and repair it.

22   If there is vacuum in the line to the actuator when that function is selected, and there is no leak detected, check for binding of the door the actuator operates. If the door operates freely, independent of the actuator, then the actuator needs replacing.

Eliminating air-conditioning odors

Refer to illustration 13.26

23   Unpleasant odors that often develop in air-conditioning systems are caused by the growth of a fungus, usually on the surface of the evaporator core. The warm, humid environment there is a perfect breeding ground for mildew to develop.

24   The evaporator core on most vehicles is difficult to access, and factory dealerships have a lengthy, expensive process for eliminating the fungus by opening up the evaporator case and using a powerful disinfectant and rinse on the core until the fungus is gone. You can service your own system at home, but it takes something much stronger than basic household germ-killers or deodorizers.

25   Aerosol disinfectants for automotive air-conditioning systems are available in most auto parts stores, but remember when shopping for them that the most effective treatments are also the most expensive. The basic procedure for using these sprays is to start by running the system in the RECIRC mode for ten minutes with the blower on its highest speed. Use the highest heat mode to dry out the system and keep the compressor from engaging by disconnecting the wiring connector at the compressor (see Section 14).

26   The disinfectant can usually comes with a long spray hose. Remove the power module (see Section 10), point the nozzle inside the hole and to the left towards the evaporator

core, and spray according to the manufacturer's recommendations. Try to cover the whole surface of the evaporator core, by aiming the spray up, down and sideways. You can also spray through the cover of the heater core, aiming towards the back of the evaporator **(see illustration)**. Follow the manufacturer's recommendations for the length of spray and waiting time between applications.

27   Once the evaporator has been cleaned, the best way to prevent the mildew from coming back again is to make sure your evaporator housing drain tube is clear **(see illustration 13.1)**.

---

14   Air conditioning compressor clutch circuit - check

Refer to illustrations 14.3 and 14.7

1   Proper operation of the compressor clutch is essential to the function of the air conditioning system. If your system doesn't seem to get cold, first check the clutch operation.

2   With the engine warmed up, set the air conditioning temperature selector on the coldest setting and the fan on high. Open the doors (to make sure the air conditioning system doesn't cycle off as soon as it cools the passenger compartment down).

3   Turn the air conditioning On and observe the front of the compressor. The clutch will make an audible click and the center of the clutch should rotate. If it doesn't, shut the engine off and disconnect the air conditioning system pressure switch **(see illustration)**. Using a jumper wire, connect the two terminals of the pressure switch connector and try the air conditioning again. If the compressor now engages, the system pressure is too high or too low. Have your system tested by a dealer service department or air conditioning shop.

4   If the clutch still didn't operate, check the appropriate fuses. Refer to the wiring diagrams at the end of Chapter 12 to see the

compressor clutch circuit and which fuses to check.

5   While you switch the air conditioning on and off, listen at the A/C relay in the underhood fuse/relay box. If the relay doesn't cycle, remove it and attach a test light between sockets 2 and 5. If the compressor clutch now clicks on, install a new relay. **Note:** *The terminals should be numbered on the relay.*

6   If the old relay did cycle in the test, then remove the relay and attach a jumper wire between sockets 1 and 4. If the compressor clutch engages, install a new relay.

7   If the clutch did not engage in Step 6, attach a jumper wire between relay sockets 1 and 5. If the clutch engages, this indicates there is an open in the circuit from the fuse panel to socket 4 for the relay. If it didn't engage, disconnect the two-pin connector from the compressor clutch **(see illustration)**. Probe the power wire terminal on the connector. If there is no voltage present, check the power circuit from the relay to the clutch connector.

8   If the power circuit to the clutch is good, check the ground side of the clutch connector. If there is power and ground to the two pins on the clutch and it doesn't operate, the clutch must be replaced.

---

15   Air conditioning accumulator-drier - removal and installation

Refer to illustration 15.3

Warning: *The air conditioning system is under high pressure. Do not loosen any hose fittings or remove any components until after the system has been discharged. Air conditioning refrigerant should be properly discharged into an EPA-approved recovery/recycling unit at a dealer service department or an automotive air conditioning repair facility. Always wear eye protection when disconnecting air conditioning system fittings.*

Caution: *When replacing entire components, additional refrigerant oil should be added equal to the amount that is removed with the*

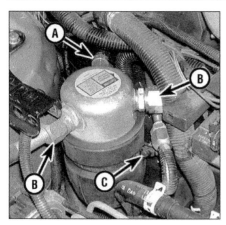

15.3 Disconnect the pressure switch (A), then use backup wrenches to disconnect the refrigerant lines (B) - loosen the accumulator clamp bolt (C) to remove the accumulator

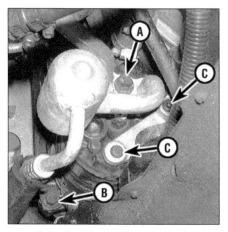

16.7c On V8 models, remove the refrigerant line bolt (A), the rear-facing front mounting bolt (B), and the two rear mounting bolts (C)

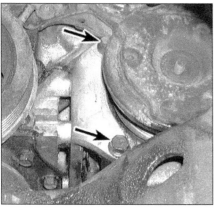

16.7d Front the front, remove the two compressor mounting bolts (arrows)

16.7a On V6 models, remove the clutch connector (A), the bolt (B) retaining the refrigerant lines, and the top mounting bolt (C)

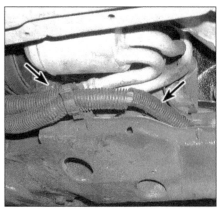

16.7b From below, remove the two lower compressor mounting bolts (arrows)

---

component being replaced. Be sure to read the container before adding any oil to the system, to make sure it is compatible with the R-12 system.

1   Have the refrigerant discharged and recovered by an air-conditioning technician.

2   Disconnect the negative battery cable.

3   Disconnect the refrigerant lines from the accumulator and cap the open fittings to prevent dirt and moisture entry **(see illustration)**. Disconnect the pressure switch connector. **Note:** *On most covered models, the accumulator is located in the right-rear corner of the engine compartment, near the evaporator case. On some models it is located just ahead of the right strut tower. Models that have the accumulator located in the left-front corner of the engine compartment, near the radiator, will not have a low-pressure switch. The low-pressure switch on these models is near the evaporator in the line from the evaporator to the compressor.*

4   Loosen the accumulator bracket clamping screw and lift the accumulator out of the vehicle.

5   Installation is the reverse of removal, using new O-rings, lubricated with refrigerant oil, where the lines connect to the accumulator. If the accumulator is being replaced, add 3.5 ounces of refrigerant oil to the new accumulator.

6   Have the system evacuated, charged and leak-tested by an air-conditioning technician.

---

16   Air conditioning compressor - removal and installation

Warning: *The air conditioning system is under high pressure. Do not loosen any hose fittings or remove any components until after the system has been discharged. Air conditioning refrigerant should be properly discharged into an EPA-approved recovery/recycling unit at a dealer service department or an automotive air conditioning repair facility. Always wear eye protection when disconnecting air conditioning system fittings.*

Caution: *When replacing entire components, additional refrigerant oil should be added equal to the amount that is removed with the component being replaced. Be sure to read*

the container before adding any oil to the system, to make sure it is compatible with the R-12 system.

Note: *If the compressor is being replaced due to damaged internal components, replace the accumulator and the orifice tube as well (refer to Sections 15 and 19).*

Removal

Refer to illustrations 16.7a, 16.7b, 16.7c and 16.7d

1   Have the air conditioning system refrigerant discharged and recovered by an air conditioning technician.

2   Disconnect the negative battery cable from the battery.

3   Set the parking brake, block the rear wheels and raise the front of the vehicle, supporting it securely on jackstands.

4   Remove the drivebelt (see Chapter 1).

5   From below, remove the plastic splash shield just below the compressor. Remove the right-side fenderwell splash shield for access to the front of the compressor.

6   Disconnect the compressor clutch wiring harness **(see illustration 14.7)**. **Note:** *On some models, there is a second electrical connector at the rear of the compressor that must be disconnected.*

7   Disconnect the refrigerant lines from the compressor. Plug the open fittings to prevent entry of dirt and moisture **(see illustrations)**.

8   Unbolt the compressor from the mounting bracket and remove it from the vehicle.

Installation

9   The clutch may have to be transferred from the old compressor to the new unit. This may require special tools and should be done by an air-conditioning technician.

10   Add the proper amount of refrigerant oil to the new compressor using the following calculations:

a) *Drain the refrigerant oil from the old compressor through the suction fitting and measure it in ounces.*

b) *Drain the new oil from the new compressor.*

c) *If the amount drained from the old compressor was less than 1 ounce, add 1 ounce of new oil to the new compressor.*

17.4 Remove the battery for access to the condenser fittings (arrows)

17.6 Remove the two condenser brackets (arrow indicates left bracket) from the radiator support

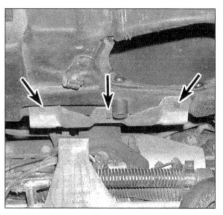

18.8 Remove the mounting screws (arrows indicate the upper three) and remove the heat shield from the blower case

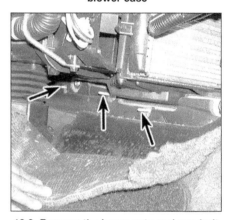

18.9 Remove the lower nuts and one bolt (arrows) retaining the evaporator/blower case on the interior side of the firewall

d) If the amount drained from the old compressor was 1 ounce or more, add that exact amount of new oil to the new compressor.

11   Installation is the reverse of removal, using new O-rings, lubricated with clean refrigerant oil, where the line fittings attach to the compressor.

12   Have the system evacuated, recharged and leak tested by an air conditioning technician.

17   Air conditioning condenser - removal and installation

Warning: *The air conditioning system is under high pressure. Do not loosen any hose fittings or remove any components until after the system has been discharged. Air conditioning refrigerant should be properly discharged into an EPA-approved recovery/recycling unit at a dealer service department or an automotive air conditioning repair facility. Always wear eye protection when disconnecting air conditioning system fittings.*
Caution: *When replacing entire components, additional refrigerant oil should be added equal to the amount that is removed with the component being replaced. Be sure to read the container before adding any oil to the system, to make sure it is compatible with the R-12 system.*

Removal

Refer to illustrations 17.4 and 17.6

1   Have the air conditioning system discharged by a dealer service department or air conditioning shop.
2   Disconnect the negative cable from battery, then remove the battery (see Chapter 5).
3   Drain the cooling system (see Chapter 1).
4   Disconnect the refrigerant lines from the condenser **(see illustration)**. Always use backup wrenches to void twisting a refrigerant line.
5   Remove the radiator (see Section 5).
6   Remove the mounting bolts from the

condenser brackets **(see illustration)**.
7   Lift the condenser out of the vehicle and plug the lines to prevent dirt and moisture from entering the system.
8   If the original condenser will be reinstalled, store it with the line fittings on top to prevent oil from draining out.

Installation

9   If a new condenser is being installed, pour one ounce of refrigerant oil into the new condenser prior to installation.
10   Reinstall the components in the reverse order of removal. Be sure the rubber pads are in place under the condenser. Also, be sure to use new O-rings on the refrigerant lines and lubricate them with a small amount of clean refrigerant oil.
11   Have the system evacuated, recharged and leak tested by the shop that discharged it.

18   Air conditioning evaporator - removal and installation

Warning: *The air conditioning system is under high pressure. Do not loosen any hose fittings or remove any components until after the system has been discharged. Air conditioning refrigerant should be properly discharged into an EPA-approved recovery/recycling unit at a dealer service department or an automotive air conditioning repair facility. Always wear eye protection when disconnecting air conditioning system fittings.*
Caution: *When replacing entire components, additional refrigerant oil should be added equal to the amount that is removed with the component being replaced. Be sure to read the container before adding any oil to the system, to make sure it is compatible with the R-12 system.*

Removal

Refer to illustrations 18.8, 18.9, 18.10 and 18.11

1   Have the air conditioning system discharged by a dealer service department or air conditioning shop.

2   Disconnect the negative cable from battery.
3   On models with a brace between the two strut towers, refer to Chapter 2, Part A or B and remove the brace.
4   Drain the cooling system (see Chapter 1) and disconnect the two heater hoses at the firewall **(see illustration 12.4)**.
5   Refer to Section 15 and remove the accumulator. On models where the accumulator is at the front of the vehicle instead of next to the evaporator case, you will be disconnect the two lines at the evaporator case.
6   Remove the cover over the engine compartment fuse/relay panel on the cowl. Remove two bolts and move the fuse/relay panel out of the way for access to the evaporator/blower housing mounting screws.
7   Refer to Section 10 and remove the blower motor and power module from the evaporator/blower case.
8   Below the blower opening is a heat shield, between the exhaust system and the evaporator/blower case **(see illustration)**. Remove the three top screws and from below remove the lower screws and the shield.
9   Remove the nuts on the studs retaining the housing to the firewall on the inside of the vehicle **(see illustration)**.

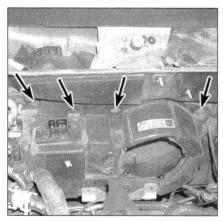

18.10 Remove the upper case screws (arrows) and access the lower case screws from below

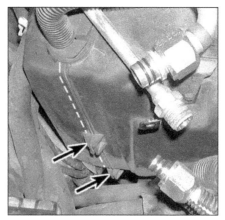

18.11 If the evaporator case is marked with a line and the words "CUT HERE" cut the case along the indicated line - arrows indicate areas where clips are installed when the case is reassembled

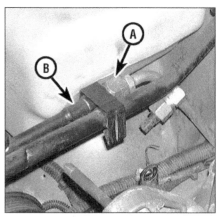

19.3a When the accumulator is located near the evaporator, the orifice tube is located in this line - (A) is the fitting disconnect, (B) is the orifice tube location

10   Remove all of the screws mounting the case to the firewall, working from the evaporator section to the driver's side **(see illustration)**. **Note:** *See Step 11, on some models, you don't have to remove the nearly-inaccessible screws on the passenger side.*
11   Carefully inspect the evaporator case. If the case is marked with a line between two flanges, cut this line with a sharp utility knife to separate the main case from the small right section **(see illustration)**. The right section stays in place on the firewall. **Warning:** *Extend the blade of the utility knife only enough to just cut through the case.*
12   Remove the evaporator case to expose and remove the evaporator core. **Note:** *If there is a temperature sensor clipped to the core, remove it carefully to transfer to the new core.*

Installation

13   Installation is the reverse of the removal process. Make sure that any sealing materials are reused and if you install a new unit that you add 3 ounces of new refrigerant oil. When joining the case sections together at

19.3b On models where the accumulator is located in the left front of the engine compartment - the junction (arrow) for the orifice tube is in the high-pressure line just before the evaporator

the line you cut in Step 11, use a weatherstripping sealant along the cut line and obtain clips to secure the sections together.
14   Have the system evacuated, leak-tested and recharged at a dealer or air conditioning shop. After the engine has run and is cooled off again, recheck the coolant level.

19 Air conditioning expansion (orifice) tube - removal and installation

Refer to illustrations 19.3a, 19.3b and 19.4
Warning: *The air conditioning system is under high pressure. DO NOT loosen any fittings or remove any components until after the system has been discharged. Air conditioning refrigerant should be properly discharged into an EPA-approved container at a dealership service department or an automotive air conditioning repair facility. Always wear eye protection when disconnecting air conditioning system fittings.*
1   Have the air conditioning system discharged and the refrigerant recovered (see **Warning** above). Disconnect the cable from the negative terminal of the battery.
2   Remove the air cleaner and duct (see Chapter 4).
3   Disconnect the refrigerant high-pressure line, using backup wrenches to avoid twisting the refrigerant lines. On models where the accumulator is on the right side of the engine,

the orifice tube is located in the line in front of the coolant reservoir, while on models where the accumulator is at the left front of the engine compartment, the orifice tube is in the high pressure line near the cowl, just ahead of the evaporator **(see illustrations)**. There is always a fitting junction next to the orifice location.
4   The expansion *(orifice)* tube is a tube with a fixed-diameter orifice and a mesh filter at each end **(see illustration)**. When you separate the pipe at the fitting you will see one end of the orifice tube inside the pipe leading to the evaporator. Use needle-nose pliers to remove the orifice tube.
5   The orifice tube acts to meter the refrigerant, changing it from high-pressure liquid to low-pressure liquid. It is possible to reuse the orifice tube if:

a) *The screens aren't plugged with grit or foreign material*
b) *Neither screen is torn*
c) *The plastic housing over the screens is intact*
d) *The brass orifice inside the plastic housing is unrestricted*

6   Installation is the reverse of removal. Be sure to insert the expansion tube with the shorter end in first, toward the evaporator. **Caution:** *Always use a new O-ring when installing the expansion tube.*
7   Retighten the fitting and refrigerant line, then have the system evacuated, recharged and leak-tested by the shop that discharged it.

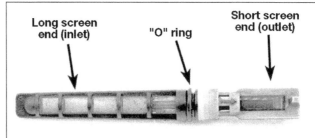

Long screen end (inlet)   "O" ring   Short screen end (outlet)

19.4 The expansion tube (orifice) contains a precise orifice and several screen filters - it should be replaced whenever a compressor is replaced - install it the same direction in the line as the original

Chapter 4
Fuel and exhaust systems

Contents

Specifications

General

Fuel pressure (key On, engine not running)	
Throttle Body Injection (TBI) system	9 to 13 psi
Multi Port Fuel Injection (MPFI) system	41 to 47 psi
Fuel injector resistance	
TBI system	1.16 to 1.36 ohms
MPFI system	11.6 to 12.4 ohms
Minimum idle speed setting	
Cadillac models	525 RPM
Buick and Oldsmobile	500 RPM in drive
ISC motor maximum extension setting (Cadillac models)	
1987 and earlier	1.05 to 1.10 volts
1988 through 1989	1.05 to 1.15 volts
1990 and later	1.15 to 1.20 volts

Torque specifications

	Ft-lbs (unless otherwise indicated)
Exhaust pipe-to-manifold nuts	15 to 22
Fuel meter cover screws (TBI)	28 in-lbs
Fuel inlet nut (TBI)	30
Fuel outlet nut (TBI)	21
Fuel rail mounting bolts	
Cadillac models	18
Buick and Oldsmobile	84 to 168 in-lbs
Throttle body bolts/nuts	132 in-lbs

1   General information

Fuel system

Refer to illustrations 1.1a, 1.1b and 1.1c

The fuel system consists of a fuel tank, an electric fuel pump located in the fuel tank, a fuel pump relay, an air cleaner assembly and either one of two fuel injection systems: Throttle Body Injection (TBI) or Multi Port Fuel Injection (MPFI). The TBI system is easily identified by the large throttle body with two injectors mounted side by side and is used on 1986 through 1989 Cadillac models only **(see illustration)**, while the MPFI system uses injectors mounted on the intake manifold, positioned to shoot fuel into the intake port of each cylinder. This type of system is used on all Buick and Oldsmobile models and 1990 and later Cadillac models **(see illustrations)**.

Throttle Body Injection (TBI) system

The throttle body system utilizes two injectors, centrally mounted in a carburetor-like housing. The injector is an electrical solenoid, with fuel delivered to the injector at a constant pressure level. To maintain the fuel pressure at a constant level, fuel is supplied at a greater pressure than required and controlled by a fuel pressure regulator, which returns excess fuel to the fuel tank.

A signal from the ECM opens the solenoid, allowing fuel to spray through the injector into the throttle body. The amount of time the injector is held open by the ECM determines the fuel/air mixture ratio.

Multi Port Fuel Injection (MPFI) system

This system utilizes injectors of a different type than the throttle body injection system. Instead of an injector mounted in a centrally located throttle body as in the TBI system, one injector is installed above each intake port. The throttle body on the MPFI system serves only to control the amount of air passing into the system. Because each cylinder is equipped with an injector mounted immediately adjacent to the intake valve, much better control of the fuel/air mixture ratio is possible.

Fuel pump and lines

Fuel is circulated from the fuel tank to the fuel injection system, and back to the fuel tank, through a pair of lines running along the underside of the vehicle. An electric fuel pump is attached to the fuel sending unit inside the fuel tank. A return system routes all vapors and excess fuel back to the fuel tank through separate return lines.

Exhaust system

The exhaust system includes an exhaust manifold fitted with an exhaust oxygen sensor, a catalytic converter, an exhaust pipe, and a muffler. The catalytic converter is an

1.1a Typical fuel injection system components (1986 through 1989 Cadillac models)

1	Fuel injector	5	Fuel pressure test port
2	Fuel pressure regulator	6	Throttle body
3	Fuel feed and return lines	7	ISC motor
4	Distributor		

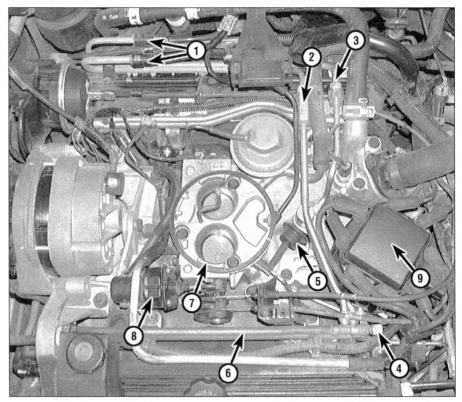

1.1b Typical fuel injection system components (1990 and later Cadillac models)

1	Fuel injector	6	Fuel rail assembly
2	Fuel feed line	7	Throttle body
3	Fuel return line	8	ISC motor
4	Fuel pressure test port	9	Distributor
5	Fuel pressure regulator		

1.1c Typical fuel injection system components (Buick and Oldsmobile models)

1	Fuel rail assembly	4	Fuel injector
2	Fuel pressure test port	5	Fuel feed line
3	Fuel pressure regulator	6	Throttle body

emission control device added to the exhaust system to reduce pollutants. A single-bed converter is used in combination with a three-way (reduction) catalyst. Refer to Chapter 6 for more information regarding the catalytic converter.

2 Fuel pressure relief procedure

Refer to illustrations 2.3a and 2.3b
Warning: *Gasoline is extremely flammable, so take extra precautions when you work on any part of the fuel system. Don't smoke or allow open flames or bare light bulbs near the work area, and don't work in a garage where a natural gas-type appliance (such as a water heater or a clothes dryer) with a pilot light is present. Since gasoline is carcinogenic, wear latex gloves when there's a possibility of being exposed to fuel, and, if you spill any fuel on your skin, rinse it off immediately with soap and water. Mop up any spills immediately and do not store fuel-soaked rags where they could ignite. The fuel system is under constant pressure, so, if any fuel lines are to be disconnected, the fuel pressure in the system must be relieved first. When you perform any kind of work on the fuel system, wear safety glasses and have a Class B type fire extinguisher on hand.*
Note: *After the fuel pressure has been relieved, it's a good idea to lay a shop towel over any fuel connection to be disassembled, to absorb the residual fuel that may leak out when servicing the fuel system.*

1   Before servicing any fuel system component, you must relieve the fuel pressure to minimize the risk of fire or personal injury.
2   Remove the fuel filler cap - this will relieve any pressure built up in the tank.
3   Most of the fuel systems covered by this manual are equipped with a fuel pressure test port located on the fuel rail assembly that will enable the system to be bled **(see illustrations 1.1a, 1.1b and 1.1c)**. On later model 3800 V6 engines the fuel pressure test port is located on the fuel pressure regulator **(see illustration)**. Install a fuel pressure gauge equipped with a bleeder hose (available at most auto parts stores) onto the test port and relieve the pressure by bleeding the fuel through the bleed-off valve and into a metal can **(see illustration)**.

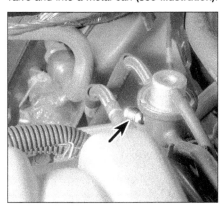

2.3a On late model 3.8L engines, the fuel pressure test port (arrow) is located on the fuel pressure regulator - unscrew the cap and install a fuel pressure gauge

Note: *Earlier model TBI systems are not equipped with a fuel pressure test port that will enable the system to be bled. On these models it will be necessary to relieve fuel pressure at the fuel filter which is located beneath the vehicle (see Chapter 1 if necessary). Position a metal pan under the fuel filter. Use shop rags or towels to cover the inlet fuel line at the fuel filter and then disconnect the fuel line to allow any residual fuel pressure to dissipate.*
4   If the specialized fuel pressure gauge is not available, follow this alternate method that will easily bleed the fuel pressure. Cover the test port fitting with shop rags or towels, then carefully depress the Schrader valve with a small screwdriver or awl. Dispose of the rags in a covered, marked container.
5   Unless this procedure is followed before servicing fuel lines or connections, fuel spray (and possible injury) may occur.

3 Fuel pump/fuel pressure - check

Warning: *Gasoline is extremely flammable, so take extra precautions when you work on any part of the fuel system. Don't smoke or allow open flames or bare light bulbs near the work area, and don't work in a garage where a natural gas-type appliance (such as a water heater or a clothes dryer) with a pilot light is present. Since gasoline is carcinogenic, wear latex gloves when there's a possibility of being exposed to fuel, and, if you spill any fuel on your skin, rinse it off immediately with soap and water. Mop up any spills immediately and do not store fuel-soaked rags where they could ignite. The fuel system is under constant pressure, so, if any fuel lines are to be disconnected, the fuel pressure in the system must be relieved first (see Section 2). When you perform any kind of work on the fuel system, wear safety glasses and have a Class B type fire extinguisher on hand.*
Note : *The following checks assume the fuel filter is in good condition. If you doubt its condition, install a new one (see Chapter 1).*
1   Check that there is adequate fuel in the fuel tank. Relieve the fuel pressure (see Section 2).

2.3b Attach a fuel pressure gauge to the test port and open the valve to drain the excess fuel out of the drain tube and into an approved fuel container

3.2 Attach the fuel pressure gauge to the test port and check the pressure with the ignition key ON (engine not running) (Cadillac model shown)

3.5 Connect a vacuum pump to the fuel pressure regulator, apply vacuum to the fuel pressure regulator and check the fuel pressure - the fuel pressure should decrease as the vacuum increases

3.6 Checking vacuum to the fuel pressure regulator using a vacuum gauge

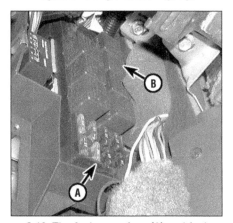

3.10 The fuel pump fuse (A) and fuel pump relay (B) are located in the interior relay center under the instrument panel (1991 Cadillac Deville shown)

Fuel pump output and pressure check

Refer to illustrations 3.2, 3.5 and 3.6

2   Install a fuel pressure gauge onto the fuel lines or the fuel rail **(see illustration)**. Each type of system has a different location on the fuel line for the fuel pressure test port **(see illustrations 1.1a, 1.1b, 1.1c and 2.3a)**. You will need a fuel pressure gauge capable of measuring high fuel pressure and equipped with the correct fitting to thread onto the test port. On earlier model TBI systems the fuel pressure gauge must be connected in-line with the fuel inlet line using a T-fitting. Since the inlet fuel line to the throttle body on TBI system is equipped with quick-connect fittings, it will be necessary to install the correct fuel line adapters between the throttle body and inlet fuel line. Consult a specialty tool company for the correct fuel pressure gauge and adapter set, or fabricate your own adapters from the proper fittings.

3   Turn the ignition switch ON (engine not running). The fuel pump should run for about two seconds - note the reading on the gauge. After the pump stops running the pressure should hold steady. It should be within the range listed in this Chapter's Specifications. If the pump does not turn On check the fuel pump electrical circuits (see Step 9).

4   Start the engine and let it idle at normal operating temperature. The pressure should be lower by 3 to 10 psi. If all the pressure readings are within the limits listed in this Chapter's Specifications, the system is operating properly.

5   If the pressure did not drop by 3 to 10 psi after starting the engine, apply 12 to 14 inches of vacuum to the pressure regulator **(see illustration)**. If the pressure drops, repair the vacuum source to the regulator. If the pressure does not drop, replace the regulator. **Note:** *The tests in Steps 5 and 6 will work only on engines where the fuel pressure regulator is easily accessible (MPFI systems). On TBI fuel systems, have the fuel pressure regulator diagnosed by a dealer service*

department or other qualified repair shop.

6   If the fuel pressure is not within specifications, check the following:

a) *If the pressure is higher than specified, check for vacuum to the fuel pressure regulator* **(see illustration)**. *Vacuum must fluctuate with the increase or decrease in the engine rpm. If vacuum is present, check for a pinched or clogged fuel return hose or pipe. If the return line is OK, replace the regulator.*

b) *If the pressure is lower than specified, change the fuel filter to rule out the possibility of a clogged filter. If the pressure is still low, install a fuel line shut-off adapter between the pressure regulator and the return line. With the valve open, start the engine (if possible) and slowly close the valve. If the pressure rises above 50 psi, replace the regulator* (see Section 14). **Warning:** *Don't allow the fuel pressure to exceed 60 psi. Also, don't attempt to restrict the return line by pinching it, as the nylon fuel line will be damaged.*

c) *If the pressure is still low with the fuel return line restricted, an injector (or injectors) may be leaking (see Section 15) or the fuel pump may be faulty.*

7   After the testing is done, relieve the fuel pressure (see Section 2) and remove the fuel pressure gauge.

8   If there are no problems with any of the above-listed components, the fuel pump and circuit are functioning correctly.

Fuel pump electrical circuit check

Refer to illustration 3.10

9   Should the fuel system fail to deliver the proper amount of fuel, or any fuel at all, inspect it as follows. Remove the fuel filler cap. Have an assistant turn the ignition key to the On position (engine not running) while you listen at the fuel filler opening. You should hear a whirring sound that lasts for a

couple of seconds.

10   If you don't hear anything, check the fuel pump fuse (see Chapter 12). If the fuse is blown, replace it and see if it blows again **(see illustration)**. If it does, trace the fuel pump circuit for a short. Refer to the wiring diagrams at the end of Chapter 12 for additional wiring schematics.

11   If the fuse is OK, remove the fuel pump relay and check for battery voltage to the fuel pump relay connector with the ignition key in the OFF position. Then turn the ignition key to the ON position and check for battery voltage from the ECM. **Note 1:** *The fuel pump relay is located in the interior relay center on all models (see Chapter 12).* **Note 2:** *The fuel pump relay is equipped with a primary and secondary voltage circuit. The primary circuit is controlled by the ECM and the secondary circuit provides battery voltage to the fuel pump as the relay is energized. With the ignition switch ON (engine not running), the ECM will ground the relay for several seconds. During cranking, the ECM grounds the fuel pump relay as long as the camshaft position sensor (CMP) sends its position signal (see Chapter 6). If there are no reference pulses, the fuel pump will shut off after two or three seconds.*

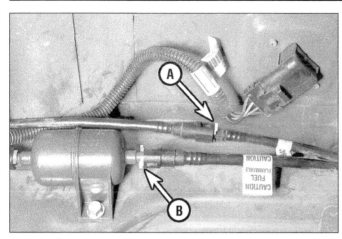

4.10 Some models are equipped with fuel lines that can be disconnected by pinching the tabs and separating each connector

A   *Fuel return line*　　　　　B   *Fuel feed line*

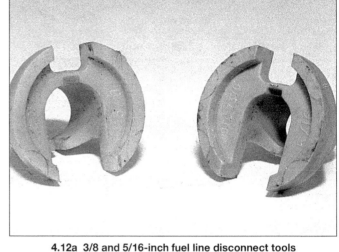

4.12a 3/8 and 5/16-inch fuel line disconnect tools

Refer to the wiring schematics at the end of Chapter 12 for additional information on the wiring color designations for the fuel pump relay. **Note 3:** *On some models, if oil pressure drops below the specified pressure level, the oil pressure switch will act as a fuel pressure cut-off device. Be sure to check the oil pressure switch and circuit in the event of a difficult problem diagnosing the fuel pump circuit (refer to the wiring diagrams at the end of Chapter 12).*

12   If there is no voltage present at the relay connector, check the fuse(s) and the wiring circuit for the fuel pump relay and or ECM (see Chapter 12). If voltage is present at the relay connector, check the relay (see Chapter 12). If the relay is OK, check for battery voltage at one terminal of the fuel pump harness connector located near the fuel tank. If there is no voltage reaching the fuel pump connector check for an open circuit between the fuel pump and the relay. If there is battery power at the fuel pump connector and the fuel pump still does not work, check for continuity to ground at one terminal of the fuel pump harness connector. If the ground is OK, replace the fuel pump.

---

4   Fuel lines and fittings - repair and replacement

Warning: *Gasoline is extremely flammable, so take extra precautions when you work on any part of the fuel system. Don't smoke or allow open flames or bare light bulbs near the work area, and don't work in a garage where a natural gas-type appliance (such as a water heater or a clothes dryer) with a pilot light is present. Since gasoline is carcinogenic, wear latex gloves when there's a possibility of being exposed to fuel, and, if you spill any fuel on your skin, rinse it off immediately with soap and water. Mop up any spills immediately and do not store fuel-soaked rags where they could ignite. The fuel system is under constant pressure, so, if any fuel lines are to*

be disconnected, the fuel pressure in the system must be relieved first (see Section 2). When you perform any kind of work on the fuel system, wear safety glasses and have a Class B type fire extinguisher on hand.

1   Always relieve the fuel pressure before servicing fuel lines or fittings on fuel-injected vehicles (see Section 2).

2   The fuel feed, return and vapor lines extend from the fuel tank to the engine compartment. The lines are secured to the underbody with clip and screw assemblies. These lines must be occasionally inspected for leaks, kinks and dents.

3   If evidence of dirt is found in the system or fuel filter during disassembly, the line should be disconnected and blown out. Check the fuel strainer on the fuel level sending unit (see Section 6) for damage and deterioration.

Steel and nylon tubing

4   Because fuel lines used on fuel-injected vehicles are under high pressure, they require special consideration.

5   If replacement of a metal fuel line or emission line is called for, use welded steel tubing meeting original equipment manufacturers standards or its equivalent. Don't use copper or aluminum tubing to replace steel tubing. These materials cannot withstand normal vehicle vibration.

6   If it becomes necessary to replace a section of nylon fuel line, replace it only with the correct part number - don't use any substitutes.

7   Most fuel lines have threaded fittings with O-rings. Any time the fittings are loosened to service or replace components:

 a) *Use a back-up wrench while loosening and tightening the fittings.*

 b) *Check all O-rings for cuts, cracks and deterioration. Replace any that appear worn or damaged.*

 c) *If the lines are replaced, always use original equipment parts, or parts that meet the manufacturer's specifications.*

Rubber hose

Warning: *These models are equipped with electronic fuel injection; use only original equipment replacement hoses or their equivalent. Others may fail from the high pressures of this system.*

8   When a rubber hose is replaced, use reinforced, fuel resistant hose meeting original equipment manufacturers standards. Other hoses could fail to meet Federal emission standards. Hose inside diameter must match line outside diameter. **Warning:** *Don't substitute rubber hose for metal line on high-pressure systems. Use only genuine factory replacement lines or lines meeting factory specifications.*

9   Don't use rubber hose within four inches of any part of the exhaust system or within ten inches of the catalytic converter. Metal lines and rubber hoses must never be allowed to chafe against the frame. A minimum of 1/4-inch clearance must be maintained around a line or hose to prevent contact with the frame.

Removal and installation

Refer to illustrations 4.10, 4.12a, 4.12b and 4.13

Note: *The following procedure and accompanying illustrations are typical for vehicles covered by this manual. On quick-disconnect (non-threaded) fittings, clean off the fittings before disconnection to prevent dirt from getting in the fittings. After disconnection, clean the fittings and apply a few drops of oil.*

10   Relieve the fuel pressure (see Section 2) and disconnect the fuel feed, return or vapor line at the fuel tank **(see illustration)**. **Note:** *Some fuel line connections may be threaded. Be sure to use a back-up wrench when separating the connections.*

11   Remove all fasteners attaching the lines to the vehicle body.

12   Detach the fitting(s) that attach the fuel hoses to the engine compartment metal lines **(see illustrations)**. Twisting them back and forth will allow them to separate more easily.

4.12b On quick-connect fuel lines, use the special disconnect tools and push them into the connector to separate the fuel lines

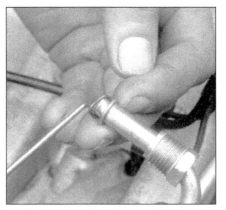

4.13 Always replace the fuel line O-rings (if equipped)

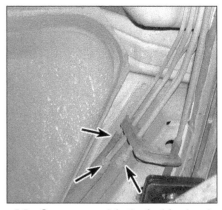

5.5a On some models the fuel feed and return lines (arrows) will be located on the side of the tank

13　Installation is the reverse of removal. Be sure to use new O-rings at the threaded fittings, if equipped **(see illustration)**.

5　Fuel tank - removal and installation

Refer to illustrations 5.5a, 5.5b, 5.5c and 5.8
Warning: *Gasoline is extremely flammable, so take extra precautions when you work on any part of the fuel system. Don't smoke or allow open flames or bare light bulbs near the work area, and don't work in a garage where a natural gas-type appliance (such as a water heater or a clothes dryer) with a pilot light is present. Since gasoline is carcinogenic, wear latex gloves when there's a possibility of being exposed to fuel, and, if you spill any fuel on your skin, rinse it off immediately with soap and water. Mop up any spills immediately and do not store fuel-soaked rags where they could ignite. The fuel system is under constant pressure, so, if any fuel lines are to be disconnected, the fuel pressure in the system must be relieved first (see Section 2). When you perform any kind of work on the fuel system, wear safety glasses and have a Class B type fire extinguisher on hand.*

1　Relieve the fuel pressure (see Section 2).
2　Detach the cable from the negative terminal of the battery.
3　If the tank is full or nearly full, use a hand-operated pump to remove as much fuel through the filler tube as possible (if no such pump is available, you can siphon the tank at the fuel feed line after raising the vehicle - see Step 6).
4　Raise the vehicle and place it securely on jackstands.
5　Disconnect the fuel lines, the vapor return line and the fuel filler neck **(see illustrations)**. **Note:** *The fuel feed and return lines and the vapor return line are three different diameters, so reattachment is simplified. If you have any doubts, however, clearly label the three lines and their corresponding pipes. Be sure to plug the hoses to prevent leakage and contamination of the fuel system. Later model vehicles are equipped with quick-disconnect fittings at the fuel feed and return lines (see Section 4 for further information).*
6　If fuel remains in the tank, siphon the fuel from the tank at the fuel feed line (not the return line). **Warning:** *DO NOT start the siphoning action by mouth! Use a siphoning kit, available at most auto parts stores.*
7　Support the fuel tank with a floor jack or other suitable means. Place a sturdy plank

between the jack head and the fuel tank to protect the tank.
8　Disconnect both fuel tank retaining straps and pivot them down until they are hanging out of the way **(see illustration)**.
9　Lower the tank far enough to disconnect the electrical wires and ground strap (if equipped) from the fuel pump/fuel gauge sending unit.
10　Remove the tank from the vehicle.
11　Installation is the reverse of removal.

6　Fuel tank cleaning and repair - general information

1　Any repairs to the fuel tank or filler neck should be carried out by a professional who has experience in this critical and potentially dangerous work. Even after cleaning and flushing of the fuel system, explosive fumes can remain and ignite during repair of the tank.
2　If the fuel tank is removed from the vehicle, it should not be placed in an area where sparks or open flames could ignite the fumes coming out of the tank. Be especially careful inside garages where a natural gas-type appliance is located, because the pilot light could cause an explosion.

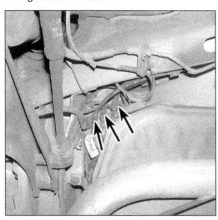

5.5b On other models the fuel feed and return lines (arrows) will be located at the front of the tank

5.5c Loosen the clamps and remove the fuel filler hose and vapor return hose (arrows)

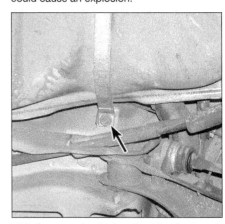

5.8 Support the fuel tank with a jack and remove the fuel tank strap bolts (arrow)

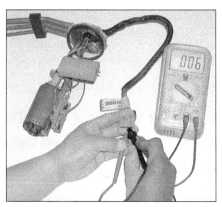

7.2 Check the resistance of the fuel tank sending unit as you move the float arm

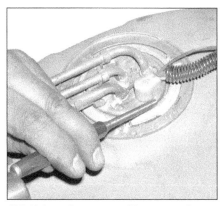

7.8 Using a hammer and brass punch, tap the lock ring counterclockwise to loosen it

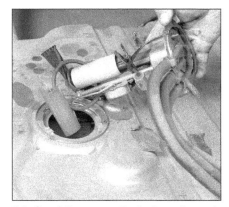

7.9 Carefully remove the fuel sending unit assembly through the opening

7 Fuel level sending unit - check and replacement

Warning: *Gasoline is extremely flammable, so take extra precautions when you work on any part of the fuel system. Don't smoke or allow open flames or bare light bulbs near the work area, and don't work in a garage where a natural gas-type appliance (such as a water heater or a clothes dryer) with a pilot light is present. Since gasoline is carcinogenic, wear latex gloves when there's a possibility of being exposed to fuel, and, if you spill any fuel on your skin, rinse it off immediately with soap and water. Mop up any spills immediately and do not store fuel-soaked rags where they could ignite. The fuel system is under constant pressure, so, if any fuel lines are to be disconnected, the fuel pressure in the system must be relieved first (see Section 2). When you perform any kind of work on the fuel system, wear safety glasses and have a Class B type fire extinguisher on hand.*

Check

Refer to illustration 7.2

1   Remove the fuel level sending unit (see below).

2   Using an ohmmeter, connect the probes to the fuel level sending unit electrical connector terminals **(see illustration)**.

3   Check the resistance of the sending unit as you move the float from the empty position to the full position. Look for a change in resistance when the sending unit float travels from one position to the other. The resistance should change smoothly with the movement of the float arm.

4   If there is no resistance or there is very little change in resistance as the float travels from full to empty, replace the fuel level sending unit assembly.

Replacement

Refer to illustrations 7.8 and 7.9

5   Disconnect the cable from the negative battery terminal.

6   Remove the fuel tank (see Section 5).

7   The fuel pump/sending unit assembly is located inside the fuel tank. It is held in place

by a cam lock ring mechanism consisting of an inner ring with three locking cams and an outer ring with three tangs. The outer ring is welded to the tank and can't be turned.

8   To unlock the fuel pump/sending unit assembly, turn the inner ring counterclockwise until the locking cams are free of the tangs **(see illustration)**. A special tool is available to loosen the lock ring, but if the tool is not available a hammer and brass punch can be used. **Warning:** *If using a punch to loosen the lock ring, use a brass punch. Do not use a steel punch - a spark could cause an explosion!*

9   Lift the fuel pump/sending unit assembly from the fuel tank **(see illustration)**. **Caution:** *The fuel level float and sending unit are delicate. Do not bump them against the tank during removal or the accuracy of the sending unit may be affected.*

10   Inspect the condition of the rubber gasket around the opening of the tank. If it is dried, cracked or deteriorated, replace it.

11   If your replacing the sending unit with a new one it will be necessary to remove the fuel pump from the sending unit assembly (see Section 8).

12   Installation is the reverse of removal.

8 Fuel pump - removal and installation

Refer to illustrations 8.4, 8.5, 8.7a and 8.7b

Warning: *Gasoline is extremely flammable, so take extra precautions when you work on any part of the fuel system. Don't smoke or allow open flames or bare light bulbs near the work area, and don't work in a garage where a natural gas-type appliance (such as a water heater or a clothes dryer) with a pilot light is present. Since gasoline is carcinogenic, wear latex gloves when there's a possibility of being exposed to fuel, and, if you spill any fuel on your skin, rinse it off immediately with soap and water. Mop up any spills immediately and do not store fuel-soaked rags where they could ignite. The fuel system is under constant pressure, so, if any fuel lines are to be disconnected, the fuel pressure in the system must be relieved first (see Section 2).*

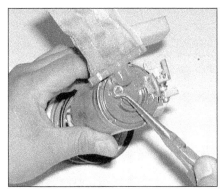

8.4 Use a small screwdriver or pliers to separate the strainer from the bottom of the fuel pump inlet - On some models it may be necessary to remove the strainer retaining clip first

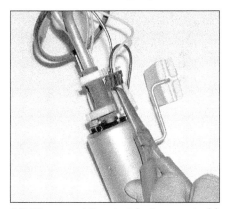

8.5 Disconnect the electrical connector(s) from the fuel pump

When you perform any kind of work on the fuel system, wear safety glasses and have a Class B type fire extinguisher on hand.

1   Disconnect the cable from the negative terminal of the battery.

2   Remove the fuel tank (see Section 5).

3   Remove the fuel pump/sending unit assembly (see Section 7).

4   Remove and inspect the strainer on the lower end of the fuel pump **(see illustration)**. If it's dirty, replace it.

5   Disconnect the electrical connector from the fuel pump **(see illustration)**.

6   Loosen the fuel line retaining clamp or fuel pulsation dampener (if equipped).

7   Remove the pump from the sending unit **(see illustration)**. **Note:** *On some models it will be necessary to push the fuel pump up into the rubber coupler and then slide the bottom of the pump away from the bottom support* **(see illustration)**. *After the pump is clear of the bottom support, pull it out of the rubber coupler. Care should be taken to prevent damage to the rubber insulator during removal.*

8   Inspect the sound insulator and all rubber components for cracks and or deterioration, replacing them if necessary. Reassemble the fuel pump onto the sending unit assembly. On vehicles equipped with a fuel pulsation dampener always install the rubber bumper onto the fuel pump first, then install the fuel pump into the pulsation dampener.

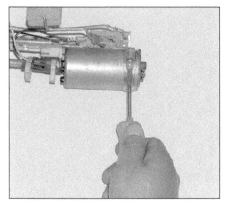

8.7a On some models the fuel pump is secured by a retaining clamp

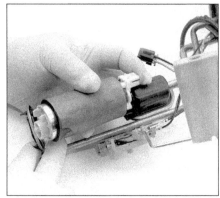

8.7b On other models it will be necessary to push the fuel pump towards the dampener to clear the retaining bracket at the bottom

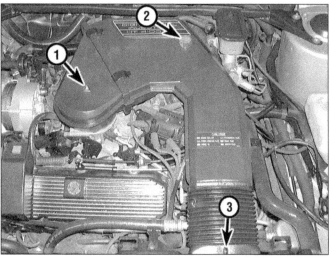

9.5 Air cleaner housing mounting details (1986 and 1987 Eldorado and Seville)

1   Air cleaner housing-to-throttle body retaining nut
2   Air cleaner housing-to-engine support bracket retaining screw
3   Air intake duct-to-radiator support retaining screw

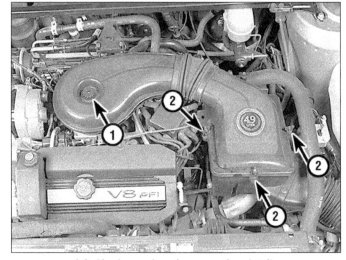

9.8 Air cleaner housing mounting details (1988 and later Cadillac models)

1   Air cleaner-to-throttle body retaining nut
2   Upper air cleaner housing-to-lower air cleaner housing retaining nuts

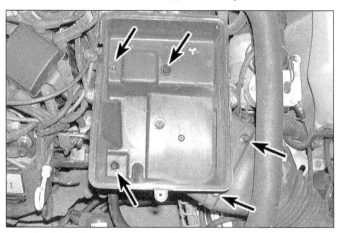

9.10 Air cleaner lower housing mounting bolts (arrows) (1988 and later Cadillac models)

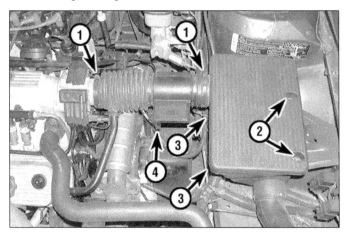

9.13 Air cleaner housing mounting details (Buick and Oldsmobile models)

1   Air intake duct retaining clamp
2   Air cleaner housing upper retaining bolts
3   Air cleaner housing lower retaining bolts (not visible)
4   MAF sensor

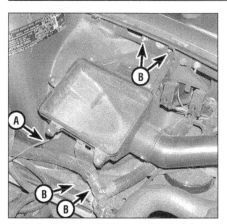

9.18 On Buick and Oldsmobile models the lower housing can be removed after disconnecting the IAT sensor (A) and the retaining bolts (B)

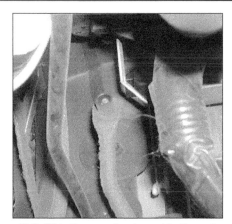

10.2 Slide the accelerator cable through the slot in the pedal arm

10.4 On Buick and Oldsmobile models detach the retaining clips securing the accelerator and the cruise control cables to the throttle lever

9   Insert the fuel pump/sending unit assembly into the fuel tank.

10   Turn the inner lock ring clockwise until the locking cams are fully engaged by the retaining tangs. **Note:** *If you have installed a new O-ring type rubber gasket, it may be necessary to push down on the inner lock ring until the locking cams slide under the retaining tangs.*

11   Install the fuel tank (see Section 5).

9   Air cleaner assembly - removal and installation

Cadillac models

1986 and 1987

Deville and Fleetwood

1   Remove the air cleaner cover and filter.

2   Remove the air intake duct and the heat riser tube from the air cleaner snorkel.

3   Remove the retaining screw securing the front of the snorkel, lift the air cleaner assembly upward and disconnect any vacuum hoses which would interfere with removal.

4   Installation is the reverse of removal.

Eldorado and Seville

Refer to illustration 9.5

5   Remove the air cleaner housing retaining screws **(see illustration)**.

6   Lift the air cleaner housing upward, disconnect any vacuum hoses which would interfere with removal and remove the heat riser tube from the bottom of the air cleaner housing.

7   Installation is the reverse of removal.

1988 and later (all models)

Refer to illustrations 9.8 and 9.10

8   Remove the upper air cleaner housing retaining screws **(see illustration)**.

9   Lift the upper air cleaner housing and disconnect any vacuum hoses which would interfere with removal. Remove the upper housing and the air filter element from the

engine compartment.

10   Remove the bolts securing the lower housing **(see illustration)**.

11   Loosen the clamp securing the heat riser tube to the bottom of the lower housing and remove it from the vehicle.

12   Installation is the reverse of removal.

Buick and Oldsmobile models

Refer to illustrations 9.13 and 9.18

13   Loosen the air intake duct from the throttle body **(see illustration)**.

14   On early models disconnect the MAF sensor connector.

15   Remove the air cleaner housing mounting bolts **(see illustration 9.13)**.

16   Remove the engine compartment support brace if equipped.

17   Lift the upper air cleaner housing and the filter element from the engine compartment.

18   Remove the bolts that retain the air cleaner lower assembly to the body **(see illustration)**.

19   Disconnect the IAT sensor and lift the lower air cleaner assembly from the engine compartment.

20   Installation is the reverse of removal.

10   Accelerator cable - removal and installation

Refer to illustrations 10.2, 10.4, 10.5, 10.6a and 10.6b

Removal

1   Detach the screws and the clip retaining the lower instrument panel trim and lower the trim (if necessary) (see Chapter 11).

2   Detach the accelerator cable from the accelerator pedal **(see illustration)**.

3   Squeeze the accelerator cable cover tangs and push the cable through the firewall into the engine compartment.

4   On Buick and Oldsmobile models detach the accelerator cable retaining clips **(see illustration)**.

5   On Cadillac models rotate the throttle lever and pass the cable through the slot in the lever **(see illustration)**.

6   Remove the cable from the accelerator cable bracket **(see illustrations)**.

7   Remove any remaining cable retaining clips and remove the cable.

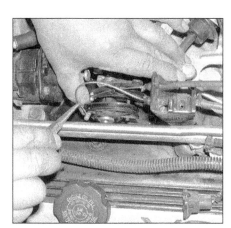

10.5 On Cadillac models rotate the throttle lever and pass the cable(s) through the slot in the lever

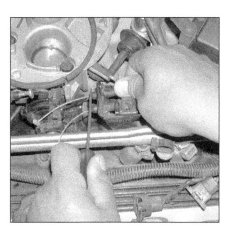

10.6a On some models it will be necessary to pry off the retaining clip to detach the cable from the cable bracket

Installation

8   Installation is the reverse of removal. **Note:** *To prevent possible interference, flexible components (hoses, wires, etc.) must not be routed within two inches of moving parts, unless routing is controlled.*

9   Operate the accelerator pedal and check for any binding condition by completely opening and closing the throttle.

10   Apply sealant around the accelerator cable at the engine compartment side of the firewall.

---

11   Fuel injection system - general information

1   These models are equipped with either the Throttle Body Injection (TBI) system (1989 and earlier Cadillac models) or the Multiport Fuel Injection (MPFI) system. All models are equipped with the OBD I self diagnostic system.

2   The fuel system consists of a fuel tank, an electric fuel pump and fuel pump relay, an air cleaner assembly and an electronic fuel injection system.

Throttle Body Injection (TBI) systems

3   The main component of the TBI system is the Throttle Body Injection (TBI) unit, which is mounted on the intake manifold just like a carburetor **(see illustration 1.1a)**. The TBI unit is made up of two major assemblies: the throttle body and the fuel metering assembly.

4   The throttle body contains a throttle valve, controlled by the accelerator pedal, similar to a carburetor. Attached to the exterior of the body are the Throttle Position Sensor (TPS), which sends throttle position information to the ECM, and the Idle Speed Control (ISC) motor, which is used by the ECM to maintain a constant idle speed during normal engine operation.

5   The fuel metering assembly contains the fuel pressure regulator and the fuel injectors. The regulator dampens the pulsations of the fuel pump and maintains a steady pressure at the injectors. The fuel injectors are controlled by the ECM through an electrically operated solenoid. The amount of fuel injected into the intake manifold is varied by the length of time the injector plunger is held open.

Multiport Fuel Injection (MPFI) system

6   Multiport Fuel Injection (MPFI) consists of an air intake manifold, the throttle body, the injectors, the fuel rail assembly, an electric fuel pump and associated plumbing **(see illustrations 1.1b and 1.1c)**.

7   Air is drawn through the air cleaner and throttle body. A Mass Airflow Sensor (Buick and Oldsmobile models) or a Manifold Absolute Pressure sensor (Cadillac models) informs the ECM of volume and pressure variations.

8   While the engine is running, the fuel

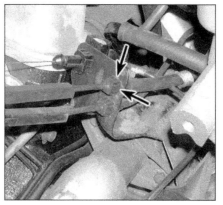

10.6b On others it will be necessary to squeeze the accelerator cable retaining tangs (arrows) and pull the cable through the accelerator cable bracket

constantly circulates through the fuel rail, which removes vapors and keeps the fuel cool while maintaining sufficient pressure to the injectors under all running conditions.

9   As with TBI, the operation of the MPFI system is controlled by the ECM so that it works in conjunction with the rest of the vehicle functions to provide optimum driveability and emissions control.

10   Because the MPFI system meters fuel and air precisely, it is important to the proper operation of the vehicle that the fuel and air filters be changed at the specified intervals.

---

12   Fuel injection system - check

Refer to illustrations 12.8, 12.9a, 12.9b and 12.10

Warning: *Gasoline is extremely flammable, so take extra precautions when you work on any part of the fuel system. Don't smoke or allow open flames or bare light bulbs near the work area, and don't work in a garage where a natural gas-type appliance (such as a water heater or a clothes dryer) with a pilot light is present. Since gasoline is carcinogenic, wear latex gloves when there's a possibility of being exposed to fuel, and, if you spill any fuel on your skin, rinse it off immediately with soap and water. Mop up any spills immediately and do not store fuel-soaked rags where they could ignite. The fuel system is under constant pressure, so, if any fuel lines are to be disconnected, the fuel pressure in the system must be relieved first (see Section 2). When you perform any kind of work on the fuel system, wear safety glasses and have a Class B type fire extinguisher on hand.*

Note: *The following procedure is based on the assumption that the fuel pump is working and the fuel pressure is adequate (see Section 3).*

General checks

1   Check to see that the battery is fully charged, as the control unit and sensors depend on an accurate supply voltage in order to properly meter the fuel.

2   Check the air filter element - a dirty or

12.8 Use a stethoscope or screwdriver to determine if the injectors are working properly - they should make a steady clicking sound that rises and falls with engine speed changes

partially blocked filter will severely impede performance and economy (see Chapter 1).

3   Check the fuel filter and replace it if necessary (see Chapter 1).

5   Check the ground wire connections on the intake manifold for tightness. Check all electrical connectors that are related to the system. Loose connectors and poor grounds can cause many problems that resemble more serious malfunctions.

5   If a blown fuse is found, replace it and see if it blows again. If it does, search for a grounded wire in the harness.

System checks

6   Check the air intake duct to the throttle body for leaks, which will result in an excessively lean mixture. Also check the condition of all vacuum hoses connected to the intake manifold.

7   Remove the air intake duct or the air cleaner housing from the throttle body and check for dirt, carbon or other residue build-up. If it's dirty, clean it with aerosol carburetor cleaner and a rag.

8   Remove the injector cover, if equipped. With the engine running, place an automotive stethoscope against each injector, one at a time, and listen for a clicking sound, indicating operation **(see illustration)**. If you don't have a stethoscope, place the tip of a screwdriver against the injector and listen through the handle. **Note:** *On the TBI system, you can actually see the fuel being sprayed in the throttle body, it is not necessary to check them for a clicking sound using an automotive stethoscope.*

9   Unplug the injector electrical connectors and test the resistance of each injector. Compare the values to the Specifications listed in this Chapter **(see illustrations)**.

10   Install an injector test light ("noid" light) into each injector electrical connector, one at a time **(see illustration)**. Crank the engine over. Confirm that the light flashes evenly on each connector. This will test for voltage to the injectors.

11   The remainder of the system checks can be found in Chapter 6.

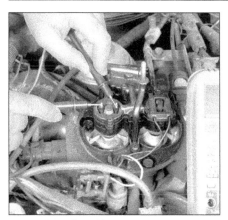

12.9a Checking the resistance of a TBI injector

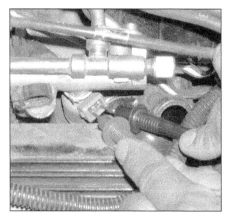

12.9b Checking the resistance of a MPFI injector

12.10 Install the "noid" light into each injector electrical connector and confirm that it blinks when the engine is cranking or running (MPFI system shown)

13 Throttle body – removal and installation

Cadillac models

Refer to illustrations 13.4a, 13.4b and 13.7

1   Remove the air cleaner (see Section 9).
2   Remove the ISC motor from the throttle body (see Chapter 6).
3   Unplug the electrical connector from the Throttle Position Sensor (TPS).
4   On TBI equipped models remove the fuel feed and return lines **(see illustration)**. Remove the fuel line O-rings from the nuts and discard them. Unplug the electrical connectors from the fuel injectors and detach the injector wiring harness grommet from the throttle body **(see illustration)**.
5   Detach the accelerator cable, throttle return spring(s), transmission detent cable (automatics) and, if equipped, cruise control cable.
6   Clearly label, then detach, all vacuum hoses.
7   Remove the throttle body mounting bolts **(see illustration)** and lift the throttle body from the intake manifold. Remove and discard the throttle body mounting gasket.
8   Installation is the reverse of removal.

Buick and Oldsmobile models

Refer to illustration 13.15

Warning: *Wait until the engine is completely cool before beginning this procedure.*

9   Disconnect the cable from the negative terminal of the battery and drain the coolant from the radiator.
10   Detach the air intake duct (see Section 9).
11   Unplug the Idle Air Control (IAC) valve and the Throttle Position Sensor (TPS) electrical connectors (see Chapter 6). Unplug the Mass Air Flow sensor (MAF) electrical connector on 1988 and later models.
12   Mark and disconnect any vacuum hoses connected to the throttle body. Also detach the breather hose, if equipped.
13   Disconnect the accelerator cable, the cruise control cable (see Section 10) and the transmission detent cable (if equipped) from the throttle lever, On 1991 and later models

remove the accelerator cable retaining bracket from the throttle body.
14   On 1987 and earlier models detach the coolant bypass hose from the intake manifold.
15   Remove the throttle body retaining bolts and detach the throttle body **(see illustration)**. **Note:** *1990 and earlier throttle bodies are fastened to the intake manifold by two bolts while 1991 and later models are fastened to the intake manifold by three bolts.*

13.4a On TBI models be sure to use a back-up wrench when disconnecting the fuel feed and return lines

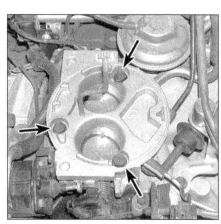

13.7 Throttle body mounting bolts (arrows) (MPFI system shown, TBI system similar)

16   Clean off all traces of old gasket material from the throttle body and the plenum.
17   Install the throttle body and a new gasket and tighten the bolts to the torque listed in this Chapter's Specifications.
18   The remainder of the installation is the reverse of removal. Be sure to refill the engine coolant (see Chapter 1).

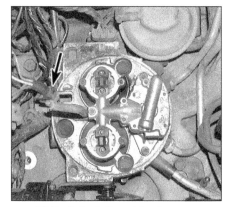

13.4b Detach the wiring harness connectors from the fuel injectors and the wiring harness grommet (arrow) from the throttle body

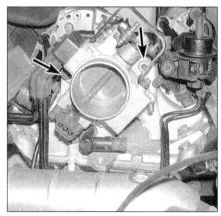

13.15 Throttle body mounting bolts (arrows) (early model Oldsmobile shown)

14 Throttle Body Injection (TBI) - component replacement

Warning: *Gasoline is extremely flammable, so take extra precautions when you work on any part of the fuel system. Don't smoke or allow open flames or bare light bulbs near the work area, and don't work in a garage where a natural gas-type appliance (such as a water heater or a clothes dryer) with a pilot light is present. Since gasoline is carcinogenic, wear latex gloves when there's a possibility of being exposed to fuel, and, if you spill any fuel on your skin, rinse it off immediately with soap and water. Mop up any spills immediately and do not store fuel-soaked rags where they could ignite. The fuel system is under constant pressure, so, if any fuel lines are to be disconnected, the fuel pressure in the system must be relieved first (see Section 2). When you perform any kind of work on the fuel system, wear safety glasses and have a Class B type fire extinguisher on hand.*

Note: *Because of its relative simplicity, the throttle body assembly does not need to be removed from the intake manifold or disassembled for component replacement. However, for the sake of clarity, the following procedures are shown with the TBI assembly removed from the vehicle.*

1   Relieve system fuel pressure (see Section 2).

2   Detach the cable from the negative terminal of the battery.

3   Remove the air cleaner housing assembly, adapter (if equipped) and gaskets.

Fuel meter cover/fuel pressure regulator assembly

Refer to illustrations 14.5, 14.6 and 14.7

Note: *The fuel pressure regulator is housed in the fuel meter cover. Whether you are replacing the meter cover or the regulator itself, the entire assembly must be replaced. The regulator must not be removed from the cover.*

4   Unplug the electrical connectors to the fuel injectors **(see illustration 14.14)**.

5   Remove the long and short fuel meter

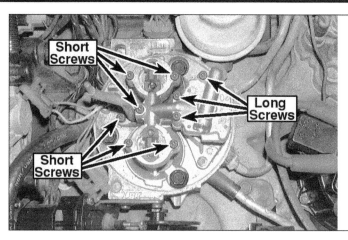

14.5 Fuel meter cover screws

cover screws and remove the fuel meter cover **(see illustration)**.

6   Remove the fuel meter outlet passage gasket, cover gasket and pressure regulator seal. Carefully remove any old gasket material that is stuck with a razor blade **(see illustration)**. **Caution:** *Do not attempt to re-use either of these gaskets.*

7   Inspect the cover for dirt, foreign material and casting warpage. If it is dirty, clean it with a clean shop rag soaked in solvent. Do not immerse the fuel meter cover in cleaning solvent - it could damage the pressure regulator diaphragm and gasket. **Warning:** *Do not remove the four screws* **(see illustration)** *securing the pressure regulator to the fuel meter cover. The regulator contains a large spring under compression which, if accidentally released, could cause injury. Disassembly might also result in a fuel leak between the diaphragm and the regulator housing. The new fuel meter cover assembly will include a new pressure regulator.*

8   Install the new pressure regulator seal, fuel meter outlet passage gasket and cover gasket.

9   Install the fuel meter cover using Loctite 262 or equivalent on the screws. **Note:** *The short screws go next to the injectors.*

10   Attach the electrical connectors to both injectors.

11   Attach the cable to the negative terminal

of the battery.

12   With the engine Off and the ignition On, check for leaks around the gasket and fuel line couplings.

13   Install the air cleaner, adapter (if equipped) and gaskets.

Fuel injector(s)

Refer to illustrations 14.14, 14.16, 14.21, 14.22, 14.23, 14.24 and 14.25

Note: *When replacing a fuel injector, be sure to use one having the identical part number. Injectors from other models are calibrated with different flow rates but can be interchanged with other Model 220 TBI units.*

14   To unplug the electrical connectors from the fuel injectors, squeeze the plastic tabs and pull straight up **(see illustration)**.

15   Remove the fuel meter cover/pressure regulator assembly. **Note:** *Do not remove the fuel meter cover assembly gasket - leave it in place to protect the casting from damage during injector removal.*

16   Use two screwdrivers **(see illustration)** to pry out the injector(s).

17   Remove the upper (larger) and lower (smaller) O-rings and filter from the injector(s).

18   Remove the steel backup washer from the top of each injector cavity.

19   Inspect the fuel injector filters for evidence of dirt and contamination. If present, check for the presence of dirt in the fuel lines

14.6 Carefully peel away the old fuel meter outlet passage gasket and fuel meter cover gasket with a razor blade

14.7 DO NOT remove the four pressure regulator screws (arrows) from the fuel meter cover

14.14 To remove either injector electrical connector, depress the two tabs on the front and rear of each connector and lift straight up

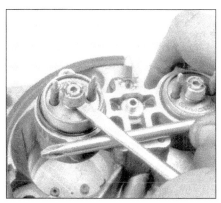

14.16 To remove an injector, slip the tip of a flat-bladed screwdriver under the lip of the lug on top of the injector and, using another screwdriver as a fulcrum, carefully pry the injector up and out

and fuel tank.

20   Be sure to replace the fuel injector with an identical part. Injectors from other models can fit in the Model 220 TBI assembly but are calibrated for different flow rates.

21   Slide the new filter into place on the nozzle of the injector **(see illustration)**.

22   Lubricate the new lower (smaller) O-ring with automatic transmission fluid and place it on the small shoulder at the bottom of the fuel injector cavity in the fuel meter body **(see illustration)**.

23   Install the steel back-up washer in the injector cavity **(see illustration)**.

24   Lubricate the new upper (larger) O-ring with automatic transmission fluid and install it on top of the steel back-up washer **(see illustration)**. **Note:** *The backup washer and the large O-ring must be installed before the injector. If they aren't, improper seating of the large O-ring could cause fuel leakage.*

25   To install an injector, align the raised lug on the injector base with the notch in the fuel meter body cavity **(see illustration)**. Push down on the injector until it is fully seated in the fuel meter body. **Note:** *The electrical terminals should be parallel with the throttle shaft.*

26   Install the fuel meter cover assembly and gasket.

27   Attach the cable to the negative terminal

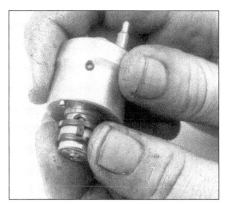

14.21 Slide the new filter onto the nozzle of the fuel injector

of the battery.

28   With the engine Off and the ignition On, check for fuel leaks.

29   Attach the electrical connectors to the fuel injectors.

30   Install the air cleaner housing assembly, adapter (if equipped) and gaskets.

---

15   Multi Port Fuel Injection (MPFI) - component replacement

---

Warning: *Gasoline is extremely flammable, so take extra precautions when you work on any part of the fuel system. Don't smoke or allow open flames or bare light bulbs near the work area, and don't work in a garage where a natural gas-type appliance (such as a water heater or a clothes dryer) with a pilot light is present. Since gasoline is carcinogenic, wear latex gloves when there's a possibility of being exposed to fuel, and, if you spill any fuel on your skin, rinse it off immediately with soap and water. Mop up any spills immediately and do not store fuel-soaked rags where they could ignite. The fuel system is under constant pressure, so, if any fuel lines are to be disconnected, the fuel pressure in the system must be relieved first (see Section 2). When you perform any kind of work on the fuel system, wear safety glasses and have a Class B type fire extinguisher on hand.*

14.22 Lubricate the lower O-ring with transmission fluid then place it on the shoulder in the bottom of the injector cavity

14.23 Place the steel back-up washer on the shoulder near the top of the injector cavity

Fuel rail and injectors

Refer to illustrations 15.1, 15.3a, 15.3b, 15.5a, 15.5b, 15.6, 15.7, 15.8 and 15.10

1   Relieve fuel system pressure and then detach the cable from the negative terminal of the battery. On late model 3800 V6 engines remove the intake manifold cover **(see illustration)**. On Cadillac V8 engines remove the fuel injector covers.

14.24 Lubricate the upper O-ring with transmission fluid then install it on top of the steel washer

14.25 Make sure that the lug is aligned with the groove in the bottom of the fuel injector cavity

15.1 On late model 3800 V6 engines detach the nut (arrow) remove the intake manifold cover

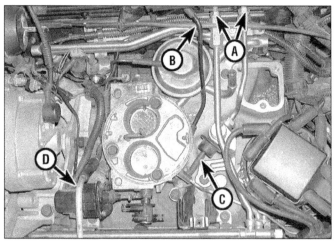

15.3a To allow removal of the fuel rail assembly on Cadillac models, detach the fuel feed and return lines (A), the **MAP** sensor vacuum line (B), the fuel pressure regulator vacuum line (C) and the **PCV** valve tube (D)

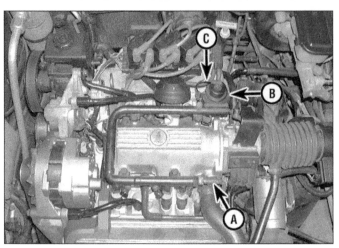

15.3b To allow removal of the fuel rail assembly on Buick and Oldsmobile models, detach the fuel feed line (A), the fuel return line (B), the fuel pressure regulator vacuum line (C) (1987 model shown)

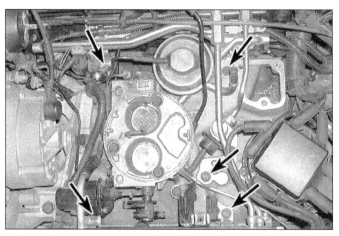

15.5a Fuel rail mounting bolts - 1990 and later Cadillac models

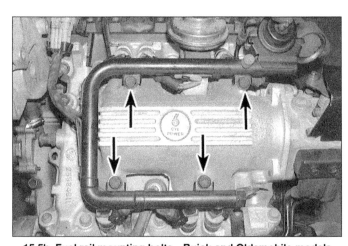

15.5b Fuel rail mounting bolts - Buick and Oldsmobile models (early model shown, late models similar)

2   Detach the vacuum lines from the fuel pressure regulator. On Cadillac models, disconnect the MAP sensor vacuum line and the PCV valve to throttle body tube.

3   Detach the fuel feed and return lines from the fuel rail **(see illustrations)**. **Note:** *1991 and later Buick and Oldsmobile models may require the use of special quick disconnect tools to separate the fuel feed and return lines from the fuel rail (see Section 4).*

4   Label and unplug the injector electrical connectors. **Note:** *On some Cadillac models it will be necessary to remove the power steering pump and the power steering pump mounting bracket.*

5   Remove the fuel rail retaining bolts **(see illustrations)**.

6   Carefully remove the fuel rail with the injectors and the pressure regulator attached **(see illustration)**. **Caution:** *Use care when handling the fuel rail assembly to avoid damaging the injectors.* **Note:** *On Cadillac models, an eight digit identification number is stamped on the side of the fuel rail assembly. Refer to this number if servicing or parts*

replacement is required.

7   To remove the fuel injectors, spread the injector retaining clip and pull the injector from the fuel rail **(see illustration)**.

15.6 Pull upward on the fuel rail to separate the injectors from the intake manifold, then carefully maneuver the fuel rail assembly off the engine (Cadillac model shown)

8   Remove the injector O-ring seals **(see illustration)**.

9   Install the new O-ring seal(s), as required, on the injector(s) and lubricate them

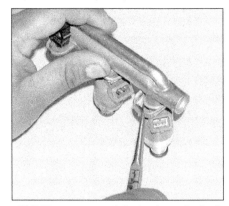

15.7 To remove an injector from the fuel rail, spread the retaining clip with a small screwdriver, then pull the injector from the fuel rail

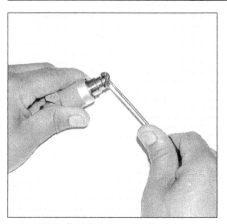

15.8 Carefully pry the O-rings off the injectors

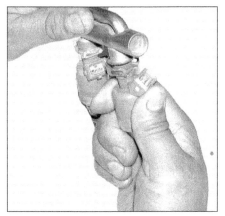

15.10 Install a new retaining clip onto the top of each injector, then push each injector into the injector cup until the retaining clip snaps into place around the cup lip

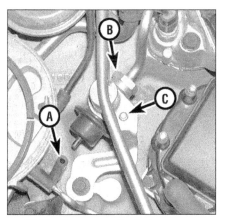

15.16a To remove the fuel pressure regulator on 1990 and later Cadillac models, detach the vacuum hose (A), unscrew the fuel return line (B) and remove the mounting bolt (C)

with a light film of engine oil.

10   Install a new retaining clip onto the top of each injector. Push each injector into the fuel rail cup until the retaining clip snaps into place around the fuel rail cup **(see illustration)**.

11   Installation is the reverse of the removal procedure.

Fuel pressure regulator

Refer to illustrations 15.16a, 15.16b and 15.17

12   Relieve the fuel system pressure (see Section 2).

13   Disconnect the cable from the negative terminal of the battery.

14   On Cadillac models remove the air cleaner (see section 9).

15   Disconnect the vacuum line from the fuel pressure regulator.

16   Remove the fuel return line from the fuel pressure regulator assembly **(see illustrations)**. On Cadillac models, remove the fuel rail (see steps 1 through 11). Remove the mounting screw(s) and separate the pressure regulator from the fuel rail.

17   On 1991 and later Buick and Oldsmobile models, compress the snap-ring with snap-ring pliers and lift the ring from the regulator **(see illustration)**. Lift the regulator while

simultaneously twisting it and remove it from the housing. Remove the filter screen from the regulator housing and clean it with compressed air. Install new O-rings and lubricate them with a light coat of oil.

18   Installation is the reverse of removal. Be sure to replace all gaskets and seals, otherwise a dangerous fuel leak may develop. When installing the seals, lubricate them with a light film of engine oil.

16 Fuel injection system - adjustments

Cadillac models

Minimum idle speed adjustment

Refer to illustrations 16.5 and 16.7

1   Warm the engine to normal operating temperature.

2   Connect a tachometer in accordance with the tool manufacturer's instructions.

3   Ground the diagnostic test terminal on the ALDL connector. Set the ignition timing (refer to Chapter 1 timing specifications and procedure).

4   Shut off the engine, remove the air cleaner, disconnect the THERMAC vacuum

15.16b To remove the fuel pressure regulator on 1990 and earlier Buick and Oldsmobile models, detach the vacuum hose (A), unscrew the fuel return line (B) and remove the mounting bolts (C)

hose (if equipped) and disable the charging system by grounding the green test connector located in the wiring harness at the rear of the alternator.

5   Using a pair of fused jumper wires, fully retract the ISC plunger by applying 12 volts to terminal C of the ISC motor connector and grounding terminal D **(see illustration)**.

15.17 On 1991 and later Buick and Oldsmobile models, remove the snap-ring (arrow) from the top of the regulator

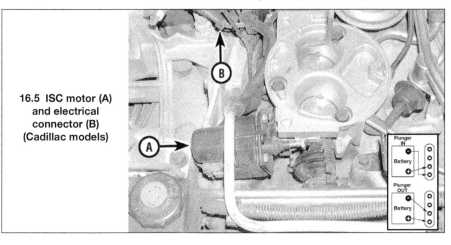

16.5 ISC motor (A) and electrical connector (B) (Cadillac models)

16.7 Minimum idle speed screw (arrow)
(Cadillac MPFI system shown)

16.10 With the ISC plunger at fully extended, turn the plunger
(arrow) until the correct voltage reading is obtained at the TPS

Note 1: *Apply finger pressure to the to the ISC plunger while connecting terminal D to ground. Do not allow the battery voltage (12V) to contact terminal C any longer than necessary to retract the ISC plunger. Prolonged contact will cause damage to the motor.* **Note 2:** *It is very important to never connect a voltage source across terminals A and B, as it may damage the internal throttle contact switch.*

6   With the ISC motor fully retracted, the plunger should not be touching the throttle lever. If the plunger is contacting the throttle lever, turn the plunger inward (clockwise) until there is clearance between the two and the throttle lever is resting on the minimum idle screw. Check that the throttle lever is not binding and moves freely through its full range of motion.

7   Start the engine and check the minimum idle speed. Compare the reading on the tachometer with the Specifications at the front of this Chapter. If the minimum idle speed is out of specification, turn the minimum idle screw in the direction needed to bring the idle to the correct specification **(see illustration)**. Remove the ground wire from the alternator test connector.

Idle Speed Control (ISC) motor maximum extension adjustment

Refer to illustration 16.10

8   Check and adjust if necessary the TPS setting (see Chapter 6) with the minimum idle speed set to specification.

9   With the voltmeter still connected to the TPS, fully extend the ISC plunger by applying battery positive voltage (12V) to terminal D of the ISC connector and grounding terminal C **(see illustration 16.5)**. **Note:** *Do not allow the battery voltage (12V) to contact terminal D any longer than necessary when extending the ISC plunger. Prolonged contact will cause damage to the motor.*

10   The TPS voltage should be between the Specifications listed at the front of this Chap-

ter with the ISC motor fully extended. Turn the ISC plunger as necessary to achieve the correct specification **(see illustration)**.

11   Disconnect all the testing instruments, and reconnect the ISC motor connector. Turn the ignition key Off for at least 15 seconds, start the engine and check for proper operation of the ISC motor.

12   The adjustment procedure may set a diagnostic code in the On-Board Diagnostic system. If the CHECK ENGINE light comes on, refer to Chapter 6, and erase the codes from the ECM memory.

Buick and Oldsmobile models

Minimum idle speed adjustment

Refer to illustration 16.17

13   Warm the engine to normal operating temperature. Block the wheels and be sure the parking brake is applied.

14   Connect a tachometer in accordance with the tool manufacturer's instructions and turn the engine Off.

15   Ground the diagnostic test terminal on the ALDL connector and set the ignition timing (refer to Chapter 1 timing specifications and procedure). Turn the engine Off and wait at least 30 seconds.

16   With the ALDL connector still grounded turn the ignition On (engine not running) and disconnect the IAC connector.

17   Have an assistant sit in the driver's seat, start the engine, apply the brakes and move the shift lever into the Drive range. Check the minimum idle speed and compare the reading on the tachometer with the Specifications at the front of this Chapter. If the minimum idle speed is out of specification, turn the minimum idle screw in the direction needed to bring the idle to the correct specification **(see illustration)**.

18   Turn the ignition Off and reconnect the IAC connector.

19   Check and adjust if necessary the TPS setting (see Chapter 6).

17   Exhaust system servicing - general information

Refer to illustrations 17.1a, 17.1b, 17.1c and 17.1d

Warning: *Inspection and repair of exhaust system components should be done only after enough time has elapsed after driving the vehicle to allow the system components to cool completely. Also, when working under the vehicle, make sure it is securely supported on jackstands.*

1   The exhaust system consists of the exhaust manifold(s), the catalytic converter, the muffler, the tailpipe and all connecting pipes, brackets, hangers and clamps **(see illustrations)**. The exhaust system is attached to the body with mounting brackets and rubber hangers. If any of the parts are improperly installed, excessive noise and vibration will be transmitted to the body.

2   Conduct regular inspections of the exhaust system to keep it safe and quiet. Look for any damaged or bent parts, open seams, holes, loose connections, excessive corrosion or other defects which could allow

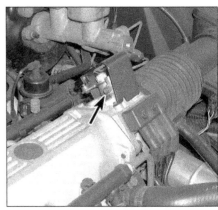

16.17 Minimum idle speed screw (arrow)
(Buick and Oldsmobile
MPFI system shown)

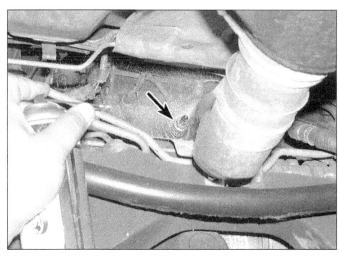

17.1a Be sure to apply penetrating lubricant to the exhaust flange nuts (arrows) before attempting to remove them

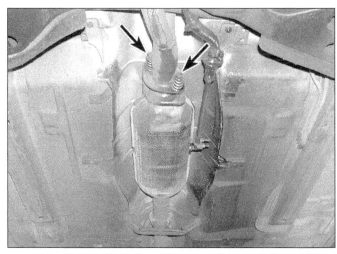

17.1b Remove the exhaust flange retaining nuts (arrows) from the center section of the exhaust system

exhaust fumes to enter the vehicle. Deteriorated exhaust system components should not be repaired; they should be replaced with new parts.

3   If the exhaust system components are extremely corroded or rusted together, welding equipment will probably be required to remove them. The convenient way to accomplish this is to have a muffler repair shop remove the corroded sections with a cutting torch. If, however, you want to save money by doing it yourself (and you don't have a welding outfit with a cutting torch), simply cut off the old components with a hacksaw. If you have compressed air, special pneumatic cutting chisels can also be used. If you do decide to tackle the job at home, be sure to wear safety goggles to protect your eyes from metal chips and work gloves to protect your hands.

4   Here are some simple guidelines to follow when repairing the exhaust system:

a) *Work from the back to the front when removing exhaust system components.*
b) *Apply penetrating oil to the exhaust system component fasteners to make them easier to remove.*
c) *Use new gaskets, hangers and clamps when installing exhaust systems components.*
d) *Apply anti-seize compound to the threads of all exhaust system fasteners during reassembly.*
e) *Be sure to allow sufficient clearance between newly installed parts and all points on the underbody to avoid overheating the floor pan and possibly damaging the interior carpet and insulation. Pay particularly close attention to the catalytic converter and heat shield.*

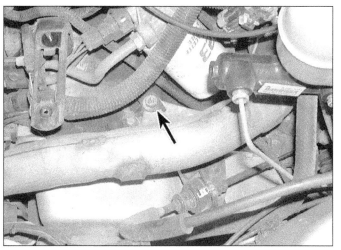

17.1c Exhaust crossover pipes are often fastened in several places - this one is fastened (arrow) to the rear of the engine block

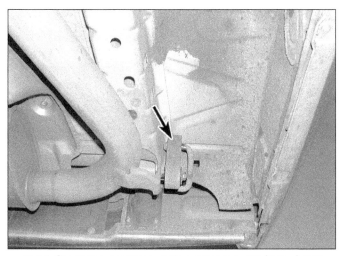

17.1d Check the condition of the rubber mounts (arrow) that support the weight of the exhaust system

Notes

Chapter 5
Engine electrical systems

Contents

Specifications

General

HEI system	
Ignition coil resistance	
Primary resistance	0.4 to 0.5 ohms
Secondary resistance	6,000 to 30,000 ohms
Pick-up coil resistance	500 to 1,500 ohms
DIS system	
Ignition coil resistance	
Primary resistance	0.3 to 0.7 ohms
Secondary resistance	5,000 to 6,500 ohms
Spark plug wire resistance	Less than 5, 000 ohms per foot
Cylinder numbers	See Chapter 2
Firing order	See Chapter 2

1   General information

Warning: *Because of the very high voltage generated by the ignition system, extreme care should be taken whenever an operation involving ignition components is performed. This not only includes the distributor, coil(s), module and spark plug wires, but related items that are connected to the systems as well, such as the plug connections, tachometer and testing equipment.*

The models covered by this manual are equipped with two different types of ignition systems. All Cadillac models are equipped with a conventional coil-in-cap High Energy

Ignition (HEI) system. All Buick and Oldsmobile models are equipped with a Distributorless Ignition System (DIS) (see Section 5).

Although these systems use different ways to generate the ignition signals, they share certain similar components, such as the ignition switch, battery, coil, primary (low tension) and secondary (high tension) wiring circuits and spark plugs.

The charging system consists of a belt-driven alternator with an integral voltage regulator and the battery. These components work together to supply electrical power for the ignition system, the lights and all accessories.

All models are equipped with either a

CS-144 (120 amp) or a CS-144 GEN II (124 amp) alternator. All types use a conventional pulley and fan. CS-144 alternators can be rebuilt but it is recommended that the home mechanic exchange the alternator for a rebuilt unit. Because of the expense and the limited availability of parts, no alternator overhaul information is included in this manual.

2   Battery - emergency jump starting

Refer to the *Booster battery (jump) starting* procedure at the front of this manual.

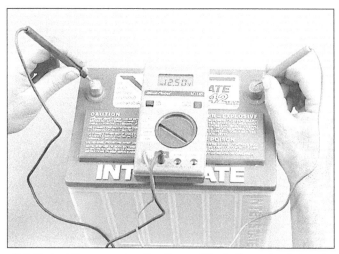

3.2 To test the open circuit voltage of the battery, simply touch the black probe of the voltmeter to the negative terminal and the red probe to the positive terminal of the battery – a fully charged battery should read between 11.5 to 12.5 volts depending on the outside air temperature

3.3 Some battery load testers are equipped with a ammeter which enables the battery load to be precisely dialed in, as shown – less expensive testers have a load switch and a voltmeter only

3   Battery – check and replacement

Warning: *Hydrogen gas is produced by the battery, so keep open flames and lighted cigarettes away from it at all times. Always wear eye protection when working around a battery. Rinse off spilled electrolyte immediately with large amounts of water.*

Caution : *Always disconnect the negative cable first and hook it up last or the battery may be shorted by the tool being used to loosen the cable clamps.*

Check

Refer to illustrations 3.2 and 3.3

1   Disconnect the negative battery cable, then the positive cable from the battery.

2   Check the battery state of charge. Visually inspect the indicator eye on the top of the battery, if the indicator eye is black in color charge the battery as described in Chapter 1. Next perform an open voltage circuit test using a digital voltmeter **(see illustration)**. **Note:** *The battery's surface charge must be removed before accurate voltage measurements can be made. Turn On the high beams for ten seconds, then turn them Off, let the vehicle stand for two minutes.* With the engine and all accessories Off, touch the negative probe of the voltmeter to the negative terminal and the positive probe to the positive terminal of the battery. The battery voltage should be 11.5 to 12.5 volts or slightly above. If the battery is less than the specified voltage, charge the battery before proceeding to the next test. Do not proceed with the battery load test unless the battery charge is correct.

3   Perform a battery load test. An accurate check of the battery condition can only be performed with a load tester (available at most auto parts stores). This test evaluates the ability of the battery to operate the starter and other accessories during periods of heavy amperage draw (load). Install a special battery load testing tool onto the terminals **(see illustration)**. Load test the battery according to the manufacturer's instructions for the particular tool. This tool utilizes a carbon pile to increase the load demand (amperage draw) on the battery. Maintain the load on the battery for 15 seconds or less and observe that the battery voltage does not drop below 9.6 volts. If the battery condition is weak or defective, the tool will indicate this condition immediately. **Note:** *Cold temperatures will cause the minimum voltage requirements to drop slightly. Follow the chart given in the tool manufacturer's instructions to compensate for cold climates. Minimum load voltage for freezing temperatures (32 degrees F) should be approximately 9.1 volts.*

Replacement

Refer to illustrations 3.5 and 3.6

4   The battery is located at the right front corner of the engine compartment. Remove the engine compartment support brace (if equipped).

5   Detach the cables from the negative and positive terminals of the battery **(see illustration)**. **Warning:** *To prevent arcing, disconnect the negative (-) cable first, then remove the positive (+) cable.*

6   Remove the hold-down clamp bolt and the clamp from the battery carrier **(see illustration)**.

7   Carefully lift the battery from the carrier. **Note:** *The battery carrier and hold-down clamp should be clean and free from corrosion before installing the battery. Make certain that there are no parts in the carrier before installing the battery.*

8   To install the battery, simply position it in the battery carrier. Don't tilt it.

9   Install the hold-down clamp and bolt. The bolt should be snug, but overtightening it may damage the battery case.

10   Install both battery cables, positive first, then the negative. **Note:** *The battery terminals and cable ends should be cleaned prior to connection* (see Chapter 1).

4   Battery cables - check and replacement

Refer to illustrations 4.4a, 4.4b, 4.4c and 4.4d

1   Periodically inspect the entire length of each battery cable for damage, cracked or burned insulation and corrosion. Poor battery cable connections can cause starting problems and decreased engine performance.

2   Check the cable-to-terminal connections at the ends of the cables for cracks, loose wire strands and corrosion. The presence of white, fluffy deposits under the insulation at the cable terminal connection is a sign that the cable is corroded and should be replaced. Check the terminals for distortion, missing mounting bolts and corrosion.

3   When replacing the cables, always disconnect the negative cable first and hook it up last or the battery may be shorted by the tool used to loosen the cable clamps. Even if only the positive cable is being replaced, be sure to disconnect the negative cable from the battery first.

4   Disconnect and remove the cable **(see illustrations)**. Make sure the replacement cable is the same length and diameter.

5   Clean the threads of the relay or ground connection with a wire brush to remove rust and corrosion. Apply a light coat of petroleum jelly to the threads to prevent future corrosion.

6   Attach the cable to the relay or ground connection and tighten the mounting nut/bolt securely.

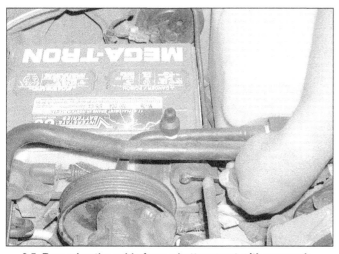

3.5 Removing the cable from a battery post with a wrench - always remove the ground cable first and hook it up last!

3.6 Remove the bolt and the battery hold down clamp (arrow) which secures the base of the battery to the battery tray

7   Before connecting the new cable to the battery, make sure that it reaches the battery post without having to be stretched. Clean the battery posts thoroughly and apply a light coat of petroleum jelly to prevent corrosion (see Chapter 1).

8   Connect the positive cable first, followed by the negative cable.

5   Ignition system - general information

Warning: *Because of the very high voltage generated by the ignition system, extreme care should be taken whenever an operation involving ignition components is performed. This not only includes the distributor, coil(s), module and spark plug wires, but related items that are connected to the systems as well, such as the plug connections, tachometer and other testing equipment.*

The ignition system consists of the ignition switch, the battery, the coil(s), the primary (low tension) and secondary (high tension) wiring circuits, the ignition module, the distributor (Cadillac models) and the spark plugs.

High Energy Ignition (HEI) system

A High Energy Ignition (HEI) distributor with Electronic Spark Timing (EST) is used on all Cadillac models. The ignition coil is mounted in the distributor cap. The HEI distributor has an internal magnetic pick-up assembly which contains a permanent magnet, a pole piece with internal teeth and a pick-up coil.

All spark timing changes in the HEI/EST distributor are carried out electronically by the Electronic Control Module (ECM), which monitors data from various engine sensors, computes the desired spark timing and signals the distributor to change the timing accordingly. A back-up spark advance system is incorporated to signal the ignition module in case of ECM failure. No vacuum or mechanical advance is used.

Direct Ignition System (DIS)

A Direct Ignition System (DIS) with Electronic Spark Timing (EST) is used on all Buick and Oldsmobile models. The DIS ignition systems use a "waste spark" method of spark distribution. Each cylinder is paired with its

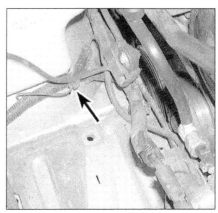

4.4a Detach any battery cable ties or retaining clips (arrow)

opposing cylinder in the firing order (1-4, 2-5, 3-6) so one cylinder under compression fires simultaneously with its opposing cylinder, where the piston is on the exhaust stroke. Since the cylinder on the exhaust stroke requires very little of the available voltage to fire its plug, most of the voltage is used to fire the plug of the cylinder on the compression stroke.

4.4b On most models, the positive battery cable is fastened to the starter (arrow) . . .

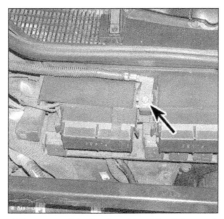

4.4c . . . and to the junction block (arrow)

4.4d Negative battery cables are usually fastened to the engine block (arrow) and to the inner fenderwell or radiator support

6.1 To use a calibrated ignition tester (available at most auto parts stores), simply disconnect a spark plug wire, attach the wire to the tester, clip the tester to a convenient ground and operate the starter - if there's enough power to fire the plug, sparks will be visible between the electrode tip and the tester body

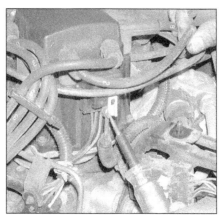

6.5 On Cadillac models, check for power at the BAT+ terminal on the HEI distributor with the ignition On

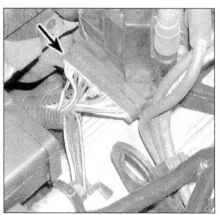

6.7 On Buick and Oldsmobile models, backprobe ignition control module connector (arrow) and check for battery power to the module with the ignition On

The DIS system includes three ignition coils, an ignition module, a single hall effect crankshaft position sensor (except 1988 and later models which are equipped with a dual hall effect crankshaft position sensor), a engine crankshaft balancer with crankshaft sensor interrupter rings, a camshaft position sensor, spark plugs, spark plug wires and the Electronic Control Module (ECM).

Conventional ignition coils have one end of the secondary winding connected to the engine ground. On DIS, neither end of the secondary winding is grounded - instead, one end of the coils secondary winding is directly attached to the spark plug and the other end is attached the spark plug of the companion cylinder.

The crankshaft reluctor ring disrupts signal voltage from the crankshaft position sensors to indicate crankshaft position and crankshaft speed. These signals are used by the Ignition Control Module (ICM) during start up and passed on to the Electronic Control Module (ECM) to adjust ignition timing.

The DIS system is also integrated with the ESC system which uses a knock sensor in connection with the Electronic Control Module (ECM) to control spark timing. The KS system allows the engine to have maximum spark advance without spark knock which also improves driveability and fuel economy.

Secondary (spark plug) wiring

The secondary (spark plug) wires are a carbon-impregnated cord conductor encased in an 8 mm (5/16-inch) diameter rubber jacket with an outer silicone covering. This type of wire will withstand very high temperatures and provides an excellent insulator for the high secondary ignition voltage. Silicone spark plug boots form a tight seal on the plug. The boot should be twisted 1/2-turn before removing (for more information on spark plug wiring refer to Chapter 1).

---

6   Ignition system - check

Refer to illustrations 6.1, 6.5 and 6.7
Warning: *Because of the very high voltage generated by the ignition system, extreme care should be taken whenever an operation is performed involving ignition components. This not only includes the coils, control module and spark plug wires, but related items connected to the system as well, such as the plug connections, tachometer and any test equipment.*
1   If the engine turns over , but won't start, check for spark at the spark plug by installing a calibrated ignition system tester to one of the spark plug wires **(see illustration)**. **Note:** *The tool is available at most auto parts stores. Be sure to get the correct tool for your particular ignition system.*
2   Connect the clip on the tester to a ground such as a metal bracket or valve cover bolt, crank the engine and watch the end of the tester for bright blue, well defined sparks.
3   If sparks occur, sufficient voltage is reaching the plugs to fire the engine. However the plugs themselves may be fouled, so remove and check them as described in Chapter 1 or replace them with new ones.

Cadillac models

4   If no sparks occur, the distributor cap, rotor or spark plug wires may be defective. Remove the distributor cap and check the cap, rotor and spark plug wires as described in Chapter 1. Replace defective parts as necessary. If moisture is present in the distributor cap, use WD-40 or something similar to dry out the cap and rotor, then reinstall the cap and repeat the spark test at the spark plug wire.
5   If no sparks occur, check the primary wire connections at the coil to make sure they are clean and tight. Check for voltage to the ignition coil **(see illustration)**. Battery voltage should be available to the ignition coil with the ignition switch ON. **Note:** *If the reading is 7 volts or less, repair the primary circuit from the ignition switch to the ignition coil.*
6   If there's still no spark, check the ignition module and the ignition coil or other internal components that may be defective (see Section 8).

Buick and Oldsmobile models

7   If no sparks occur, check the primary wire connections at the coil to make sure they are clean and tight. Refer to the wiring diagrams (see Chapter 12) and check for battery voltage to the ignition coil from the IGN fuse **(see illustration)**. Battery voltage should be available to the ignition coil with the ignition switch ON. **Note:** *If no voltage is available check the ignition system fuses (see Chapter 12). If the reading is 7 volts or less, repair the primary circuit from the ignition switch to the ignition coil.*
8   If voltage is available at the ignition coil and there's still no spark, check the ignition module and the ignition coil (see Section 7).

---

7   Distributorless Ignition System (DIS) - component check and replacement

Component checks

Ignition module

Refer to illustration 7.2
1   Disconnect the ignition module harness connector and verify that battery voltage is available to the ignition control module as described in section 6.
2   Next, check for a trigger signal from the ignition module. Remove the coil pack from the ignition module to expose the module terminals. Connect a test light between each of the module terminals and have an assistant crank the engine over **(see illustration)**. This test checks the triggering circuit in the ignition module. A blinking test light indicates the module is triggering. The test light should blink quickly and constantly as each coil

7.2 Remove the coil pack from the module assembly and check for a trigger signal on the module terminals with a test light while an assistant cranks the engine over

pack is triggered to fire by the switching signal from the ignition module.

3   If there is no blinking light, the ignition module is most likely the problem but not always. The ECM should also relay reference signals to fire the fuel injectors if all the infor-

mation is being received from the camshaft and crankshaft sensors. It is best to install a SCAN tool and monitor the ECM activity to differentiate the ignition system problems (see Chapter 6).

4   A slowly blinking light, at this point, indicates the ECM is not seeing a crank sensor signal (see Chapter 6). At this point, the problem is in the camshaft or crankshaft sensor(s), sensor circuits or the ignition module. **Note:** *Refer to Chapter 6 for additional information and testing procedures for the camshaft sensor and the crankshaft sensors. It will be necessary to verify that the crankshaft sensors and camshaft sensor is operating correctly before changing the ignition module. A defective ignition module can only be diagnosed by process of elimination.*

Ignition coil packs

Refer to illustrations 7.6 and 7.7

5   Refer to Steps 6 through 8 to determine if the coils are receiving the correct trigger signal from the ignition module.

6   Use an ohmmeter and check the secondary resistance for each coil pack **(see illustration)**. Refer to the Specifications listed at the beginning of this Chapter for the correct amount of resistance.

7   Next, remove each coil pack (see Step 13) and check the primary resistance for each individual coil pack **(see illustration)**. Refer to the Specifications listed at the beginning of this Chapter for the correct amount of resistance.

8   If the indicated resistance is less than the resistance listed in this Chapter's Specifications, replace the coil pack.

Component replacement

Ignition coils and control module

Refer to illustrations 7.12a, 7.12b, 7.13a and 7.13b

9   Detach the cable from the negative terminal of the battery.

10   Unplug the electrical connectors from the module **(see illustration 6.7)**.

11   If the plug wires are not numbered, label them and detach the plug wires at the coil assembly.

12   Remove the module/coil assembly mounting nuts and lift the assembly from the vehicle. **Note:** *On early models it will be necessary to remove the module/coil assembly mounting bracket from the engine first, after removal of the bracket the module retaining nuts can be accessed* **(see illustrations)**.

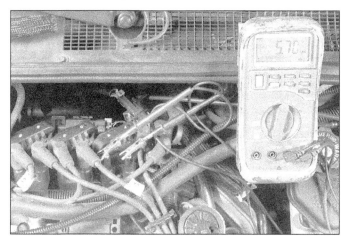

7.6 Check the coil secondary resistance by connecting the ohmmeter leads to the towers of each coil pack

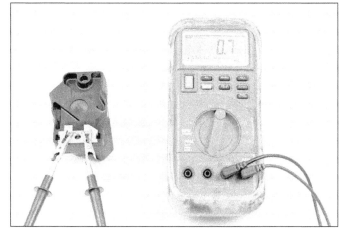

7.7 With the coil pack removed from the module assembly, check the coil primary resistance with an ohmmeter

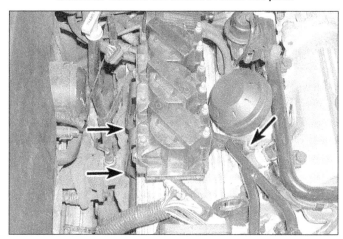

7.12a 1987 and earlier ignition module/coil pack mounting bracket bolts (arrows)

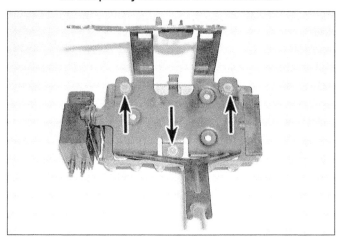

7.12b Ignition module to engine mounting bracket retaining nuts (arrows) (1987 and earlier shown, all others similar)

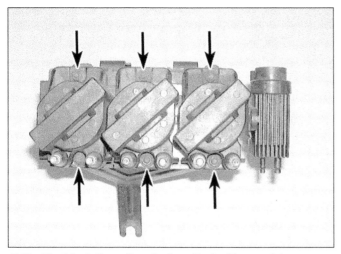

7.13a To detach the coil packs from the ignition module, remove the screws (arrows) . . .

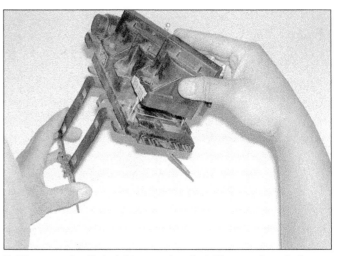

7.13b . . . and pull straight up - when installing a coil pack, line up the blade terminals and press downward - make sure it's fully seated before tightening the retaining screws

13   To remove the coils from the ignition module, simply remove the mounting screws and unplug the coil from the ignition module terminals **(see illustrations)**. **Note:** *It is not necessary to remove the ignition control module from the engine to remove the ignition coils.*
14   Installation is the reverse of removal. **Note:** *When installing the coils, make sure they are properly seated on the ignition module.*

8   High Energy Ignition (HEI) system - component check and replacement

Component checks

Ignition coil
Refer to illustrations 8.2 and 8.3
1   Remove the distributor cap.
2   Attach the leads of an ohmmeter across the two primary coil terminals as shown **(see illustration)**. The indicated resistance should be less than one ohm.
3   Using the high scale, attach one lead of the ohmmeter to the SECONDARY terminal in the middle of the distributor and the other lead to the GROUND terminal and note the resistance **(see illustration)**. Next attach the ohmmeter leads across the SECONDARY and TACH terminals and note the resistance. If both readings indicate infinite resistance, replace the coil.

Distributor pick-up coil
Refer to illustrations 8.6 and 8.7
4   Disconnect the cable from the negative battery terminal.
5   Unplug the electrical connectors, disengage the latches and remove the distributor cap and rotor.
6   To test the pick-up coil, remove the pick-up coil leads from the module. Connect the negative ohmmeter probe to the distributor body and the positive probe to the pick-up coil electrical connector **(see illustration)**. Resistance should be infinite. If continuity is indicated, there is a short to ground in the pick-up coil. Replace the pick-up coil.
7   Next, measure the pick-up coil resistance across the two pick-up coil terminals. It should be 500 to 1,500 ohms **(see illustration)**. If the resistance is incorrect, replace the pick-up coil.
8   Flex the pick-up coil leads and pull on the connectors (with the ohmmeter connected to the pick-up coil leads). Any change in resistance would indicate a broken wire in the pick-up coil, replace the pick-up coil.

Ignition module
Refer to illustration 8.10
9   Remove the cap from the distributor (see Chapter 1). Verify that the ignition coil and the distributor pick-up coil are function-

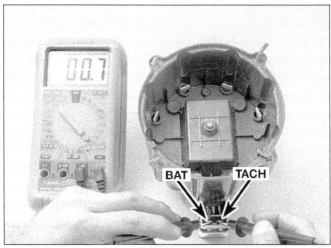

8.2 To test the HEI type coil, attach the leads of an ohmmeter to the primary terminals and verify that the indicated resistance is zero or very near zero . . .

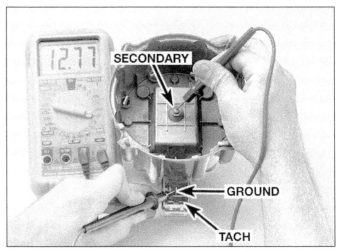

8.3 . . . then, using the high scale, attach one lead to the high tension terminal and the other to the GROUND terminal and then the TACH terminal. Verify that both of the readings are not infinite - if the indicated resistance is not as specified, replace the coil

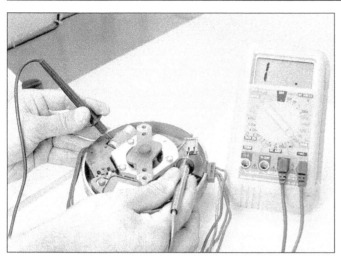

8.6 Connect the ohmmeter to the pick-up coil terminal and the distributor body. If continuity is indicated, there is a short from the pick-up coil wiring to the distributor body

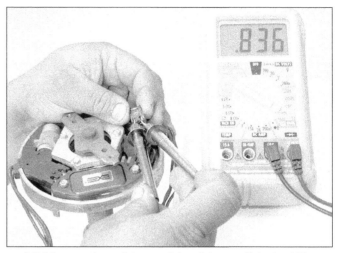

8.7 Measure the resistance of the pick-up coil. It should be between 500 and 1,500 ohms

ing properly as described above in Steps 1 through 8.

10   With the ignition coil leads connected to the distributor, check for battery voltage to

8.10 Check for battery voltage to the ignition module with the ignition key ON (engine not running)

the ignition module **(see illustration)**. **Note:** *The ignition wiring to the distributor cap is short and will not allow much movement. Use a jumper wire from the BAT terminal on the distributor cap to the pink wire.*

11   If the module is receiving power and the pick-up coil, ignition coil and secondary circuit are good, most likely the ignition module is defective. Have the module checked at a dealer service department.

Component replacement

Ignition coil

Refer to Illustrations 8.14, 8.16 and 8.17

12   Detach the cable from the negative terminal of the battery.

13   Disconnect the electrical connectors from the distributor cap.

14   Remove the coil cover screws and the cover **(see illustration)**.

15   Remove the screws securing the coil to the cap.

16   Note the position of each wire, marking them if necessary. Remove the coil ground wire, then push the terminals from the under-

side of the cap. Remove the coil from the distributor cap **(see illustration)**.

17   Installation is the reverse of the removal procedure. Be sure that the center electrode is in good condition **(see illustration)** and that the leads are connected to their original positions.

Ignition module

Refer to illustration 8.23

Note: *It's not necessary to remove the distributor from the engine to replace the ignition module.*

18   Disconnect the cable from the negative terminal of the battery.

19   Remove the air cleaner to provide room to work around the distributor.

20   Remove the distributor cap and wires as an assembly and position them out of the way (see Chapter 1).

21   Remove the rotor.

22   Carefully detach the electrical connectors from the module terminals.

23   Remove the screws and lift out the module **(see illustration)**.

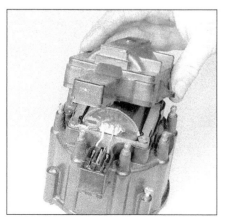

8.14 To access the HEI ignition coil, remove the coil cover screws and the cover

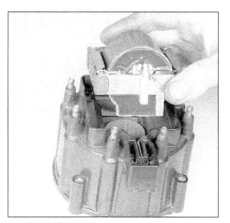

8.16 To separate the coil from the distributor cap, clearly mark the wires, detach the coil ground wire and push the terminals from the underside of the cap

8.17 Before installing a new coil, make sure that the center electrode is in good condition

24   Installation is the reverse of removal. Be sure to apply the silicone dielectric grease supplied with the new module to the distributor base - DO NOT use any other type of grease! If the grease isn't used, the module will overheat and destroy itself.

Distributor pick-up coil

Refer to illustrations 8.27 and 8.28

25   Remove the distributor (see Section 9).
26   Place the distributor in a vise, using blocks of wood to protect it.
27   Mark the relative positions of the gear and shaft. Drive the roll pin out **(see illustration)**. Remove the gear and pull the shaft from the distributor housing.
28   Remove the aluminum shield (if equipped) for access to the pick-up coil. Carefully pry the retaining clip from the distributor shaft and remove the pick-up coil and pole piece assembly **(see illustration)**.
29   Reassembly is the reverse of the disassembly procedure. After reassembly, spin the distributor shaft to make sure there is no contact between the pick-up coil and the pick-up teeth. Loosen and retighten the teeth to eliminate the contact, if necessary.

---

9   Distributor (Cadillac models) - removal and installation

Removal

Refer to illustrations 9.5 and 9.6

1   Position the engine at number one Top Dead Center (see Chapter 2). Disconnect the cable from the negative terminal of the battery.
2   Remove the air cleaner assembly (see Chapter 4).
3   Disconnect the electrical connectors from the distributor and the coil.
4   Remove the distributor cap (see Chapter 1).
5   Make a mark on the distributor body directly in-line with the tip of the rotor **(see illustration)**.
6   Mark the position of the distributor body in relation to the engine **(see illustration)**.
7   Remove the distributor hold-down bolt

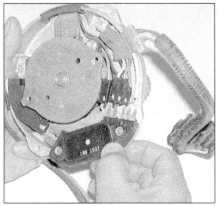

8.23 To remove the ignition module, remove the two mounting screws, unplug the connectors and lift the module out

and clamp.
8   Remove the distributor from the engine. Do not rotate the engine with the distributor removed, or the mark you made in Step 5 will be useless.

Installation

Crankshaft not turned after distributor removal

9   Position the rotor in the exact location it was in when the distributor was removed.
10   Lower the distributor into the engine. To mesh the gears at the bottom of the distributor it may be necessary to turn the rotor slightly. It is possible that the distributor may not seat down fully against the block because the lower part of the distributor shaft has not properly engaged the oil pump shaft. Make sure the distributor and rotor are properly aligned with the marks made earlier, then use a large socket and breaker bar on the crankshaft bolt to turn the engine in the normal direction of rotation until the two shafts engage and the distributor drops down against the block.
11   With the base of the distributor seated against the engine block turn the distributor housing to align the marks made on the distributor base and the engine block. Make sure the rotor is pointing to its mark.

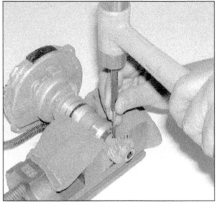

8.27 To remove the distributor shaft, drive the roll pin out of the shaft with a hammer and punch, remove the gear, shim and tanged washer from the shaft, inspect the shaft for any burrs that might prevent its removal, then pull it out (be careful not to lose the washer at the upper end of the shaft)

12   Place the hold-down clamp in position and loosely install the hold-down bolt.
13   Reconnect the distributor wiring harness connectors.
14   Install the distributor cap.
15   Reconnect the coil connector.
16   With the distributor in its original position, tighten the hold-down bolt.
17   Check the ignition timing (see Chapter 1).

Crankshaft turned after distributor removal

18   Remove the number one spark plug.
19   Place your finger over the spark plug hole while turning the crankshaft with a wrench on the pulley bolt at the front of the engine.
20   When you feel compression, continue turning the crankshaft slowly until the timing mark on the vibration damper is aligned with the "0" on the engine timing indicator.
21   Position the rotor between the number one and eight spark plug terminals on the cap.

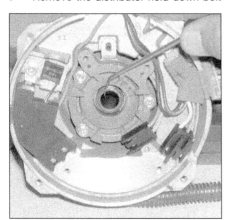

8.28 The pick-up coil and pole piece assembly can be removed after carefully prying out the C-clip

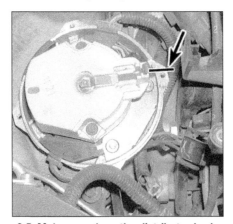

9.5 Make a mark on the distributor body (arrow) to show the direction the rotor is pointing before removing the distributor

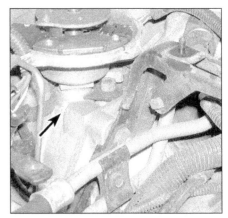

9.6 Mark the position of the distributor in relation to the engine (arrow) before loosening the distributor hold-down clamp

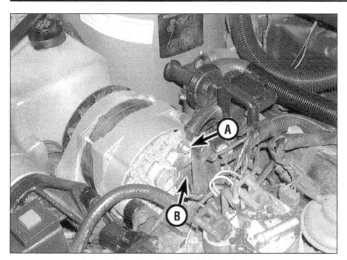

11.3 Alternator terminal identification

A B+ terminal
B L terminal

11.5 To measure charging voltage, attach the voltmeter leads to the battery terminals, start the engine and record the voltage reading

22   Lower the distributor into the engine. To mesh the gears at the bottom of the distributor, it may be necessary to turn the rotor slightly. If the distributor does not drop down flush against the block it is because the distributor shaft has not mated to the oil pump shaft. Place a large socket and breaker bar on the crankshaft bolt and turn the engine over slightly in the normal direction of rotation until the two shafts engage properly, allowing the distributor to seat flush against the block.

23   With the base of the distributor properly seated against the engine block, turn the distributor housing to align the marks made on the distributor base and the engine block.

24   Place the hold-down clamp in position and loosely install the hold-down bolt.

25   Reconnect the distributor wiring harness connectors.

26   Install the distributor cap. If the secondary wiring harness was removed from the cap, reinstall it.

27   Reconnect the coil connector.

28   With the distributor in its original position, tighten the hold-down bolt and check the ignition timing.

10 Charging system - general information and precautions

The charging system consists of a belt-driven alternator with an integral voltage regulator and the battery. These components work together to supply electrical power for the ignition system, the lights and all accessories.

All models are equipped with either a CS-144 (120 amp) or a CS-144 GEN II (124 amp) alternator. All engines use a conventional pulley and fan. CS-144 alternators can be rebuilt but it is recommended that the home mechanic exchange the alternator for a rebuilt unit. Because of the expense and the limited availability of parts, no alternator overhaul information is included in this manual.

The purpose of the voltage regulator is to limit the alternator's voltage to a preset value. This prevents power surges, circuit overloads, etc., during peak voltage output. On all models with which this manual is concerned, the voltage regulator is contained within the alternator housing.

The charging system does not ordinarily require periodic maintenance. The drivebelt, electrical wiring and connections should, however, be inspected at the intervals suggested in Chapter 1.

Take extreme care when making circuit connections to a vehicle equipped with an alternator and note the following. When making connections to the alternator from a battery, always match correct polarity. Before using arc welding equipment to repair any part of the vehicle, disconnect the wires from the alternator and the battery terminals. Never start the engine with a battery charger connected. Always disconnect both battery leads before using a battery charger.

The charging indicator light on the dash lights up when the ignition switch is turned on and goes out when the engine starts. If the light stays on or comes on once the engine is running, a charging system problem has occurred. See Section 11 for the proper diagnosis procedure of the alternator.

11 Charging system - check

Refer to illustrations 11.3 and 11.5

1   If a malfunction occurs in the charging circuit, do not immediately assume that the alternator is causing the problem.

2   First, check the following items:

  a) *Make sure the battery cable connections at the battery are clean and tight.*

  b) *The battery electrolyte specific gravity (if possible). If it is low, charge the battery.*

  c) *Check the external alternator wiring and connections. They must be in good condition.*

  d) *Check the drivebelt condition and tension (see Chapter 1).*

  e) *Make sure the alternator mounting bolts are tight.*

  f) *Run the engine and check the alternator for abnormal noise (may be caused by a loose drive pulley, loose mounting bolts, worn or dirty bearings, defective diode or defective stator).*

3   Check the charge light bulb and circuit as follows:

  a) *With the ignition key ON and the engine running, the lamp should be OFF. If it remains ON while running, stop the engine and remove the lead from the L terminal on the alternator* **(see illustration)**.

  b) *If the lamp on the dash goes OFF, the alternator is defective.*

  c) *If the lamp on the dash remains ON, there is a grounded L terminal in the wiring harness*

4   Using a voltmeter, check the battery voltage with the engine off. It should be approximately 12-volts.

5   Check the charging voltage with the engine running. Start the engine and raise the engine speed to 1500 rpm. It should now be approximately 14 to 15-volts **(see illustration)**.

6   Load the battery and observe the charging voltage. Turn on the high beam headlights, the A/C blower on HIGH, the windshield wipers and the radio. The voltage should drop and then come back up as each accessory is selected. If the charging system is working properly the voltage should stay above 13.5 volts. If the voltage drops below 13 volts, the charging system is weak or defective.

7   Lower the engine rpm back to idle and apply the brakes while observing the charging voltage. The charging voltage should not drop below 13 volts with the decrease in engine rpm. **Note:** *Some smaller amperage alternators may drop below 13 volts but, if they are in good condition, they will regulate the charging voltage to normal.*

12.4a Alternator mounting bolts (arrows) (Buick and Oldsmobile models)

12.4b Alternator mounting bolts (arrows) (1988 and earlier Cadillac models)

8   Turn off all the electrical loads (high beam headlights, the A/C blower, the windshield wipers and the radio), run the engine at 1500 rpm and watch the charging voltage rise. It should not rise above 15 volts. **Note:** *Cold temperatures will cause the voltage readings to increase slightly while hot temperatures will lower the charging system voltage readings.*

9   If the charging voltage does not exhibit distinct changes when engine rpm increases and accessory loads are added, the voltage regulator is defective. If the charging voltages are low and the drivebelts and battery are all in good condition, the alternator is defective. In this situation, replace the alternator and voltage regulator as a single unit.

12   Alternator - removal and installation

Refer to illustrations 12.4a, 12.4b, 12.4c and 12.5
1   Detach the cable from the negative terminal of the battery.

2   Remove the serpentine drivebelt (see Chapter 1).
3   Loosen the BAT+ terminal retaining nut **(see illustration 11.3)**.
4   Remove the alternator mounting bolts **(see illustrations)**.
5   Clearly label, if necessary, then remove the electrical connectors from the rear of the alternator **(see illustration)**.
6   Remove the alternator from the engine.
7   Installation is the reverse of removal.

13   Starting system - general information and precautions

The sole function of the starting system is to turn over the engine quickly enough to allow it to start.

The starting system consists of the battery, the starter motor, the starter solenoid and the wires connecting them. The solenoid is mounted directly on the starter motor.

The solenoid/starter motor assembly is installed on the upper part of the engine, next to the transaxle bellhousing.

When the ignition key is turned to the Start position, the starter solenoid is actuated through the starter control circuit. The starter solenoid then connects the battery to the starter. The battery supplies the electrical energy to the starter motor, which does the actual work of cranking the engine.

The electrical circuitry of the vehicle is arranged so that the starter motor can only be operated when the transmission selector lever is in Park or Neutral.

Always observe the following precautions when working on the starting system:

a) *Excessive cranking of the starter motor can overheat it and cause serious damage. Never operate the starter motor for more than 15 seconds at a time without pausing to allow it to cool for at least two minutes.*
b) *The starter is connected directly to the battery and could arc or cause a fire if mishandled, overloaded or shorted out.*
c) *Always detach the cable from the negative terminal of the battery before working on the starting system.*

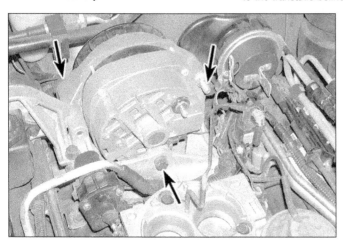

12.4c Alternator mounting bolts (arrows) (1989 and later Cadillac models)

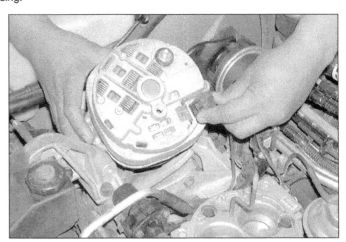

12.5 On most models it will be necessary to remove the alternator mounting bolts first, then rotate the rear of the alternator upward to access the electrical connections

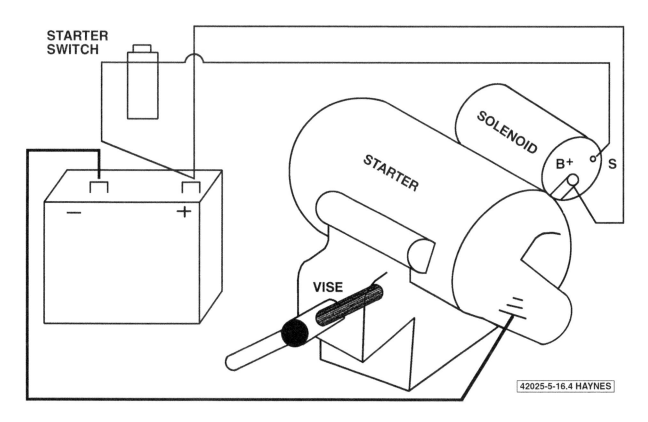

STARTER
SWITCH

SOLENOID

STARTER

B+   S

VISE

42025-5-16.4 HAYNES

14.4 Starter motor bench testing details

14   Starter motor and circuit - check

Refer to illustration 14.4

1   If a malfunction occurs in the starting circuit, do not immediately assume that the starter is causing the problem. First, check the following items:

a) *Make sure the battery cable clamps, where they connect to the battery, are clean and tight.*

b) *Check the condition of the battery cables (see Section 4). Replace any defective battery cables with new parts.*

c) *Test the condition of the battery (see Section 3). If it does not pass all the tests, replace it with a new battery.*

d) *Check the starter solenoid wiring and connections. Refer to Chapter 12 wiring diagrams.*

e) *Check the starter mounting bolts for tightness.*

f) *Check the fusible links (if equipped) exiting the engine compartment fuse box (see Chapter 12). If they're burned, determine the cause and repair the circuit. Also, check the ignition switch circuit for correct operation (see Chapter 12).*

g) *Check the operation of the Park/Neutral position switch. Refer to Chapter 7 for the Park/Neutral position switch check and adjustment procedure. These systems must operate correctly to provide battery voltage to the starter solenoid.*

2   If the starter does not activate when the ignition switch is turned to the start position, check for battery voltage to the solenoid. This will determine if the solenoid is receiving the correct voltage signal from the ignition switch. Install a voltmeter to the starter solenoid positive terminal while an assistant turns the ignition switch to the start position, observe the voltmeter. It should be approximately 12.6 volts. If voltage is not available, refer to the wiring diagrams in Chapter 12 and check all the fuses and relays in series with the starting system. If voltage is available but there is not movement from the starter motor, remove the starter from the engine (see Section 15) and bench test the starter (see Step 4).

3   If the starter turns over slowly, check the starter cranking voltage and the current draw from the battery. This test must be performed with the starter assembly on the engine. Crank the engine over (for 10 seconds or less) and observe the battery voltage. It should not drop below 8.5 volts. Also, observe the current draw using an amp meter. Typically a starter should not exceed 300 amps. If the starter motor exceeds these values, have it tested by a dealer service department or other qualified repair shop. There are several conditions that may affect the starter cranking potential. The battery must be in good condition and the battery cold-cranking rating must not be under-rated for the particular application. Be sure to check the battery specifications carefully. The battery terminals and cables must be clean and not corroded. Also, in cases of extreme cold temperatures, make sure the battery and/or engine block is warmed before performing the tests.

4   If the starter is receiving voltage but does not activate, remove and check the starter/solenoid assembly on the bench. Most likely the solenoid is defective. In some rare cases, the engine may be seized so be sure to try and rotate the crankshaft pulley (see Chapter 2A or 2B) before proceeding. With the starter/solenoid assembly mounted in a vise on the bench, install one jumper cable from the negative terminal (-) to the body of the starter **(see illustration)**. Install another jumper cable from the positive terminal (+) on the battery to the B+ terminal on the starter. Install a starter switch and apply battery voltage to the solenoid S terminal (for 10 seconds or less) and observe the solenoid plunger, shift lever and overrunning clutch extend and rotate the pinion drive. If the pinion drive extends but does not rotate, the solenoid is operating but the starter motor is defective. If there is no movement but the solenoid clicks, the solenoid and/or the starter motor is defective. If the solenoid plunger extends and rotates the pinion drive, the starter/solenoid assembly is working properly.

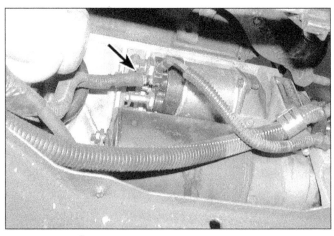

**15.4 Remove the starter electrical connections
on the solenoid (arrow)**

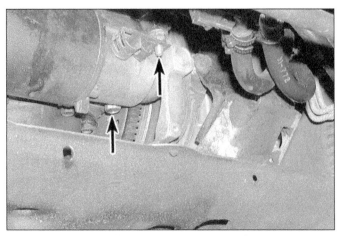

**15.5a Remove the starter mounting
bolts (arrows)**

15 Starter motor - removal and installation

Removal and installation

Refer to illustrations 15.4, 15.5a and 15.5b

1   Disconnect the cable from the negative terminal of the battery.
2   Raise the front of the vehicle and support it securely on jackstands.
3   From under the vehicle, remove the flexplate inspection cover and the starter motor shield (if equipped). **Note:** *On some Cadillac models it may be necessary to remove the crossover pipe from the exhaust system to allow removal of the starter* (see Chapter 4).
4   Disconnect the solenoid wire and battery cable from the terminals on the solenoid **(see illustration)**.
5   Remove the starter motor bolts **(see illustrations)**.
6   Remove the starter motor. Note the location of the spacer shim(s), if equipped.
7   Installation is the reverse of removal. Be sure to install the spacer shim(s) in exactly the same location, if equipped (see Steps 8 through 13 for the correct shim adjustment procedure).

Shim adjustment procedure

8   Remove the flywheel housing cover (see Chapter 7).
9   Inspect the flywheel or driveplate for signs of unusual wear such as chipped or missing gear teeth.
10   Using a wire type feeler gauge (round diameter), measure the clearance between the top of the flywheel ring gear tooth and the bottom of the pinion tooth on the starter. Normal clearance should be 0.010 to 0.060 inch.
11   If the starter clearance is less than 0.020 inches and the starter whined after engagement, the starter is too close to the flywheel. Add a 0.040 shim to gain clearance. Do not use more than two shims.
12   If the starter clearance is more than 0.060 inches and the starter whines during engagement then the starter is far from the flywheel. Remove a 0.040 shim to gain clearance. Do not remove more than two shims.

13   To install a shim, loosen the inside bolt, remove the outer bolt and slide the shim between the engine and the starter without removing the starter from the engine block.

16 Starter motor solenoid - removal and installation

Refer to illustration 16.3

1   Remove the starter unit (see Section 15).
2   Disconnect the connector strap from the starter motor terminal. Remove the two screws which secure the solenoid housing to the end-frame assembly.
3   Twist the solenoid in a clockwise direction to disengage the flange key and then withdraw the solenoid **(see illustration)**.
4   Install the solenoid to the starter motor by first checking that the return spring is in position on the plunger and then insert the solenoid body into the drive housing and turn the body counterclockwise to engage the flange key.
5   Install the two solenoid securing screws and connect the starter motor connector strap.

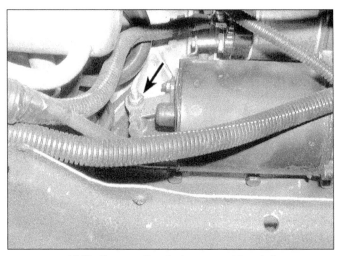

**15.5b Remove the starter support bracket
retaining nut (arrow) (if equipped)**

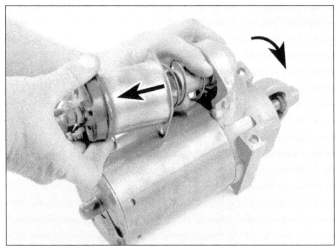

**16.3 To disengage the solenoid from the starter,
turn it in a clockwise direction**

Chapter 6
Emissions and engine control systems

Contents

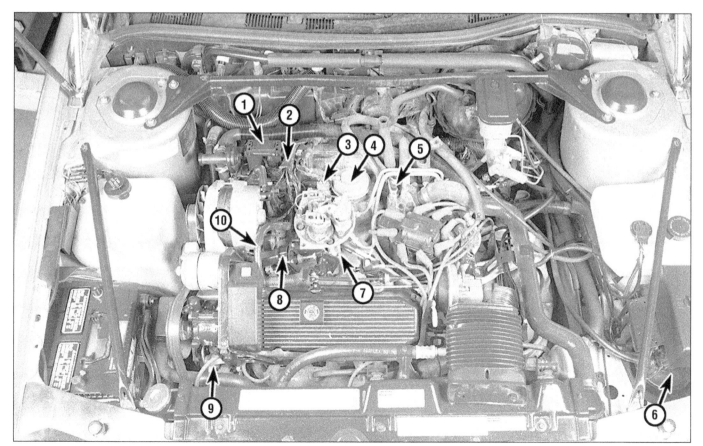

1.1a Typical emissions and engine control system components (4.1L engine)

1	Manifold Absolute Pressure (MAP) sensor
2	Intake Air Temperature (IAT) sensor
3	Throttle Position Sensor (TPS)
4	Exhaust Gas Recirculation (EGR) valve
5	Engine Coolant Temperature (ECT) sensor

6	Evaporative emissions (EVAP) canister
7	Early Fuel Evaporation (EFE) heater (under throttle body)
8	Idle Speed Control (ISC) motor
9	Air Injection Reactor (AIR) pump (not visible)
10	PCV valve and hose

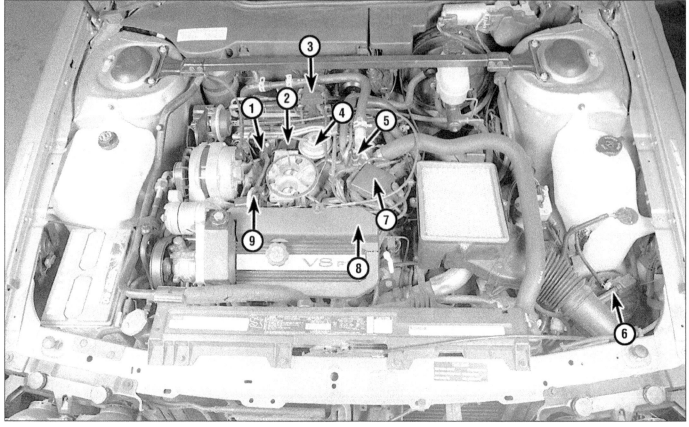

1.1b Typical emissions and engine control system components (4.9L engine)

1 Intake Air Temperature (IAT) sensor	6 Evaporative emissions (EVAP) canister
2 Throttle Position Sensor (TPS)	7 Camshaft Position (CMP) sensor (inside distributor)
3 Manifold Absolute Pressure (MAP) sensor	8 PCV valve (under valve cover shroud)
4 Exhaust Gas Recirculation (EGR) valve	9 Idle Speed Control (ISC) motor
5 Engine Coolant Temperature (ECT) sensor	

1   General information

Refer to illustrations 1.1a, 1.1b, 1.1c and 1.7

To prevent pollution of the atmosphere from incompletely burned and evaporating gases, and to maintain good driveability and fuel economy, a number of emission control systems are incorporated **(see illustrations)**. They include the:

On-Board Diagnostic (OBD) system
Electronic Control Module (ECM)
Electronic Spark Timing (EST)
Electronic Spark Control (ESC)
Positive Crankcase Ventilation (PCV)
  system
Exhaust Gas Recirculation (EGR) system
Evaporation Control System (EVAP)/
  Evaporative Emissions Control System
  (EECS)
Air Injection Reactor (AIR) system
Early Fuel Evaporation (EFE) system
Thermostatic Air Cleaner (THERMAC)
Catalytic converter

All of these systems are linked, directly or indirectly, to the emission control system.

The Sections in this Chapter include general descriptions, checking procedures within the scope of the home mechanic and component replacement procedures (when possible) for each of the systems listed above.

Before assuming that an emissions control system is malfunctioning, check the fuel and ignition systems carefully. The diagnosis of some emission control devices requires specialized tools, equipment and training. If checking and servicing become too difficult or if a procedure is beyond your ability, consult a dealer service department. Remember, the most frequent cause of emissions problems is simply a loose or broken vacuum hose or wire, so always check the hose and wiring connections first.

This doesn't mean, however, that emission control systems are particularly difficult to maintain and repair. You can quickly and easily perform many checks and do most of the regular maintenance at home with common tune-up and hand tools. **Note:** *Because of a Federally mandated extended warranty which covers the emission control system components, check with your dealer about warranty coverage before working on any emissions-related systems. Once the warranty has expired, you may wish to perform some of the component checks and/or*

replacement procedures in this Chapter to save money.

Pay close attention to any special precautions outlined in this Chapter. It should be noted that the illustrations of the various systems may not exactly match the system installed on the vehicle you're working on because of changes made by the manufacturer during production or from year-to-year.

A Vehicle Emissions Control Information (VECI) label is located in the engine compartment **(see illustration)**. This label contains important emissions specifications and adjustment information, as well as a vacuum hose schematic with emissions components identified. When servicing the engine or emissions systems, the VECI label in your particular vehicle should always be checked for up-to-date information.

2   On Board Diagnosis (OBD) system and trouble codes

Diagnostic tool information

Refer to illustrations 2.1 and 2.2

1   A digital multimeter is necessary for checking fuel injection and emission related

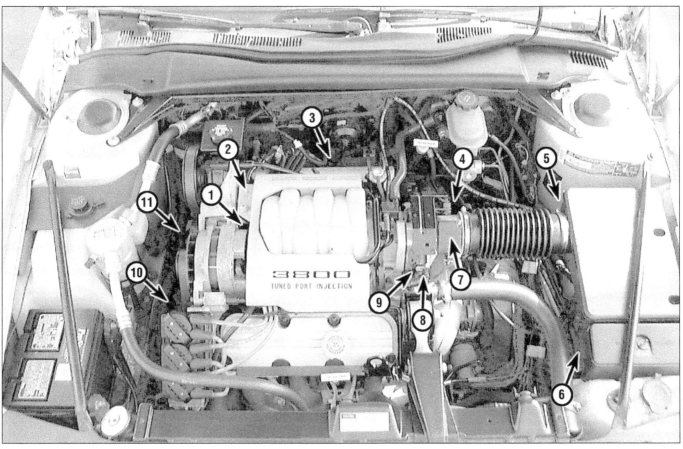

1.1c Typical emissions and engine control system components (V6 engine)

1   Knock Sensor (KS) (not visible)
2   PCV valve
3   Oxygen (O2) sensor (in rear exhaust manifold)
4   Engine Coolant Temperature (ECT) sensor (not visible)
5   Intake Air Temperature (IAT) sensor (in air cleaner housing)
6   Evaporative emissions (EVAP) canister (under fuse/relay panel)

7   Mass Airflow (MAF) sensor
8   Throttle Position Sensor (TPS)
9   Idle Air Control (IAC) valve
10 Crankshaft Position (CKP) sensor (behind harmonic balancer)
11 Camshaft Position (CMP) sensor (on front of timing cover)

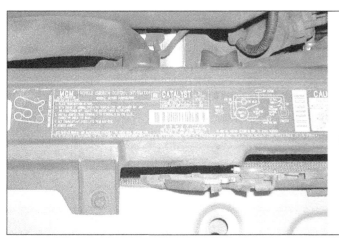

1.7 The Vehicle Emission Control Information (VECI) label is located in the engine compartment and contains information on the emission devices on your vehicle, vacuum line routing, etc.

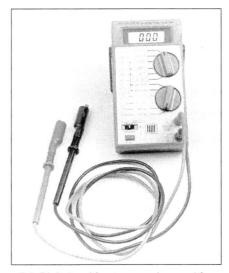

2.1 Digital multimeters can be used for testing all types of circuits; because of their high impedance, they are much more accurate than analog meters for measuring low-voltage computer circuits

components **(see illustration)**. A digital volt-ohmmeter is preferred over the older style analog multimeter for several reasons. The analog multimeter cannot display the volts-ohms or amps measurement in hundredths and thousandths increments. When working with electronic circuits which are often very low voltage, this accurate reading is most important. Another good reason for the digital multimeter is the high impedance circuit. The digital multimeter is equipped with a high resistance internal circuitry (10 million ohms). Because a voltmeter is hooked up in parallel with the circuit when testing, it is vital that none of the voltage being measured should be allowed

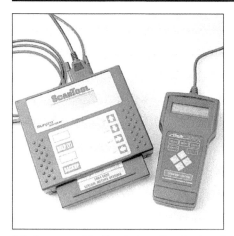

2.2 Scanners like the Actron Scantool and the AutoXray XP240 are powerful diagnostic aids - programmed with comprehensive diagnostic information, they can tell you just about anything you want to know about your engine management system

to travel the parallel path set up by the meter itself. This dilemma does not show itself when measuring larger amounts of voltage (9 to 12 volt circuits) but if you are measuring a low voltage circuit such as the oxygen sensor signal voltage, a fraction of a volt may be a significant amount when diagnosing a problem. Obtaining the diagnostic trouble codes is one exception where using an analog voltmeter is necessary.

2   Hand-held scanners are the most powerful and versatile tools for analyzing engine management systems used on later model vehicles **(see illustration)**. Early model scanners handle codes and some diagnostics for many OBD-I systems. Each brand scan tool must be examined carefully to match the year, make and model of the vehicle you are working on. Often interchangeable cartridges are available to access the particular manufacturer (Ford, GM, Chrysler, etc.). Some manufacturers will specify by continent (Asia, Europe, USA, etc.).

OBD system general description

3   All vehicles described in this manual are equipped with an On Board Diagnostic (OBD-I) system. The OBD-I system consists of several onboard computers, known as the Electronic Control Module (ECM) and the Body Control Module (BCM) and information sensors, which monitor various functions of the engine and send data to the ECM. Based on the data and the information programmed into the computer's memory, the ECM generates output signals to control various engine functions via control relays, solenoids and other output actuators.

4   The ECM is the "brain" of the electronic engine control system and is interlinked with the BCM to control various engine and body subsystems. It receives data from a number

of sensors and other electronic components (switches, relays, etc.). Based on the information it receives, the ECM generates output signals to control various relays, solenoids and other actuators. The ECM is specifically calibrated to optimize the emissions, fuel economy and driveability of the vehicle.

5   Both computers have self diagnostic capabilities that can be accessed through the Electronic Climate Control Panel or CRT display and the Data Link Connector (DLC) with the use of a scan tool.

Information sensors

6   **Oxygen sensors (O2)** - The O2 sensor generates a voltage signal that varies with the difference between the oxygen content of the exhaust and the oxygen in the surrounding air.

7   **Crankshaft position (CKP) sensor -** The CKP sensor provides a signal to the ignition control module which the ECM uses to determine crankshaft position and the engine speed. The ECM and the Ignition control module use this information to control ignition and fuel injection timing.

8   **Camshaft position (CMP) sensor -** The CMP sensor provides information on camshaft position and the speed to the ECM. The ECM uses this information to trigger the fuel injectors in the proper sequence.

9   **Engine coolant temperature (ECT) sensor -** The ECT monitors engine coolant temperature and sends the ECM a voltage signal that affects ECM control of the fuel mixture, ignition timing, and EGR operation.

10   **Intake Air Temperature (IAT) sensor -** The IAT provides the ECM with intake air temperature information. The ECM uses this information along with the MAP or MAF sensor to calculate the quantity of air flowing into the engine. This sensor may also be referred to as the Manifold Air temperature (MAT) sensor.

11   **Throttle Position Sensor (TPS) -** The TPS senses throttle movement and position, then transmits a voltage signal to the ECM. This signal enables the ECM to determine when the throttle is closed, in a cruise position, or wide open.

12   **Mass airflow (MAF) sensor -** The MAF sensor measures the molecular mass of the intake airflow entering the engine. The MAF sensor, along with the IAT sensor, provide mass airflow and air temperature information for the most precise fuel metering.

13   **Manifold Absolute Pressure (MAP) sensor -** This sensor is used to monitor intake manifold pressure and ambient barometric pressure. The ECM uses this input signal along with the IAT sensor signal to control fuel injector pulse width (On time) and ignition timing.

14   **Vehicle speed sensor (VSS) -** The vehicle speed sensor provides information to the ECM to indicate vehicle speed.

15   **Knock sensor (KS) -** The knock sensor is a piezoresistive microphone that detects

the sound of engine detonation, or "pinging". The ECM uses the input signal from the knock sensor to recognize detonation and retard spark advance to avoid engine damage.

16   **Power steering pressure (PSP) switch** - The PSP sensor is used to detect excessive line pressure in the power steering system. The ECM uses this input signal to adjust the idle speed under increased engine loads during low-speed vehicle maneuvers.

17   **Park Neutral Position (PNP) switch –** The PNP switch is mounted on the transaxle and senses the position of the gear selector as chosen by the driver. Depending on the gear selector position, current is routed through the switch contacts and back to the ECM. The ECM and BCM use this signal for a number of engine and transaxle control operations.

18   **A/C selector switch -** During air conditioning operation, the ECM controls the air conditioning clutch control relay to delay clutch engagement after the air conditioning is turned ON to allow the IAC valve to adjust the idle speed of the engine to compensate for the additional load.

Output actuators

19   **Fuel pump relay -** The fuel pump relay is activated by the ECM with the ignition switch in the Start or Run position. When the ignition switch is turned on, the relay is activated to supply initial line pressure to the system. Refer to Chapter 12 or your owner's manual for more information on relay location. For more information on fuel pump check and replacement, refer to Chapter 4.

20   **Fuel injectors -** The ECM opens the fuel injectors individually in firing order sequence. The ECM also controls the time the injector is open, called the "pulse width." The pulse width of the injector (measured in milliseconds) determines the amount of fuel delivered. For more information on the fuel delivery system and the fuel injectors, including injector replacement, refer to Chapter 4.

21   **Ignition Control Module (ICM) -** The ICM on V8 engines amplifies the current generated by the distributor pick-up coil and intermittently grounds the primary circuit to the ignition coil which generates high voltage in the secondary circuit, thus sending spark from the ignition coil to the distributor spark plug wires. The ICM on V6 engines works in conjunction with the Crankshaft Position (CKP) sensor to determine the correct ignition coil firing sequence and to fire the ignition coil(s) at the proper time. The ICM also sends a crankshaft reference signal to the ECM, which determines proper ignition timing. Refer to Chapter 5 for more information on the Ignition Control Module.

22   **Idle Air Control (IAC) valve -** The IAC valve controls the amount of air to bypass the throttle plate when the throttle valve is closed or at idle position. The IAC valve opening and the resulting airflow is controlled by the ECM.

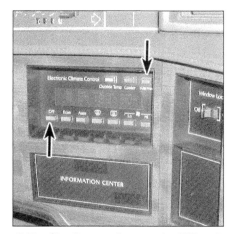

2.28a Simultaneously press the OFF and WARMER buttons on the climate control panel to enter the diagnostic service mode

23 **Idle Speed Control (ISC) motor** – The ISC motor is mounted to throttle body and is actuated by the ECM to control engine idle speed under varying conditions. The ISC motor also contains a throttle switch (ON/OFF). The position of the switch tells the ECM whether or not to send idle speed commands to the ISC motor.

24 **EGR and EVAP vacuum control solenoids** - The EGR and EVAP control solenoids are controlled by the ECM to help regulate the opening of the vacuum-operated EGR valve and to help control the purging of the EVAP system.

25 **Torque Converter Clutch (TCC) solenoid** – The TCC solenoid is located in the transaxle and is controlled by the ECM. The TCC solenoid when actuated increases line pressure to the torque converter clutch thereby reducing the converter slippage and improving fuel economy.

26 **Body Control Module (BCM)** - The BCM receives input signals from various sensors and switches such as the vehicle speed sensor, park neutral position switch to control various vehicle functions. The BCM is a separate control module from the ECM although both control modules are interlinked to share information, which gives the self diagnostic system greater control capabilities.

Obtaining and clearing OBD system codes

Refer to illustrations 2.28a, 2.28b, 2.28c and 2.29

27 The ECM will illuminate the CHECK ENGINE light (also known as the SERVICE ENGINE SOON light) on the dash if it recognizes a component fault for two consecutive drive cycles. It will continue to set the light until the codes are cleared or the ECM does not detect any malfunction for three or more consecutive drive cycles. **Note:** *Some diagnostic trouble codes will not set the CHECK ENGINE or SERVICE ENGINE SOON light, it*

ENTERING SERVICE MODE	Turn ignition key to the ON position. Simultaneously press the OFF and WARMER buttons on the electronic climate control panel for three seconds.
SEGMENT CHECK	Visually inspect that "**-1.8.8**" illuminates on the electronic climate control panel and that "**8.8.8**" illuminates on the fuel data center after the OFF and WARMER buttons are pressed for three seconds.
DIAGNOSTIC TROUBLE CODE DISPLAY (ECM TROUBLE CODES)	After the service mode has been entered all ECM diagnostic trouble codes will automatically be displayed. "**.. E**" designates the start of the ECM history codes. "**. E . E**" designates the start of the ECM current codes. Write down all trouble codes as they're being displayed, if any trouble codes are missed turn the ignition key OFF and restart the procedure.
DIAGNOSTIC TROUBLE CODE DISPLAY (BCM TROUBLE CODES)	After the ECM trouble codes, the BCM trouble codes will automatically be displayed. "**.. F**" designates the start of the BCM history codes. "**. F . F**" designates the start of the BCM current codes. Write down all trouble codes as they're being displayed, if any trouble codes are missed turn the ignition key OFF and restart the procedure.
SYSTEM READY MODE	After all diagnostic trouble codes have been displayed or if no codes are present , the fuel data center will display a "**.7.0**" this indicates that the system is ready for further instructions or testing.
CLEARING CODES	Enable to clear codes, the fuel data center must first display a system ready mode ("**.7.0**"). To clear ECM codes, simultaneously press the OFF and HI buttons on the electronic climate control panel until "**E.0.0**" is displayed. To clear BCM codes, simultaneously press the OFF and LO buttons on the electronic climate control panel until "**F.0.0**" is displayed.

2.28b Trouble code extraction process - quick reference chart (Deville and Fleetwood models)

is always a good idea to access the self diagnostic system and look for any trouble codes which may exist that have not set the CHECK ENGINE light.

28 The diagnostic codes for OBD-I systems can be extracted from the ECM using two methods. The first method requires accessing the computers self diagnostic mode with the use of the Electronic Climate Control Panel or CRT display. To extract the diagnostic trouble codes using this method proceed as follows:

Cadillac Deville and Fleetwood models

a) *Turn the ignition key ON (engine not running).* **Note:** *Before you begin the code extraction process, be sure to obtain a pad of paper and a pencil to write down any stored trouble codes as they're displayed.*

b) *Simultaneously press the OFF and WARMER buttons on the electronic climate control panel until "-1.8.8" illuminates on the climate control panel and "8.8.8" illuminates on the fuel data center* **(see illustrations 2.28a and 2.28b)**. **Note:** *If all of the panel segments do not illuminate, it will be necessary to remove the affected display panel and have it serviced at a dealer service department or other qualified repair facility. Burnt out or broken panel segments will lead to incorrect trouble code displays, thereby leading to misdiagnosis of a particular system or component.*

c) *After the segment check has been performed, "8.8.8" will be displayed on the fuel data center for approximately two more seconds, this indicates the beginning of the diagnostic trouble codes.*

d) *ECM trouble codes will now be displayed numerically from the lowest denomination to the highest on the fuel data center in two separate passes. The first pass will contain the ECM history codes and the second pass will contain the ECM current codes. All ECM trouble codes will begin with an "E" prefix.*

e) *The display ". E" designates the start of ECM history codes (first pass).*

f) *The display ". E . E" designates the start of ECM current codes (second pass).*

g) *If no ECM codes are present, the self diagnostic system will bypass the ECM code symbols (". E" or ". E . E") and begin to display the BCM codes.*

h) *After the ECM trouble codes, the self diagnostic system will now begin to display the BCM trouble codes numerically from the lowest denomination to the highest on the fuel data center in two separate passes. The first pass will contain the BCM history codes and the second pass will contain the BCM current codes. All BCM trouble codes will begin with an "F" prefix.*

i) *The display ". F" designates the start of BCM history codes (first pass).*

j) *The display ". F. F" designates the start of BCM current codes (second pass).*

k) *When all ECM and BCM codes are fin-*

*ished being displayed or if no codes are present ".7.0" will be displayed. This code indicates a system ready status and the beginning of a switch test mode for various components within the self diagnostic system. **Note:** Because of the complexity of the remaining OBD system, all further testing of the self diagnostic system is considered beyond the scope of the home mechanic and should be performed by a dealer service department or other qualified repair facility.*

l) To clear ECM trouble codes, simultaneously press the OFF and HI buttons on the electronic climate control panel until "E.0.0" is displayed.

m) To clear BCM trouble codes, simultaneously press the OFF and LO buttons on the electronic climate control panel until "F.0.0" is displayed.

n) To exit the self diagnostic system at any time, press the "AUTO" button and turn the ignition key OFF.

Cadillac Eldorado, Seville, Buick Riviera and Oldsmobile Toronado models

*a) Turn the ignition key ON (engine not running). **Note:** Before you begin the code extraction process, be sure to obtain a pad of paper and a pencil to write down any stored trouble codes as they're displayed.*

*b) Simultaneously press the OFF and WARMER buttons (OFF and TEMP UP buttons on 1990 and later Buick models) on the electronic climate control panel until "-1.8.8" illuminates on the climate control panel and all the lights illuminate on the instrument panel cluster **(see illustrations 2.28a and 2.28c). Note 1:** If all of the panel segments do not illuminate, it will be necessary to remove the affected display panel and have it serviced at a dealer service department or other qualified repair facility. Burnt out or broken panel segments will lead to incorrect trouble code displays, thereby leading to misdiagnosis of a particular system or component. **Note 2:** On vehicles equipped with a CRT display, scroll the CRT menu and select the "CLIMATE CONTROL" page, press the OFF and WARMER buttons until the CRT beeps twice or the "SERVICE MODE" page is present on the CRT screen.*

*c) After the segment check has been performed, ECM trouble codes will now be displayed numerically from the lowest denomination to the highest on the climate control panel or the instrument panel cluster. **Note:** All Cadillac models display the trouble codes on the climate control panel. All Buick and Oldsmobile models display the trouble codes on the CRT (if equipped) or the instrument panel cluster on models without a CRT.*

d) All ECM trouble codes will begin with an "EO" prefix and will be followed by the suffix "C " (current) or "H" (history). Suf-

ENTERING SERVICE MODE	Turn ignition key to the ON position. Simultaneously press the OFF and WARMER buttons on the electronic climate control panel or the climate control page on CRT equipped models for three seconds.
SEGMENT CHECK (NON-CRT MODELS)	Visually inspect that "-1.8.8" illuminates on the electronic climate control panel and all the lights on the instrument panel cluster illuminate after the OFF and WARMER buttons are pressed for three seconds.
DIAGNOSTIC TROUBLE CODE DISPLAY	After the service mode has been entered all diagnostic trouble codes will automatically be displayed. Write down all trouble codes as they're being displayed, if any trouble codes are missed turn the ignition key OFF and restart the procedure.
SYSTEM LEVEL MODE (ECM or BCM)	To select a system level, press the fan LO button to scroll to the desired system level, then press the fan HI button to enter your selection. A system level must be selected before the computer can proceed to the next menu. On CRT equipped models use the NO pad on the "SERVICE MODE" page to scroll and the YES pad to enter your selection.
CLEARING CODES	To clear codes, press the fan LO button to scroll to the "CLEAR CODES" menu, then press the fan HI button to enter your selection. A "CODES CLEAR" message should appear for approximately three seconds. On CRT equipped models use the NO pad on the "SERVICE MODE" page to scroll and the YES pad to enter your selection.

2.28c Trouble code extraction process - quick reference chart (Eldorado, Seville, Riviera and Toronado models)

fix "C" (current) after the trouble code indicates that the fault was still present the last time the ECM performed a self diagnostic check. Suffix "H" (history) after the trouble code indicates that the fault was not present the last time the ECM performed a self diagnostic check.

e) If no ECM codes are present, the self diagnostic system will display a "NO ECM CODES" message or a "NO X CODE" message.

f) After the ECM trouble codes, the self diagnostic system will now begin to display the BCM trouble codes numerically from the lowest denomination to the highest. All BCM trouble codes will begin with a "B" prefix and again will be followed by the suffix "C " (current) or "H" (history) as described in above.

g) If no BCM codes are present, the self diagnostic system will display a "NO BCM CODES" message or a "NO X CODE" message.

*h) When all ECM and BCM codes are finished being displayed the self diagnostic system will display codes for other subsystems such as the CRT, instrument panel cluster and the SIR system. **Note 1:** Because of the complexity of the remaining OBD system, All further trouble code diagnosis and system testing on the CRT, instrument panel and SIR systems is considered beyond the scope of this manual and should be performed by a dealer service department or other qualified repair facility. **Note 2:** If a "NO X DATA" message appears on the display screen at any time during the trouble code extraction process it signifies that the BCM lost communication with that particular system and cannot retrieve any codes until the communication line or circuit is fixed.*

i) To clear trouble codes, you must first select the system level to be cleared. Press the fan LO button on the climate control panel until the system to be cleared ("ECM" or "BCM") is displayed, then press the fan HI button to select (enter) the system.

*j) Second you must select the test type. Again press the fan LO button until the "CLEAR CODES" menu is displayed, then press the fan HI button to select (enter) the test type. A "CODES CLEAR" message should appear for three seconds. **Note 1:** This procedure must be performed for each system level to be cleared. Example: clearing both the ECM and the BCM codes requires*

2.29 The Data Link Connector (DLC) (arrow) is located to the left or right of the steering column depending on the model year of the vehicle

performing this procedure twice, once for the ECM system codes and a second time for the BCM system codes. **Note 2:** *On models equipped with a CRT display it will be necessary to press the NO button on the "SERVICE MODE" page to scroll through the system levels and the test types and then press the YES button to select (enter) the system levels and "CLEAR CODES" menu.*

29   The second method uses a special SCAN tool that is programmed to interface with the OBD-I system by plugging into the DLC **(see illustration)**. When used, the SCAN tool has the ability to diagnose in-depth driveability problems and it allows data to be retrieved from the ECM stored memory.

If the tool is not available and intermittent driveability problems exist, have the vehicle checked at a dealer service department or other qualified repair shop.

All models

30   Always clear the codes from the ECM before a new electronic emission control component is installed onto the engine. The ECM will often store trouble codes during sensor malfunctions. The ECM will also record new trouble codes if a new sensor is allowed to operate before the parameters from the old sensor have been erased. Clearing the codes will allow the computer to relearn the new operating parameters relayed by the new component. During the

computer relearning process, the engine may experience a rough idle or slight driveability changes. This period of time, however, should last no longer than 15 to 20 minutes. **Caution:** *Do not disconnect the battery from the vehicle to clear the codes. This will erase stored operating parameters from the memory and cause the engine to run rough for a period of time while the computer relearns the information. If necessary, have the codes cleared by a dealer service department or other qualified repair facility.* **Note:** *If using a OBD-I SCAN tool, scroll the menu for the function that describes "CLEARING CODES" and follow the prescribed method for that particular SCAN tool.*

ECM Trouble codes

Trouble code	Code identification
12	Distributor signal fault
13	Oxygen (02) sensor or circuit fault (right 02 sensor on 1990 and later Cadillac models)
14	Engine Coolant Temperature (ECT) sensor circuit fault (short)
15	Engine Coolant Temperature (ECT) sensor circuit fault (open)
16	Alternator voltage out of range
17 (Cadillac models)	Left Oxygen (02) sensor or circuit fault (1990 and later models)
17 (Buick and Oldsmobile models)	Spark reference signal fault
18	Starter solenoid signal to ECM circuit fault (open)
19	Fuel pump circuit fault (short)
20	Fuel pump circuit fault (open)
21	Throttle Position Sensor (TPS) circuit fault (short)
22	Throttle Position Sensor (TPS) circuit fault (open)
23 (Cadillac models)	Electronic Spark timing (EST) signal fault
23 (Buick and Oldsmobile models)	Intake Air Temperature (IAT) sensor circuit fault (open)
24	Vehicle Speed Sensor (VSS) circuit fault
25	Intake Air Temperature (IAT) sensor circuit fault (short)
26 (Cadillac models)	Idle Speed Control (ISC) motor/throttle switch circuit fault (short)
26 (Buick and Oldsmobile models)	ECM failure
27 (Cadillac models)	Idle Speed Control (ISC) motor/throttle switch circuit fault (open)
27 (Buick and Oldsmobile models)	Second gear switch or circuit fault
28 (Cadillac models)	Third or fourth gear switch fault (short)
28 (Buick and Oldsmobile models)	Third gear switch or circuit fault
29	Fourth gear switch or circuit fault
30	Idle Speed Control (ISC) motor rpm out of range
31 (Cadillac models)	Manifold Absolute Pressure (MAP) sensor circuit fault (short)
31 (Buick and Oldsmobile models)	Park Neutral Position (PNP) switch or circuit fault
32 (Cadillac models)	Manifold Absolute Pressure (MAP) sensor circuit fault (open)
32 (Buick and Oldsmobile models)	Exhaust Gas Recirculation (EGR) vacuum control system fault
33	Mass Airflow (MAF) sensor or circuit fault
34 (Cadillac models)	Manifold Absolute Pressure (MAP) sensor signal out of range
34 (Buick and Oldsmobile models)	Mass Airflow (MAF) sensor or circuit fault
36	Transaxle shift control solenoid fault
37	Intake Air Temperature (IAT) sensor circuit fault (short)
38	Intake Air Temperature (IAT) sensor circuit fault (open)
38 (1988 and later Buick and Oldsmobile models)	Brake light switch circuit fault
39	Torque Converter Clutch (TCC) engagement fault
40	Power Steering Pressure (PSP) switch fault (open)
41	Camshaft Position (CMP) sensor or circuit fault
42 (Cadillac models)	Left Oxygen (02) sensor signal (lean) (1990 and later models)
42 (Buick and Oldsmobile models)	Electronic Spark timing (EST) circuit fault

ECM Trouble codes (continued)

Trouble code	Code identification
43 (Cadillac models)	Left Oxygen (02) sensor signal (rich) (1990 and later models)
43 (Buick and Oldsmobile models)	Electronic Spark Control (ESC) circuit fault
44	Oxygen (02) sensor signal (lean) (right 02 sensor on 1990 and later Cadillac models)
45	Oxygen (02) sensor signal (rich) (right 02 sensor on 1990 and later Cadillac models)
46 (Cadillac models)	Fuel imbalance fault (left to right bank)
46 (Buick and Oldsmobile models)	Power Steering Pressure (PSP) switch fault (open)
47	Body Control Module (BCM) to Electronic Control Module (ECM) data fault
48 (Cadillac models)	Exhaust Gas Recirculation (EGR) system fault
48 (Buick and Oldsmobile models)	Misfire detected
49	Air Injection Reaction (AIR) system fault
51	MEM-CAL or PROM fault
52 (Cadillac models)	Electronic Control Module (ECM) memory reset
52 (Buick and Oldsmobile models)	Calpak fault
53	Distributor signal fault (interrupted)
55 (Cadillac models)	Throttle Position Sensor (TPS) out of adjustment
55 (Buick and Oldsmobile models)	Electronic Control Module (ECM) fault
58	Personal Automotive Security (PASS-KEY) or Vehicle Anti Theft (VAT) system fuel enable circuit fault
59	Torque Converter Clutch (TCC) temperature sensor circuit fault
60	Cruise control system engaged with transaxle in park or neutral
61	Cruise control system fault (vent control solenoid or circuit)
62	Cruise control system fault (vacuum control solenoid or circuit)
63	Cruise control system fault (vehicle speed increased more than 20 MPH over the set speed)
63 (1990 and earlier Buick and Oldsmobile)	EGR system fault
64	Cruise control system fault (vehicle speed increased more than 4 MPH in a .25 second
64 (1990 and earlier Buick and Oldsmobile)	EGR system fault
65	Cruise control system fault (servo position sensor or circuit)
65 (1990 and earlier Buick and Oldsmobile)	EGR system fault
66	Cruise control system fault (engine RPM increased to 4800 RPM or greater for longer than a .25 second)
67	Cruise control system fault (set/resume switch short to power)
68	Cruise control system command fault (servo position sensor stroke out of range)
70	Throttle Position Sensor (TPS) signal fault (intermittent)
71	Manifold Absolute Pressure (MAP) sensor signal fault (intermittent)
73	Engine Coolant Temperature (ECT) sensor signal fault (intermittent)
74	Intake Air Temperature (IAT) sensor signal fault (intermittent)
75	Vehicle Speed Sensor (VSS) signal fault (intermittent)
80	Fuel injection system fault (rich)
85	Throttle body service required (throttle valve not closing)
90	Torque Converter Clutch (TCC) brake switch signal fault
91	Park Neutral Position (PNP) switch or circuit fault
92	Heated windshield circuit fault
96	Torque converter overstress fault (driver induced symptom caused by applying the brake and the accelerator simultaneously)
97	Transaxle shift fault (driver induced symptom caused by shifting the transaxle into gear from park or neutral when engine speed is greater than 2000 RPM)
98	Transaxle shift fault (driver induced symptom caused by shifting the transaxle into gear from park or neutral when engine speed is between 1200 and 2000 RPM)
99	Cruise control system engagement fault (servo motor still engaged when cruise control system is off)

Note: *All models covered by this manual will display an "E" or"E0" prefix before the ECM trouble code number. Trouble code numbers and definitions are the same for all models except where specified in the code list above. Examples: Code 13 on a Cadillac model has the same definition as code 13 on a Buick or Oldsmobile model. Code 42 on the other hand has a different definition for Cadillac models than it does for Buick and Oldsmobile models.*

BCM Trouble codes

Cadillac Deville and Fleetwood models

Trouble code	Code identification
10	Outside temperature sensor or circuit fault
11	Air conditioning temperature sensor or circuit fault (high side)
12	Air conditioning temperature sensor or circuit fault (low side)
13	In-car temperature sensor or circuit fault
15	Sunload temperature sensor or circuit fault
30	Climate control panel to Body Control Module (BCM) fault (loss of communication)
31	Fuel data center to Body Control Module (BCM) fault (loss of communication)
32	Electronic Control Module (ECM) to Body Control Module (BCM) fault (loss of communication)
40	Heating, Ventilation and Air Conditioning (HVAC) system fault (air mix door)
43	Heated windshield circuit fault
46	Air Conditioning system fault (low refrigerant)
47	Air Conditioning system fault (very low refrigerant)
48	Air Conditioning system fault (low refrigerant pressure)
49	Air Conditioning system fault (high side temperature too high)
51	Body Control Module (BCM) PROM fault

Note: *All Cadillac Deville and Fleetwood models covered by this manual will display an "F" prefix before the BCM trouble code number.*

Cadillac Eldorado, Seville, Buick Riviera and Oldsmobile Toronado models

Trouble code	Code identification
110	Outside temperature sensor or circuit fault
111	Air conditioning temperature sensor or circuit fault (high side)
112	Air conditioning temperature sensor or circuit fault (low side)
113	In-car temperature sensor or circuit fault
115	Sunload temperature sensor or circuit fault
118	Door jamb/door ajar switch or circuit fault
119	Twilight photocell or circuit fault
120	Twilight delay pot or circuit fault
121	Twilight enable switch or circuit fault
122	Instrument panel dimmer switch or circuit fault
123	Courtesy light switch or circuit fault
124	Vehicle Speed Sensor (VSS) or circuit fault
127	Park Neutral Position (PNP) switch fault
128	Door jamb/door ajar switch or circuit fault
131	Engine oil pressure switch or circuit fault
132	Engine oil pressure sensor or circuit fault
140	Cellular phone system fault
331	Tape deck to CRT fault (loss of communication)
332	Compass module to Body Control Module (BCM) fault (loss of communication)
333	SIR (airbag) to Body Control Module (BCM) system fault (loss of communication)
334	Electronic Control Module (ECM) to Body Control Module (BCM) fault (loss of communication)
335	Climate control panel or CRT to Body Control Module (BCM) fault (loss of communication)
336	Instrument Panel Cluster (IPC) to Body Control Module (BCM) fault (loss of communication)
337	Heating, Ventilation and Air Conditioning (HVAC) programmer to Body Control Module (BCM) fault (loss of communication)
339	Radio to CRT fault (loss of communication)
409	Charging system or circuit fault (alternator)
410	Charging system or circuit fault (regulator)
411	Battery voltage fault (low)
412	Battery voltage fault (high)
418	Climate control panel or CRT to Body Control Module (BCM) fault (loss of crank signal)

BCM Trouble codes (continued)

Cadillac Eldorado, Seville, Buick Riviera and Oldsmobile Toronado models

Trouble code	Code identification
420	Headlight and courtesy lamp relay(s) or circuit fault
440	Heating, Ventilation and Air Conditioning (HVAC) system fault (air mix door)
441	Cooling fans or circuit fault
445	Air Conditioning clutch or circuit fault
446	Air Conditioning system fault (low refrigerant)
447	Air Conditioning system fault (very low refrigerant)
448	Air Conditioning system fault (low refrigerant pressure)
449	Air Conditioning system fault (high side temperature too high)
450	Air Conditioning system fault (engine coolant temperature too high)
482	Anti-lock brake (ABS) system pressure fault
552	Body Control Module (BCM) memory reset
553	CRT memory reset
556	EE prom fault
660	Cruise control system fault (cruise engaged with vehicle in Park, Neutral or Reverse)
663	Cruise control system fault (vehicle speed increased or decreased more than 20 MPH over the set speed)
664	Cruise control system fault (vehicle speed increased more than 4 MPH in a .25 second
665	Cruise control system fault (engine coolant temperature too high)
666	Cruise control system fault (engine exceeds 5000 RPM while cruise is engaged)
667	Cruise control system fault (set/coast or resume/accel switch shorted)
671	Cruise control system fault (servo position sensor or circuit)
672	Cruise control system fault (vent control solenoid or circuit)
673	Cruise control system fault (vacuum control solenoid or circuit)
710	CRT switches or circuit fault

Note: *All Cadillac Eldorado, Seville, Buick Riviera and Oldsmobile Toronado models covered by this manual will display an "B" prefix before the BCM trouble code number.*

3 Electronic Control Module (ECM) – general information and replacement

Warning: *Some models covered by this manual are equipped with Supplemental Inflatable Restraint systems (SIR), more commonly known as airbags. Always disable the airbag system before working in the vicinity of the impact sensors, steering column or instrument panel to avoid the possibility of accidental deployment of the airbag, which could cause personal injury (see Chapter 12).*

General information

1   The Electronic Control Module (ECM) is often referred too as the Powertrain Control Module (PCM) on 1991 and later vehicles and is the same component. For clarification purposes this manual will use the terminology Electronic Control Module (ECM) when referring to this component.
2   The ECM is highly reliable component and rarely requires replacement. Because the ECM is the most expensive part in the On-Board Diagnostic (OBD-I) system, you should be absolutely positive it has failed before replacing it. If in doubt have the system tested by a dealer service department or other qualified repair facility.

3   On 1987 and earlier models the ECM is equipped with a replaceable PROM and CALPAK that must be installed into the new ECM when exchanged.
4   On 1988 and later models the ECM is equipped with a replaceable MEM-CAL that must be installed into the new ECM when exchanged.
5   If the PROM, CALPAK OR MEM-CAL unit(s) become defective and must be replaced, it will require removing the ECM from the vehicle to allow access to the calibration units. Simply follow the procedures outlined below.

Replacement

Refer to illustrations 3.10a, 3.10b, 3.10c, 3.13a and 3.13b
Caution 1: *To avoid electrostatic discharge damage to the ECM, handle the ECM only by its case. Do not touch the electrical terminals during removal and installation. If available, ground yourself to the vehicle with a anti-static ground strap, available at computer supply stores.*
Caution 2: *To prevent damage to the ECM, the ignition switch must be turned Off when disconnecting or connecting the ECM connectors.*
6   The ECM on all models is located in the right front corner of the passenger compart-

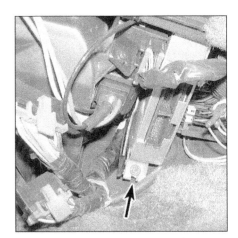

3.10a On some models the ECM is mounted vertically in the vehicle to the heater box . . .

ment under the instrument panel.
7   Disconnect the cable from the negative terminal of the battery.
8   Remove the right hand sound insulator panel from beneath the dashboard (see Chapter 11).
9   Unplug the electrical connectors from the ECM. Each connector is color coded to fit its respective receptacle in the ECM.

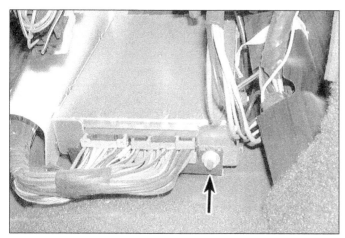

3.10b ... On other models the ECM is mounted horizontally in the vehicle to the firewall - In either case, simply remove the retaining nut (arrow) and guide the ECM out of its mounting bracket

3.10c On some models it may be easier to lower the ECM from its mounting bracket and disconnect the electrical connectors

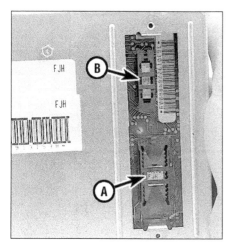

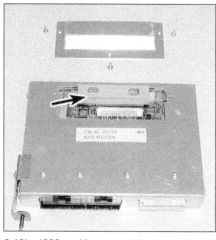

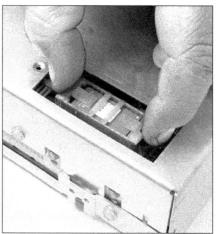

3.13a 1987 and earlier models are equipped with a **PROM (A)** and a **CALPAK (B)** chip

3.13b 1988 and later models are equipped with a **MEM-CAL** chip (arrow)

3.14 Grasp the **PROM** carrier at the narrow ends and gently rock it until the **PROM** is disconnected from its socket

10   Remove the ECM retaining nut **(see illustrations)**.

11   Carefully lift the ECM from its mounting bracket without damaging the electrical connectors and wiring harness to the computer. **Note:** It is not necessary to remove the ECM mounting bracket when removing the ECM from the vehicle.

12   All models are equipped with a memory calibration chip(s) (see Step 4 and 5). New ECMs are not supplied with new calibration chips, therefore it is necessary to remove the original and install it into the new ECM. Use care not to damage the PROM, CALPAK or MEM-CAL when working with these delicate electrical components.

13   Remove the screws and detach the access cover from the ECM to allow access to the memory calibration chip(s) **(see illustrations)**.

1987 and earlier models (PROM and CALPAK replacement)

Refer to illustrations 3.14, 3.16 and 3.17

14   Using a PROM removal tool (if avail-

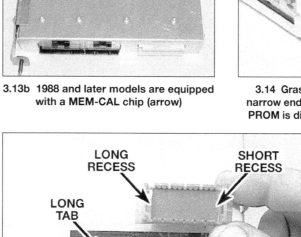

3.16 The notch in the PROM carrier is matched up with the small tang on the socket

able), grasp the PROM carrier at the narrow ends **(see illustration)**. Gently rock the carrier from end to end while carefully pulling up. **Note:** *PROM and CALPAK removal tools usually come supplied with new PROMs and ECMs.*

15   The PROM carrier and PROM should lift out of the PROM socket easily.

16   Note the reference end of the PROM carrier before setting it aside, so you will be able to install the prom with the same orientation **(see illustration)**.

17   Position the PROM/PROM carrier assembly squarely over the PROM socket

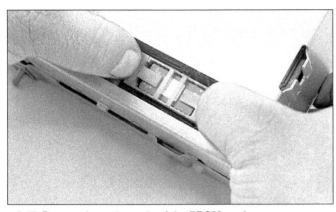

3.17 Press only on the ends of the PROM carrier - pressure on the center could result in bent or broken pins or damage to the PROM

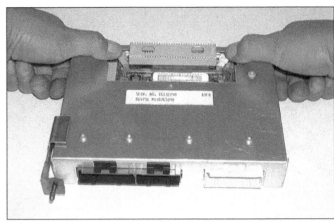

3.20 Press the retaining tabs out to release the MEM-CAL

with the small notched end of the carrier aligned with the small notch in the socket at the pin on one end. Press on the PROM carrier until it seats firmly in the socket **(see illustration)**. **Caution 1:** *Do not press on the PROM - press only on the carrier.* **Caution 2:** *If the PROM is installed backwards and the ignition switch is turned on, the PROM will be destroyed.*

18   If you are replacing the old PROM, CAL-PAK or ECM with a new one, always check the service numbers on the new parts to make sure that it is the same as the number of the old part before installing the new parts.
19   The CALPAK is removed and installed in the same manner as the PROM. Repeat steps 14 through 18 and reinstall the CAL-PAK into the ECM.

1988 and later models (MEM-CAL replacement)

Refer to illustrations 3.20 and 3.21
20   Using two fingers, carefully push both retaining clips back away from the MEM-CAL and lift the unit straight out of the ECM **(see illustration)**.
21   Installation is the reverse of removal. Be sure to use the alignment notches in the seat

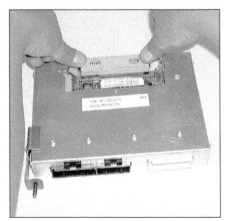

3.21 Carefully align the MEM-CAL with the notches in the seat, then press on the edges of the MEM-Cal carrier until the unit snaps into place

area of the ECM when sliding the MEM-CAL back in place. Press only on the ends of the PROM assembly until it snaps back into place **(see illustration)**.

All models

22   Install the memory calibration chip access cover on the ECM.
23   Install the ECM in the reverse order of removal.
24   Start the engine and enter the diagnostic mode (see Section 2). If no trouble codes occur, the PROM, CALPAK or MEM-CAL is correctly installed.
25   If a Trouble Code occurs, or if the CHECK ENGINE/SERVICE ENGINE SOON light comes on and remains constantly lit, the memory calibration chip(s) may not be not fully seated, has bent pins or is defective.
26   If the memory calibration chip is not fully seated, pressing firmly on both ends of the carrier should correct the problem.
27   If the pins have been bent, remove the chip, straighten the pins and reinstall it. If the bent pins break or crack when you attempt to straighten them, discard the chip and replace it with a new one. **Note:** *After this procedure some vehicles may experience a rough idle or a slight driveability problem until the computer completes its relearning process, This period of time, however, should last no longer than 15 to 20 minutes after normal driving is resumed.*

4   Throttle Position Sensor (TPS) – check, replacement and adjustment

Check

Refer to illustrations 4.2a, 4.2b and 4.2c
Note: *If the following tests indicate that a sensor is good, and not the cause of a driveability problem or DTC, check the wiring harness and connectors between the sensor and the ECM for an open or short circuit. If no problems are found, have the vehicle checked by a dealer service department or other qualified repair shop.*
1   The Throttle Position Sensor (TPS) is a variable-resistance potentiometer, mounted on the side of the throttle body and connected to the throttle shaft **(see illustrations 1.1a, 1.1b and 1.1c)**. By monitoring the output voltage from the TPS, the ECM can determine fuel delivery based on throttle valve angle (driver demand). A broken or loose TPS can cause intermittent bursts of fuel from the injector and an unstable idle because the ECM thinks the throttle is moving. A problem with the TPS and or circuit will set a diagnostic trouble code.
2   Check the reference voltage from the ECM to the TPS. Disconnect the TPS harness connector and install the probes of a volt-

4.2a TPS terminal guide (1989 and earlier Cadillac models)

SIG | GND | REF

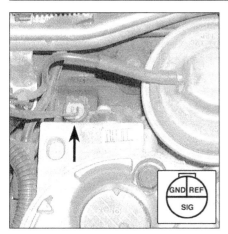

4.2b TPS terminal guide (1990 and later Cadillac models)

4.2c TPS terminal guide (Buick and Oldsmobile models)

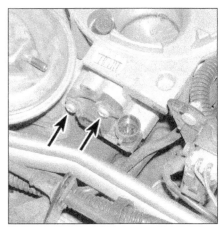

4.7 Detach the TPS mounting screws (arrows) (Cadillac model shown, others similar)

meter on the REF terminal (+) (gray wire) and the GND terminal (-) (black or black and pink wire). With the ignition key ON (engine not running), it should read approximately 5.0 volts **(see illustrations)**. If reference voltage is not available, there is an open circuit from the ECM or a defective ECM.

3   Next, probe the SIG terminal (+) (dark blue wire) and the GND terminal (-) (black or black and pink wire) on the TPS. Check the signal voltage of TPS with the throttle fully closed, then gradually open the throttle to full throttle. There should be approximately 0.44 to 0.55 volts with the throttle closed and 4.0 to 4.5 volts at full throttle. The voltage should increase smoothly as the throttle is opened.

4   If the readings are incorrect, the TPS may need adjustment or the sensor may be defective.

Replacement

Refer to illustration 4.7

5   Disconnect the cable from the negative terminal of the battery.

6   Remove the air intake duct (Buick and Oldsmobile) or the air cleaner assembly (Cadillac models) (see Chapter 4) and disconnect the TPS electrical connector.

7   Remove the two retaining screws **(see illustration)** and separate the TPS from the throttle body.

8   Install the new TPS leaving the mounting screws loose. Connect the electrical connector.

9   Adjust the TPS as described below (see Step 12) and tighten the screws securely.

Adjustment

10   Remove the air intake duct (Buick and Oldsmobile) or the air cleaner assembly (Cadillac models) (see Chapter 4).

11   Loosen the TPS mounting screws.

12   Set the minimum idle speed adjustment to specification (see Chapter 4).

13   Using a digital voltmeter, backprobe the electrical connector SIG terminal (+) (dark blue wire) and the GND terminal (-) (black or black and pink wire). Be very careful not to damage the wiring harness **(see illustrations**

4.2a, 4.2b and 4.2c). With the ignition key ON (engine not running) and the throttle fully closed, the voltage should read between 0.45 to 0.55 volts on Cadillac models and between 0.38 to 0.42 volts on Buick and Oldsmobile.

14   If the readings are incorrect, rotate the TPS body until the correct voltage is attained and tighten the mounting screws.

---

5   Mass Airflow (MAF) sensor (Buick and Oldsmobile models) - check and replacement

Check

Refer to illustration 5.3

1   The Mass Airflow (MAF) sensor is installed in the air intake duct **(see illustration 1.1c)**. This sensor uses a hot-wire sensing element to measure the molecular mass (or weight) of air entering the engine. The air passing over the hot wire causes it to cool, and the sensor converts this temperature change into an analog voltage signal to the ECM. The ECM in turn calculates the required fuel injector pulse width to obtain the necessary air/fuel ratio. A defective MAF sensor can cause surging, stalling, rough idle and other driveability problems. The On-board Diagnostic (OBD-I)

system can detect several different MAF sensor problems and set trouble codes to indicate the specific fault.

2   To check for power to the MAF sensor, disconnect the MAF sensor electrical connector.

3   Connect the positive (+) lead of your voltmeter to the B+ terminal (pink and black wire) of the harness connector; connect the meter negative (-) lead to the sensor connector GND terminal (-) (black and white wire) **(see illustration)**.

4   Turn the ignition On but do not start the engine. The meter should read more than 10 volts or close to battery voltage.

5   Connect the positive (+) lead of your voltmeter to the REF (+) terminal (yellow wire) of the harness connector; connect the meter negative (-) lead to the sensor connector GND terminal (-) (black and white wire). The meter should read 4 to 6 volts. This will check the reference voltage from the ECM.

6   If the voltage readings are correct, and the On-board Diagnostic (OBD-I) system detects a MAF sensor trouble code, the MAF sensor may be defective. **Note:** *Have the sensor tested by a dealer service department or other qualified repair shop. A MAF sensor can develop voltage signal problems that can't be detected by conventional test equip-*

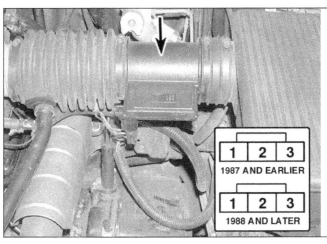

5.3 The MAF sensor (arrow) is located in the air intake duct or on the throttle body- check for battery voltage to the B+ terminal of the MAF sensor connector (key ON engine not running) (1987 and earlier model shown)

5.11 On 1991 and later models, unplug the electrical connector and remove the MAF sensor retaining screws (arrows)

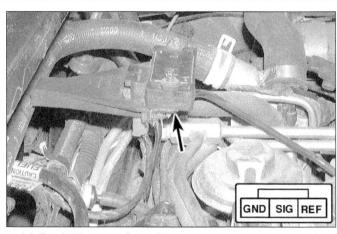

6.2 The MAP sensor (arrow) is mounted between the throttle body and the rear valve cover (1991 Deville shown)

ment such as a voltmeter. The ECM can see such signal faults and a driveability problem will result.

7   If the voltage readings are incorrect, use an ohmmeter to check for continuity between the MAF GND terminal and a known good body ground. If continuity does not exist repair the ground circuit and perform the voltage tests again. If the ground circuit is OK, refer to the wiring diagrams and check the wiring harness for open circuits or a damaged harness.

Replacement

Refer to illustration 5.11
8   Disconnect the electrical connector from the MAF sensor.
9   On 1987 and earlier models, loosen the clamps that secure the sensor to the intake air duct and the air cleaner housing and remove the sensor from the vehicle **(see illustration 5.3)**.
10   On 1988 through 1990 models, detach the air intake duct from the throttle body, remove four screws securing the MAF assembly to the throttle body and remove the MAF assembly from the vehicle. **Note:** *The plastic MAF sensor body on these models and the metal air duct on which it is mounted are an assembly that must be replaced as a unit. Do not try to separate the sensor body from the metal duct.*
11   On 1991 and later models, remove the screws and lift the MAF sensor from the throttle body **(see illustration)**.
12   Installation is the reverse of removal.

---

6   Manifold Absolute Pressure (MAP) sensor (Cadillac models) - check and replacement

---

Check

Refer to illustrations 6.2 and 6.3
Note: *If the following tests indicate that a sensor is good, and not the cause of a driveability problem or DTC, check the wiring harness and connectors between the sensor and*

VOLTAGE RANGE	ALTITUDE
3.8 - 5.5V	Below 1,000
3.6 - 5.3V	1,000 - 2,000
3.5 - 5.1V	2,000 - 3,000
3.3 - 5.0V	3,000 - 4,000
3.2 - 4.8V	4,000 - 5,000
3.0 - 4.6V	5,000 - 6,000
2.9 - 4.5V	6,000 - 7,000
2.8 - 4.3V	7,000 - 8,000
2.6 - 4.2V	8,000 - 9,000
2.5 - 4.0V	9,000 - 10,000 FEET

LOW ALTITUDE = HIGH PRESSURE = HIGH VOLTAGE

24071-6-4.09 HAYNES

6.3 Typical Manifold Absolute Pressure (MAP) sensor altitude (pressure) vs. voltage values

the ECM for an open or short circuit. If no problems are found, have the vehicle checked by a dealer service department or other qualified repair shop.
1   The Manifold Absolute Pressure (MAP) sensor monitors the intake manifold pressure changes resulting from changes in engine load and speed and converts the information into a voltage output. The ECM uses the MAP sensor to control fuel delivery and ignition timing. The ECM will receive information as a voltage signal that will vary from 0.5 to 1.0 volts at closed throttle (high engine vacuum) and 4.0 to 4.5 volts at wide open throttle (low engine vacuum). The voltage range values will vary slightly according to changes in altitude. A failure in the MAP sensor circuit will set a trouble code. The MAP sensor is attached to a bracket located between the throttle body and the rear valve cover.
2   To check the sensor, begin by checking the reference voltage from the ECM. Disconnect the electrical connector and connect a voltmeter to the REF terminal (+) (gray wire) and GND terminal (-) (black and pink or orange and black wire) of the harness connector **(see illustration)**. Turn the ignition key On, approximately 5.0 volts should be present. If reference voltage is not present, check the circuits from the sensor to the ECM. If the

GND | SIG | REF

6.5 MAP sensor retaining screws (arrows)

circuits are good the ECM is defective.
3   If reference voltage is available, turn the ignition key Off and reconnect the connector to the sensor. Using suitable probes, backprobe the connector and connect the voltmeter to the SIG terminal (+) (light green wire) and GND terminal (-) (black and pink or orange and black wire) at the sensor. Turn the ignition Key ON (engine not running) and compare your voltage reading with the following chart **(see illustration)**.
4   If the voltage reading is within the correct range, disconnect the vacuum hose at the sensor, connect a hand held vacuum pump to the sensor fitting and apply 10 to 15 in-Hg of vacuum to the sensor (or start the engine). With vacuum applied to the sensor, the voltage should decrease sharply to 1.2 to 2.3 volts less than the voltage recorded in Step 3. If the voltage readings are not correct, replace the sensor.

Replacement

Refer to illustration 6.5
5   To replace the sensor, detach the vacuum hose, disconnect the electrical connector and remove the mounting screws **(see illustration)**.
6   Installation is reverse of removal.

Temperature (degrees-F)	Resistance (ohms)
212	176
194	240
176	332
158	458
140	668
122	972
112	1182
104	1458
95	1800
86	2238
76	2795
68	3520
58	4450
50	5670
40	7280
32	9420

7.2 Intake Air Temperature (IAT) and Engine Coolant Temperature (ECT) sensors approximate temperature vs. resistance values

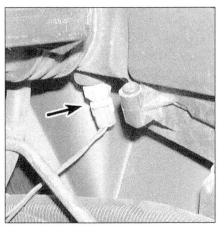

7.6 On Buick and Oldsmobile models, disconnect the electrical connector, then carefully pull the IAT sensor (arrow) from the air cleaner housing

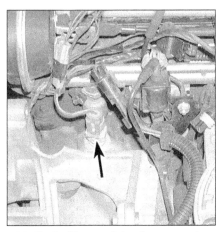

7.7 On Cadillac models, disconnect the electrical connector and unscrew the IAT sensor (arrow) from the intake manifold (alternator remove for clarity)

---

7   Intake Air Temperature (IAT) sensor - check and replacement

Check

Refer to illustration 7.2

Note: *If the following tests indicate that a sensor is good, and not the cause of a driveability problem or DTC, check the wiring harness and connectors between the sensor and the ECM for an open or short circuit. If no problems are found, have the vehicle checked by a dealer service department or other qualified repair shop.*

1   The Intake Air Temperature (IAT) sensor is sometimes referred to as the Manifold Air Temperature (MAT) sensor on Cadillac models and is the same component. For clarification purposes this manual will use the terminology IAT sensor when referring to this component. The IAT sensor is a thermistor that changes resistance as temperature changes. This sensor is installed in the intake air duct (Buick and Oldsmobile models) or the intake manifold (Cadillac models) to sense air temperature **(see illustrations 1.1a, 1.1b and 1.1c)**. As temperature increases, sensor resistance decreases and vice versa. The ECM uses this information to compute the intake temperature and fine tune fuel metering. A problem in the IAT sensor circuit will set a trouble code. The fault may be in the circuit wiring or connections or in the sensor itself.

2   To check the sensor resistance, disconnect the harness connector from the IAT sensor and measure the resistance across the two terminals on the sensor. Check its resistance with the engine completely cold, then start the engine and monitor the resistance change as the engine warms-up. Compare your readings with the following chart **(see illustration)**.

3   If the resistance reading is not within the correct range, replace the sensor.

4   If the resistance values of the sensor are

correct, check for reference voltage from the ECM to the sensor connector. The open-circuit voltage at the sensor connector should be approximately 5 volts.

Replacement

Refer to illustrations 7.6 and 7.7

5   Disconnect the electrical connector from the IAT sensor.

6   On Buick and Oldsmobile models carefully remove the IAT sensor from the air cleaner housing **(see illustration)**. Be careful not to damage any of the plastic parts.

7   On Cadillac models unscrew the IAT sensor from the intake manifold **(see illustration)**. Before installing the new sensor, wrap the threads with Teflon sealing tape to prevent vacuum leaks.

8   Install the sensor and reconnect the electrical connector.

---

8   Engine Coolant Temperature (ECT) sensor - check and replacement

Check

Refer to illustration 8.2

Note 1: *If the following tests indicate that a sensor is good, and not the cause of a driveability problem or DTC, check the wiring harness and connectors between the sensor and the ECM for an open or short circuit. If no problems are found, have the vehicle checked by a dealer service department or other qualified repair shop.*

Note 2: *Before condemning an ECT sensor, check the coolant level in the system.*

1   Like the IAT sensor, the ECT sensor is a thermistor, which is a variable resistor that changes its resistance as temperature changes. The sensor is typically installed in a cooling system passage on the intake manifold or the thermostat housing **(see illustrations 1.1a, 1.1b and 1.1c)** to sense coolant

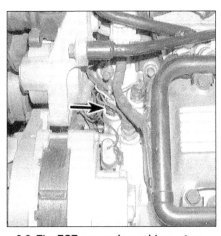

8.2 The ECT sensor (arrow) has a two-wire connector and is identified by yellow and black wires (1987 and earlier Buick and Oldsmobile shown)

temperature. As coolant temperature increases, sensor resistance decreases and vice versa. The ECM uses this information to compute the engine operating temperature. A problem in the ECT sensor circuit will set a trouble code. The fault may be in the circuit wiring or connections or in the sensor itself.

2   1987 and earlier Buick and Oldsmobile engines have two almost identical coolant temperature sensor units **(see illustration)** mounted next to each other. One is the sender for the instrument panel temperature gauge, the other is the ECT sensor for the On-board diagnostic (OBD-I) system. The temperature sender connector for the instrument panel gauge has a single wire connector, the ECT sensor connector has a two wire connector.

3   Disconnect the ECT sensor and use an ohmmeter to measure resistance across the two terminals of the sensor. At 68 degrees F, resistance should be approximately 3,200 to 3,500 ohms **(see illustration 7.2)**.

4   Next, start the engine and warm it up until it reaches operating temperature. The

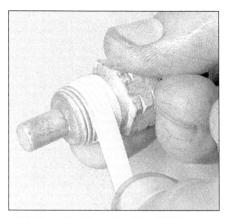

8.6 Wrap the threads of the ECT sensor with Teflon tape before installing it

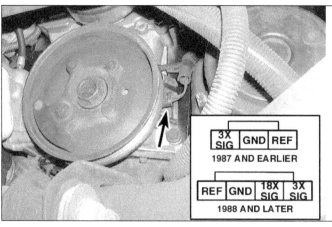

9.2 The CKP sensor (arrow) is mounted to the front timing cover behind the crankshaft balancer

resistance should be lower. For example, at 176-degrees F resistance should be 300 to 330 ohms.

5   If the resistance values of the sensor are correct, check for reference voltage from the ECM to the sensor connector. The open-circuit voltage at the sensor connector should be approximately 5 volts.

Replacement

Refer to illustration 8.6

Warning: *Wait until the engine is completely cool before performing this procedure.*

6   Before installing the new sensor, wrap the threads with Teflon sealing tape to prevent leakage and thread corrosion **(see illustration)**.

7   Disconnect the electrical connector, then unscrew the ECT sensor from the engine. Install the new sensor as quickly as possible to minimize coolant loss. Tighten the sensor securely and reconnect the electrical connector.

8   Check the coolant level as described in Chapter 1, adding some, if necessary. Start the engine and allow it to reach normal operating temperature, then check for coolant leaks. Check the coolant level in the expansion tank after the engine has warmed up and then cooled down again.

9   Crankshaft position (CKP) sensor (Buick and Oldsmobile models)- check and replacement

Check

Refer to illustration 9.2

Note: *If the following tests indicate that a sensor is good, and not the cause of a driveability problem or DTC, check the wiring harness and connectors between the sensor and the ECM for an open or short circuit. If no problems are found, have the vehicle checked by a dealer service department or other qualified repair shop.*

1   The CKP sensor consists of a single hall effect switch or a dual hall effect switch, and a magnet which are separated by an air gap.

Hall effect switches complete a ground circuit when a magnetic field is present. Interrupter ring(s) located on the back of the crankshaft balancer disrupt the ground signal from the sensor to produce ON OFF signals to the ECM as the engine rotates. This reference signal is used to determine crankshaft position and engine RPM. 1987 and earlier models are equipped with single hall effect crankshaft sensor and a 3X (slot) interrupter ring on the back of the balancer. This type of sensor produces three ON OFF signals per crankshaft revolution. 1988 and later models are equipped with a dual hall effect crankshaft position sensor which consists of two hall-effect switches and a shared magnet mounted between them. On these sensors the crankshaft balancer is equipped with two "interrupter rings [a 3X (slot) and 18X (slot)] which create two separate ON OFF signal patterns that can be used for the timing and the ignition sequence. Dual hall effect crankshaft position sensors produce three ON OFF signals per crankshaft revolution and eighteen ON OFF signals per crankshaft revolution. Always be sure to check the interrupter ring for damage while diagnosing a faulty crankshaft position sensor.

2   To check hall effect type sensors, select the DC volts function on your multi-meter, carefully backprobe the harness connector terminals using pins and position the voltmeter probes onto the pins. Connect the negative probe (-) to the ground connection and the positive probe to the 3X SIGNAL terminal **(see illustration)**. With the ignition key in the ON position and the engine OFF, rotate the engine crankshaft by hand (one full turn) and observe the voltmeter. There should be three ON OFF signals per crankshaft revolution and 10 to 12 volts of signal voltage at the terminal when the sensor is in the ON cycle.

3   On 1988 and later models with dual hall effect type sensors, connect the positive probe to the second 18X SIGNAL terminal, rotate the engine crankshaft again by hand (one full turn) and observe the voltmeter. There should be eighteen ON OFF signals per crankshaft revolution and 10 to 12 volts of signal voltage at the 18X terminal when the sensor is in the ON cycle.

4   If SIGNAL voltage does not exist, test

the system for the proper reference voltage from the computer. Simply disconnect the harness connector at the sensor, select the voltage range on the rotary switch on the volt/ohmmeter and probe the correct terminals on the harness for the reference signal. With ignition switch in the ON position (engine not running) there should be 10 to 12 volts of reference voltage.

Replacement

Refer to illustrations 9.8, 9.9 and 9.10

Note: *This procedure requires a special alignment tool (available at aftermarket tool manufacturers) to align the harmonic balancer vanes with the slots on the sensor. Attempts to perform this procedure without the use of this tool may result in damage to the sensor and or the harmonic balancer.*

5   Refer to the front timing cover removal procedure in Chapter 2A and remove the harmonic balancer. If you're working on a 1989 and later vehicle remove the torque axis mount and brackets also following the procedure outlined in chapter 2A.

6   Remove the CKP sensor deflector shield if equipped.

7   Disconnect the electrical connector from the CKP sensor.

8   Remove the crankshaft sensor pinch

9.8 Remove the pinch bolt (arrow) and detach the crankshaft sensor from the sensor mounting bracket

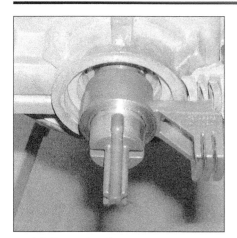

9.9 Place the alignment tool on the crankshaft snout, align the tool with the slots in the sensor and tighten the pinch bolt

9.10 Position the opposite end of the tool onto the harmonic balancer and rotate it 360 degrees – the interrupter ring should not touch the slots in the tool

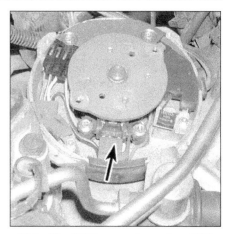

10.1a The CMP sensor (arrow) on 1990 and later Cadillac models is located in the distributor

bolt and detach the sensor from the mounting bracket (see illustration).
9   To Install the sensor, position the sensor on the sensor mounting bracket and loosely install the pinch bolt. Place the alignment tool on the crankshaft snout and align the tool with the slots in the sensor (see illustration). Tighten the pinch bolt to 30 to 35 in-lbs.
10   Remove the tool from the crankshaft and insert it into the bore of the harmonic balancer (see illustration). Rotate the tool 360 degrees and check the alignment between the harmonic balancer vanes and the slots in the tool. If any of the balancer vanes contact the tool, they're bent and the balancer must be replaced.
11   Install the harmonic balancer onto the crankshaft and tighten the bolt to the torque listed in Chapter 2A Specifications. Be careful not to disturb the positioning of the crankshaft sensor.
12   The remainder of the installation is the reverse of removal.

10   Camshaft position (CMP) sensor - check and replacement

Refer to illustrations 10.1a and 10.1b
Note: *If the following tests indicate that a sensor is good, and not the cause of a driveability problem or DTC, check the wiring harness and connectors between the sensor and the ECM for an open or short circuit. If no problems are found, have the vehicle checked by a dealer service department or other qualified repair shop.*
1   The camshaft position sensor consists of a single hall effect type switch and a magnet which are separated by an air gap. The hall effect switch completes the ground circuit when a magnetic field is present. As the engine rotates, these sensors will produce ON OFF signals to the ECM to determine camshaft position and speed. Hall effect camshaft sensors typically use a rotating magnet on the camshaft sprocket or in the

sensor housing to complete the hall effect switch ground signal. The ECM uses the sensors to control fuel delivery and ignition timing. The camshaft sensor typically produces only one ON OFF signal per camshaft revolution. 1990 and later Cadillac models are equipped with a camshaft sensor located inside the HEI distributor (see illustration). All Buick and Oldsmobile models are equipped with a camshaft sensor that is mounted to the front timing cover (see illustration). **Note:** *1989 and earlier Cadillac models are not equipped with a camshaft sensor.*

1990 and later Cadillac models
Check
Refer to illustration 10.3
2   To test the switch it must first be removed (see Step 6).
3   Connect a 12-volt power supply and voltmeter as shown (see illustration). Check for polarity markings carefully before making any connections.

10.1b The CMP sensor (arrow) on Buick and Oldsmobile models is mounted to the front timing cover

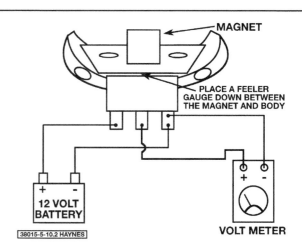

10.3 To test a CMP sensor on Cadillac models, connect a 12-volt power supply and voltmeter as shown - always check for polarity markings carefully before making any connections

10.12 CMP sensor retaining bolt (arrow) (Buick and Oldsmobile models)

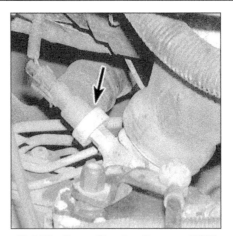

11.4 The Power Steering Pressure (PSP) switch (arrow) is mounted to the steering gear

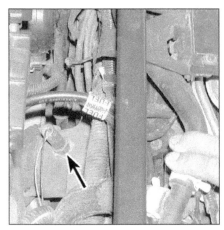

12.1 Typical oxygen sensor installation

4   When the feeler gauge is not inserted as shown, the voltmeter should read less than 0.5 volts. If the reading is more, the camshaft sensor is faulty and must be replaced with a new one.

5   With the feeler gauge inserted, the voltmeter should read within 0.5 volts of battery voltage. Replace the camshaft sensor with a new one if the reading is greater than battery voltage.

Replacement

6   Remove the distributor cap and rotor and position it aside (see Chapter 5 if necessary).

7   Disconnect the electrical connector and detach the sensor mounting screws and remove camshaft sensor from the distributor **(see illustration 10.1a)**

8   Installation is the reverse of removal.

Buick and Oldsmobile models
Check

9   To check a camshaft sensor on Buick or Oldsmobile models, select the DC volts function on the multi-meter, carefully backprobe the harness connector terminals using pins and connect the voltmeter probes to the pins. Connect the negative probe (-) to the GND terminal and the positive probe (+) to the SIGNAL (+) terminal **(see illustration 10.1b)**. With the ignition key in the ON position and the engine OFF, rotate the engine crankshaft by hand (two full turns) and observe the voltmeter. If the sensor is operating properly it should trigger ON OFF voltage readings on the voltmeter while the engine is being rotated. Depending on the make and model of the vehicle, there should be 10 to 12 volts of signal voltage. **Note:** *When testing camshaft sensors it may be necessary to have an assistant crank the engine over in short bursts at the ignition key (always disable the fuel system first to prevent the engine from starting see Chapter 4) while observing the voltmeter. It could take as many as two complete crankshaft revolutions before a camshaft sensor signal is detected.*

Replacement
Refer to illustration 10.12

10   Loosen the water pump pulley bolts several turns, then remove the engine drivebelt (see Chapter 1 if necessary).

11   Remove the water pump pulley from the front of the engine and disconnect the sensor electrical connector.

12   Remove the sensor retaining bolt **(see illustration)**. Using a pulling and twisting motion remove the camshaft sensor from the timing cover.

13   Installation is the reverse of removal. Always lubricate the sensor O-ring with clean engine oil before installing it into the timing cover.

---

11   Power Steering Pressure (PSP) switch - check and replacement

---

Check
Refer to illustration 11.4

1   The power steering pressure (PSP) switch is a normally-closed switch, mounted on the steering gear. When steering system pressure reaches a high-pressure setpoint, the PSP switch opens and sends a signal to the ECM that the ECM uses to maintain engine idle speed during parking maneuvers. The On-Board Diagnostic (OBD-I) system can detect switch problems and set trouble codes to indicate specific faults.

2   Check the operation of the PSP switch if the engine stalls during parking or if the engine idles continuously at high rpm.

3   Refer to the wiring diagrams at the end of this manual to identify the functions of connector terminals.

4   Disconnect the PSP switch connector and connect an ohmmeter to the terminals on the switch body **(see illustration)**.

5   Start the engine and let it idle.

6   Turn the steering wheel to point the front wheels straight ahead and read the ohmmeter. It should indicate a closed circuit (conti-

nuity of close to zero ohms).

7   Turn the steering wheel to either side and watch the ohmmeter. The PSP switch should open as the wheel nears the steering stop on either side, and the meter should indicate an open circuit (infinite resistance).

8   If the switch fails either test, replace it. If the switch is OK, troubleshoot the engine idle control operation if high idle speed or stalling problems continue.

Replacement

9   Disconnect the electrical connector from the switch.

10   Position a suitable drain pan below the switch to catch any fluid draining from the steering gear.

10   Using a wrench, unscrew the switch from the steering gear.

11   Install and connect the new switch. Refer to Chapter 10 and bleed air from the power steering system. Add fluid as required (see Chapter 1).

---

12   Oxygen sensor (O2S) - check and replacement

Refer to illustration 12.1

1   The oxygen sensor, which is located in the exhaust pipe **(see illustration)**, monitors the oxygen content of the exhaust gas stream. The oxygen content in the exhaust reacts with the oxygen sensor to produce a voltage output from 100 millivolts (high oxygen, lean mixture) to 900 millivolts (low oxygen, rich mixture). The ECM monitors this voltage output to determine optimum the air/fuel mixture. The ECM alters the air/fuel mixture ratio by controlling the pulse width (open time) of fuel injectors. A mixture of 14.7 parts air to 1 part fuel is the ideal ratio for minimizing exhaust emissions, thus allowing the catalytic converter to operate at maximum efficiency. It is this ratio of 14.7 to 1 which the ECM and the oxygen sensor attempt to maintain at all times.

2   The oxygen sensor must be hot to operate properly (approximately 600-degrees F).

During the initial warm-up period, the ECM operates in open loop mode - that is, it controls fuel delivery in accordance with a programmed default value instead of feedback information from the oxygen sensor.

3   The proper operation of the oxygen sensor depends on four conditions:

a) **Electrical** - *The low voltages and low currents generated by the sensor depend upon good, clean connections which should be checked whenever a malfunction of the sensor is suspected or indicated.*

b) **Outside air supply** - *The sensor is designed to allow air circulation to the internal portion of the sensor. Whenever the sensor is removed and installed or replaced, make sure the air passages are not restricted.*

c) **Proper operating temperature** - *The ECM will not react to the sensor signal until the sensor reaches approximately 600-degrees F. This factor must be taken into consideration when evaluating the performance of the sensor.*

d) **Unleaded fuel** - *The use of unleaded fuel is essential for proper operation of the sensor. Make sure the fuel you are using is of this type.*

4   In addition to observing the above conditions, special care must be taken whenever the sensor is serviced.

a) *The oxygen sensor has a permanently attached pigtail and connector which should not be removed from the sensor. Damage or removal of the pigtail or connector can adversely affect operation of the sensor.*

b) *Grease, dirt and other contaminants should be kept away from the electrical connector and the louvered end of the sensor.*

c) *Do not use cleaning solvents of any kind on the oxygen sensor.*

d) *Do not drop or roughly handle the sensor.*

e) *The silicone boot must be installed in the correct position to prevent the boot from being melted and to allow the sensor to operate properly.*

Check

5   The On-Board Diagnostic (OBD-I) system can detect O2 sensor problems and set trouble codes to indicate specific faults. **Note:** *1990 and later Eldorado and Seville models are equipped with two O2 sensors, a right side and a left side. Always check both sensors for proper operation when diagnosing sensor problems.*

6   The sensor can also be checked with a high-impedance digital voltmeter. Warm up the engine to normal operating temperature, then turn the engine off.

7   Without disconnecting the oxygen sensor electrical connector, backprobe the sensor wire with a suitable probe and connect the positive lead of the voltmeter to the probe. **Caution:** *Don't let the sensor wire or*

the voltmeter lead touch the exhaust pipe or manifold. Connect the negative lead of the meter to a good chassis ground. Place the meter on the millivolt setting. **Note:** *Some models are equipped with two wire O2 sensors, in this case it will be necessary to refer to the wiring diagrams at the end of this manual to identify the functions of connector terminals. Then carefully backprobe the harness connector terminals using suitable probes and connect the voltmeter leads to the probes. Connect the negative probe (-) to the GND terminal and the positive probe (+) to the SIGNAL (+) terminal and perform the remaining tests.*

8   Start the engine and monitor the voltage output of the sensor. When the engine is cold the sensor should produce a steady voltage of approximately 100 to 200 millivolts. After a period of approximately two minutes the engine should reach operating temperature and the voltage should begin to fluctuate between 100 and 900 millivolts. This indicates the system has reached closed loop and the computer is controlling fuel delivery. If the system fails to reach closed loop mode in a reasonable amount of time, the sensor may be defective or a fuel system problem could be the cause. If the voltage is fixed (or fluctuating very slowly), create a rich condition by opening the accelerator sharply and snapping it shut and then a lean condition by disconnecting a vacuum hose, if the sensor voltage does not respond quickly with the proper voltage fluctuations, the sensor is probably defective.

Replacement

Refer to illustration 12.12

Note: *Because the oxygen sensor is located in the exhaust pipe, it may be too tight to remove when the engine is cold. If you find it difficult to loosen, start and run the engine for a minute or two, then shut it off. Be careful not to burn yourself during the following procedure.*

12.12 A special socket (available at many auto parts stores) that allows clearance for the wiring harness is required for oxygen sensor removal

9   Disconnect the cable from the negative terminal of the battery.

10   If necessary for clearance, raise the vehicle and place it securely on jackstands.

11   Disconnect the electrical connector from the sensor.

12   Note the position of the silicone boot, if equipped, and carefully unscrew the sensor from the exhaust manifold or exhaust pipe **(see illustration)**. **Caution:** *Excessive force may damage the threads.*

13   Anti-seize compound must be used on the threads of the sensor to facilitate future removal. The threads of a new sensor will already be coated with this compound, but if an old sensor is removed and reinstalled, recoat the threads.

14   Install the sensor and tighten it to the specified torque.

15   Reconnect the electrical connector of the pigtail lead to the main engine wiring harness.

16   Lower the vehicle and reconnect the cable to the negative terminal of the battery.

13   Vehicle Speed Sensor (VSS) - check and replacement

Check

Refer to illustrations 13.1a and 13.1b

Note: *If the following tests indicate that a sensor is good, and not the cause of a driveability problem or DTC, check the wiring harness and connectors between the sensor and the ECM for an open or short circuit. If no problems are found, have the vehicle checked by a dealer service department or other qualified repair shop.*

1   The vehicle speed sensor (VSS) is a pickup coil (variable-reluctance) sensor mounted on the transaxle case **(see illustrations)**. It produces an AC voltage sine wave, the frequency of which is proportional to vehicle speed. The ECM and BCM use the

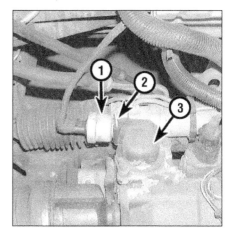

13.1a Retaining-clip type Vehicle Speed Sensor (VSS) mounting details

1   *Vehicle speed sensor*
2   *Retaining clip*
3   *Governor housing*

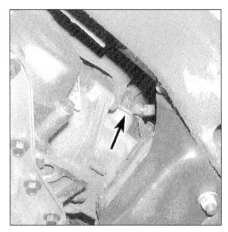

13.1b Location of a bolt-on type Vehicle Speed Sensor (VSS)

14.3a IAC valve location (1987 and earlier models)

14.3b IAC valve location (1991 and later models)

sensor input signal for several different engine and transmission control functions. A defective VSS can cause various driveability and transmission problems. The On- Board Diagnostic (OBD-I) system can detect sensor problems and set trouble codes to indicate specific faults.

2   Refer to the wiring diagrams in Chapter 12 to identify the functions of connector terminals.

3   Disconnect the VSS connector and turn the ignition ON but do not start the engine. Use a voltmeter to check for voltage between the sensor connector and ground as shown on the wiring diagrams. Approximately 1 to 5 volts should be present on one of the sensor wires with the key ON and the engine OFF.

4   Raise the front of the vehicle and place it securely on jackstands. Block the rear wheels and place the transaxle in Neutral. Place the voltmeter on the AC scale and connect the electrical connector to the VSS, turn the ignition to ON (engine not running) and backprobe the VSS connector yellow wire terminal with a voltmeter positive lead. Connect the negative lead of the meter to a chassis ground. Securely hold or block one of the front wheels, rotate the other wheel by hand. The voltmeter should pulse between 0.5 and 6 volts.

5   If no pulsing voltage signal is produced, replace the sensor. **Note:** *1989 and earlier Deville and Fleetwood models are equipped with an externally mounted VSS buffer that is not incorporated into the ECM. On these models always check the sensor first, if the sensor tests OK refer to the wiring diagrams in Chapter 12 and probe the positive terminal of the voltmeter to the brown wire of the VSS buffer and the negative lead to chassis ground. This checks the output signal from the buffer to the ECM. There should 6 to 12 volts present with the ignition key On (engine not running) and the wheel being turned.*

Replacement

6   Raise the vehicle and support it securely on jackstands.

7   Disconnect the electrical connector

from the VSS.

8   Remove the hold-down bolt and clamp or retaining clip and remove the VSS from the transaxle **(see illustrations 13.1a and 13.1b)**.

9   Inspect the O-ring on the sensor and replace it if damaged. If you are installing a new sensor, use a new O-ring.

10   Installation is the reverse of removal.

---

14   Idle Air Control (IAC) valve (Buick and Oldsmobile models) – check and replacement

Check

Refer to illustrations 14.3a and 14.3b

1   The Idle Air Control (IAC) valve controls the amount of air that bypasses the throttle valve, which controls the engine idle speed. This output actuator is mounted on the throttle body and is controlled by voltage pulses sent from the ECM (computer). The IAC valve within the body moves in or out, allowing more or less intake air into the system. To increase idle speed, the ECM extends the IAC valve from the seat and allows more air to bypass the throttle bore. To decrease idle speed, the ECM retracts the IAC valve towards the seat, reducing the air flow. A fault in this system will not be detected by the OBD system.

2   Start engine and allow the engine to reach normal operating temperature, turn the ignition off and connect a tachometer.

3   Disconnect the electrical connector from the IAC valve and start the engine **(see illustrations)**. If the idle speed increases, turn the ignition off and reconnect the IAC valve electrical connector. Restart the engine and check the idle, if the idle speed returns to normal, the IAC valve is functioning properly.

4   If the idle speed didn't respond as described, disconnect the IAC valve electrical connection and turn the ignition switch On (engine not running).

5   Using a test light connected to ground, probe each of the four terminals in the har-

ness connector. The test light should light at each IAC valve terminal indicating the ECM is attempting to control the IAC valve. If there is no control signal at the connector, check for an open wire, short to ground, faulty connection at the ECM or possible faulty ECM.

6   Finally, measure the resistance across each pair of terminals in the IAC valve. There should be a minimum of 20 ohms between the IAC valve terminals opposite the connector terminals A to B and C to D (the terminals are marked on the harness connector). If the resistance is not as specified, replace the IAC valve.

Replacement

Refer to illustrations 14.9, 14.10 and 14.11

7   Remove the air intake duct (if necessary) and disconnect the IAC electrical connector.

8   On 1987 and earlier models simply, unscrew the IAC valve and withdraw the valve from the throttle body **(see illustration 14.3a)**. Clean the mounting surface and discard the old gasket.

9   On 1988 and later models, remove the mounting screws and detach the IAC from the throttle body **(see illustration).** Check the condition of the O-ring; if it's damaged replace it with a new one.

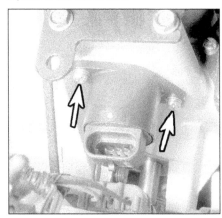

14.9 On 1988 and later models, remove the screws (arrows) and separate the IAC from the throttle body

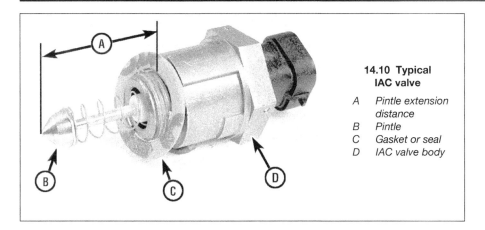

14.10 Typical IAC valve

A   Pintle extension
     distance
B   Pintle
C   Gasket or seal
D   IAC valve body

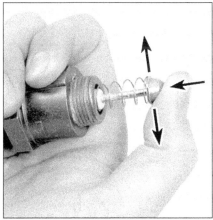

14.11 To adjust an IAC valve, retract the pintle by exerting firm pressure while moving it from side-to-side very slightly

10   Before installing the valve, measure the distance between the tip of the pintle and the housing mounting surface **(see illustration)**. If the distance is greater than 1-1/8 inch, it must be reduced to prevent damage to the valve.

11   To adjust the valve, retract the pintle by exerting firm pressure while moving it from side-to-side slightly **(see illustration)**.

12   Installation is the reverse of removal.

13   Start the engine and allow the engine to reach operating temperature, then turn it off. The IAC valve is reset by the ECM when the engine is turned off.

15   Idle Speed Control (ISC) motor (Cadillac models) - check and replacement

Check

Refer to illustration 15.3

1   The ISC motor is mounted to throttle body and is actuated by the ECM to control engine idle speed under varying conditions. The ISC motor also contains a throttle (ON/OFF) switch. The position of the switch tells the ECM whether or not to send idle speed commands to the ISC motor. The On-Board Diagnostic (OBD-I) system can detect ISC motor problems and set trouble codes to indicate specific faults.

2   To check the ISC motor, unplug the connector from the ISC motor.

3   Using a pair of fused jumper wires, fully retract the ISC plunger by applying 12 volts to terminal C of the ISC motor connector and grounding terminal D **(see illustration)**. **Note 1:** *Apply finger pressure to the to the ISC plunger while connecting terminal D to ground. Do not allow the battery voltage (12V) to contact terminal C any longer than necessary to retract the ISC plunger. Prolonged contact will cause damage to the motor.* **Note 2:** *It is very important to never connect a voltage source across terminals A and B, as it may damage the internal throttle contact switch.*

4   Fully extend the plunger by applying battery positive voltage (12V) to terminal D of the ISC connector and grounding terminal C **(see illustration 15.3)**. **Note:** *Do not allow the battery voltage (12V) to contact terminal D any longer than necessary when extending the ISC plunger. Prolonged contact will cause damage to the motor.*

5   If the ISC plunger does not respond when battery voltage is applied to the C and D terminals, it is faulty and should be replaced.

6   Check the operation of the closed throttle switch. Using an ohmmeter, check for continuity between terminals A and B of the ISC connector. There should be continuity between terminals A and B with the ISC plunger fully retracted and no continuity between terminals A and B with the ISC plunger fully extended.

7   If the closed throttle switch is faulty the entire ISC motor must be replaced.

Replacement

Refer to illustration 15.9

8   Unplug the electrical connector from the ISC motor.

9   Remove the ISC mounting nuts and detach the ISC motor from the mounting bracket **(see illustration)**.

10   Installation is the reverse of removal. If the ISC motor is being replaced with a new one, perform a minimum idle speed adjustment (see Chapter 4).

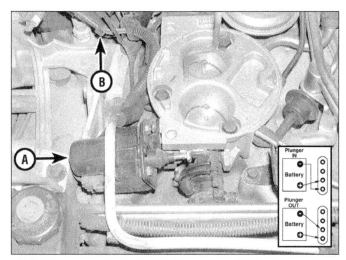

15.3 ISC motor (A) and electrical connector (B) (Cadillac models)

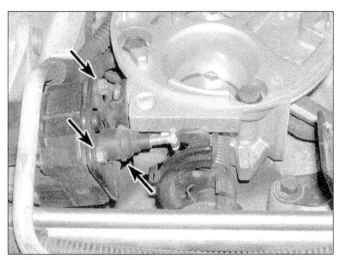

15.9 ISC motor retaining nuts (arrows) (lower nut not visible)

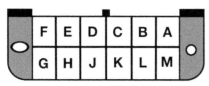

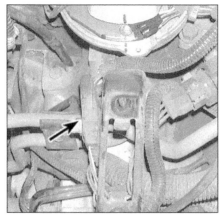

16.3 Connect a jumper wire between terminals A and B of the Assembly Line Data Link (ALDL) connector to enter the diagnostic mode (with the key On, engine Off)

16.4 The EST connector (arrow) is located in the wiring harness near the distributor

17.3 The ESC module (arrow) on 1987 and earlier models is mounted to the firewall

16   Electronic Spark Timing (EST) system

General description

Refer to illustrations 16.3 and 16.4

1   To provide improved engine performance, fuel economy and control of exhaust emissions, the Electronic Control Module (ECM) controls spark advance (ignition timing) with the Electronic Spark Timing (EST) system.

2   On Cadillac models the ECM receives a reference pulse from the distributor, which indicates both engine rpm and crankshaft position. The ECM then determines the proper spark advance for the engine operating conditions and sends an EST pulse to the distributor. On Buick and Oldsmobile models the ECM receives reference pulses from the ignition module which monitors signals from the crankshaft and camshaft position sensors. On these models the ignition module controls spark timing during start up (under 400 rpm). At 400 rpm and over, the ECM takes control of the ignition and fuel injection timing.

  The On-Board Diagnostic (OBD-I) system can detect EST system problems and set trouble codes to indicate specific faults.

Check

3   The ECM will set EST at a specified value when the diagnostic test terminal is grounded. To check for EST operation, the timing should be checked at 2000 rpm with the terminal ungrounded. Then ground the test terminal **(see illustration)**. If the timing changes at 2000 rpm, the EST is operating. Refer to Chapter 1 for the complete set timing procedure.

4   On Cadillac models, check the distributor 4-wire connector for corrosion, bent or broken pins **(see illustration)**. If the connector is damaged, repair the connector and retest the system.

5   For further information regarding the testing and component replacement procedures for the HEI/EST distributor, refer to Chapter 5.

17   Electronic Spark Control (ESC) system (Buick and Oldsmobile models)

General description

Refer to illustration 17.3

1   Irregular octane levels in modern gasoline can cause detonation in an engine. Detonation is sometimes referred to as "spark knock."

2   The Electronic Spark Control (ESC) system is designed to retard spark timing up to 10-degrees to reduce spark knock in the engine. This allows the engine to use maximum spark advance to improve driveability and fuel economy.

3   Models covered by this manual are equipped with two different types of ESC systems. The first system is used on 1987 and earlier models. This system uses an externally mounted ESC module which is located on the firewall **(see illustration)**. When the knock sensor detects abnormal vibration (spark knock) on these models, the ESC module turns off the circuit to the ECM. The ECM then retards the EST until spark knock is eliminated. Failure of the ESC knock sensor signal or loss of ground at the ESC module will cause the signal to the ECM to remain high. This condition will result in the ECM controlling the EST as if no spark knock is occurring. Therefore, no retard will occur and spark knock may become severe under heavy engine load conditions.

4   The second system is used on 1988 and later models which incorporates the ESC module into the MEM-CAL chip and is located in the ECM. On this system the ECM supplies 5 volts to the knock sensor which contains a resistor that pulls available line voltage down to 2.5 volts under a no knock condition. When the knock sensor detects abnormal vibration (spark knock), the knock sensor sends a wave forming AC signal back to the ECM, thus indicating a spark knock condition.

Check

5   Connect a timing light to the engine according to the manufacturers instructions.

With the engine running at approximately 1500 rpm carefully tap on the engine block with a hammer near the knock sensor while monitoring the ignition timing with the light. The timing should retard with each tap of the hammer.

6   If the timing does not retard on 1987 and earlier models, disconnect the electrical connector to the knock sensor. Connect the lead of a 12-volt test light to the positive terminal of the battery and touch the terminal of the knock sensor harness connector with the tip of the test light while monitoring the ignition timing with the timing light. The timing should retard each time the terminal is touched with the test light. If it does, the knock sensor is defective. If timing still did not retard, the ESC module may be defective, check the following wiring and connections at the ESC harness connector **(see illustration 17.3)** and repair as necessary:

   a) *Check for battery voltage with the key On at the pink/black wire terminal.*
   b) *Check for continuity to ground at the black/white wire terminal.*
   c) *Check for continuity to the knock sensor at the dark blue terminal.*
   d) *Check for continuity to the ECM at the yellow/black wire terminal.*

7   If the timing does not retard on 1988 and later models, disconnect the electrical connector to the knock sensor. Place the voltmeter to the lowest AC voltage scale and connect the negative probe (-) to ground and the positive probe to the sensor terminal on the knock sensor. With the voltmeter connected, tap sharply on the engine block with a hammer or similar device (this simulates the knock from the engine) and observe voltage fluctuations on the meter. If no voltage fluctuations can be detected, the sensor is bad and should be replaced with a new part. If the knock sensor is OK, check for a 5 volt DC reference signal from the ECM with the ignition key On (engine not running). There should be 5 volts. If reference voltage is not present at the knock sensor the ESC portion of the MEM-CAL chip may be defective.

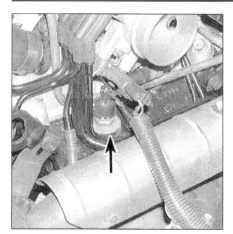

17.8 Knock sensor (arrow) location (1987 and earlier Buick and Oldsmobile models)

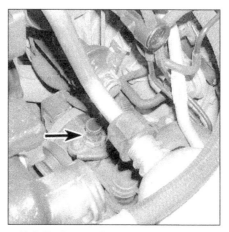

18.6 Typical A.I.R control valve (arrow) (Eldorado and Seville models)

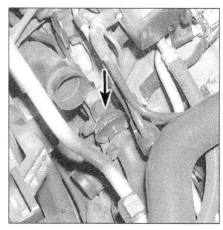

18.7 Typical one way check valve (arrow)

Component replacement

Knock sensor

Refer to illustration 17.8

8   On 1987 and earlier models, the knock sensor is located on the engine block below the throttle body **(see illustration)**. On 1988 and later models it is located at the lower right side of the engine block, above the right hand driveaxle.

9   Disconnect the electrical connector and remove the knock sensor from the block.

10   Installation is the reverse of the removal.

ESC module

11   On 1987 and earlier models, disconnect the electrical connector, remove the module mounting screws and remove the module **(see illustration 17.3)**.

12   Installation is the reverse of removal.

13   On 1988 and later models it will be necessary to remove the ECM and replace the MEM-CAL chip (see Section 3).

---

18 Air Injection Reaction (AIR) system (Cadillac models)

General description

1   The AIR system helps reduce hydrocarbons and carbon monoxide levels in the exhaust by injecting air into the exhaust ports of each cylinder during cold engine operation, or directly into the catalytic converter during normal operation. It also helps the catalytic converter reach proper operating temperature quickly during warm-up.

2   The AIR system uses an air pump to force the air into the exhaust stream. An air management valve, controlled by the vehicle's Electronic Control Module (ECM), directs the air to the correct location, depending on engine temperature and driving conditions. During certain situations, such as deceleration, the air is diverted to the air cleaner to prevent backfiring from too much oxygen in the exhaust stream. One-way check valves are also used in the AIR system

tubing to prevent exhaust gases from being forced back through the system.

3   The AIR system components include an engine driven air pump, air control, air switching and divert management valves, air flow and control hoses, check valves and a catalytic converter.

Check

Refer to illustrations 18.6 and 18.7

4   Because of the complexity of this system, it is difficult for the home mechanic to make an accurate diagnosis. If the system is malfunctioning, individual components can be checked.

5   Begin any inspection by carefully checking all hoses, vacuum lines and wires. Be sure they are in good condition and that all connections are tight and clean. Also make sure that the pump drivebelt is in good condition and properly adjusted (see Chapter 1).

6   To check the pump, allow the engine to reach normal operating temperature and run it at about 1500 rpm. Locate the air pump output hose running from the air pump **(see illustration)** to the control valve assembly and squeeze it to check for pulsations. Have an assistant increase the engine speed and check for a parallel increase in air flow. If an increase is noted, the pump is functioning properly. If not, there is a fault in the pump.

7   Each check valve can be inspected after removing it from the hose **(see illustration)**. Make sure the engine is completely cool before attempting this. Try to blow through the check valve from both ends. Air should pass through it only in the direction of normal air flow. If it is stuck either open or closed, the valve should be replaced.

8   To check the air control valve assembly, disconnect the vacuum signal hose at the valve. With the engine running, check for vacuum at the hose. If none is felt, the line is clogged or the solenoid may be malfunctioning.

Component replacement

Drivebelt

9   The A.I.R system on these models shares a single serpentine drivebelt that is used to run all engine accessories at the front of the engine. Drivebelt tension is maintained automatically by a spring loaded tensioner. Refer to Chapter 1 for the drivebelt removal and installation procedures.

AIR pump pulley and filter

Refer to illustrations 18.12, 18.13 and 18.15

10   Compress the drivebelt to keep the pulley from turning and loosen the pulley bolts.

11   Remove the drivebelt as described in Chapter 1.

12   Remove the mounting bolts and lift off the pulley **(see illustration)**.

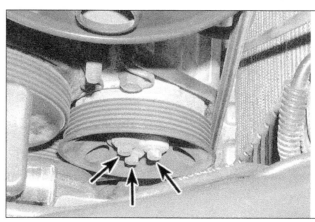

18.12 Loosen the pulley bolts (arrows) with the drivebelt and tensioner compressed

18.13 On Deville and Fleetwood models, the filter can be pulled out of the air pump with a pair of needle-nose pliers

18.15 On Eldorado and Seville models, simply pry up the plastic cover (arrow) and remove the filter

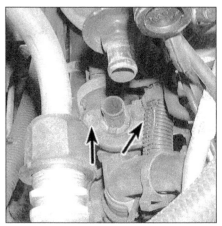

18.24 A.I.R control valve mounting bolts (Eldorado and Seville models shown)

13   On Deville and Fleetwood models, if the fan-like filter must be removed, grasp it firmly with a pair of needle-nose pliers **(see illustration)** and pull it from the pump. **Note:** *Do not insert a screwdriver between the filter and pump housing as the edge of the housing could be damaged. The filter will usually be distorted when pulled off. Be sure no fragments fall into the air intake hose.*

14   The new filter is installed by placing it in position on the pump, placing the pulley over it and tightening the pulley bolts evenly to draw the filter into the pump. Do not attempt to install a filter by pressing or hammering it into place. **Note:** *It is normal for the new filter to have an interference fit with the pump housing and, upon initial operation, it may squeal until worn in.*

15   On Eldorado and Seville models the filter is easily accessible by prying up the cover and removing the filter from the A.I.R air cleaner assembly **(see illustration)**. **Note:** *On these models it is not necessary to remove the pulley to change the filter, although it will be necessary to remove the pulley if the A.I.R pump is to be removed.*

16   To install the pulley on any model, reposition it back onto the air pump and hand tighten the bolts. Install the drivebelt and, while compressing the belt, tighten the pulley bolts securely.

Hoses and tubes

17   To replace a tube or hose, always note how it is routed first, either with a sketch or with numbered pieces of tape.

18   Remove the defective hose or tube and replace it with a new one of the same material and size and tighten all connections.

Check valve

19   Disconnect the pump outlet hose at the check valve.

20   Unscrew the check valve from the pipe assembly. Be careful not to bend or twist the assembly.

21   Install a new valve after making sure that it is a duplicate of the part removed, then tighten all connections.

Air control valve

Refer to illustration 18.24

22   Disconnect the negative battery cable at the battery. Remove the air cleaner if necessary.

23   Disconnect the vacuum signal line from the valve. Disconnect the air hoses and wiring connectors.

24   If the mounting bolts are retained by tabbed lockwashers, bend the tabs back, then remove the mounting bolts and lift the valve off the adapter or bracket **(see illustration)**.

25   Installation is the reverse of the removal procedure. Be sure to use a new gasket when installing the valve.

Air pump

26   Remove the air management valve and adapter, if so equipped.

27   Loosen the pulley retaining bolts must be prior to removing the drivebelt. Remove the drivebelt and the A.I.R pump pulley.

28   Detach the inlet and outlet hoses or pipes from the rear of the pump.

29   Remove the pump mounting bolts and separate the pump from the engine.

30   Installation is the reverse of removal. **Note:** *Do not tighten the pump mounting bolts until all components are installed.*

31   Following installation, install pump pulley and the drivebelt (see Chapter 1).

---

19   Early Fuel Evaporation (EFE) system (Cadillac models)

General description

1   This system is used on 1986 and 1987 models to rapidly heat the intake air supply to promote evaporation of the fuel when the engine is started cold.

2   The components involved in the EFE's operation include a ceramic electronic grid mounted beneath the throttle body, a control relay, the IAT sensor, the coolant temperature sensor and the ECM.

3   Current is applied to the EFE heater grid when engine is cold and the following conditions exist:

a) *Battery voltage is above 10 volts*
b) *The IAT sensor is less than 165 degrees F*
c) *The coolant temperature sensor is less than 220 degrees F*

4   After one to ten minutes the EFE operation should turn off when the engine warms to normal operating temperature. If the EFE system is not functioning, poor cold engine performance will result. If the system is not shutting off when the engine is warmed up, the engine will run as if it is out of tune and also may overheat.

Check

Refer to illustration 19.5

5   With the engine cold, disconnect the electrical connector from the heater grid **(see illustration)**. Check for battery voltage between the terminals on the wiring harness side of the connector. If power does not exist, check for continuity to ground on the black wire of the connector. If the ground is OK, check the relay and or the circuit from the ECM. If the relay and the circuit from the

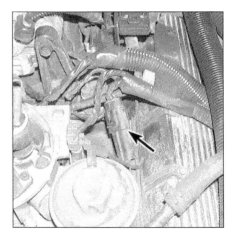

19.5 The EFE heater grid connector (arrow) is located between the throttle body and the rear valve cover

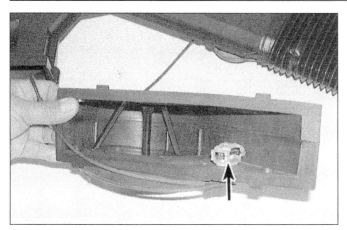

20.2 The THERMAC temperature sensor (arrow) is located inside the air cleaner housing

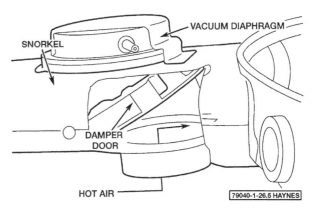

20.6 When the engine is cold, the damper door closes off the snorkel passage, allowing air warmed by the exhaust to enter the throttle body

ECM are good, check the operation of the coolant temperature and the IAT sensors.

6   If voltage is present, check for continuity between the heater grid terminals. Continuity should exist. If continuity is not present, the heater is defective and should be replaced.

Heater replacement

7   Disconnect the electrical connector.
8   Remove the throttle body (see Chapter 4).
9   Remove the heater grid from the intake manifold.
10   Installation is the reverse of removal. Always be sure to use new gaskets when installing the throttle body.

20   Thermostatic Air Cleaner (THERMAC) system (Cadillac models)

General description

Refer to illustration 20.2

1   The thermostatic air cleaner (THERMAC) system improves engine efficiency and driveability under varying climatic conditions by controlling the temperature of the air coming into the air cleaner. A uniform incoming air temperature allows leaner air/fuel ratios during warm-up, which reduces hydrocarbon emissions.
2   The system uses a damper assembly, located in the snorkel of the air cleaner housing, to control the ratio of cold and warm air directed into the throttle body. This damper is controlled by a vacuum motor which is, in turn, modulated by a temperature sensor in the air cleaner **(see illustration)**. On some engines a check valve is used in the sensor, which delays the opening of the damper when the engine is cold and the vacuum signal is low.
3   It is during the first few miles of driving (depending on outside temperature) that this system has its greatest effect on engine performance and emissions output. When the

engine is cold, the damper blocks off the air cleaner inlet snorkel, allowing only warm air from around the exhaust manifold to enter the engine. As the engine warms up, the damper gradually opens the snorkel passage, increasing the amount of cold air allowed into the air cleaner. By the time the engine has reached its normal operating temperature, the damper opens completely, allowing only cold, fresh air to enter.
4   Because of this "cold engine only" function, it is important to periodically check this system to prevent poor engine performance when cold, or overheating of the fuel mixture once the engine has reached operating temperatures. If the air cleaner damper sticks in the "no heat" position the engine will run poorly, stall and waste gas until it has warmed up on its own. A valve sticking in the "heat" position causes the engine to run as if it is out of tune due to the constant flow of hot air to the throttle body.

Check

Refer to illustrations 20.6 and 20.7

5   Detach the flexible air duct attached to the end of the snorkel, leading to an area behind the grille, disconnect it at the snorkel. This will enable you to look through the end of the snorkel and see the damper inside.

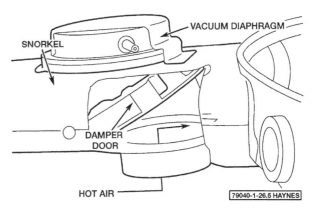

20.7 As the engine warms up, the damper door moves down to close off the heat stove passage and open the snorkel passage so outside air can enter the throttle body

6   The check should be done when the engine is cold. Start the engine and look through the snorkel at the damper, which should move to a closed position. With the damper closed, air cannot enter through the end of the snorkel, but instead enters the air cleaner through the flexible duct attached to the exhaust manifold and the heat stove passage **(see illustration)**.
7   As the engine warms up to operating temperature, the damper should open to allow air through the snorkel end **(see illustration)**. Depending on outside temperature, this may take 10 to 15 minutes. To speed up this check you can reconnect the snorkel air duct, drive the vehicle, then check to see if the damper is completely open.
8   If the damper does not close off snorkel air when the cold engine is started, disconnect the vacuum hose at the snorkel vacuum motor and place your thumb over the hose end, checking for vacuum. Replace the vacuum motor if the hose routing is correct and vacuum is present.
9   If there was no vacuum going to the motor in the above test, check the hoses for cracks, crimped areas and proper connection. If the hoses are clear and in good condition, replace the temperature sensor inside the air cleaner housing.

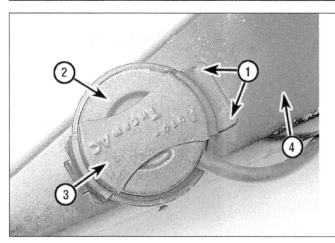

20.11 To remove the vacuum motor, drill out the spot welds and detach the retaining strap

1   *Spot welds*
2   *Vacuum diaphragm motor*
3   *Retaining strap*
4   *Snorkel*

20.17 On Eldorado and Seville models simply depress the motor retaining clip (arrow) and detach the motor from the air cleaner housing

Component replacement

Note: *Check on parts availability before attempting component replacement. Some components may no longer be available for all models, requiring replacement of the complete air cleaner assembly.*

Air cleaner vacuum motor

10   Remove the air cleaner assembly from the engine and disconnect the vacuum hose from the motor.

Deville and Fleetwood models

Refer to illustration 20.11

11   Drill out the two spot welds which secure the vacuum motor retaining strap to the snorkel tube **(see illustration)**.

12   Remove the motor retaining strap.

13   Lift up the motor, cocking it to one side to unhook the motor linkage at the damper assembly.

14   To install, drill a 7/64-inch hole in the snorkel tube at the center of the vacuum motor retaining strap.

15   Insert the vacuum motor linkage into the damper assembly.

16   Using the sheet metal screw supplied with the motor service kit, attach the motor and retaining strap to the snorkel. Make sure the sheet metal screw does not interfere with the operation of the damper door. Shorten the screw if necessary.

Eldorado and Seville models

Refer to illustration 20.17

17   Depress the motor retaining strap and disengage it from the air cleaner housing **(see illustration)**.

18   Lift up the motor, cocking it to one side to unhook the motor linkage at the damper assembly.

19   To install, place the actuator in position in the housing and engage the spring.

20   Connect the vacuum hose to the motor and install the air cleaner assembly.

Air cleaner temperature sensor

Refer to illustration 20.23

21   Remove the air cleaner from the engine and disconnect the vacuum hoses at the sensor.

22   Carefully note the position of the sensor. The new sensor must be installed in exactly the same position.

23   Pry up the tabs on the sensor retaining clip and remove the sensor and clip from the air cleaner **(see illustration)**.

24   Install the new sensor with a new gasket in the same position as the old one.

25   Press the retaining clip onto the sensor. Do not damage the control mechanism in the center of the sensor.

26   Connect the vacuum hoses and attach the air cleaner to the engine.

21   Positive Crankcase Ventilation (PCV) system

General description

Refer to illustration 21.1

1   The Positive Crankcase Ventilation (PCV) system reduces hydrocarbon emissions by circulating fresh air through the crankcase to pick up blow-by gases, which are then rerouted through a PCV valve to the throttle body or intake manifold **(see illustration)** to be burned in the engine.

2   The main components of the PCV system are vacuum hoses and the PCV valve, which regulates the flow of gases according to engine speed and manifold vacuum.

Check

3   The PCV system can be checked quickly and easily for proper operation. It should be checked regularly because carbon and gunk deposited by blow-by gases will eventually clog the PCV valve and system hoses. The common symptoms of a plugged or clogged PCV valve include a rough idle, stalling or a slow idle speed, oil leaks, oil in the air cleaner or sludge in the engine.

20.23 The sensor retainer must be pried off with a screwdriver

21.1 Typical PCV valve (arrow) location (Cadillac models)

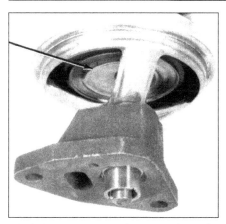

22.8 Push up on the diaphragm (arrow) when checking the EGR valve - if the diaphragm is stuck or binds in the up position, replace the valve with a new one

4   Refer to Chapter 1 for PCV valve checking and replacement procedures.

22   Exhaust Gas Recirculation (EGR) system

General description

1   The EGR system meters exhaust gases into the engine induction system through passages cast into the intake manifold. From there the exhaust gases pass into the fuel/air mixture for the purpose of lowering combustion temperatures, thereby reducing the amount of oxides of nitrogen (NOx) formed.
2   The amount of exhaust gas admitted is regulated by a negative or positive backpressure controlled (EGR) valve in response to engine operating conditions. The EGR valve is under the control of the EGR vacuum control solenoid valve, which, in turn, is under the control of the PCM. The vacuum signal to the EGR valve is controlled by varying the duty cycle (on-time) of the solenoid valve. The duty cycle is calculated by the computer using information from the ECT, MAP or MAF and the TPS sensors.

3   1988 through 1990 Buick and Oldsmobile models use a digital EGR valve that feeds small amounts of exhaust gas back into the intake manifold and then into the combustion chamber independent of intake manifold vacuum. The digital EGR valve operates without intake manifold vacuum thereby allowing accurate (finer) amounts of exhaust gas to be recirculated back into the intake system. A PCM-controlled pintle within the digital EGR valve passes exhaust gasses through a small orifice upon command from the PCM. This EGR valve operates similar to the stepper motor type particular to IAC valves that control idle quality. By sending a 5 volt reference signal to the EGR valve, the PCM detects and sets the pintle position with the feedback signal from the EGR valve.
4   The On-Board Diagnostic (OBD-I) system can detect EGR system problems and set trouble codes to indicate specific faults.
5   Common engine problems associated with the EGR system are rough idling or stalling at idle, rough engine performance during light throttle application and stalling during deceleration.

Check

Non-digital type

Refer to illustration 22.8

6   Start the engine, warm it up to normal operating temperature and allow the engine to idle. Manually lift the EGR diaphragm. The engine should stall, or at least the idle should drop considerably.
7   If the EGR valve appears to be in proper operating condition, carefully check all hoses connected to the valve for breaks, leaks or kinks. Replace or repair the valve/hoses as necessary.
8   With the engine idling at normal operating temperature, disconnect the vacuum hose from the EGR valve and connect a vacuum pump **(see illustration)**. When vacuum is applied, the engine should stumble or die, indicating the vacuum diaphragm is operating properly. **Note:** *These models use a backpressure-type EGR valve. Backpressure must be created in the exhaust system before the*

vacuum pump will actuate the valve. To create backpressure, install a large socket into the tailpipe and clamp it to the pipe to prevent it from dropping out during the test. The 1/2 inch hole in the socket will allow the engine to idle but still create backpressure on the EGR valve. Don't restrict the exhaust system any longer than necessary to perform this test. Replace the EGR valve with a new one if the test does not affect the idle.
9   Check the operation of the EGR control solenoid. Check for vacuum to the solenoid. If vacuum exists, check for battery voltage to the solenoid.
10   Further testing of the EGR system, vacuum solenoid and the PCM will require a SCAN tool to access computer information that directly controls the EGR system. Have the vehicle checked by a dealer service department or other qualified repair facility.

Digital type (1988 through 1990 Buick and Oldsmobile models)

11   Special electronic diagnostic equipment is needed to check this valve and should be left to a dealer service department or other qualified repair facility.

Component replacement

EGR valve

Refer to illustrations 22.13a and 22.13b

12   Disconnect the vacuum hose (non-digital type) or e electrical connector (digital type) from the EGR valve.
13   Remove the nuts or bolts which secure the valve to the intake manifold or adapter **(see illustrations)**.
14   Lift the EGR valve from the engine.
15   Clean the mounting surfaces of the EGR valve. Remove all traces of gasket material.
16   Place the new EGR valve, with a new gasket, on the intake tube or adapter and tighten the attaching nuts or bolts.
17   Connect the vacuum signal hose or the electrical connector.

EGR vacuum solenoid

Refer to illustrations 22.19a and 22.19b

18   Remove the hoses from the EGR vac-

**22.13a EGR valve mounting bolt (arrow)
(1986 Oldsmobile model shown)**

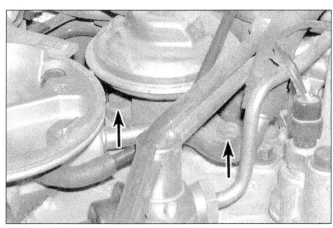

**22.13b EGR valve mounting bolts (arrows)
(1991 Cadillac model shown)**

22.19a Typical EGR valve solenoid location (arrow) (1986 and 1987 3.8L V6 engine)

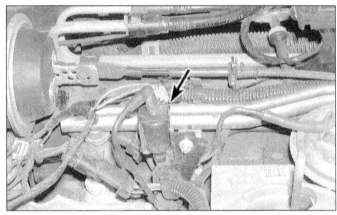

22.19b Typical EGR valve solenoid location (arrow) (V8 engines)

uum control solenoid, labeling them to ensure proper installation.
19   Remove the solenoid and replace it with the new one **(see illustrations)**.

---

23   Evaporative Emission Control System (EECS)

General description

Refer to illustration 23.2
1   This system is designed to trap and store fuel vapors that evaporate from the fuel tank, throttle body and intake manifold.
2   The Evaporative Emission Control System (EECS) consists of a charcoal-filled canister and the lines connecting the canister to the fuel tank, ported vacuum and intake manifold vacuum **(see illustration)**.
3   Fuel vapors are transferred from the fuel tank, throttle body and intake manifold to a canister where they are stored when the engine is not operating. When the engine is running, the fuel vapors are purged from the canister by intake air flow and consumed in the normal combustion process.
4   On later models there is a purge valve in the canister and, on some models, a tank pressure control valve. On some models the purge valve is operated by the ECM.
5   An indication that the system is not operating properly is a strong fuel odor. Poor idle, stalling and poor driveability can be caused by an inoperative purge valve, a damaged canister, split or cracked hoses or hoses connected to the wrong tubes.

Check

6   Checking and maintenance procedures for the EECS system canister and hoses are included in Chapter 1.
7   To check the purge valve, apply a short length of hose to the lower tube of the purge valve assembly and attempt to blow through it. Little or no air should pass into the canister (a small amount of air will pass because the canister has a constant purge hole).
8   With a hand vacuum pump, apply vacuum through the control vacuum signal tube

to the purge valve diaphragm. If the diaphragm does not hold vacuum for at least 20 seconds, the diaphragm is leaking and the canister must be replaced.
9   If the diaphragm holds vacuum, again try to blow through the hose while vacuum is still being applied. An increased flow of air should be noted. If it isn't, replace the canister.

Canister replacement

10   Disconnect the hoses and electrical connections (if equipped) from the canister.
11   Remove the pinch bolt and loosen the clamp.
12   Remove the canister.
13   Installation is the reverse of removal.

---

24   Catalytic converter

Note: *Because of a Federally-mandated extended warranty which covers emissions-related components such as the catalytic converter, check with a dealer service department before replacing the converter at your own expense.*

General description

1   The catalytic converter is an emission control device added to the exhaust system to reduce pollutants from the exhaust gas stream. A single-bed converter design is used in combination with a three-way (reduction) catalyst. The catalytic coating on the three-way catalyst contains platinum and rhodium, which lowers the levels of oxides of nitrogen (NOx) as well as hydrocarbons (HC) and carbon monoxide (CO).

Check

2   The test equipment for a catalytic converter is expensive and highly sophisticated. If you suspect that the converter on your vehicle is malfunctioning, take it to a dealer or authorized emissions inspection facility for diagnosis and repair.
3   Whenever the vehicle is raised for servicing of underbody components, check the converter for leaks, corrosion, dents and other

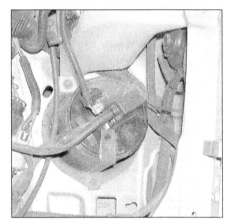

23.2 Typical charcoal-filled canister

damage. Check the welds/flange bolts that attach the front and rear ends of the converter to the exhaust system. If damage is discovered, the converter should be replaced.
4   Although catalytic converters don't break too often, they can become plugged. The easiest way to check for a restricted converter is to use a vacuum gauge to diagnose the effect of a blocked exhaust on intake vacuum.

 a) *Connect a vacuum gauge to an intake manifold vacuum source.*
 b) *Warm the engine to operating temperature, place the transaxle in park and apply the parking brake.*
 c) *Note and record the vacuum reading at idle.*
 d) *Open the throttle until the engine speed is about 2000 rpm.*
 e) *Release the throttle quickly and record the vacuum reading.*
 f) *Perform the test three more times, recording the reading after each test.*
 g) *If the reading after the fourth test is more than one in-Hg lower than the reading recorded at idle, the catalytic converter, muffler or exhaust pipes may be plugged or restricted.*

Component replacement

5   Refer to the exhaust system servicing section in Chapter 4.

Chapter 7
Automatic transaxle

Contents

Specifications

General

Fluid type and capacity	See Chapter 1

Torque specifications

Ft-lbs (unless otherwise indicated)

Brake/transaxle interlock assembly bolts	144 in-lbs
Floor shift mounting bolts	18
Transaxle lever nut-to-Park/Neutral switch	15
Oil cooler bracket bolts	15
Park/Neutral switch-to-transaxle case bolts	20
Shift cable-to-transaxle lever nut	20
Shift cable bracket bolts (at transaxle)	18
Subframe assembly	
Left subframe member-to-right subframe member bolts	40
Subframe-to-chassis bolts	74
Transaxle-to-engine bolts	55
Torque converter-to-driveplate bolts	46
Torque converter cover bolts	60 in-lbs
Transaxle cradle mounting bolts and nuts	
1990 and earlier	
Upper mounting bolts	70
Lower mounting bolts	50
Lower mounting nuts	35
1991 and later	
Upper mounting bolts	
Horizontal bolt	55
Vertical bolt	43
Lower mounting bolts	47
Lower mounting nuts	30
Throttle valve cable-to-transaxle case bolt	72 in-lbs

1.1 An underside view of an automatic transaxle and its related components (1987 Seville shown)

1	*Driveplate*	*2*	*Left subframe member*	*3*	*Transaxle fluid pan*	*4* *Right subframe member*

1   General information

Refer to illustration 1.1

The vehicles covered by this manual are equipped with a four-speed Turbo Hydra-Matic automatic transaxle **(see illustration)**. Three models of the Turbo Hydra-Matic automatic transaxles are used on these vehicles: the THM 440-T4, THM 4T60 and the THM 4T60-E.

The Turbo Hydro-Matic transaxle does not drive the transaxle geartrain directly. The torque converter drives two sprockets and a chain that transfers the power from the engine to the geartrain positioned below the crankshaft centerline. The drive chain is located at the end of the transaxle case on the driver's side of the vehicle.

The Turbo Hydra-Matic transaxle can be easily identified by the identification tag riveted onto the right rear of the transaxle case near the right rear driveshaft. Refer to the Introductory Pages for the location and illustration of the transaxle identification tag.

1991 and later Turbo Hydra-Matic transaxles are designated 4T60-E. These updated transaxles are equipped with electronic control solenoids that activate shift sequences in response to computer control. Four solenoids respond to the PCM signals; the SHIFT A and SHIFT B solenoids control transaxle shifting while the VCC APPLY solenoid and the VCC PWM solenoid controls the application of the torque converter clutch. Refer to Chapter 6 for additional information on the self-diagnosis system and accessing trouble codes.

Cadillac models are equipped with a Viscous Converter Clutch (VCC) instead of the typical Torque Converter Clutch (TCC) system. The VCC system has different operating parameters than an automatic transaxle with TCC, The "chuggle" or "surge" normally experienced with the TCC systems, is eliminated by the VCC MOD solenoid which controls the apply of the converter clutch using a pulse width modulated (PWM) grounding signal. This allows smoother transition into lock-up.

Due to the complexity of the clutches and the hydraulic control system, and because of the special tools and expertise required to perform an automatic transaxle overhaul, it should not be undertaken by the home mechanic. Therefore, the procedures in this Chapter are limited to general diagnosis, routine maintenance, adjustment and transaxle removal and installation.

If the transaxle requires major repair work, it should be left to a dealer service department or an automotive or transmission repair shop. You can, however, remove and install the transaxle yourself and save the expense, even if the repair work is done by a transmission shop (but be sure a proper diagnosis has been made before removing the transaxle).

Replacement and adjustment procedures the home mechanic can perform include those involving the throttle valve cable and the shift linkage.

Caution: *Never tow a disabled vehicle with an automatic transaxle at speeds greater than 35 mph or for distances over 50 miles if the front wheels are on the ground.*

2 Diagnosis - general

Note: *Automatic transaxle malfunctions may be caused by five general conditions: poor engine performance, improper adjustments, hydraulic malfunctions, mechanical malfunctions or malfunctions in the computer or its signal network. Diagnosis of these problems should always begin with a check of the easily repaired items: fluid level and condition (see Chapter 1) and shift linkage adjustment. Next, perform a road test to determine if the problem has been corrected or if more diagnosis is necessary. If the problem persists after the preliminary tests and corrections are completed, additional diagnosis should be done by a dealer service department or transmission repair shop. Refer to the Troubleshooting section at the front of this manual for information on symptoms of transaxle problems.*

Preliminary checks

1   Drive the vehicle to warm the transaxle to normal operating temperature.
2   Check the fluid level as described in Chapter 1:
   a) *If the fluid level is unusually low, add enough fluid to bring the level within the designated area of the dipstick, then check for external leaks (see below).*
   b) *If the fluid level is abnormally high, drain off the excess, then check the drained fluid for contamination by coolant. The presence of engine coolant in the automatic transaxle fluid indicates that a failure has occurred in the internal radiator walls that separate the coolant from the transaxle fluid (see Chapter 3).*
   c) *If the fluid is foaming, drain it and refill the transaxle, then check for coolant in the fluid or a high fluid level.*
3   Check the engine idle speed as described in Chapter 1. **Note:** *If the engine is malfunctioning, do not proceed with the preliminary checks until it has been repaired and runs normally.*
4   Check the shift cable (see Section 4). Make sure it's properly adjusted and operates smoothly.
5   On 1990 and earlier models, check the throttle valve cable (see Section 5). Make sure it's properly adjusted and operates smoothly. Check the vacuum modulator and its base for leaks (see Section 8).

Fluid leak diagnosis

6   Most fluid leaks are easy to locate visually. Repair usually consists of replacing a seal or gasket. If a leak is difficult to find, the following procedure may help.
7   Identify the fluid. Make sure it's transaxle fluid and not engine oil or brake fluid (automatic transaxle fluid is a deep red color).
8   Try to pinpoint the source of the leak. Drive the vehicle several miles, then park it over a large sheet of cardboard. After a minute or two, you should be able to locate the leak by determining the source of the fluid dripping onto the cardboard.
9   Make a careful visual inspection of the suspected component and the area immediately around it. Pay particular attention to gasket mating surfaces. A mirror is often helpful for finding leaks in areas that are hard to see.
10   If the leak still cannot be found, clean the suspected area thoroughly with a degreaser or solvent, then dry it.
11   Drive the vehicle for several miles at normal operating temperature and varying speeds. After driving the vehicle, visually inspect the suspected component again.
12   Once the leak has been located, the cause must be determined before it can be properly repaired. If a gasket is replaced but the sealing flange is bent, the new gasket will not stop the leak. The bent flange must be straightened.
13   Before attempting to repair a leak, check to make sure that the following conditions are corrected or they may cause another leak. **Note:** *Some of the following conditions cannot be fixed without highly specialized tools and expertise. Such problems must be referred to a transmission repair shop or a dealer service department.*

Gasket leaks

14   Check the pan periodically. Make sure the bolts are tight, no bolts are missing, the gasket is in good condition and the pan is flat (dents in the pan may indicate damage to the valve body inside).
15   If the pan gasket is leaking, the fluid level or the fluid pressure may be too high, the vent may be plugged, the pan bolts may be too tight, the pan sealing flange may be warped, the sealing surface of the transaxle housing may be damaged, the gasket may be damaged or the transaxle casting may be cracked or porous. If sealant instead of gasket material has been used to form a seal between the pan and the transaxle housing, it may be the wrong type of sealant.

Seal leaks

16   If a transaxle seal is leaking, the fluid level or pressure may be too high, the vent may be plugged, the seal bore may be damaged, the seal itself may be damaged or improperly installed, the surface of the shaft protruding through the seal may be damaged or a loose bearing may be causing excessive shaft movement.
17   Make sure the dipstick tube seal is in good condition and the tube is properly seated. Periodically check the area around the speedometer gear or sensor for leakage. If transaxle fluid is evident, check the O-ring for damage.

Case leaks

18   If the case itself appears to be leaking, the casting is porous and will have to be repaired or replaced.
19   Make sure the oil cooler hose fittings are tight and in good condition.

Fluid comes out vent pipe or fill tube

20   If this condition occurs, the transaxle is overfilled, there is coolant in the fluid, the case is porous, the dipstick is incorrect, the vent is plugged or the drain-back holes are plugged.

3   Shift lever (floor shift models) - removal and installation

Refer to illustration 3.6
Warning: *On models equipped with a Supplemental Inflatable Restraint (SIR) system (more commonly known as airbags), always disable the airbag system before working in the vicinity of any airbag system component to avoid the possibility of accidental deployment of the airbag, which could cause personal injury (see Chapter 12).*
1   Disconnect the negative cable from the battery.
2   Remove the console (see Chapter 11). **Note:** *On some models, it will be necessary to remove the handle retainer clip and separate the handle from the shift lever.*
3   Disconnect the shift cable from the shift lever (see Section 4).
4   Disconnect the park lock cable from the shift lever (see Section 6)
5   Disconnect any harness connectors from the console area.
6   Remove the retaining nuts and lift the floor shift control assembly out of the vehicle **(see illustration)**.

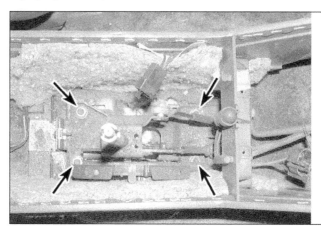

3.6 Remove the four mounting nuts (arrows) to separate the floorshift assembly from the body

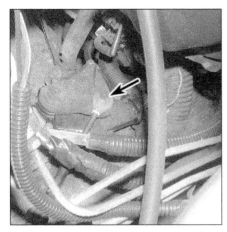

4.3 Use a screwdriver to pry up on the cable end (arrow) - snap-fit type shift cable shown

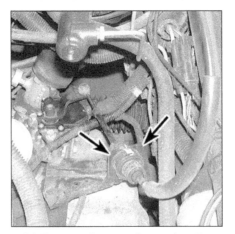

4.4 Squeeze the tangs (arrows) on the backside of the cable housing to release the grommet from the bracket

4.6 Separate the shift cable from the lever pin on the shift lever assembly

7   To install the shift lever assembly, place the floor shift control assembly in position on the mounting studs and install the nuts. Tighten the nuts to the torque listed in this Chapter's Specifications.

8   Connect the shift lever and the park lock cables.

9   Install the console (see Chapter 11).

10   Reconnect the negative battery cable.

4   Shift cable - replacement and adjustment

Warning: *On models equipped with a Supplemental Inflatable Restraint (SIR) system (more commonly known as airbags), always disable the airbag system before working in the vicinity of any airbag system component to avoid the possibility of accidental deployment of the airbag, which could cause personal injury (see Chapter 12).*

Floor shift models

Refer to illustrations 4.3, 4.4, 4.6 and 4.7

1   Disconnect the cable from the negative battery terminal.

2   Position the shift lever into Neutral position.

3   Working in the engine compartment, remove the nut and disconnect the shift cable from the lever on the transaxle. **Note:** *Some models are equipped with a snap-fit over a ball stud which can be removed using a dull bladed screwdriver and carefully prying the cable end* **(see illustration)**.

4   To detach the shift cable from the transaxle bracket, remove the U-clip retainer (if used), then squeeze the tangs on the cable housing grommet and pull the cable and grommet through the bracket **(see illustration)**.

5   Working inside the passenger compartment, remove the center console (see Chapter 11).

6   Working in the center console area in the passenger compartment, pull the retain-

ing clip from the end of the shift cable located on the shift lever pin **(see illustration)**.

7   Remove the clip at the cable housing end, then pry out the grommet **(see illustration)** from the front of the shift lever assembly and pull out the shift cable through the hole in the mount. You may have to pull back the carpeting to expose the cable. **Note:** *There are two different types of grommets that retain the shift cable to the shift lever assembly. The retaining clip grommet uses a single clip to lock the cable housing to the bracket while the spring grommet uses a spring to pressurize the locating tangs.*

8   To install the shift cable, guide the new cable through the hole in the floor and install the grommet.

9   Connect the shift cable to the shift lever.

10   Install the center console (see Chapter 11).

11   Place the shift control lever (inside the vehicle) in Neutral. Make sure it remains in the Neutral position until the shift cable is installed.

12   Attach the shift cable to the transaxle bracket.

13   Place the shift lever on the transaxle in

the Neutral position by rotating it counter-clockwise from Park through Reverse into Neutral.

14   Connect the shift cable end to the shift lever on the transaxle, and install new cable-ties where needed.

15   Adjust the shift cable (see Steps 25 through 28).

Column shift models

Refer to illustrations 4.21 and 4.22

16   Disconnect the cable from the negative battery terminal.

18   Working in the engine compartment, disconnect the shift cable from the lever on the transaxle **(see illustration 4.3)**.

19   To detach the shift cable from the transaxle bracket, remove the U-clip retainer (if used), then squeeze the tangs on the cable housing grommet and pull the cable and grommet through the bracket **(see illustration 4.4)**.

20   Remove the lower sound insulator panel and the lower steering column cover (see Chapter 11).

21   Working on the steering column, disconnect the cable end from the shift control lever **(see illustration)**.

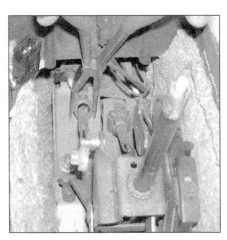

4.7 Pry the grommet from the floorshift bracket using pliers

4.21 Pry the shift cable from the lever on the steering column (arrow) - shift interlock solenoid removed for clarity

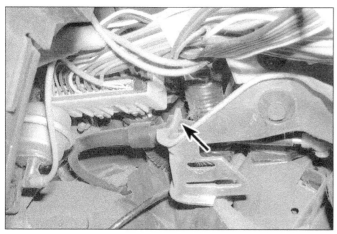

4.22 Squeeze the tabs (arrow) then pull the shift cable through the bracket housing

5.2 To detach the throttle valve cable from the throttle lever pin, grasp it firmly and lift up

22   Pry the cable grommet from the U-shaped mount on the steering column **(see illustration)**.
23   Pull the cable through the hole in the firewall to the engine side.
24   Installation is the reverse of removal. After installation, adjust the cable (see Steps 25 through 28).

Adjustment

Note: *Some models are equipped with a nut on the cable end at the transaxle lever that must be loosened while the correct shifting position is located, while other models are equipped with a lock tab on the shift cable grommet at the transaxle bracket. This lock tab, when pressed, locates the correct shift position and locks the cable in place.*

Nut type

25   The shift linkage should be adjusted so that the engine starts when the shift lever is positioned in Park or Neutral. First, position the shift lever (column or floor) into the Neutral position.
26   Loosen the cable retaining nut on the transaxle lever.
27   Make sure the shift lever is correctly

positioned in the Park or Neutral detent.
28   Tighten the shift cable nut to the torque listed in this Chapter's Specifications.

Lock tab type

29   The shift linkage should be adjusted so that the engine starts when the shift lever is positioned in Park or Neutral. First, position the shift lever into the Neutral position.
30   Make sure the transaxle lever is correctly positioned in the Neutral detent.
31   Push the lock tab on the cable grommet assembly to lock the shift cable into place.

---

5   Throttle valve cable (1986 through 1990 models) - replacement and adjustment

---

Note: *On 1991 and later models, the 4T60-E electronic controlled transaxle is not equipped with the throttle valve cable assembly.*

Replacement

Refer to illustrations 5.2, 5.3, 5.5 and 5.6
1   Detach the cable from the negative battery terminal.

2   Disconnect the throttle valve cable from the throttle lever by grasping the connector, pulling it forward to disconnect it and then lifting up and off the lever pin **(see illustration)**.
3   If equipped with a bracket, disconnect the throttle valve cable housing from the bracket by compressing the tangs and pushing the housing back through the bracket **(see illustration)**.
4   Disconnect any clips or straps retaining the cable to the transaxle.
5   Remove the bolt retaining the cable housing to the transaxle **(see illustration)**.
6   Pull up on the housing until the end of the cable can be seen, then disconnect it from the transaxle throttle valve link **(see illustration)**. Remove the cable from the vehicle.
7   To install the cable, connect it to the transaxle throttle valve link, push the housing into the transaxle and install the bolt. Tighten the bolt to the specified torque. Route the cable to the top of the engine, push the housing through the bracket until it clicks into place, place the connector over the throttle lever pin and pull back to lock it. Secure the cable with any retaining clips or straps.

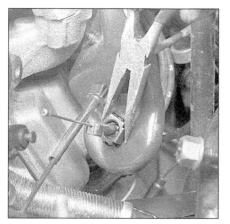

5.3 Use needle-nose pliers to compress the throttle valve cable tangs, then push the housing back through the bracket

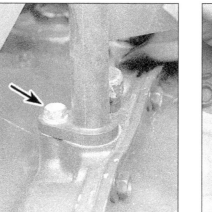

5.5 Remove the throttle valve cable bolt (arrow) and pull the housing out of the transaxle

5.6 Disconnect the throttle valve link from the cable end

5.9a To adjust the throttle valve cable, press down on the adjustment tab, push the cable housing into the adjuster assembly until it stops and release the adjustment tab - rotate the throttle lever to the wide open position and the cable housing will ratchet out of the adjuster assembly the proper distance

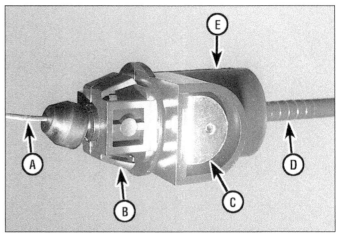

5.9b Details of the throttle valve cable adjuster assembly

A	Throttle valve cable	D	Cable casing
B	Locking lugs	E	Adjuster assembly
C	Release tab		

Adjustment

Refer to illustrations 5.9a and 5.9b

8   The engine MUST NOT be running during this adjustment.

9   Depress the adjustment tab and push the cable housing into the adjuster assembly (away from the throttle lever) as far as it will go **(see illustrations)**.

10   Release the adjustment tab.

11   Manually turn the throttle lever to the "wide open throttle" position, the cable housing will ratchet out of the adjuster assembly, then release the throttle lever. The cable is now adjusted. **Note:** *Don't use excessive force at the throttle lever to adjust the throttle valve cable. If great effort is required to adjust the cable, disconnect the cable at the transaxle end and check for free operation. If it's still difficult, replace the cable. If it's now free, suspect a bent throttle valve link in the transaxle or a problem with the throttle lever.*

12   Reattach the cable to the negative battery terminal.

---

6   Park/lock cable (floor shift models) - removal and installation

Warning: *On models equipped with a Supplemental Inflatable Restraint (SIR) system (more commonly known as airbags), always disable the airbag system before working in the vicinity of any airbag system component to avoid the possibility of accidental deployment of the airbag, which could cause personal injury (see Chapter 12).*

Removal

1   Floor shift assemblies are equipped with a park/lock cable. The column park/lock system is not equipped with a park/lock cable. The column park/lock system is integrated into the steering column. The column shift lock system stops the lock cylinder from moving into the Lock or Accessory positions until the key is installed and the shift selector is in Park. The floor shift assemblies are equipped with a slightly different park/lock system. This system uses a flexible park/lock cable actuator which is attached at one end of the shift lever and the other end to the column mounted ignition switch where it actuates a locking pin. The locking pin engages a sliding contact when the shift lever is in Park, Neutral, Drive. When the shift lever is in Park, the pin disengages from the slider and goes into the lock position. The locking pin engages a cam and prevents the shift lever from being moved to another position.

2   Disconnect the negative cable from the battery.

3   Remove the driver's side kick panel and pull back the carpet near the lower section of the steering column.

4   Place the shift lever in Park and the ignition switch in the Run position.

5   Remove the steering column mounting bolts and partially lower the steering column (see Chapter 10).

6   Insert a screwdriver blade into the slot in the ignition switch inhibitor, depress the cable latch and detach the cable.

7   Remove the console (see Chapter 11).

8   Working in the center console area in the passenger compartment, pull the retaining clip from the end of the park/lock cable located on the park/lock lever pin.

9   Push the cable connector lock button (located at the shift control base) to the UP position and detach the cable from the park lock lever pin. Depress the two cable connector latches and remove the cable from the shift control base.

10   Remove the cable clips and separate the park/lock cable from the console area.

Installation

11   Make sure the cable lock button is in the UP position and the shift lever is in Park. Snap the cable connector into the shift control base.

12   With the ignition key in the Run position (this is very important), snap the cable into the inhibitor housing. Do *not* attempt to insert the cable with the key in any other position.

13   Turn the ignition key to the Lock position.

14   Snap the end of the cable onto the shifter park/lock pin.

15   Push the nose of the cable connector forward to remove the slack.

16   With no load on the connector nose, snap down the cable connector lock button.

17   Check the operation of the park/lock cable as follows.

a)   With the shift lever in Park and the key in Lock, make sure the shift lever cannot be moved to another position and the key can be removed.

b)   With the key in Run and the shift lever in Neutral, make sure the key cannot be turned to Lock.

18   If it operates as described above, the park/lock cable system is properly adjusted.

19   If the park/lock system doesn't operate as described, return the cable connector lock to the UP position and repeat the adjustment procedure. Push the cable connector down and recheck the operation.

---

7   Brake/transaxle shift interlock system - description, check and adjustment

Warning: *On models equipped with a Supplemental Inflatable Restraint (SIR) system (more commonly known as airbags), always disable the airbag system before working in the vicinity of any airbag system component to avoid the possibility of accidental deployment of the airbag, which could cause personal injury (see Chapter 12).*

Description

1   Later models are equipped with the brake/transaxle shift interlock system. The brake transaxle shift interlock system prevents the (column or floor-mounted) shift lever from being moved out of Park unless the brake pedal is depressed simultaneously. When the car is started, a solenoid is energized, locking the shift lever in Park; when the brake pedal is depressed, the solenoid is de-energized, unlocking the shift lever so that it can be moved out of Park. **Note:** *Before making the following checks, refer to Chapter 9 and verify that the brake light switch is functioning properly because it is part of the interlock circuit.*

Check

2   Remove the trim panel below the steering column and the knee bolster (if equipped). Using a flashlight, locate the brake/transaxle shift interlock solenoid - it's mounted to the left of the column, just ahead of the column mount.

3   Verify that the brake/transaxle shift interlock solenoid operates as follows:

a) *When the ignition key is in the Lock position, the solenoid plunger should be in and you should not be able to move the shift lever, even with the brake pedal applied.*

b) *With the ignition key is turned to the Off position, the solenoid plunger should be popped out and you should be able to move the shift lever to any gear position without applying the brake pedal.*

c) *When the ignition key is turned to the Run position, the solenoid should go back into the solenoid and you should not be able to move the shift lever; except with the brake pedal applied, the solenoid plunger should be released (pop out) and you should be able to move the shift lever out of Park into any gear.*

d) *Place the key in the Lock position, then turn the key to the Acc (accessory) position. The solenoid plunger should go in a little further than it does when the key is turned to Lock, and you should not be able to move the shift lever (even with the brake applied).*

4   If the solenoid doesn't operate as described above, unplug the electrical connector from the solenoid, apply battery voltage to the solenoid and verify that it "clicks" on, then open the circuit and verify that the solenoid clicks off.

a) *If the solenoid doesn't operate as described, replace it.*

b) *If the solenoid is operating properly, troubleshoot the circuit between the battery and the solenoid, and between the solenoid and ground. Refer to the wiring schematics at the end of Chapter 12.*

c) *If the solenoid circuit is okay, take the vehicle to a dealer service department.*

8.10 Remove the Park/Neutral switch mounting bolts (arrows)

Adjustment

5   Check the Park/Neutral switch for correct operation and adjustment (see Section 8).

6   Loosen the bolts that retain the interlock assembly to the steering column bracket.

7   Make sure the locking button is released to allow movement of the interlock solenoid.

8   Move the interlock assembly toward the steering column approximately 1/8 inch.

9   Tighten the bolt(s) lightly and engage the locking button. Check the operation of the brake/transaxle shift interlock assembly (see Steps 3 and 4).

10   If the solenoid operates correctly, tighten the mounting bolts to the torque listed in this Chapter's Specifications.

8   Park/Neutral switch - check, replacement and adjustment

Warning: *On models equipped with a Supplemental Inflatable Restraint (SIR) system (more commonly known as airbags), always disable the airbag system before working in the vicinity of any airbag system component to avoid the possibility of accidental deployment of the airbag, which could cause personal injury (see Chapter 12).*

Check

1   The Park/Neutral switch prevents the starter motor from operating when the automatic transaxle is in gear by opening the starter motor circuit. When the shift lever is put into Reverse with the engine running, the switch closes the back-up light circuit and turns on the back-up lights.

2   To check the Neutral start function of the switch, verify that the engine will start only in Neutral. If the engine starts with the transaxle in gear, verify that the switch is properly adjusted (see below). If the switch still fails to work properly after it's been adjusted, replace it.

3   To check the back-up light function of the switch, place the shift lever in Reverse

8.15 Insert a 3/32-inch drill bit (arrow) into the service adjustment hole and rotate the switch assembly until the drill bit drops into the carrier tang hole.

with the ignition key On. The back-up lights should come on.

4   If the back-up lights don't come on, either a fuse is blown, the bulbs are burned out, there's an open in the circuit, or the back-up light switch is defective.

5   Always begin any electrical troubleshooting procedure by checking the fuses (see Chapter 12). If a fuse is blown, replace it. If it blows again, find out why.

6   Check the bulbs for the back-up light circuit. If either bulb is bad, replace it (see Chapter 12). If the bulbs are good, check the circuit. Using a voltmeter, verify that voltage is getting to the switch and that, when the shift lever is placed in Reverse, there is voltage to the bulb. If the bulbs are okay and the circuit is okay, check the continuity of the switch.

Replacement

Refer to illustration 8.10

7   Disconnect the cable from the negative battery terminal.

8   Remove the fuel line bracket and bolt. If necessary, remove the air cleaner housing and air intake ducts for access to the Park/Neutral switch (see Chapter 4).

9   Unplug the electrical connector from the Park/Neutral switch.

10   Remove the switch retaining screws **(see illustration)** and remove the switch.

11   Be sure to adjust the switch (see below) before tightening the switch retaining screws. Installation is otherwise the reverse of removal.

Adjustment

Refer to illustration 8.15

12   Place the shift lever in Neutral.

13   Insert the tang on the switch assembly into the slot on the shift lever.

14   Install the switch retaining screws loosely.

15   Insert a 3/32-inch drill bit into the service adjustment hole **(see illustration)** and rotate the switch assembly until the drill bit

9.1 Disconnect the vacuum line from the modulator (arrow) and install a vacuum gauge

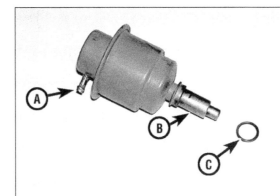

drops into the carrier tang hole. **Note: A new switch should be equipped with a plastic adjustment pin in place.** *If the screw holes will not align with the holes in the shift lever assembly, verify that the shift lever is in Neutral. Do NOT rotate the switch. If the switch has been rotated and the pin is broken, adjust the switch using a drill bit as described.*

16   Tighten the switch retaining screws securely to the torque listed in this Chapter's Specifications.

9   Vacuum modulator - check and replacement

Check

Refer to illustration 9.1

1   The vacuum modulator is connected to engine manifold vacuum to rapidly respond to changes in engine loading, and has an important effect on shift quality. If your vehicle exhibits shifting problems such as slipping or shifts that are either too soft or too harsh, check the vacuum supply to the modulator **(see illustration)**. With the engine running, connect a vacuum gauge to the modulator's vacuum line (at the modulator). Anything less than normal engine vacuum here of 13 to 17 inches (idling hot in Drive with the brakes applied) could cause shifting problems. If vacuum is too low, find the engine problem, hose kink or hose leak that is causing the low vacuum signal.

2   Connect a hand-held vacuum pump to the vacuum connection on the modulator (removed from vehicle) and apply 15 to 20 inches of vacuum while watching the plunger. If the plunger isn't drawn in as vacuum is applied, the modulator should be replaced. The modulator should be able to hold this vacuum for at least 30 seconds.

3   When the modulator is withdrawn from the transaxle, turn it so the vacuum pipe is down. If any oil, water or other fluid drips out, the modulator should be replaced. **Note:** *A*

9.7 Use a screwdriver to remove the O-ring from the transaxle

vehicle with a modulator whose diaphragm has a leak may exhibit excessive smoking at the tailpipe, due to transaxle fluid being drawn into the engine and burned.

4   The body of the modulator can be checked for leaks by coating the outside with soapy water and blowing (by mouth, no more than 6 psi) into the vacuum connector, using a short length of vacuum hose. Bubbles on the outside of the modulator or along the seam indicate a leak.

Replacement

Refer to illustrations 9.6, 9.7 and 9.8

5   The modulator is located on the front (radiator) side of the transaxle, below the exhaust crossover pipe.

6   Disconnect the vacuum line, and remove the mounting bolt or stud, then withdraw the modulator **(see illustration)**.

7   Remove the O-ring from the modulator cavity, using a small screwdriver or hook **(see illustration)**.

8   Use a small magnet to remove the modulator valve from the transaxle **(see illustration)**. Inspect the valve for signs of abnormal wear or scoring.

9   Installation is the reverse of the removal procedure. Always use a new O-ring and make sure the modulator valve is installed the same way it came out.

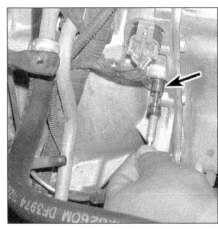

9.8 A magnet can be used to reach in and extract the modulator valve (arrow)

10   Driveaxle oil seals - replacement

Refer to illustrations 10.3 and 10.6

1   Raise the vehicle and support it securely on jackstands.

2   Remove the driveaxle(s) (see Chapter 8).

3   Use a hammer and chisel to pry up the outer lip of the seal to dislodge it so it can be pried out of the housing **(see illustration)**.

4   Compare the new seal to the old one to make sure they're the same.

5   Coat the lips of the new seal with

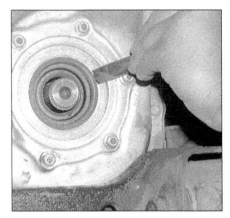

10.3 Dislodge the metal-type differential seal by working around the outer edge with a chisel and hammer

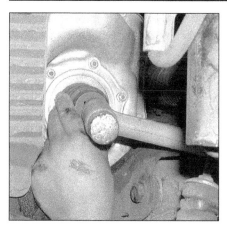

10.6 Tap the new seal into place using a large socket or section of pipe and hammer - use a socket or section of pipe with a diameter slightly smaller than the outside diameter of the seal itself

transaxle fluid.

6   Place the new seal in position and tap it into the bore with a hammer and a large socket or a piece of pipe that's the same diameter as the outside edge of the seal **(see illustration)**.

7   Reinstall the various components in the reverse order of removal.

---

11   Auxiliary oil cooler - removal and installation

1   On some models, an auxiliary oil cooler is provided for the automatic transaxle fluid. The cooler looks like a small radiator and is mounted in front of the engine radiator and air-conditioning condenser, just behind the grille. Transaxle fluid flow comes from the transaxle to the auxiliary cooler through the transaxle fluid output line, and then from the auxiliary cooler to the standard cooler in the right-hand tank of the engine radiator.

2   To remove the cooler, place a suitable drain pan under the cooler to catch the fluid and use backup wrenches to remove the lower fluid line from the cooler.

3   When no more fluid comes out, remove the fitting on the upper fluid line.

4   Remove the four screws (two into the upper radiator support, two into the lower radiator support) and remove the cooler. **Caution:** *If transaxle fluid spills on painted body surfaces, clean it off immediately to avoid damaging the finish.*

5   Installation is the reverse of the removal procedure. **Caution:** *Do not tighten the mounting bolts until you are sure the cooling lines are properly threaded into the cooler, with no cross-threading.*

---

12   Automatic transaxle - removal and installation

Warning: *This is a difficult procedure for the home mechanic, requiring the use of*

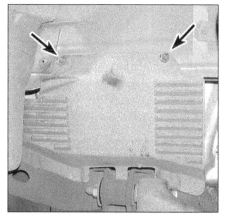

12.2 Remove the two screws (arrows) and take off the left inner splash shield for access to the transaxle mount

several specialized tools, including a transaxle jack and a three-bar engine support fixture. Such tools can be rented, but safely raising the vehicle enough for the transaxle and subframe to be pulled out from underneath is a problem without a vehicle hoist.

Removal

Refer to illustrations 12.2, 12.5, 12.17a, 12.17b, 12.20, 12.21a, 12.21b and 12.21c

1   Disconnect the negative cable from the battery.

2   Raise the vehicle and support it securely on jackstands. Remove the left side inner splash shield to access the transaxle **(see illustration)**.

3   Drain the transaxle fluid (Chapter 1).

4   If equipped, disconnect the engine oil cooler line bracket.

5   Remove the torque converter cover. Mark the torque converter-to-driveplate relationship so they can be installed in the same relative position **(see illustration)**.

6   Remove the torque converter-to-driveplate bolts. Turn the crankshaft pulley bolt for access to each bolt.

12.17a Use a flare nut wrench on the tube nut and an open end wrench on the fitting adapter when detaching the transaxle cooler lines (arrows) from the transaxle

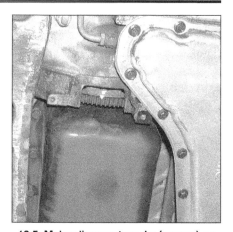

12.5 Make alignment marks (arrows) on the flywheel and torque converter so they can be reinstalled in the same relative positions

7   Remove the air cleaner assembly (see Chapter 4).

8   Remove the starter motor (see Chapter 5).

9   Disconnect the vacuum modulator (see Section 9).

10   Disconnect the electrical connectors from the transaxle, TCC converter clutch, cruise control, Park/Neutral switch and shift solenoids (4T60-E).

11   Disconnect the exhaust crossover pipe (see Chapter 4).

12   Disconnect the accelerator cable from the throttle body (see Chapter 4).

13   If equipped, disconnect the throttle valve cable from the transaxle (see Section 5).

14   Disconnect the shift cable from the transaxle (see Section 4).

15   If equipped, disconnect the air injection crossover pipe fitting. Reposition the pipe away from the engine.

16   Remove the driveaxles (see Chapter 8).

17   Disconnect the transaxle cooler lines from the transaxle **(see illustrations)** and separate the lines.

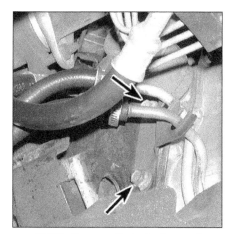

12.17b Remove the bolts (arrows) retaining the transaxle cooler bracket to the side of the transaxle

12.20 Remove the transaxle cradle mounting bolts (arrows) - upper and lower mounting bolts shown, the lower mounting nuts are shown in Chapter 2B at the engine mount

12.21a Remove the two bolts (arrows) that secure the left subframe member to the right subframe member at the left, rear corner of the subframe

12.21b Remove the subframe-to-chassis mounting bolt (arrow)

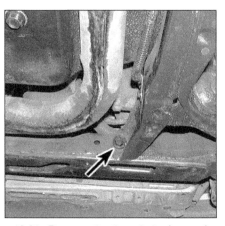

12.21c Remove the three bolts (arrows) that secure the left subframe member to the right subframe member at the right, front corner of the subframe

18   Support the engine using a three-bar support fixture. The engine must remain supported during the entire procedure.

19   Support the transaxle with a jack - preferably a special jack made for this purpose. Safety chains will help steady the transaxle on the jack.

20   On models equipped with a transaxle cradle, remove the transaxle cradle mounting nuts and bolts and the left side engine mount **(see illustration)** (see Chapter 2A or 2B). **Note:** *The transaxle cradle is a one-piece metal brace that supports the transaxle to the engine mount, located on the subframe. This cradle stabilizes the transaxle and protects the transaxle cover from impact.*

21   Remove the subframe. These models are equipped with a two-piece subframe. Remove the bolts that retain the left side subframe **(see illustrations)** and lower the section from the bottom of the engine compartment. The subframe mounting bolts for the left section are located in the front right and the left rear corners of the left section. **Note:** *The lower control arm will be retained onto the left section subframe but it will be necessary to disconnect the hoses and lines from the subframe before removal.*

22   Install a floorjack under the transaxle. Position a piece of wood between the transaxle and the floorjack to prevent damaging the transaxle pan.

23   Remove the upper transaxle-to-engine mounting bolts. **Note:** *Make notes or a schematic for each bolt regarding the length and location to insure correct installation.*

24   Remove the front left and rear left side engine mounts that secure the transaxle to the subframe (see Chapters 2A or 2B).

25   If equipped, remove the torque brace from the upper engine compartment (see Chapters 2A or 2B).

26   Remove the lower transaxle-to-engine mounting bolts. **Note:** *On some models it will be necessary to use a three foot extension*

and socket to access the rear throughbolt. **Note:** *Make notes or a chart for each bolt regarding the length and location to insure correct installation.*

27   Lower the transaxle slightly and disconnect and plug the transaxle cooler lines. Make sure all of the fluid has drained out into a suitable container.

28   Move the transaxle back to disengage it from the engine block dowel pins and make sure the torque converter is detached from the driveplate. Secure the torque converter to the transaxle so it will not fall out during removal. Lower the front subframe and transaxle from the vehicle. Remove the transaxle mount from the subframe and separate the transaxle.

Installation

29   Prior to installation, make sure the torque converter is fully engaged in the transaxle. To do this, rotate the converter while pushing it toward the transaxle. If it wasn't already fully in place, you'll feel it "clunk" into position as it engages with the input shaft and front pump. It may even "clunk" more than once. Lubricate the torque converter hub with multi-purpose grease.

30   With the transaxle and subframe secured to the jack, raise it evenly into position. Be sure to keep it level so the torque converter does not slide out. **Warning:** *Do not place any part of your body under the subframe until the subframe bolts are back in place.*

31   Rotate the torque converter to align the bolt holes with the holes in the driveplate. The mark on the torque converter and the driveplate made in Step 5 must align.

32   Move the transaxle forward carefully until the dowel pins and the torque converter are engaged.

33   Install the transaxle housing-to-engine bolts. Tighten them securely.

34   Install the torque converter-to-driveplate

bolts. Tighten the bolts to the specified torque.

35   Install the subframe and chassis components which were removed. Tighten the bolts and nuts to the specified torque.

36   Remove the jacks supporting the transaxle and the engine.

37   Install the dipstick tube.

38   Install the starter motor (Chapter 5).

39   Connect the vacuum hose(s) (if equipped).

40   Connect the throttle valve cable.

41   Plug in the transaxle electrical connectors.

42   Install the torque converter cover.

43   Install the driveaxles (Chapter 8).

44   If equipped, install the transaxle cradle.

45   Install the shift cable (see Section 4).

46   Install any exhaust system components that were removed or disconnected.

47   Lower the vehicle. Adjust the throttle valve cable and shift cable.

48   Fill the transaxle (Chapter 1), run the vehicle and check for fluid leaks.

49   Have the wheel alignment checked by a qualified alignment shop as soon as possible.

Chapter 8   Driveaxles

Contents

Specifications

Driveaxles

Inner CV joint standard length (dimension A in **illustration 3.3u**)

1986 through 1990 models	5-1/16 inch
1991 and later models...	4-29/32 inch

Torque specifications

	Ft-lbs
Driveaxle/hub nut	
Oldsmobile and Cadillac Seville/Eldorado models	185
Buick models	
1986 and 1987	192
1988 through 1993..	183
Cadillac Deville/Fleetwood models	
1986 through 1990..	183
1991	192
1992 and 1993	
RPO J55 Heavy Duty Brakes..............................	130
All others..	107
Wheel lug nuts	See Chapter 1

1   General information

Power is transmitted from the transaxle to the front wheels by two driveaxles, which consist of splined axleshafts with constant velocity (CV) joints at each end. There are two types of inner CV joints used. Early models are equipped with a "tri-pot" design using needle bearings and round bearing surfaces. Later models are equipped with inner tri-pot bearings with square bearing design that incorporate the bearings within a metal case. Tri-pot bearings are identified by the tri-pot housing which will have three major indentations in it. Both the round and square tri-pot designs incorporate an inner stop-ring and outer retainer ring (clip) on the axleshaft. All outer CV joints are the double-offset type, and allow angular, but not axial, movement. The outer bearings are referred to as the ball-and-cage style CV joint design.

The CV joints are protected by rubber boots, which are retained by clamps so the joints are protected from water and dirt. The boots should be inspected periodically (see Chapter 1). Damaged CV joint boots must be replaced immediately or the joints can be damaged. Boot replacement involves removing the driveaxles (Section 2). It's a good idea to disassemble, clean, inspect and repack the CV joint whenever replacing a CV joint boot to make sure the joint isn't contaminated with moisture or dirt, which would cause premature failure of the CV joint.

The most common symptom of worn or damaged CV joints, besides lubricant leaks, are a clicking noise in turns, a clunk when accelerating from a coasting condition or vibration at highway speeds.

2.1 Loosen the driveaxle/hub nut with a long breaker bar

2.6a A two-jaw puller works well for pushing the stub axle out of the hub

2.6b Support the outer end of the driveaxle with a piece of wire

2.7 Use a large prybar positioned as shown to "pop" the inner CV joint out of the transaxle - it may be necessary to tap the prybar with a hammer if the driveaxle is stuck

2.9 A large punch or screwdriver, positioned in the groove on the CV joint housing, can be used to seat the joint in the transaxle

2   Driveaxles - removal and installation

Refer to illustrations 2.1, 2.6a, 2.6b, 2.7 and 2.9

Removal

1   Remove the wheel cover and loosen the hub nut **(see illustration)**. Loosen the wheel lug nuts, raise the front of the vehicle and support it securely on jackstands. Apply the parking brake and block the rear wheels to keep the vehicle from rolling off the jackstands. Remove the front wheel.
2   Remove the driveaxle/hub nut. To prevent the hub from turning, insert a screwdriver through the caliper and into the cooling vanes of the brake disc, then unscrew the nut and remove the washer.
3   Place a drain pan under the transaxle near the differential to catch any transaxle lubricant that may leak out when the driveaxle is removed.
4   Remove the brake caliper and disc and support the caliper out of the way with a piece of wire (see Chapter 9).

5   Remove the control arm-to-steering knuckle balljoint stud nut and separate the lower arm from the steering knuckle (see Chapter 10 if necessary).
6   Push the driveaxle out of the hub with a puller **(see illustration)**, then support the outer end of the driveaxle with a piece of wire to prevent damage to the inner CV joint **(see illustration)**.
7   Carefully pry the inner end of the driveaxle out of the transaxle, using a large prybar positioned between the transaxle housing and the CV joint housing **(see illustration)**.
8   Support the CV joints and carefully remove the driveaxle from the vehicle.

Installation

9   Lubricate the differential seal with multi-purpose grease, raise the driveaxle into position while supporting the CV joints and insert the splined end of the inner CV joint into the differential side gear. Seat the shaft in the side gear (or onto the side gear shaft, depending on design) by positioning the end of a screwdriver in the groove in the CV joint

and tapping it into position with a hammer **(see illustration)**.
10   Apply a light coat of multi-purpose grease to the outer CV joint splines, pull out on the strut/steering knuckle assembly and install the stub axle in the hub.
11   Insert the control arm balljoint stud into the steering knuckle and tighten the nut to the torque listed in the Chapter 10 Specifications. Be sure to use a new cotter pin (refer to Chapter 10).
12   Install the brake disc and caliper (see Chapter 9 if necessary).
13   Install the driveaxle/hub nut and washer. Lock the disc so it can't turn, using a screwdriver or punch inserted through the caliper into a disc cooling vane, and tighten the driveaxle/hub nut to approximately 70 ft-lbs. This is an initial torque setting. The final torque will be applied after the vehicle has been lowered to the ground.
14   Grasp the inner CV joint housing (not the driveaxle) and pull out to make sure the axle has seated securely in the transaxle.
15   Install the wheel and lug nuts and lower the vehicle.

3.3a Cut off the boot retaining clamps, using wire cutters or a chisel and hammer

3.3b Slide the housing off the spider assembly

3.3c Slide the boot towards the center of the driveaxle (square tri-pot bearing block design shown)

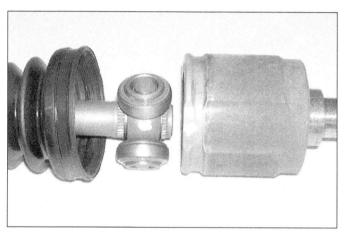

3.3d Mark the relationship of the tri-pot assembly to the outer race

3.3e Spread the ends of the stop-ring apart and slide it towards the center of the shaft

16   Tighten the driveaxle/hub nut to the torque listed in this Chapter's Specifications. Tighten the wheel lug nuts to the torque listed in the Chapter 1 Specifications, then install the wheel cover.
17   Check the transaxle lubricant level, adding as required (see Chapter 1).

3   Driveaxle boot - replacement

Note: *If the CV joint boots must be replaced, explore all options before beginning the job. Complete rebuilt driveaxles are available on an exchange basis, which eliminates much time and work. Whichever route you choose to take, check on the cost and availability of parts before disassembling the vehicle.*
1   Remove the driveaxle (see Section 2).
2   Place the driveaxle in a vise lined with rags to avoid damage to the axleshaft. Check the CV joint for excessive play in the radial direction, which indicates worn parts. Check for smooth operation throughout the full range of motion for each CV joint. If a boot is torn, disassemble the joint, clean the components and inspect for damage due to loss of lubrication and possible contamination by foreign matter.

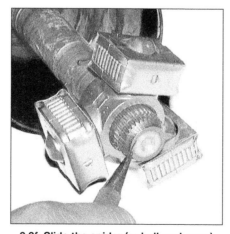

3.3f Slide the spider (or ball-and-cage) assembly back to expose the retaining ring and pry off the ring

Inner CV joint

Refer to illustrations 3.3a through 3.3w
Note: *Later models use a square tri-pot bear-ing design instead of the typical round bear-ing design on the inner CV joints. The square bearing type houses the bearings within a*

3.3g Remove the spider from the driveaxle. Hold the bearings in place with your hand; even better, use tape or a cloth wrapped around the spider bearing assembly to retain them

metal cage. Both the square and round tri-pot bearing designs incorporate an inner stop-ring and outer retainer (clip) on the driveaxle.
3   To replace the inner boot, refer to the accompanying photo sequence (**see illustra-tions 3.3a through 3.3w**).

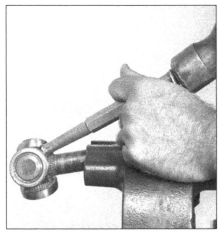

3.3h If the spider sticks to the shaft, tap it off the axleshaft with a hammer and brass punch

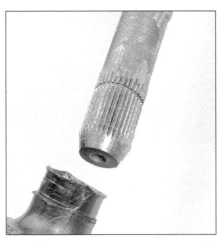

3.3i Slide the stop-ring and the boot off the axleshaft

3.3j Clean all of the old grease out of the housing and spider assembly, then remove each bearing, one at time

3.3k On models with needle bearings, carefully disassemble each section of the spider assembly, clean the needle bearings with solvent and inspect the rollers, spider cross, bearings and housing for scoring, pitting and other signs of abnormal wear

3.3l Apply CV joint grease to hold the needle bearings in place and slide the bearing over them

3.3m On square tri-pot bearing assemblies, install the bearing blocks onto the posts of the spider, then turn them 90-degrees to lock them in place

3.3n Wrap the axleshaft splines with tape to avoid damaging the boot, then slide the small clamp and boot onto the axleshaft

3.3o Remove the tape and slide the stop-ring onto the axleshaft, past the groove in which it seats

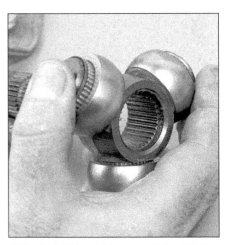

3.3p Install the spider assembly with the recess in the counterbore facing the end of the driveaxle

3.3q Install the retaining ring, then slide the spider (or ball-and-cage) assembly against it and install the stop-ring in its groove

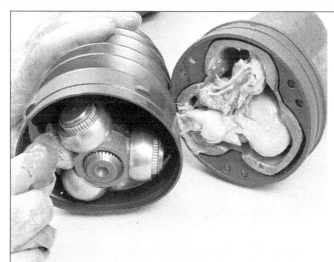

3.3r Pack the housing with half of the grease furnished with the new boot and place the remainder in the boot – on square tri-pot bearing design, be sure to hold the bearing blocks in alignment so they can be inserted into the joint housing without cocking to the side

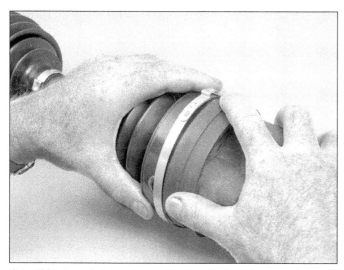

3.3s With the retaining clamps in place (but not tightened), install the joint housing

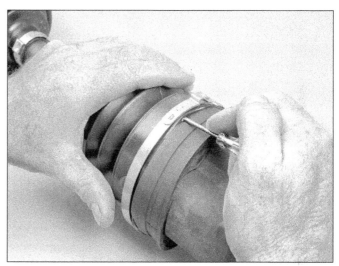

3.3t Seat the boot in the housing and axle seal grooves - a small screwdriver can make the job easier (make sure the boot isn't dimpled, stretched or out of shape)

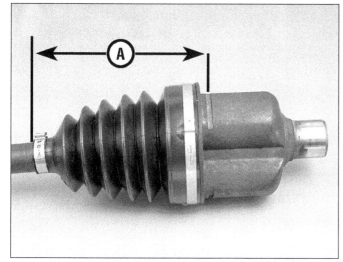

3.3u Adjust the length of the joint to the dimension listed in this Chapter's Specifications

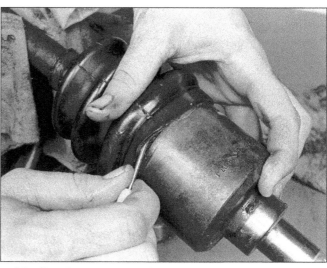

3.3v Equalize the pressure inside the boot by inserting a *dull* screwdriver between the boot and the outer race . . .

3.3w . . . then secure the boot clamps with special pliers (available at most auto parts stores)

3.4a Cut off the band retaining the boot to the shaft, then slide the boot toward the center of the shaft

3.4b Using snap-ring pliers, spread the ears of the snap-ring apart, then slide the joint off the shaft and remove the old boot

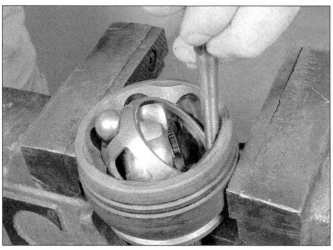

3.4c Press down on the inner race far enough to allow a ball bearing to be removed - if it's difficult to tilt, gently tap the cage and inner race with a brass punch and hammer

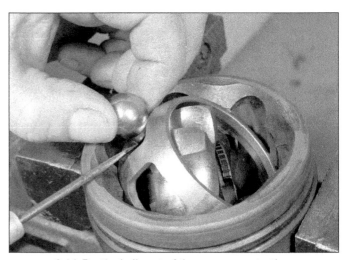

3.4d Pry the balls out of the cage, one at a time

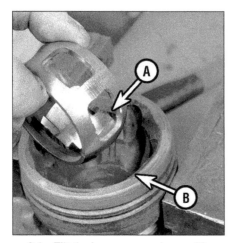

3.4e Tilt the inner race and cage 90-degrees, then align the windows in the cage (A) with the lands of the housing (B) and rotate the inner race up and out of the outer race

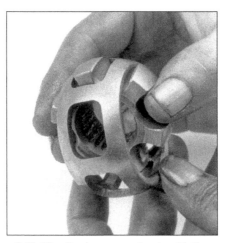

3.4f Align the inner race lands with the cage window and rotate the inner race out of the cage

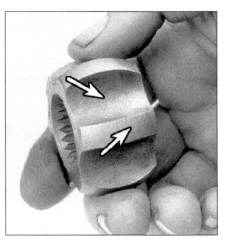

3.4g After cleaning the components with solvent, check the inner race lands and grooves for pitting and score marks

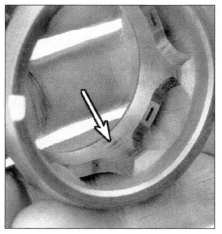

3.4h Check the cage for cracks, pitting and score marks - shiny spots are normal and don't affect operation

3.4i With the race and cage tilted 90-degrees, lower the assembly into the housing

3.4j Rotate the assembly by gently tapping with a hammer and brass punch . . .

3.4k . . . then press the balls into the cage windows, repeating until all of the balls are installed

3.4l Use needle-nose pliers to lower a new snap-ring into the groove . . .

Outer CV joint

Refer to illustrations 3.4a through 3.4r

4   Refer to the accompanying photo sequence and perform the outer CV joint boot replacement procedure **(see illustrations 3.4a through 3.4r).**

3.4m . . . then seat it into the groove with snap-ring pliers

3.4n Apply grease through the splined hole, then insert a wooden dowel (with a diameter slightly less than that of the axle) through the splined hole and push down - the dowel will force the grease into the joint - repeat until the bearing is completely packed

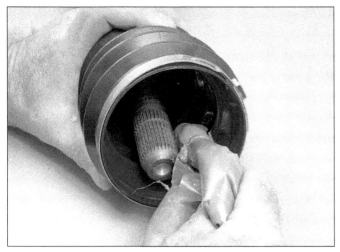

3.4o Install the small clamp and the boot on the driveaxle and apply grease to the inside of the axle boot until . . .

3.4p . . . the level is up to the end of the axle

3.4q Position the CV joint assembly on the driveaxle, aligning the splines, then use a soft-face hammer to drive the joint onto the driveaxle until the snap-ring is seated in the groove

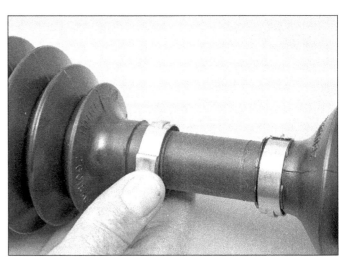

3.4r Seat the inner end of the boot in the groove and install the retaining clamp, then do the same on the other end of the boot - equalize the pressure in the boot (see illustration 3.3v), then tighten the boot clamps (see illustration 3.3w)

Chapter 9   Brakes

Contents

Specifications

General

Brake fluid type...	See Chapter 1
Brake pedal travel	
Non ABS models...	2-1/2 inches
ABS models	2-1/4 inches

Disc brakes

Brake pad lining minimum thickness...	See Chapter 1
Disc (front)	
Minimum thickness	Refer to the dimension stamped into the disc
Runout (maximum)	
1992 and 1993 Riviera and Toronado/Trofeo models	0.002 inch
All other models	0.004 inch
Thickness variation (parallelism) limit...............................	0.0005 inch
Disc (rear)	
Minimum thickness	Refer to the dimension stamped into the disc
Runout (maximum)	
1986 through 1991 models...	0.003 inch
1992 and 1993 models	0.002 inch
Thickness variation limit..	0.0005 inch

Drum brakes

Brake shoe lining minimum thickness	See Chapter 1
Brake drum	
Maximum diameter	Refer to the dimension stamped into the drum
Runout (maximum)	0.006 inch

Torque specifications
Ft-lbs (unless otherwise indicated)

Bleeder valve	
Front	115 in-lbs
Rear	
Caliper..	108 in-lbs
Wheel cylinder	
1986 through 1991 models	60 in-lbs
1992 and 1993 models...	88 in-lbs
Caliper mounting bolts	
Front	
Deville/Fleetwood models..............................	38
Seville/Eldorado models	63
Riviera and Toronado/Trofeo models	
1986 through 1991 models	63
1992 and 1993 models...	38
Rear	
Seville/Eldorado models	63
Riviera and Toronado/Trofeo models	
1986 through 1991 models	63
1992 and 1993 models...	20
Caliper mounting bracket bolts (front and rear).....................	83
Booster-to-firewall mounting nuts..	20
Brake hose-to-caliper bolt..	15 to 24
Master cylinder-to-booster mounting nuts............................	22
Parking brake lever-to-caliper nut (1986 to 1991 models)	35
Wheel cylinder mounting bolts	132 in-lbs
Wheel lug nuts	See Chapter 1

1 General information

General

The vehicles covered by this manual are equipped with hydraulically operated front and rear brake systems. All front brake systems are disc type, while the rear brakes are either disc or drum. Only the Cadillac Deville/Fleetwood models are equipped with rear drum brakes. 1986 Cadillac Deville/Fleetwood models are equipped with the anchor plate drum brake systems while 1987 through 1993 Cadillac Deville/Fleetwood models are equipped with the leading/trailing drum brake design. All brakes are self adjusting. Front and rear disc brakes automatically compensate for pad wear. Anchor plate drum brake systems apply the brake adjustment when the vehicle is driven in reverse. Leading/trailing drum brakes incorporate an adjusting mechanism which is activated any time the service brakes are applied.

Hydraulic system

The hydraulic system consists of two separate circuits. The master cylinder has separate reservoirs for the two circuits and in the event of a leak or failure in one hydraulic circuit, the other circuit will remain operative. A visual warning of circuit failure or air in the system is given by a warning light activated on the dash.

Proportioner valves

The proportioner valves are designed to provide better front to rear braking balance with heavy brake application. These valves allow more pressure to be applied to the front brakes (under certain braking operations) due to the fact the rear of the vehicle is lighter and does not require as much braking force.

Power brake booster

The power brake booster, utilizing engine manifold vacuum and atmospheric pressure to provide assistance to the hydraulically operated brakes, is mounted on the firewall in the engine compartment.

Parking brake

The parking brake operates the rear brakes only, through cable actuation. It's activated by a pedal mounted on the left side kick panel.

Service

After completing any operation involving disassembly of any part of the brake system, always test drive the vehicle to check for proper braking performance before resuming normal driving. When testing the brakes, perform the tests on a clean, dry flat surface. Conditions other than these can lead to inaccurate test results.

Test the brakes at various speeds with both light and heavy pedal pressure. The vehicle should stop evenly without pulling to one side or the other. Avoid locking the brakes because this slides the tires and diminishes braking efficiency and control of the vehicle.

Tires, vehicle load and front-end alignment are factors which also affect braking performance.

2 Anti-lock Brake System (ABS) - general information and trouble codes

General information

Refer to illustration 2.5

Anti-lock Brake Systems (ABS) maintain vehicle maneuverability, directional stability, and optimum deceleration under severe braking conditions on most road surfaces. It does so by monitoring the rotational speed of the wheels and controlling the brake line pressure to the wheels during braking. This prevents the wheels from locking up on slippery roads or during hard braking.

Under normal driving conditions, the ABS system functions the same as non-ABS systems. The Teves Mark II ABS system incorporates the booster, hydraulic actuator and master cylinder as one unit. On these models, the power assist for braking is supplied by hydraulic pressure from the booster portion of the hydraulic actuator.

If a wheel sensor detects a lock-up condition from severe braking, the ABS computer will modulate hydraulic pressure in the individual wheel circuits to prevent the wheel from grabbing or locking up. A separate hydraulic line and solenoid valves are designated for each front wheel. The rear wheels share one solenoid valve. The proportioning valve splits the hydraulic circuit to the rear wheels.

The ABS computer can decrease or hold pressure in each hydraulic circuit depending upon the information sent from

2.5 Location of the Electronic Brake Control Module (EBCM) (arrow) - 1991 Deville shown

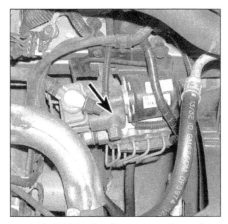

2.11 ABS hydraulic actuator (arrow) on a 1991 Deville with the Teves Mark IV ABS system

the wheel speed sensors on the rotational stability of each wheel. The driver will normally feel slight brake pedal pulsations during hard stopping. A slight ticking or popping noise may be heard during brake applications with the ABS system. This is a normal ABS sound.

1986 through 1990 models are equipped with the Teves Mark II ABS system. The hydraulic unit is mounted on the firewall near the master cylinder. The wheel speed sensors are mounted externally to the ABS ring. The master cylinder, the power brake booster and the hydraulic unit are all one assembly. The ABS computer on Deville/Fleetwood is located in the dash under the center console **(see illustration)**. The ABS computer on Seville/ElDorado, Riviera and Toronado is located in the trunk on the left wheelhouse.

The Teves Mark II ABS system, when left sitting for over 7 or 8 hours, will depressurize and require charging. The BRAKE light and the ANTI-LOCK warning light will illuminate for 40 seconds or longer during pressurizing. This is all normal for the Teves Mark II ABS system when the ABS pump is restoring pressure within the hydraulic accumulator. These systems when "active" will produce 2,000 to 2,600 psi pressure within the accumulator. When the wheel cylinders, brake lines. etc. must be serviced on the Teves Mark II ABS system, the brake system must be depressurized before any brake line or component is loosened. Refer to the depressurization procedure in this section.

1991 through 1993 Deville and Fleetwood models are equipped with the Teves Mark IV ABS system. This ABS system controls individual modulation of the braking force and also controls the traction each drive wheel (this feature is standard on the Fleetwood and optional on the Deville). On models equipped with this system, wheel spin at the drive wheels is reduced to prevent wheel lock-up and prevent slippage of the tire during slippery road conditions. All Teves Mark IV ABS models incorporate the wheel speed sensor into the hub/bearing assembly as a complete unit. The hydraulic actuator is

near the front of the engine compartment away from master cylinder.

This more sophisticated ABS system often displays unusual differences than earlier ABS systems. The pressure modulator valve (PMV) will often be heard running at slow speeds as it cycles through the ABS self-test mode. Also, it is possible to feel the pressure modulator valve running through the brake pedal when the brake is only applied lightly. The vehicle may decrease slightly in speed as the PMV applies the fluid pressure to the wheels. This is all normal for the Teves Mark IV ABS system during slow speed operation.

Models that are equipped with the Traction Control System (TCS) will display a TRACTION OFF warning light, when the ABS light is illuminated. Also, to avoid brake overheating, the computer will disable the TCS system for approximately 15 to 20 minutes while the brakes are cooling and during this period, the TRACTION OFF light will be lit. This is a normal condition and should not be confused with a defective TCS component.

1991 through 1993 Seville/Eldorado, Riviera and Toronado/Trofeo models are equipped with the Bosch Anti-Lock Brake system. Some of the 1993 models are also equipped with a Traction Control System (TCS) along with the ABS. The TCS system prevents the wheels from spinning excessively during acceleration and turning corners at high speeds. The wheel speed sensors are mounted integral to the hub assembly. The hydraulic unit is referred to as the Pressure Modulator Valve.

Hydraulic actuators

Refer to illustration 2.11

The hydraulic actuator assembly is mounted in various places depending upon the type of ABS system. On the Teves Mark II system, it is mounted on the side of the master cylinder and brake booster assemblies. This master cylinder, brake booster and hydraulic actuator assembly is recommended to be replaced as a complete unit. On the Teves Mark IV ABS system, the hydraulic actuator is mounted on the left, side engine compartment

(see illustration). This assembly is referred to as the Pressure Modulator Valve (PMV). It incorporates the ABS pump, the motor, the valve body and the fluid reservoir. On the Bosch ABS systems, the hydraulic actuator is located in the center, front of the engine compartment. This assembly is referred to as the Brake Pressure Modulator valve (BPM). All hydraulic actuators for these models must be replaced as a single unit. Consult a dealer parts department or other brake repair facility for parts information.

Electronic Brake Control Module (EBCM)

The Electronic Brake Control Module (EBCM) is located under the dash or in the trunk. The EBCM monitors the ABS system and controls the anti-lock valve solenoids. It accepts and processes information received from the brake switch and wheel speed sensors to control the hydraulic line pressure and avoid wheel lock up. It also monitors the system and stores fault codes which indicate specific problems.

Wheel speed sensors

Refer to illustrations 2.14a and 2.14b

Each type of ABS system is equipped with a speed sensor at each wheel. A wheel speed sensor measures wheel speed by monitoring the rotation of the toothed ring. As the teeth of the ring move through the magnetic field of the sensor, an AC voltage signal is generated. This signal frequency increases or decreases in proportion to the speed of the wheel. The EBCM monitors these three signals for changes in wheel speed; if it detects the sudden deceleration of a wheel, i.e. wheel lockup, the EBCM activates the ABS system.

Each wheel speed sensor assembly consists of a variable reluctance sensor mounted adjacent to a "toothed ring" with an air gap between them **(see illustrations)**. The air gap between the sensors and the rings are not adjustable, and the sensors themselves are not rebuildable. If a sensor malfunctions, the sensor must be replaced. If a ring malfunctions, it must be driven off the hub/bearing unit and a new one pressed on.

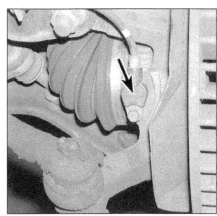

2.14a Location of the front wheel speed sensor (arrow)

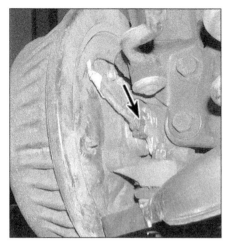

2.14b Location of the rear wheel speed sensor (arrow)

2.18 Location of the 12-pin test connector on a 1987 Seville with the Teves Mark II ABS system

Warning lights

The ABS system has self-diagnostic capabilities. There are three warning lights on the instrument panel, a red BRAKE light, an amber ANTI-LOCK light and a TRACTION OFF light, each with their own functions. During starting, the red BRAKE warning light should come on briefly then go out. If the red BRAKE light stays on, it indicates a problem with the main braking system, such as low fluid level detected or the parking brake is still on. If the lights stays on after the parking brake is released, check the brake fluid level in the master cylinder reservoir (see Chapter 1).

The amber ANTI-LOCK light indicates a problem with the ABS system, not the main or basic brake system. If the light stays on steadily, it indicates that there is a problem with the ABS system, but the main system is still working. If you have a steady ANTI-LOCK light, drive to a dealer service department or other qualified repair shop for diagnosis and repair. However, if the ANTI-LOCK light *flashes*, there is a more serious fault in the ABS system which may have affected the regular braking system. Pull over immediately and have the vehicle towed to a dealer or other qualified repair shop for service.

Obtaining ABS trouble codes

Teves Mark II system

Refer to illustration 2.18

Normally, the ABS indicator light should come on when the engine is started, then go off immediately. Under certain conditions, however, the indicator light may remain on for an extended period of time. If this occurs, the ABS computer has stored a diagnostic trouble code because it has detected a problem in the ABS system. ABS trouble codes can be accessed on the 12-pin Data Link Connector (DLC) in the passenger compartment. Using a jumper wire or a paper clip, install the tool into the specified terminals as described in the following text.

On 1986 through 1990 models with the Teves Mark II ABS system, to access the ABS self diagnosis system, connect a jumper wire from terminal G to A (1986 through 1988) or H to A (1989 and 1990) **(see illustration)**. Turn the ignition key ON (engine not running) and observe the ANTI-LOCK light on the dash. The diagnostic code is the number of flashes indicated on the ANTI-LOCK light on the dash. If no codes are stored, the ANTI-LOCK light will come on briefly and then go out. If any malfunction has been detected, the light will blink the first digit of the code at a long interval. For example, a code 41 (wheel speed sensor), will first blink four long flashes (first digit) and then pause and blink one short flash (second digit). After the second digit has flashed, the ANTI-LOCK light will stay illuminated. With the ignition key still ON (engine not running), remove the jumper wire from terminal H, reconnect the jumper wire into terminal H and observe the second stored code. Count only the flashes as the trouble code. When the ANTI-LOCK light stays illuminated, disregard counting this as a code and use it as a pass signal to the next code. Repeat the disconnect and reconnect of terminal H until no new codes are displayed. The system is capable of storing seven ABS trouble codes. Write all the stored trouble codes, turn the ignition key OFF and remove the jumper wire.

The trouble codes can be erased by driving the vehicle over 18 mph for a period of five to ten minutes. Recheck for stored trouble codes and repair the necessary components if another trouble code is recorded.

The following code chart lists the ABS codes for the most common Teves Mark II Anti-lock Brake System problems. Most ABS repairs must be performed by a dealer service department or other qualified automotive repair facility.

ABS Trouble Codes for the Teves Mark II system

Code	Probable cause
11	EBCM problem
12	EBCM problem
21	Main valve defective; hydraulic actuator problem
22	Left front solenoid valve circuit
23	Left front outlet valve circuit
24	Right front solenoid valve circuit
25	Right front outlet valve circuit
26	Rear solenoid valve circuit
27	Rear outlet valve circuit
31	Left front wheel speed sensor
32	Right front wheel speed sensor
33	Right rear wheel speed sensor
34	Left rear wheel speed sensor
35	Left front wheel sensor
36	Right front wheel sensor
37	Right rear wheel speed sensor
38	Left rear wheel speed sensor
41	Left front wheel speed sensor
42	Right front wheel speed sensor
43	Right rear wheel speed sensor
44	Left rear wheel speed sensor
45	Missing wheel speed sensor signals (left front)
46	Missing wheel speed sensor signals (right front)
47	Missing wheel speed sensor signals (rear)
48	Missing wheel speed sensor signals (all three)
51	Left front outlet valve
52	Right front outlet valve
53	Rear outlet valve
54	Rear outlet valve
55	Left front wheel speed sensor
56	Right front wheel speed sensor
57	Right rear wheel speed sensor
58	Left rear wheel speed sensor
61	EBCM circuit problem in low fluid/pressure warning loop
71	Left front outlet valve
72	Right front outlet valve
73	Rear outlet valve
74	Rear outlet valve
75	Left front wheel speed sensor
76	Right front wheel speed sensor
77	Right rear wheel speed sensor
78	Left rear wheel speed sensor

Teves Mark IV system

The trouble codes for 1991 through 1993 models with the Teves Mark IV ABS system must be accessed using a scan tool. Refer to Chapter 6 for additional information on SCAN tools and accessing diagnostic trouble codes. Most ABS system repairs must be performed by a dealer service department or other qualified automotive repair facility.

Bosch system

On 1991 through 1993 models with the Bosch ABS system, to access the ABS self diagnosis system, connect a jumper wire from terminal H to terminal A **(see illustration 2.18)**. With the ignition key ON (engine not running), observe the ANTI-LOCK light on the dash. The diagnostic code is the number of flashes indicated on the ANTI-LOCK light on the dash. The first code displayed will be Code 12. This initialization code indicates the start of the diagnostic code accessing procedure. Code 12 will flash three times. If any malfunction has been detected, the diagnostic codes will follow. The ANTI-LOCK light will blink the first digit of the code at a long interval. For example, a code 41 (solenoid valve), will first blink four long flashes (first digit) and then pause and blink one short flash (second digit). This code will flash three times and then the next code will be displayed. Write each code down on paper. If there are no codes stored, only the initialization code 12 will flash. The codes will repeat until the jumper wire is removed.

The trouble codes can be erased by grounding terminal H on the 12-pin Test Connector three times within a 10 second period **(see illustration 2.18)**. Turn the ignition key ON, wait for the ABS warning light to turn OFF, then insert a jumper wire between terminals A and H until the ABS warning light turns ON, then remove the jumper wire from the H terminal. Repeat this step two more times within a 10 second time period. The warning light should turn ON each time the jumper wire is inserted into the H terminal and it should turn off when the jumper wire is removed from the H terminal. The ABS trouble codes should now be cleared. Only Code 12 will flash after all the other codes are erased.

The following code chart lists the ABS codes for the most common Bosch ABS Anti-lock Brake System problems. Most ABS repairs must be performed by a dealer service department or other qualified automotive repair facility.

ABS Trouble Codes for the Bosch system

Code	Probable cause
12	Initialization code
21	Right front wheel sensor (open/short to body ground or short to power)
22	Right front speed sensor toothed-wheel frequency error
25	Left front wheel sensor (open/short to body ground or short to power)
26	Left front speed sensor toothed-wheel frequency error
31	Right rear wheel sensor (open/short to body ground or short to power)
32	Right rear speed sensor toothed-wheel frequency error
35	Left rear wheel sensor (open/short to body ground or short to power)
36	Left rear speed sensor toothed wheel frequency error
41	Right front solenoid valve circuit
45	Left front solenoid valve circuit
55	Left rear solenoid valve circuit
61	Hydraulic actuator motor pump circuit fault
63	Solenoid valve relay problem
71	ABS module problem
72	Serial Data Link problem

Depressurization procedure for the Teves Mark II ABS system

1986 through 1990 models equipped with the Teves Mark II ABS system require the brake system depressurized before any brake system component (wheel cylinders, master cylinder, brake lines, etc) is loosened for removal. This ABS system when active, is pressurized to 2,000 to 2,600 psi.

First, make sure the ignition key is OFF. Then disconnect the negative battery cable. Pump the brake pedal a minimum of 25 times to depressurize the hydraulic system. Use at least 50 ft-lbs of force on the brake pedal while pushing. The brake pedal should change sufficiently and become harder to push. When the brake pedal is much more difficult to push, continue with about 10 more pumps of the pedal. The brake depressurizing procedure is complete.

3.5 To make room for the new pads, use a C-clamp to depress the piston into its bore before removing the caliper

3 Disc brake pads - replacement

Warning: *Disc brake pads must be replaced on both front or rear wheels at the same time - never replace the pads on only one wheel. Also, the dust created by the brake system may contain asbestos, which is harmful to your health. Never blow it out with compressed air and don't inhale any of it. An approved filtering mask should be worn when working on the brakes. Do not, under any circumstances, use petroleum-based solvents to clean brake parts. Use brake system cleaner only!*

Note: *This procedure applies to front and rear disc brakes.*

1   Remove the cap from the brake fluid reservoir. Using a suction gun, remove approximately two-thirds of the fluid from the reservoir.

2   Loosen the wheel lug nuts, raise the front, or rear, of the vehicle and support it securely on jackstands.

3   Remove the front, or rear, wheels. Work on one brake assembly at a time, using the assembled brake for reference if necessary.

4   Inspect the brake disc carefully as outlined in Section 5. If machining is necessary, follow the information in that Section to remove the disc, at which time the calipers and pads can be removed as well.

Front pads

Refer to illustrations 3.5 and 3.6a through 3.6i

5   Wash the brake assembly with brake system cleaner, then push the piston back into the bore to provide room for the new brake pads. A C-clamp can be used to accomplish this **(see illustration)**. As the piston is depressed to the bottom of the caliper bore, the fluid in the master cylinder will rise. Make sure it doesn't overflow. If necessary, drain off some more of the fluid.

6   Follow the accompanying photos, beginning with **illustration 3.6a**, for the actual pad replacement procedure. Be sure to stay in order and read the caption under

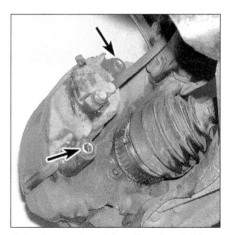

3.6a On most models, a Torx head driver must be used to remove the two caliper mounting bolts (arrows)

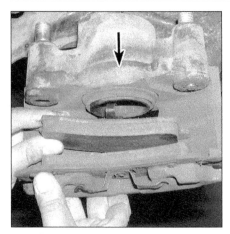

3.6b   Remove the inner pad by snapping it out of the piston

3.6c   Remove the outer brake pad

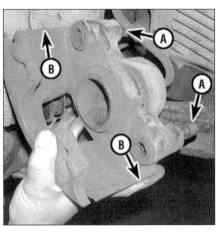

3.6d   Inspect the caliper bolts and bushings (A) for damage and the contact surfaces (B) for corrosion

3.6e   Carefully peel back the edge of the piston boot and check for leaking fluid

3.6f   Lubricate the caliper mounting bolts with high-temperature grease

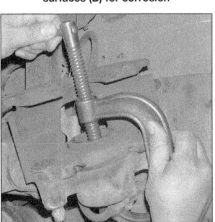

3.6g   Be sure to bottom the piston using a C-clamp - make sure the brake fluid doesn't overflow onto the inner fenderwell of the engine compartment

each illustration. **Note:** *It's a good idea to apply a thin layer of anti-squeal compound to the backing plates of the new pads. Allow the compound to dry before installing the pads.*

7   Proceed to Step 20.

Rear pads

1986 through 1991 models

Refer to illustrations 3.8a through 3.8g

8   Wash the brake assembly with brake

system cleaner, then follow the accompanying photos, beginning with **illustration 3.8a**, for the actual pad replacement procedure. Be sure to stay in order and read the caption

3.6h   Snap the inner pad into place in the piston

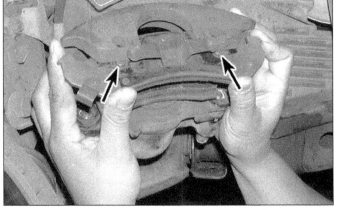

3.6i   Install the outer brake pad in the caliper, making sure the locating lugs on the pad engage with the holes in the caliper, then swing the caliper down into place. Install the caliper mounting bolts and tighten them to the torque listed in this Chapter's Specifications

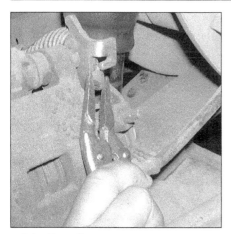

3.8a Use pliers to remove the parking brake cable from the notch in the parking brake lever

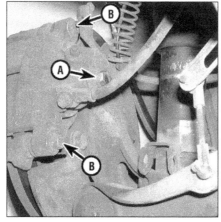

3.8b Remove the nut (A) and detach the parking brake lever, then remove the caliper mounting bolts (B) and separate the caliper from the mounting bracket

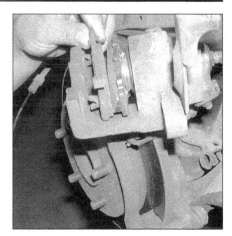

3.8c Remove the outer brake pad from the caliper

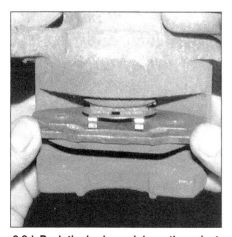

3.8d Push the brake pad down then pivot the pad out and away from the retainer

3.8e Use a C-clamp or special compressing tool to bottom the piston in the caliper bore

actuating screw, making sure the anti-friction washer and seal are in place. Install the nut and tighten it to the torque listed in this Chapter's Specifications. Rotate the lever back against its stop, then reconnect the parking brake cable.
10   Proceed to Step 20.

1992 and 1993 models

Refer to illustrations 3.15a, 3.15b, 3.16a, 3.16b, 3.17a and 3.17b

11   Wash the brake assembly with brake system cleaner, then remove the bolt securing the parking brake cable bracket to the caliper.
12   Remove the lower caliper mounting bolt and pivot the caliper up for access to the brake pads.
13   Remove the pads and retainers from the caliper mounting bracket.
14   Apply a thin coat of disc brake anti-squeal compound, in accordance with the manufacturer's recommendations, on the backing plates of the new pads.
15   Using pair of needle-nose pliers, rotate

under each illustration. **Note:** *It's a good idea to apply a thin layer of anti-squeal compound to the backing plates of the new pads. Allow*

the compound to dry before installing the pads.
9   Install the parking brake lever to the

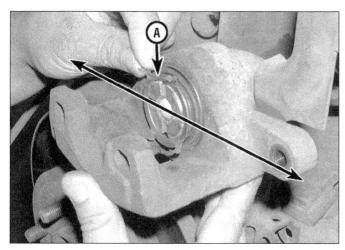

3.8f The D-shaped notches must be kept in alignment with the mounting bolt holes (arrows) - make sure the brake pad retainer (A) is positioned 90-degrees to the alignment notches

3.8g Install the inner pad in the retainer, then install the outer pad. Position the caliper in its mounting bracket and tighten the bolts to the torque listed in this Chapter's Specifications

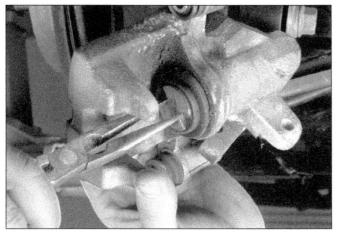

3.15a   Use a pair of needle-nose pliers with the tips engaged in the cutouts of the piston face to turn the piston into the cylinder bore

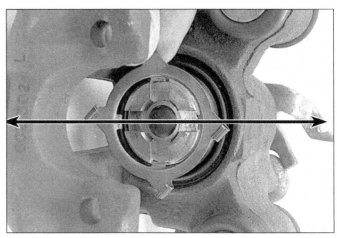

3.15b   Make sure the slots in the end of the piston are positioned 90-degrees to the mounting bolts

3.16a   Install the upper pad retainer . . .

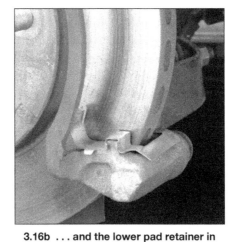

3.16b   . . . and the lower pad retainer in the caliper mounting bracket - make sure they are both fully seated

3.17a   Install the inner pad . . .

3.17b   . . . and the outer pad in the caliper mounting bracket, then position the caliper into place. Install the caliper mounting bolts and tighten them to the torque listed in this Chapter's Specifications

the piston to bottom it in its bore **(see illustration)**. Make sure the cutouts in the piston are positioned at a 90-degree angle to the caliper mounting bolt holes **(see illustration)**.

The buttons on the back side of the brake pads must engage with these alignment cutouts after the caliper has been installed onto the anchor plate.

16   Install the pad retainers in the caliper anchor plate (mounting bracket) **(see illustrations)**. Lubricate the retainers with a thin film of high-temperature grease.

17   Install the new pads to the caliper mounting bracket **(see illustrations)**.

18   Pivot the caliper down into position, making sure the tab on the inner brake pad mates with the cutout in the piston face. Install the mounting bolt and tighten it to the torque listed in this Chapter's Specifications.

19   Reconnect the parking brake cable bracket and tighten the bolt securely. Proceed to the next Step.

Front or rear pads

20   Install the wheel and lug nuts, lower the vehicle and tighten the lug nuts to the torque specified in Chapter 1.

21   Apply and release the brake pedal several times to bring the pads into contact with the brake discs. Check the operation of the brakes in an isolated area before driving the

vehicle in traffic.

22   Check the brake fluid level and add fluid, if necessary (see Chapter 1).

23   If you replaced the rear pads, check the operation of the parking brake.

---

4   Brake caliper - removal and installation

Warning: *1986 through 1990 models equipped with the Teves Mark II ABS system require the brake system to be depressurized before any brake component (master cylinder, wheel cylinder, caliper, brake line, etc.) is removed for servicing. Refer to the Teves Mark II ABS system depressurization procedure in Section 2.*

Warning 2: *Dust created by the brake system may contain asbestos, which is harmful to your health. Never blow it out with compressed air and don't inhale any of it. An approved filtering mask should be worn when working on the brakes. Do not, under any circumstances, use petroleum-based solvents to clean brake parts. Use brake system cleaner only!*

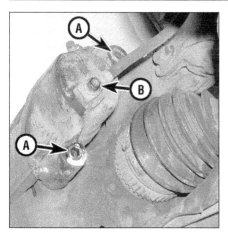

4.2a Location of the front caliper mounting bolts (A) and the brake hose banjo fitting bolt (B)

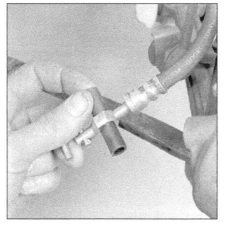

4.2b Using a short piece of rubber hose of the appropriate diameter, plug the brake line banjo fitting

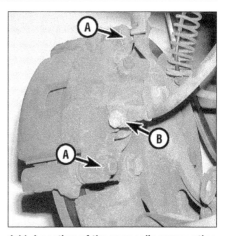

4.11 Location of the rear caliper mounting bolts (A) and the brake line banjo fitting bolt (B)

Note: *Always replace the calipers in pairs - never replace just one of them.*

Front

Removal

Refer to illustration 4.2a and 4.2b

1   Loosen - but don't remove - the lug nuts on the front wheels. Raise the front of the vehicle and place it securely on jackstands. Remove the front wheels.

2   Wash the brake assembly with brake system cleaner. Disconnect the brake line from the caliper **(see illustrations)** and plug it to keep contaminants out of the brake system and to prevent losing any more brake fluid than is necessary. **Note:** *If you are simply removing the caliper for access to other components, don't disconnect the brake hose.*

3   Remove the caliper mounting bolts **(see illustration 4.2a)**.

4   Detach the caliper from its mounting bracket.

Installation

5   Install the caliper by reversing the removal procedure. Remember to replace the sealing washers on either side of the brake line fitting with new ones. Tighten the caliper mounting bolts and the banjo bolt to the torque listed in this Chapter's Specifications.

6   Bleed the brake system (see Section 11).

7   Install the wheels and lug nuts and lower the vehicle. Tighten the wheel lug nuts to the torque listed in the Chapter 1 Specifications.

Rear

Removal

Refer to illustration 4.11

8   Loosen - but don't remove - the lug nuts on the rear wheels. Raise the rear of the vehicle and place it securely on jackstands. Remove the rear wheels.

9   Release the parking brake and remove the end of the parking brake cable **(see illustration 3.8a)** from the bracket.

5.2 Suspend the caliper with a piece of wire whenever it's necessary to reposition it - DO NOT let it hang by the brake hose!

10   Separate the parking brake cable from the bracket on the caliper.

11   Unscrew the banjo bolt and detach the brake line from the caliper **(see illustration)**. Plug the fitting to prevent fluid loss and contamination **(see illustration 4.2b)**. **Note:** *If you are simply removing the caliper for access to other components, don't disconnect the brake hose.*

12   Remove the caliper mounting bolts **(see illustration 4.11)**.

13   Detach the caliper from its mounting bracket.

Installation

14   Install the caliper by reversing the removal procedure. Remember to replace the sealing washers on either side of the brake line fitting with new ones. Tighten the caliper mounting bolts and the banjo bolt to the torque listed in this Chapter's Specifications.

15   Bleed the brake system (see Section 11).

16   Install the wheels and lug nuts. Lower the vehicle and tighten the lug nuts to the torque listed in the Chapter 1 Specifications.

5.3 The brake pads on this vehicle were obviously neglected as they wore down completely and cut deep grooves into the disc - wear this severe will require replacement of the disc

5   Brake disc - inspection, removal and installation

Refer to illustrations 5.2, 5.3, 5.4a, 5.4b, 5.5a and 5.5b

Inspection

1   Loosen the wheel lug nuts, raise the front of the vehicle and support it securely on jackstands. Apply the parking brake and block the rear wheels to keep the vehicle from rolling off the jackstands. Remove the wheel and install two lug nuts to hold the disc in place.

2   Remove the brake caliper as outlined in Section 4. You don't have to disconnect the brake hose. After removing the caliper bolts, suspend the caliper out of the way with a piece of wire - DO NOT let it hang by the hose **(see illustration)**.

3   Visually inspect the disc surface for score marks and other damage. Light scratches and shallow grooves are normal and may not be detrimental to brake opera-

5.4a Check for runout with a dial indicator - mount it with the indicator needle about 1/2-inch from the outer edge

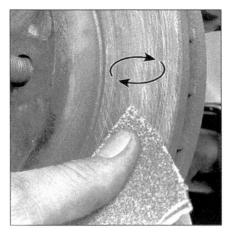

5.4b If you don't have the discs machined, at the very least be sure to break the glaze on the disc surface with sandpaper or emery cloth

5.5a The minimum wear thickness (arrow) is cast into the inside of the disc

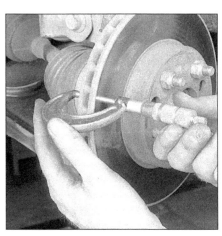

5.5b Measure the thickness of the disc at several points with a micrometer

tion, but deep score marks - over 0.015-inch deep - require disc removal and refinishing by an automotive machine shop. Be sure to check both sides of the disc **(see illustration)**. If pulsating has been felt during application of the brakes, suspect excessive disc runout.

4   To check disc runout, mount a dial indicator with the stem resting at a point about 1/2-inch from the outer edge of the disc **(see illustration)**. Set the indicator to zero and turn the disc. The indicator reading should not exceed the specified allowable runout limit. If it does, the disc should be refinished by an automotive machine shop. **Note:** *The discs should be resurfaced, regardless of the dial indicator reading, to impart a smooth finish and ensure perfectly flat brake pad surfaces (which will eliminate pedal pulsations. At the very least, if you don't have the discs resurfaced, remove the glaze with sandpaper or emery cloth using a swirling motion* **(see illustration)**.

5   Never machine the disc to a thickness less than the specified minimum allowable refinish thickness. The minimum wear (or discard) thickness is cast into the inside of the disc **(see illustration)**. The disc thickness can be checked with a micrometer **(see illustration)**.

Removal

6   If you're removing the front disc on an Eldorado or Seville, unbolt the caliper mounting bracket from the steering knuckle. On all models, remove the two lug nuts that were put on to hold the disc in place and remove the disc from the hub. If you're removing a rear disc, you'll also have to remove the two bolts and detach the caliper mounting bracket from the rear knuckle.

Installation

7   If you're working on an Eldorado or Seville (front or rear) or a rear disc on any other model, install the caliper mounting bracket and tighten the bolts to the torque

listed in this Chapter's Specifications. **Warning:** *The manufacturer states that these bolts must always be replaced with new ones whenever they are loosened or removed.* Place the disc in position over the threaded studs.

8   Install the caliper and brake pad assembly over the disc and position it on the steering knuckle (refer to Section 4 for the caliper installation procedure, if necessary). Tighten the caliper bolts to the torque listed in this Chapter's Specifications.

9   Install the wheel, then lower the vehicle to the ground. Depress the brake pedal a few times to bring the brake pads into contact with the disc. Bleeding of the system won't be necessary unless the brake hose was disconnected from the caliper. Check the operation of the brakes carefully before driving the vehicle in traffic.

---

6   Drum brake shoes - inspection and replacement

Warning 1: *1986 through 1990 models equipped with the Teves Mark II ABS system require the brake system to be depressurized before any brake component is removed for servicing. Refer to the Teves Mark II ABS system depressurization procedure in Section 2.*
Warning 2: *Drum brake shoes must be replaced on both rear wheels at the same time - never replace the shoes on only one wheel. Also, the dust created by the brake system contains asbestos, which is harmful to your health. Never blow it out with compressed air and don't inhale any of it. An approved filtering mask should be worn when working on the brakes. Do not, under any circumstances, use petroleum-based solvents to clean brake parts. Use brake system cleaner only!*
Caution: *Whenever the brake shoes are replaced, the return and hold-down springs should also be replaced. Due to the continu-*

ous heating/ cooling cycle the springs are subjected to, they lose tension over a period of time and may allow the shoes to drag on the drum and wear at a much faster rate than normal.

1   Loosen the wheel lug nuts, raise the rear of the vehicle and support it securely on jackstands. Block the front wheels to keep the vehicle from rolling off the jackstands.

2   Release the parking brake.

3   Remove the wheel. **Note:** *All four rear shoes must be replaced at the same time, but to avoid mixing up parts, work on only one brake assembly at a time.*

1986 Deville/Fleetwood models

Refer to illustrations 6.4a through 6.4y

4   The 1986 Deville/Fleetwood models are equipped with the anchor plate drum brake (or duo-servo) design. Remove the brake drum. Refer to the accompanying photographs and perform the brake shoe inspection and, if necessary, the replacement procedure. Start with **illustration 6.4a** and be sure to read each caption. After the inspec-

6.4a If the brake drum won't come off, it may be necessary to remove the plug with a hammer and chisel, then turn the adjuster screw to move the brake shoes away from the drum

6.4b Before removing anything, clean the brake assembly with brake system cleaner - DO NOT use compressed air to blow the dust out of the brake assembly!

tion or replacement procedures have been performed, proceed to Step 8. **Note:** *If the brake drum is stuck, make sure the parking brake is completely released, then apply some penetrating oil to the hub-to-brake drum joint. Allow the oil to soak in, then try to pull the drum off. If the drum still won't come off, the brake shoes will have to be retracted. This is done by removing the plug from the brake drum with a hammer and chisel* **(see illustration 6.4a)**. *With the plug removed, pull the lever off the adjusting star wheel with a hooked tool while turning the star wheel with a small screwdriver or brake adjuster tool, moving the shoes away from the drum. The drum should now come off.*

1987 through 1993 Deville/Fleetwood models

Refer to illustrations 6.5, 6.6a through 6.6y and 6.7a through 6.7h

5   The 1987 through 1993 Deville/Fleetwood models are equipped with two different types of leading/trailing drum brake designs. 1987 through 1991 models use coil type return springs, while 1992 and 1993 models are equipped with a single horseshoe type return spring. Remove the brake drum. If it's difficult to remove, back off the parking brake cable, remove the access hole plug from the backing plate, insert a screwdriver through the hole and press in to push the parking brake lever off its stop **(see illustration)**. **Note:** *On some models the access hole plug will have to knocked out with a hammer and punch. After reassembly, be sure to obtain and install a replacement rubber plug in the hole in the backing plate. This will allow the brake shoes to retract slightly. Insert a punch through the hole at the bottom of the splash shield and tap gently on the punch to loosen the drum. Avoid using excessive force.*

6   Clean the brake assembly with brake system cleaner **(see illustration 6.4b)** - DO NOT use compressed air to blow the dust out of the brake assembly. Refer to the accom-

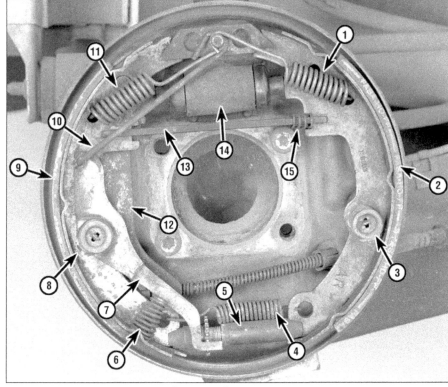

6.4c Anchor plate (duo-servo) type drum brake components

1	Return spring	9	Secondary shoe
2	Primary shoe	10	Actuator link
3	Hold-down spring	11	Return spring
4	Adjuster screw spring	12	Parking brake lever
5	Adjuster screw assembly	13	Parking brake strut
6	Return spring	14	Wheel cylinder
7	Actuator lever	15	Strut spring
8	Hold-down spring		

panying photographs for the actual shoe replacement procedure. If you're working on 1987 through 1991 models with coil return springs, follow **illustrations 6.6a through 6.6y**. Be sure to stay in order and read the caption under each illustration.

7   If you're working on a 1992 or 1993 model equipped with a single horseshoe-shaped return spring, follow **illustrations 6.7a through 6.7h.** After the inspection or replacement procedures have been performed proceed to Step 8.

6.4d Remove the return springs with a brake spring tool

6.4e Remove the hold-down springs and pins by pushing the retainer in and turning it 90-degrees - the tool shown here is available at most auto parts stores

6.4f Lift up on the actuator lever and remove the actuating link from the anchor pin pivot along with the actuator lever and return spring (arrows)

6.4g Spread the shoes apart at the top and remove the parking brake strut

6.4h With the shoe assembly spread to clear the hub flange, lift it away from the backing plate

6.4i Disconnect the parking brake lever from the cable and remove the shoe assembly

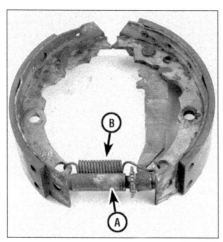

6.4j Remove the adjusting screw (A) and spring (B) from the shoe assembly - be sure to note how they're positioned

6.4k Remove the parking brake lever by prying off the C-clip

6.4l Install the parking brake lever on the new brake shoe and press the C-clip into place with needle-nose pliers

6.4m Lubricate the contact surfaces of the backing plate with high-temperature brake grease

6.4n Lubricate the threads and end of the adjuster screw with high-temperature grease, then install it and the lower return spring to the shoes

6.4o Connect the parking brake lever to the cable

6.4p Spread the brake assembly apart sufficiently to clear the hub flange and raise it into position

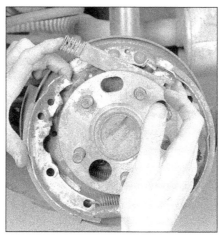

6.4q Install the parking brake strut and spring

6.4r Make sure the parking brake strut is positioned in the shoes properly (arrows)

6.4s Install the primary brake shoe hold-down pin and spring

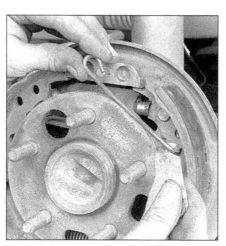

6.4t Attach the actuator link and lever to the secondary brake shoe

6.4u Install the actuator lever return spring

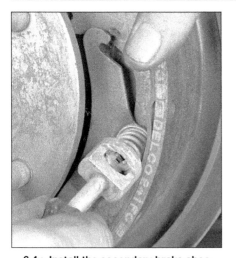

6.4v   Install the secondary brake shoe hold-down pin and spring

6.4w   Install the return springs

6.4x   Center the brake shoe assembly so the drum will slide over it

6.4y   Turn the adjusting screw so the drum fits snugly over the shoes; then, after the drum has been installed, back-off the adjustment so the drum doesn't drag

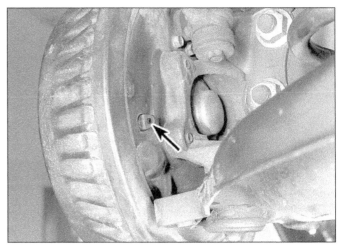

6.5   If you have trouble removing the drum, remove the access hole plug (arrow), insert a screwdriver through the hole and press in to push the parking brake lever off its stop - this will allow the brake shoes to retract slightly

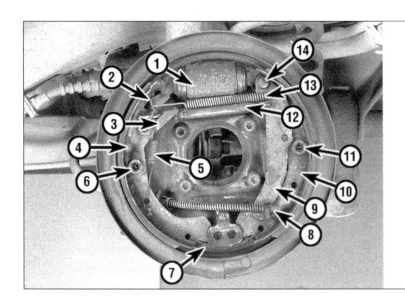

6.6a   1987 through 1991 rear drum brake components

1   Wheel cylinder
2   Spring connecting link
3   Adjuster actuator
4   Leading brake shoe
5   Actuator spring
6   Hold-down spring, pin and retainer
7   Lower return spring
8   Parking brake cable
9   Parking brake lever
10   Trailing brake shoe
11   Hold-down spring, pin and retainer
12   Adjuster screw assembly
13   Upper return spring

6.6b Unhook the upper return spring with a pair of pliers

6.6c Remove the hold-down spring and pin from the leading shoe (the tool shown here is available at most auto parts stores)

6.6d To remove the front shoe, pull it down . . .

6.6e . . . and detach the lower spring

6.6f Remove the adjuster (be sure to note the how the adjuster socket and the parking brake lever relate)

6.6g Remove the rear shoe hold-down spring, then pull down the rear shoe and parking brake assembly

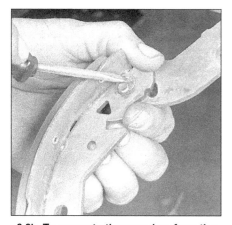

6.6h To separate the rear shoe from the parking brake lever, remove the C-clip and spring washer, then press out the lever pin - unless you're replacing the parking brake cable, it's not necessary to disconnect the cable and lever

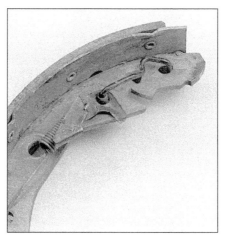

6.6i Place the front shoe assembly on a workbench and note the relationship of the actuator spring, the spring connecting link and the adjuster actuator . . .

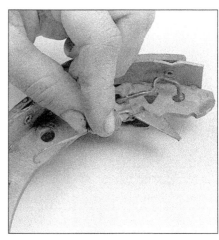

6.6j . . . then remove the actuator spring . . .

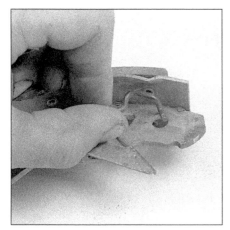

6.6k ... the spring connecting link ...

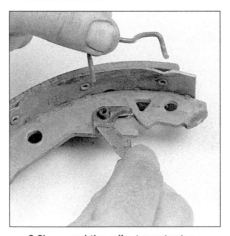

6.6l ... and the adjuster actuator - transfer these parts to the new front shoe (see illustration 6.6y)

6.6m Lubricate the raised contact surfaces of the backing plate with high-temperature grease

6.6n Install the parking brake lever on the new rear brake shoe, push the lever pin into place, install the spring washer and pop the C-clip into place with a pair of pliers (make sure the concave side of the spring washer faces toward the parking brake lever)

6.6o Clean the adjuster screw, check the threads for smooth rotation over their full length, then lubricate the adjuster threads, the inside surface of the socket and the socket face with high-temperature grease

6.6p Reattach the parking brake cable, if you disconnected it, then place the rear brake shoe in position . . .

6.6q . . . and install the hold-down spring

6.6r Attach the lower return spring to the brake shoes as shown - make sure the spring runs behind the anchor plate, not in front of it

6.6s Place the front brake shoe in position as shown . . .

6.6t . . . and install the hold-down spring

6.6u Install the adjuster assembly - make sure the parking brake lever is seated properly against the adjuster socket . . .

6.6v . . . and the adjuster actuator is properly seated against the adjuster screw

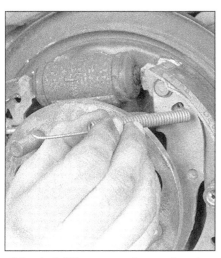

6.6w Install the upper return spring - be sure the rear end of the spring (the angled hook end) is installed through the parking brake lever and the rear shoe

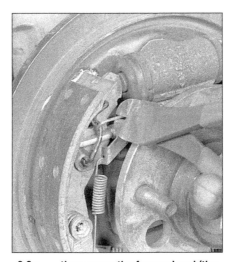

6.6x . . . then grasp the forward end (the long straight section) with a pair of pliers and hook it over the crook in the spring connecting link

6.6y Proper installation of the adjuster actuator, actuator spring and connecting link

6.7a 1992 and 1993 drum brake components with a single horseshoe type return spring

1 Actuator spring
2 Trailing shoe
3 Return spring
4 Leading shoe
5 Adjusting screw
6 Actuator lever
7 Wheel cylinder
8 Backing plate

6.7b Use a pair of needle-nose pliers to remove the actuator spring

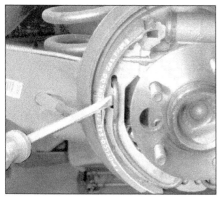

6.7c Wedge a flat-bladed screwdriver under the return spring and pry it out of the leading brake shoe, then remove the shoe, adjusting screw and actuator lever

6.7d Lift the retractor spring from the trailing brake shoe and swing the shoe out of the hub area to gain access to the parking brake cable

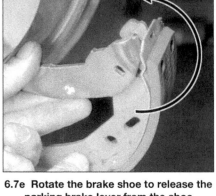

6.7e Rotate the brake shoe to release the parking brake lever from the shoe

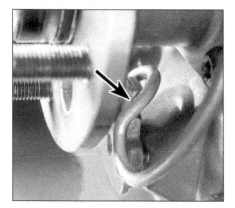

6.7f Use a screwdriver to pry the return spring over the alignment peg (arrow)

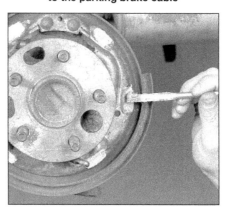

6.7g Lubricate the contact surfaces of the backing plate with high-temperature grease

6.7h Lubricate the adjuster screw with high-temperature grease prior to installation - to install new shoes on these models, reverse the removal steps, center the brake shoe assembly so the drum will slide over it, adjust the star wheel so the shoes will barely drag on the drum, then back-off the adjustment until they don't drag

All models

Refer to illustrations 6.8a and 6.8b

8   Before reinstalling the drum, check it for cracks, score marks, deep scratches and hard spots, which will appear as blue discolored areas. If the hard spots can't be removed with fine emery cloth or if any of the other conditions listed above exist, the drum must be taken to an automotive machine shop to have it turned. **Note:** *The drums should be resurfaced, regardless of the surface appearance, to impart a smooth finish and ensure a perfectly round drum (which will*

eliminate brake pedal pulsations related to out-of-round drums). At the very least, if you don't have the drums resurfaced, remove the glaze from the surface with medium-grit emery cloth using a swirling motion (**see illustration**). If the drum won't "clean up" before the maximum service limit is reached in the machining operation, install a new one. The maximum wear diameter is cast into each brake drum **(see illustration)**.

9   Install the brake drum on the axle flange. Using a screwdriver or brake adjuster tool, turn the adjuster star wheel so the shoes drag

on the drum slightly as the drum is turned. Then, back-off the star wheel until the shoes don't drag.

10   Mount the wheel, install the lug nuts, then lower the vehicle. Tighten the lug nuts to the torque listed in this Chapter's Specifications.

11   If the vehicle is equipped with anchor plate (duo servo) type brakes (1986 Deville/Fleetwood), make a number of forward and reverse stops to adjust the brakes until satisfactory pedal feel is obtained. If it's equipped with leading/trailing type brakes (1987 through 1993 Deville/Fleetwood), apply and release the brake pedal 30 to 35 times using normal pedal force. Pause about one second between pedal applications. After adjustment, make sure that both wheels turn freely. **Note:** *Make sure the brake shoes have been properly adjusted at the wheel adjusters before making the final rear brake adjustment.*

---

7   Wheel cylinder - removal and installation

Refer to illustrations 7.5a and 7.5b

Warning: *1986 through 1990 models equipped with the Teves Mark II ABS system require the brake system to be depressurized*

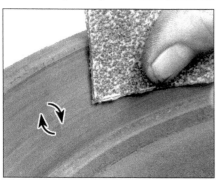

6.8a Remove the glaze from the drum with sandpaper or emery cloth - use a swirling motion

6.8b The drum has a maximum permissible diameter cast into it (arrow)

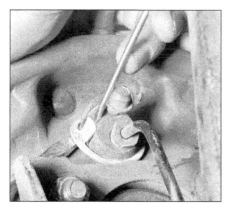

7.5a On some models the wheel cylinder is retained by a spring clip - to remove it, pry it off with two screwdrivers

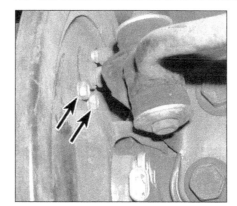

7.5b On later models the wheel cylinder is retained by two bolts (arrows)

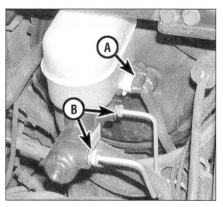

8.3 Unplug the fluid level sensor connector (A) and unscrew the brake line fittings (B)

before any brake component is removed for servicing. Refer to the Teves Mark II ABS system depressurization procedure in Section 2. **Note:** *If replacement is indicated (usually because of fluid leakage or sticky operation), it is recommended that new wheel cylinders be installed. Never replace only one wheel cylinder - always replace both of them at the same time.*

Removal

1   Raise the rear of the vehicle and support it securely on jackstands. Block the front wheels to keep the vehicle from rolling off the jackstands.
2   Remove the brake shoe assembly (Section 6).
3   Carefully clean the area around the wheel cylinder on both sides of the backing plate.
4   Unscrew the brake line fitting, but don't pull the line away from the wheel cylinder. **Note:** *If available, use a flare-nut wrench to avoid rounding off the corners of the fittings.*
5   Remove the wheel cylinder mounting bolts or retaining clip **(see illustrations)**.
6   Remove the wheel cylinder from the brake backing plate. Immediately plug the brake line to prevent fluid loss and contamination. Golf tees or rubber vacuum caps work well for plugging or capping flared metal lines. **Note:** *If the brake shoe linings are con-

8.7 Remove the master cylinder mounting nuts (arrows)

taminated with brake fluid, install new brake shoes and clean the drums with brake system cleaner.*

Installation

7   Apply a thin layer of RTV sealant to the surface of the backing plate where the wheel cylinder mounts, then place the wheel cylinder in position.
8   Start the brake line fitting by hand, being careful not to cross-thread it. Don't tighten it yet.
9   Install the wheel cylinder bolts and tighten them to the torque listed in this Chapter's Specifications.
10   Tighten the brake line fitting securely.
11   Install the brake shoes (see Section 6).
12   Bleed the brakes (see Section 11).

---

8   Master cylinder - removal and installation

Warning: *1986 through 1990 models equipped with the Teves Mark II ABS system require the brake system to be depressurized before any brake component is removed for servicing. Refer to the Teves Mark II ABS system depressurization procedure in Section 2.*
Caution: *On models equipped with an Antilock Brake System (ABS), have some plugs ready to cap the metal lines that connect the master cylinder to the hydraulic actuator. Failure to do so will allow air into the ABS actuator and can allow dirt and moisture to enter the ABS system.*
Note: *On 1986 through 1990 models equipped with the Teves Mark II ABS system, the master cylinder, hydraulic actuator, ABS pump and brake booster are integrated into one unit. Have the master cylinder diagnosed and repaired by a dealer service department or other qualified repair shop.*

Removal

Refer to illustrations 8.3 and 8.7
1   The master cylinder is located in the engine compartment, on the left (driver's) side.
2   Remove as much fluid as you can from the reservoir with a syringe, such as an old turkey baster. **Warning:** *If a baster is used,

never again use it for the preparation of food. Reinstall the reservoir cover.*
3   Detach the cable from the negative battery terminal. Unplug the fluid level sensor switch connector **(see illustration)**.
4   Place rags under the line fittings and prepare caps or plastic bags to cover the ends of the lines once they're disconnected. **Caution:** *Brake fluid will damage paint. Cover all painted parts and be careful not to spill fluid during this procedure.*
5   Loosen the fittings at the ends of the brake lines where they enter the master cylinder. To prevent rounding off the flats on the fittings, use a flare-nut wrench, which wraps around the hex.
6   Pull the brake lines away from the master cylinder and plug the ends to prevent contamination.
7   Remove the two mounting nuts **(see illustration)** and detach the master cylinder from the vehicle.
8   Remove the reservoir cover and reservoir diaphragm, then discard any remaining fluid in the reservoir.

Installation

Refer to illustration 8.10
9   Bench bleed the new master cylinder before installing it. Mount the master cylinder in a vise, with the jaws of the vise clamping on the mounting flange.

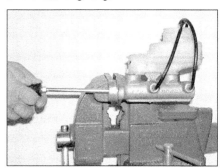

8.10 The best way to bleed air from the master cylinder before installing it on the vehicle is with a pair of bleeder tubes that direct brake fluid into the reservoir during bleeding

10   Attach a pair of master cylinder bleeder tubes to the outlet ports of the master cylinder **(see illustration)**.

11   Fill the reservoir with brake fluid of the recommended type (see Chapter 1).

12   Slowly push the pistons into the master cylinder (a large Phillips screwdriver can be used for this) - air will be expelled from the pressure chambers and into the reservoir. Because the tubes are submerged in fluid, air can't be drawn back into the master cylinder when you release the pistons.

13   Repeat the procedure until nor more air bubbles are present.

14   Remove the bleed tubes, one at a time, and install plugs in the open ports to prevent fluid leakage and air from entering. Install the reservoir cap.

15   Install the master cylinder over the studs on the power brake booster and tighten the attaching nuts only finger tight at this time.

16   Thread the brake line fittings into the master cylinder. Since the master cylinder is still a bit loose, it can be moved slightly in order for the fittings to thread in easily. Do not strip the threads as the fittings are tightened.

17   Fully tighten the mounting nuts, then the brake line fittings. Tighten the nuts to the torque listed in this Chapter's Specifications.

18   Fill the master cylinder reservoir with fluid, then bleed the master cylinder and the brake system as described in Section 11. To bleed the cylinder on the vehicle, have an assistant depress the brake pedal and hold the pedal to the floor. Loosen the fitting nut to allow air and fluid to escape. Repeat this procedure on both fittings at the master cylinder until the fluid is clear of air bubbles. **Caution:** *Have plenty of rags on hand to catch the fluid - brake fluid will ruin painted surfaces. After the bleeding procedure is completed, rinse the area under the master cylinder with clean water.*

19   Test the operation of the brake system carefully before placing the vehicle into normal service. **Warning:** *Do not operate the vehicle if you are in doubt about the effectiveness of the brake system.*

---

9   Power brake booster - check, removal and installation

Note: *On 1990 and earlier models equipped with the Teves ABS system, the master cylinder, hydraulic actuator, ABS pump and brake booster are integrated into one unit. Have the system diagnosed and repaired by a dealer service department or other qualified brake repair facility.*

Operating check

1   Depress the pedal and start the engine. If the pedal goes down slightly, operation is normal.

2   Depress the brake pedal several times with the engine running and make sure there is no change in the pedal reserve distance

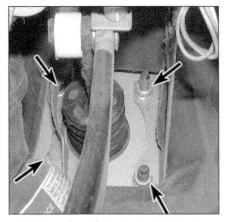

9.10 Location of the four brake booster attaching nuts (arrows)

(the distance between the pedal and the floor).

Airtightness check

3   Start the engine and turn it off after one or two minutes. Depress the brake pedal several times slowly. If the pedal goes down farther the first time but gradually rises after the second or third depression, the booster is airtight.

4   Depress the brake pedal while the engine is running, then stop the engine with the pedal depressed. If there is no change in the pedal reserve travel after holding the pedal for 30 seconds, the booster is airtight.

Removal

Refer to illustration 9.10

5   Power brake booster units should not be disassembled. They require special tools not normally found in most automotive repair stations or shops. They are fairly complex and because of their critical relationship to brake performance it is best to replace a defective booster unit with a new or rebuilt one.

6   To remove the booster, first unbolt the brake master cylinder from the booster as described in Section 8 (but without disconnecting the brake lines), then carefully move the master cylinder forward to provide clearance for booster removal.

7   Disconnect the vacuum hose leading from the engine to the booster. Be careful not to damage the hose when removing it from the booster fitting.

8   Locate the pushrod clevis pin connecting the booster to the brake pedal.

9   Remove the retaining clip (and spacer, if applicable) and slide the pushrod off its peg on the brake pedal.

10   Remove the four mounting nuts holding the brake booster to the firewall **(see illustration)**. You may need a light to see them, because they're up under the dash area.

11   Slide the booster straight out from the firewall until the studs clear the holes, turn the booster to the right (towards the center of the vehicle) so the clevis clears the firewall,

then remove the booster from the engine compartment.

Installation

12   Installation procedures are basically the reverse of those for removal. Lubricate the peg on the brake pedal arm with multi-purpose grease before installing the pushrod. Tighten the booster mounting nuts to the torque listed in this Chapter's Specifications. If the pushrod retaining clip is not in good condition or doesn't fit tightly, replace it.

13   When installing the master cylinder, be sure to tighten the mounting nuts to the torque listed in this Chapter's Specifications.

---

10   Brake hoses and lines - inspection and replacement

Refer to illustrations 10.2 and 10.11

Warning: *1990 and earlier models equipped with the Teves Mark II ABS system require the brake system to be depressurized before any brake component is removed for servicing. Refer to the Teves Mark II ABS system depressurization procedure in Section 2.*

1   About every six months, raise the vehicle and support it securely on jackstands, then check the flexible hoses that connect the steel brake lines to the front and rear brake assemblies. Look for cracks, chafing of the outer cover, leaks, blisters and other damage. The hoses are important and vulnerable parts of the brake system and the inspection should be thorough. A light and mirror will be helpful to see into restricted areas. If a hose exhibits any of the above conditions, replace it with a new one.

Front brake hose

2   Using a back-up wrench, disconnect the brake line from the hose fitting, being careful not to bend the frame bracket or brake line **(see illustration)**.

3   Use pliers to remove the U-clip from the female fitting at the bracket, then remove the

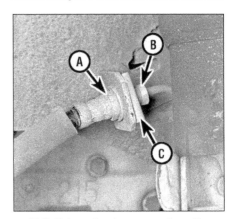

10.2 Using a back-up wrench on the flexible hose side of the fitting (A), loosen the tube nut (B) with a flare nut wrench and remove the U-clip (C) from the hose fitting

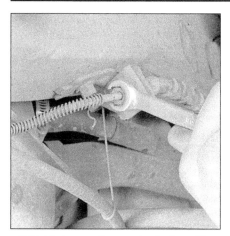

10.11 To loosen the rear brake hose/line fitting, simply loosen the nut and pull the clip off the bracket

11.8 When bleeding the brakes, a hose is connected to the bleeder valve at the caliper (or wheel cylinder) and then submerged in brake fluid - air will be seen as bubbles in the container or in the tube (all air must be expelled before continuing to the next wheel)

hose from the bracket. **Note:** *On some models the bracket is part of the hose assembly and must be unbolted from the chassis.*

4   At the caliper end of the hose, remove the bolt from the fitting block, then remove the hose and the copper gaskets on either side of the fitting block.

5   When installing the hose, always use new copper gaskets on either side of the fitting block and lubricate all bolt threads with clean brake fluid before installation.

6   With the fitting flange engaged with the caliper locating ledge, attach the hose to the caliper.

7   Without twisting the hose, install the female fitting in the hose bracket. It'll fit the bracket in only one position.

8   Install the U-clip retaining the female fitting to the frame bracket.

9   Using a back-up wrench, attach the brake line to the hose fitting.

10   When the brake hose installation is complete, there shouldn't be any kinks in the hose. Make sure the hose doesn't contact any part of the suspension. Check it by turning the wheels to the extreme left and right positions. If the hose makes contact, remove the hose and correct the installation as necessary.

Rear brake hose

11   Using a back-up wrench, if necessary, disconnect the hose at both ends, being careful not to bend the bracket or steel lines **(see illustration)**.

12   Remove the two U-clips with pliers and separate the female fittings from the brackets.

13   Unbolt the hose retaining clip and remove the hose.

14   Without twisting the hose, install the female ends in the frame brackets. It'll fit the bracket in only one position.

15   Install the U-clips retaining the female end to the bracket.

16   Using a back-up wrench, attach the steel line fittings to the female fittings. Again, be careful not to bend the bracket or steel line.

17   Make sure the hose installation didn't loosen the frame bracket. Tighten the bracket if necessary.

18   Fill the master cylinder reservoir and bleed the system (refer to Section 9).

Metal brake lines

19   When replacing brake lines, be sure to buy the correct replacement parts. Don't use copper or any other tubing for brake lines.

20   Auto parts stores and brake supply houses carry various lengths of prefabricated brake line. These sections can be bent with a tubing bender.

21   When installing the new line, make sure it's securely supported in the brackets with plenty of clearance between moving or hot components.

22   After installation, check the master cylinder fluid level and add fluid as necessary. Bleed the brake system as outlined in the next Section and test the brakes carefully before driving the vehicle in traffic.

---

11   Brake Hydraulic system bleeding

Refer to illustration 11.8

Warning: *Wear eye protection when bleeding the brake system. If you get fluid in your eyes, rinse them immediately with water and seek medical attention.*

Note: *Bleeding the brakes is necessary to remove air that manages to find its way into the system when its been opened during removal and installation of a hose, line, caliper or master cylinder.*

All except 1990 and earlier models with the Teves ABS system

1   It'll probably be necessary to bleed the

system at all four brakes if air has entered the system due to low fluid level, or if the brake lines have been disconnected at the master cylinder.

2   If a brake line was disconnected at only one wheel, then only that caliper or wheel cylinder must be bled.

3   If a brake line is disconnected at a fitting located between the master cylinder and any of the brakes, that part of the system served by the disconnected line must be bled.

4   Remove any residual vacuum from the power brake booster by applying the brake several times with the engine off.

5   Remove the master cylinder reservoir cover and fill the reservoir with brake fluid. Reinstall the cover. **Note:** *Check the fluid level often during the bleeding procedure and add fluid as necessary to prevent the level from falling low enough to allow air bubbles into the master cylinder.*

6   Have an assistant on hand, as well as a supply of new brake fluid, an empty, clear plastic container, a length of plastic, rubber or vinyl tubing to fit over the bleeder valve and a wrench to open and close the bleeder valve.

7   Beginning at the right rear wheel (all except 1991 and later Seville/Eldorado, Toronado/Trofeo and Riviera models), loosen the bleeder valve slightly, then tighten it to a point where it's snug but can still be loosened quickly and easily. **Note:** *On 1991 and later Seville/Eldorado, Toronado/Trofeo and Riviera models, start with the left front wheel for the bleeding procedure.*

8   Place one end of the tubing over the bleeder valve and submerge the other end in brake fluid in the container **(see illustration)**.

9   Have your assistant pump the brakes slowly a few times to get pressure in the system, then hold the pedal down firmly.

10   While the pedal is held down, open the bleeder valve just enough to allow fluid to flow out of the valve. Watch for air bubbles to exit the submerged end of the tube. When the fluid slows after a couple of seconds, close the valve and have your assistant release the pedal.

11   Repeat Steps 9 and 10 until no more air is seen leaving the tube, then tighten the bleeder valve. Follow the proper sequence to bleed the remaining wheel cylinders and/or calipers:

   a)   *On 1986 through 1990 models, bleed the left front wheel, the left rear wheel and the right front wheel, in that order, and perform the same procedure.*

   b)   *On 1991 through 1993 Deville/Fleetwood models, bleed the left rear wheel, the right front wheel and the left front wheel, in that order, and perform the same procedure.*

   c)   *On 1991 and later Seville/Eldorado, Toronado/Trofeo and Riviera models, bleed the right front wheel, the left rear wheel and the right rear wheel, in that order, and perform the same procedure. Be sure to check the fluid in the master cylinder reservoir frequently.*

12   Never use old brake fluid. It contains moisture which can boil, rendering the brakes useless.

13   Refill the master cylinder with fluid at the end of the operation.

14   Check the operation of the brakes. The pedal should feel firm when depressed. If necessary, repeat the entire procedure. **Warning:** *Don't operate the vehicle if you're in doubt about the effectiveness of the brake system.*

1990 and earlier models with the Teves ABS system

15   The rear calipers on models equipped with the Teves ABS system require a slightly different bleeding procedure. Turn the ignition key to the On position, which will activate the hydraulic pump and charge up the accumulator (when the system is charged the pump motor will stop running).

16   Connect a length of tubing to the right rear caliper or wheel cylinder bleeder valve **(see illustration 11.8).** Place the other end of the bleeder tube in a container partially filled with clean brake fluid.

17   Have an assistant slowly depress the brake pedal and hold it in the applied position. SLOWLY open the bleeder valve approximately 1/2-turn and allow the fluid to flow for a few seconds. **Warning:** *The brake fluid in the rear calipers is under extremely high pressure, and careless opening of the bleeder valves may cause the fluid to shoot out with great force.*

18   Close the bleeder valve and have your assistant slowly release the brake pedal. Check the fluid level and add some, if necessary. Don't let the reservoir run dry.

19   Repeat Steps 17 and 18 until clean, bubble-free brake fluid flows from the bleeder valve.

20   Perform Steps 16 through 19 on the left rear caliper.

21   The front brakes may be bled using the standard brake bleeding procedure described in Steps 8 through 14.

22   After bleeding, be sure to fill the brake fluid reservoir to the appropriate level with the recommended fluid (see Chapter 1). Check the operation of the brakes in an isolated area before returning the vehicle to normal service. **Warning:** *Don't operate the vehicle if you're in doubt about the effectiveness of the brake system.*

12   Parking brake - adjustment

1986 Deville/Fleetwood models

Refer to illustration 12.5

1   Adjust the brakes (see Step 9 in Section 6).

2   Depress the parking brake pedal 1-13/32 inches.

3   Raise the vehicle and support it securely on jackstands.

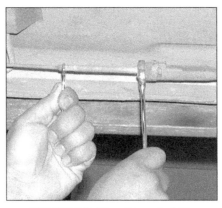

12.5 While holding the end of the threaded rod from moving, turn the adjusting nut until the right rear wheel can just barely be turned backwards but not forward

4   Before adjusting the parking brake, make sure the equalizer nut groove is lubricated with multi-purpose grease.

5   Tighten the adjusting nut **(see illustration)** until the right rear wheel can barely be turned backwards with two hands, but locks when turned forward.

6   Release the parking brake pedal and check to make sure the rear wheels turn freely in both directions.

7   Lower the vehicle.

1987 through 1993 Deville/Fleetwood models

Refer to illustration 12.13

8   Adjust the brakes (see Step 9 in Section 6).

9   Apply and release the parking brake six times to ten ratchet clicks.

10   Make sure the parking brake releases completely by turning the ignition to ON and noting whether the Brake warning light is off. If it's on even though the brake appears to be released, operate the pedal release lever and pull down on the front parking brake cable to remove slack from the assembly. Check both rear wheels to make sure they still turn freely.

11   Raise the vehicle and support it securely on jackstands.

12   Remove the access hole plug from the backing plate.

13   Turn the parking brake cable adjuster nut **(see illustration 12.5)** until you can insert a 1/8-inch drill bit - but not a 1/4-inch bit - through the access hole into the space between the shoe web and the parking brake **(see illustration)**. **Note:** *The drill bit must be perpendicular (at a right angle) to the backing plate.*

14   Apply the parking brake pedal four clicks. You should not be able to turn the wheels in a forward direction; when you try to rotate them in the rearward direction, they should either drag or not turn.

15   Release the parking brake and verify that both wheels rotate freely.

16   Replace the access hole plug.

17   Lower the vehicle.

12.13 Turn the parking brake cable adjuster until you can insert a 1/8-inch drill bit through the access hole in the backing plate and between the parking brake lever and the trailing brake shoe (drum removed for clarity)

All other models

18   Apply and release the parking brake to ten ratchet clicks (approximately 125 lbs of pressure on the pedal). Repeat this step six times.

19   Check the parking brake pedal assembly for full release by turning the ignition to On and noting whether the BRAKE warning light is off. If it's on even though the brake appears to be released, operate the pedal release lever and pull down on the front parking brake cable to remove slack from the assembly. Check both rear wheels to make sure they still turn freely.

20   Raise the vehicle and support it securely on jackstands.

21   Check the parking brake levers on both rear calipers. Levers should be against the stops on the back of the caliper housing. Check for cable binding if levers are not resting against the stops. This is the OFF position for the parking brake system.

22   Tighten the parking brake cable at the adjuster until the either lever starts to move off its stop.

23   Back off the cable adjuster until the lever that moved off its stop comes back to rest against its top. Both levers should now touch equally.

24   Activate the parking brake several times and check adjustment.

25   Release the parking brake and verify that both wheels rotate freely.

26   Lower the vehicle.

13   Parking brake cables - replacement

1   Detach the cable from the negative battery terminal.

2   If you're going to remove or replace a rear cable, loosen the wheel lug nuts.

3   Raise the vehicle and support it securely on jackstands.

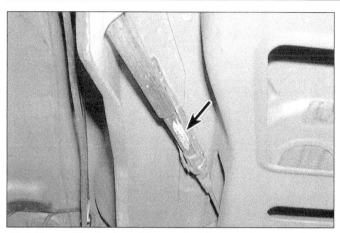

13.4 The parking brake cable equalizer is located on the left side of the rear suspension crossmember

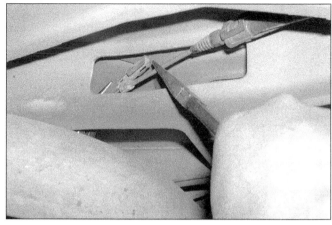

13.5 To disconnect the front or intermediate cable from the equalizer assembly, simply slide the cable out of the recess on the equalizer

13.6 Before the front cable will slide through the floor, you'll have to unscrew this nut from the underbody

13.10 To detach the intermediate cable from this mounting bracket, pinch the tabs on the cable housing with a pair of pliers

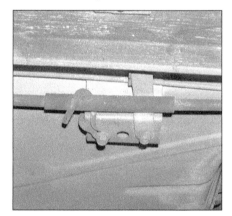

13.11 To detach the intermediate cable from the underbody, detach the guide and cable clip

Front cable

Refer to illustrations 13.4, 13.5 and 13.6

4   Locate the equalizer assembly along the left side of the underbody **(see illustration)**.
5   Loosen the parking brake adjuster **(see illustration 12.5)** and detach the front cable from the equalizer **(see illustration)**.
6   Remove the nut from the front cable **(see illustration)**.
7   Working inside the vehicle, detach the cable housing and cable from the parking brake lever assembly.
8   Installation is the reverse of removal.

Intermediate cable

Refer to illustrations 13.10, 13.11, 13.12a and 13.12b

9   Disconnect the intermediate cable from the forward equalizer assembly **(see illustration 13.5)**.
10   Detach the intermediate cable housing from the bracket **(see illustration)**.
11   Detach the intermediate cable guide and clip from the underbody **(see illustration)**.
12   Disconnect the intermediate and right rear cables **(see illustration)**, then disconnect the intermediate cable from the left rear

cable **(see illustration)**.
13   Installation is the reverse of removal.

Rear cables

Refer to illustrations 13.18 13.19a and 13.19b

14   Back off the parking brake adjuster nut until there is slack in the cable **(see illustration 12.5)**.
15   Remove the rear wheel.

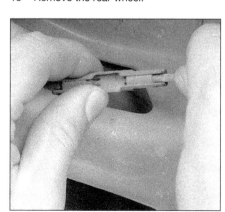

13.12a Disconnect the intermediate and right rear cable . . .

16   On rear drum systems, remove the brake drum (see Section 6). On rear disc models, disconnect the cable from the lever and bracket **(see illustration 3.8a)**.
17   On rear drum systems, remove the rear brake shoe and the parking brake lever as an assembly and detach the cable from the lever (see Section 6).
18   On rear drum systems, detach the cable

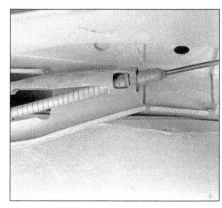

13.12b . . . then disconnect the intermediate cable from the left rear cable

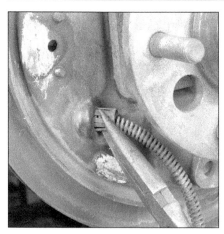

13.18 To detach the cable from the rear drum brake assembly, pinch the tabs on the housing with a pair of pliers and slide it through the hole in the brake backing plate

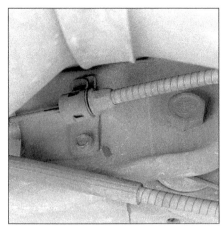

13.19a To remove the left rear cable from the underbody, detach this cable clip

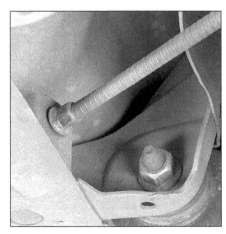

13.19b To remove the right rear cable from the underbody, pinch the tabs on this housing (not visible in this photo, they're on the other side of the housing bracket) with a pair of pliers and slide the housing and cable out of the bracket

fitting from the backing plate **(see illustration)**.

19   Detach the cable(s) from the frame **(see illustrations)**.

20   Disconnect the intermediate and right rear cable, then disconnect the left rear cable from the intermediate cable **(see illustrations 13.12a and 13.12b)**.

21   Installation is the reverse of removal.

All cables

22   Be sure to adjust the parking brake cable after installation (see Section 12).

14 Brake pedal travel check

1   If the vehicle is equipped with rear drum brakes, adjust the brakes (see Step 9 in Section 6).

2   Apply 150 ft-lbs of pressure to the brake pedal. If the BRAKE warning light illuminates, check the brake system for leaks, low fluid, broken lines, etc.

3   Turn the ignition key OFF, then pump the pedal several times to deplete the booster vacuum.

4   Place a yardstick or measuring tape on the pad of the brake pedal.

5   Measure the distance from the brake pedal to the bottom of the steering wheel.

6   Next, press the brake pedal with 100 lbs of pressure and measure the distance from the steering wheel to the brake pedal.

7   The difference between these two mea-surements is the brake pedal travel.

8   Refer to the Specifications listed in this Chapter for the correct brake pedal travel.

9   Most brake pedal travel problems occur because of air in the brake hydraulic lines. Refer to the bleeding procedure in Section 11 to correct this condition. The brake pedal must be firm before the vehicle is ready to operate.

15 Brake light switch - removal, installation and adjustment

Refer to illustration 15.3

Removal

1   The brake light switch is located on a bracket at the top of the brake pedal.

2   Remove the under dash cover and dis-connect the wiring to the courtesy light in the panel.

3   Locate the switch at the top of the brake pedal **(see illustration)**.

4   Disconnect the negative battery cable from the battery.

5   Detach the wiring connectors at the brake light switch.

6   Depress the brake pedal and pull the switch out of the clip. The switch appears to be threaded, but it's designed to be pushed into and out of the clip, not turned.

Installation and adjustment

7   With the brake pedal depressed, push the new switch into the clip. Note that audible clicks will be heard as this is done.

8   Pull the brake pedal all the way to the rear, against the pedal stop until the clicking sounds can no longer be heard. This action will automatically move the switch the proper amount and no further adjustment will be required. **Caution:** *Don't apply excessive force during this adjustment procedure, as power booster damage may result.*

9   Connect the wiring at the switch and the battery. Make sure the brake lights are func-tioning properly.

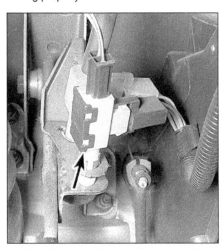

15.3 The brake light switch (arrow) is located to the right of the steering column at the end of the mounting bracket

Chapter 10
Suspension and steering systems

Contents

Specifications

Torque specifications

Ft-lbs (unless otherwise indicated)

Front suspension

Balljoint-to-steering knuckle nut
    Deville/Fleetwood
        Step 1 .... 88 in-lbs
        Step 2 .... Tighten an additional 120 degrees
    Seville/Eldorado, Riviera and Toronado
        Step 1 .... 84 in-lbs
        Step 2 .... Tighten an additional 120 degrees
Balljoint-to-control arm nuts 50
Brake line bracket bolt .... 156 in-lbs
Control arm
    Deville/Fleetwood
        Front pivot bolt/nut 140
        Rear pivot bolt/nut 91
    Seville/Eldorado, Riviera and Toronado
        Brake reaction rod nut (front) 56
        Pivot bolt/nut (rear) 95
Hub retaining bolts 70
Stabilizer bar
    Stabilizer bar link nuts
        Deville/Fleetwood models .... 156 in-lbs
        Seville/Eldorado, Riviera and Toronado
            1989 and earlier 43
            1990 and later 48
    Stabilizer bar bushing bracket bolts
        1991 and earlier 35
        1992 18
Steering knuckle-to-strut assembly bolts 140
Strut assembly upper mounting nuts 18
Tie-rod end nut 35

Rear suspension

Adjustment link-to-knuckle nut (Deville/Fleetwood)	33
Adjustment link-to-control arm nut (Deville/Fleetwood)	63
Balljoint stud-to-control arm nut (Deville/Fleetwood)	22
Control arm mounting nuts	
Deville/Fleetwood	85
Seville/Eldorado, Riviera and Toronado	66
Hub and bearing assembly bolts	52
Knuckle pivot bolt (Seville/Eldorado, Riviera and Toronado)	59
Leaf spring (Seville/Eldorado, Riviera and Toronado)	
Upper insulator nuts	21
Retainer bolts	21
Stabilizer bar	
Stabilizer bar-to-knuckle nut (Deville/Fleetwood)	156 in-lbs
Stabilizer bar-to-strut nuts (Seville/Eldorado, Riviera and Toronado)	43
Stabilizer bar pinch clamp bolts	156 in-lbs
Stabilizer bar bushing assembly nut	37
Strut-to-knuckle nuts (Deville/Fleetwood)	144
Strut lower pinch bolt (Seville/Eldorado, Riviera and Toronado)	40
Strut-to-suspension support nut (Seville/Eldorado, Riviera and Toronado)	65
Strut-to-body mounting nuts (Deville/Fleetwood)	19
Strut upper mount-to-strut bolt (Deville/Fleetwood)	62

Steering

Airbag module screws	27 in-lbs
Intermediate shaft pinch bolt	35
Power steering pump mounting bolts	20
Tie-rod end castellated nut	35
Tie-rod outer jam nut	50
Steering column mounting bolts	20
Steering gear mounting bolts	50
Steering wheel nut	30
Wheel lug nuts	See Chapter 1

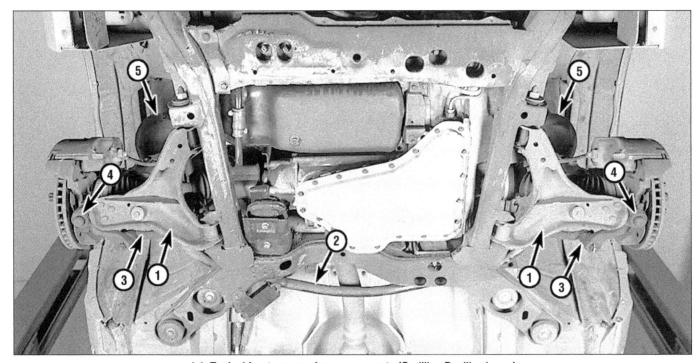

1.1　Typical front suspension components (Cadillac Deville shown)

1	Control arm	3	Outer tie-rod	5	Strut/coil spring assembly
2	Stabilizer bar	4	Balljoint		

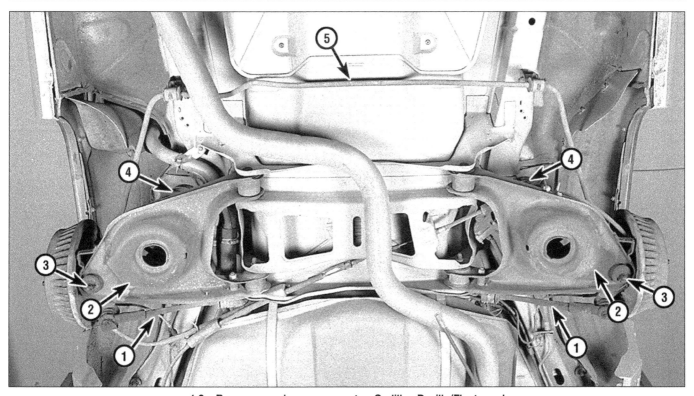

1.2a Rear suspension components - Cadillac Deville/Fleetwood

| 1 | Adjustment link | 2 | Control arm | 3 | Balljoint | 4 | Coil spring | 5 | Stabilizer bar |

1 General information

Refer to illustrations 1.1, 1.2a and 1.2b
Warning: *Whenever any of the suspension or steering fasteners are loosened or removed, they must be inspected and, if necessary, replaced with new ones of the same part number or of original equipment quality and design. Torque specifications must be followed for proper reassembly and component retention. Never attempt to heat or straighten*

any suspension or steering components. Instead, replace any bent or damaged part with a new one.

The front suspension is a combination strut and spring design. The steering knuckles are located by lower control arms which are mounted to longitudinally positioned, removable frame members. The lower end of the steering knuckle pivots on a balljoint riveted to the control arm. The balljoint is fastened to the steering knuckle with a castellated nut. The control arms are connected by

a stabilizer bar, which reduces body lean during cornering **(see illustration)**.

The rear suspension on the Deville/Fleetwood models differ from the Seville/Eldorado, Riviera and Toronado models. The rear suspension on the Deville/Fleetwood models is fully independent with each suspension knuckle supported by a lower control arm, coil spring and strut **(see illustration)**. A stabilizer bar minimizes body roll. Each control arm is equipped with a suspension adjustment link to provide for toe adjust-

1.2b Rear suspension components - Cadillac Seville/Eldorado, Buick Riviera and Oldsmobile Toronado

| 1 | Control arm | 2 | Strut pinch bolt | 3 | Strut/coil spring assembly | 4 | Leaf spring |

2.2 Detach the front brake line bracket (arrow) from the strut

2.4a Mark the strut-to-steering knuckle relationship and draw a line around the nuts with paint or a scribe

2.4b Before you remove the three strut-to-shock tower nuts, be sure to mark their relationship to the body

ment and to minimize alignment variation with suspension movement. The rear control arm is attached to the suspension knuckle through a balljoint to reduce friction.

The rear suspension on the Seville/Eldorado, Riviera and Toronado models is fully independent with a transverse mounted leaf spring **(see illustration)**. This single leaf is constructed of durable, lightweight composite material. The spring is fully isolated from the rear suspension support. The transverse leaf spring is mounted to the suspension support by a left and right insulator mounts which allows each end of the leaf spring to support the contact surface of the control arm. The knuckle pivots on the end of the control arm and is mounted by a bolt and bushings instead of a balljoint assembly.

All vehicles covered by this manual have power rack-and-pinion steering systems. The components making up the system are the steering wheel, steering column, rack-and-pinion assembly, tie-rods and tie-rod ends. The power steering system has a belt-driven pump to provide hydraulic pressure.

In the power steering system, the motion of turning the steering wheel is transferred through the column to the pinion shaft in the rack-and-pinion assembly. Teeth on the pinion shaft are meshed with teeth on the rack, so when the shaft is turned, the rack is moved left or right in the housing. A rotary control valve in the rack-and-pinion unit directs hydraulic fluid under pressure from the power steering pump to either side of the integral rack piston, which is connected to the rack, thereby reducing manual steering force. Depending on which side of the piston this hydraulic pressure is applied to, the rack will be forced either left or right, which moves the tie-rods, etc. If the power steering system loses hydraulic pressure it will still function manually, though with increased effort.

The steering column is a collapsible, energy-absorbing type, designed to compress in the event of a front end collision to minimize injury to the driver. The column also houses the ignition switch lock, key warning buzzer, turn signal controls, headlight dimmer control and windshield wiper controls

(some models). The ignition and steering wheel can both be locked while the vehicle is parked. Due to the column's collapsible design, it's important that only the specified screws, bolts and nuts be used as designated and that they're tightened to the specified torque. Other precautions particular to this design are noted in appropriate Sections.

The power rack-and-pinion steering system is located behind the engine/transaxle assembly on the firewall and actuates the tie-rods, which transmit steering inputs to the steering knuckles. The steering column is connected to the steering gear through an insulated coupler. The steering column is designed to collapse in the event of an accident.

Note: *These vehicles have a combination of standard and metric fasteners on the various suspension and steering components, so it would be a good idea to have both types of tools available when beginning work.*

---

2   Strut/coil spring assembly (front) - removal, inspection and installation

---

Removal

Refer to illustrations 2.2, 2.4a, 2.4b, 2.6 and 2.8

1   Apply the parking brake and block the rear wheels to keep the vehicle from rolling. Loosen the wheel lug nuts, raise the front of the vehicle and support it securely on jackstands. Remove the wheel.

2   Remove the brake line bracket from the strut **(see illustration)**.

3   If the vehicle is equipped with an Anti-lock Brake System (ABS), disconnect the front sensors. If the vehicle is equipped with the Computer Command Ride (CCR), disconnect the CCR electrical connector from the front strut.

4   Using white paint or a scribe, mark the strut-to-steering knuckle relationship and make a line around the strut-to-steering knuckle nuts. Also mark the relationship of

2.6 Remove the strut-to-knuckle nuts and bolts

the upper strut mounting studs to the body **(see illustrations)**. **Note:** *If the strut is replaced with a new part, be sure to paint the alignment marks onto the new strut in exactly the same position as the original to insure correct alignment.*

5   Separate the tie-rod end from the steering arm as described in Section 19.

6   Remove the strut-to-knuckle nuts **(see illustration)** and knock the bolts out with a soft-face hammer.

7   On late model Seville/Eldorado, Riviera and Toronado models, disconnect the stabilizer bar from the strut.

8   Separate the strut from the steering knuckle **(see illustration)**. Be careful not to overextend the inner CV joint or stretch the brake hose. If necessary, remove the driveaxle hub nut and separate the driveaxle from the hub (see Chapter 8).

9   Have an assistant support the strut assembly. Remove the three strut-to-shock tower nuts. Remove the assembly out through the fenderwell.

Inspection

10   Check the strut body for leaking fluid, dents, cracks and other obvious damage which would warrant repair or replacement.

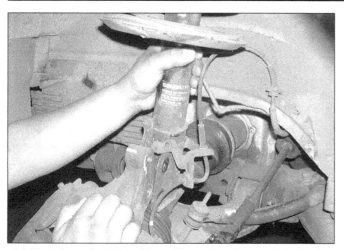

2.8 Push in on the strut while pulling out on the top of the brake disc to separate the knuckle and strut

3.4 After the spring has been compressed, remove the damper shaft nut

11   Check the coil spring for chips and cracks in the spring coating (this will cause premature spring failure due to corrosion). Inspect the spring seat for hardening, cracks and general deterioration.

12   If wear or damage is evident, replace the strut. Proceed to Section 3 for the strut/coil spring disassembly procedure.

Installation

13   Install the strut/coil spring assembly. Once the three studs protrude from the shock tower, install the nuts so the strut/coil spring assembly won't fall back through. This may require an assistant, since the strut is quite heavy and awkward. Be sure to align the marks you made on disassembly.

14   Slide the steering knuckle into the strut flange and insert the two bolts. They should be positioned with the flats situated horizontally. Install the nuts, align the marks and tighten the nuts to the torque listed in this Chapter's Specifications.

15   Install the tie-rod end to the steering knuckle and tighten the castellated nut to the torque listed in this Chapter's Specifications. Install a new cotter pin. If the cotter pin won't pass through, tighten the nut a little more, but

just enough to align the hole in the stud with a castellation on the nut (don't loosen the nut).

16   Install the wheel, lower the vehicle and tighten the lug nuts to the torque listed in Chapter 1 Specifications.

17   Tighten the three upper mounting nuts to the torque listed in this Chapter's Specifications.

18   Have the front end alignment checked and, if necessary, adjusted.

3   Strut or coil spring (front) - replacement

Refer to illustrations 3.4, 3.5a, 3.5b, 3.6, 3.7a and 3.7b

Warning: *Disassembling a strut is a dangerous job. Be very careful and follow the instructions closely or serious injury may result. Use only a high-quality spring compressor and carefully follow the manufacturer's instructions furnished with the tool. After removing the coil spring from the strut assembly, set it aside in a safe, isolated area.*

1   If the struts exhibit the telltale signs of wear (leaking fluid, loss of dampening capa-

bility) explore all options before beginning any work. Rebuilt or new strut assemblies (some complete with springs) may be available on an exchange basis which eliminates much time and work. Check on the cost and availability of parts before removing the strut.

2   Remove the strut and spring assembly following the procedure described in Section 2. Mount the strut assembly in a vise. Cushion the vise jaws with rags or blocks of wood.

3   Following the tool manufacturer's instructions, install the spring compressor (which can be obtained at most auto parts stores or equipment yards on a daily rental basis) on the spring and compress it sufficiently to relieve all pressure from the spring seat. This can be verified by wiggling the spring seat.

4   Loosen the damper shaft nut while using the proper tool to hold the shaft stationary **(see illustration)**.

5   Lift the bearing cap, upper spring seat and upper insulator off the damper shaft **(see illustrations)**. Inspect the bearing in the spring seat for smooth operation and replace it if necessary.

6   Carefully remove the compressed spring assembly **(see illustration)** and set it

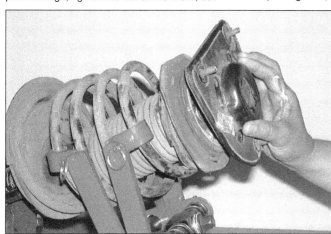

3.5a Remove the bearing cap . . .

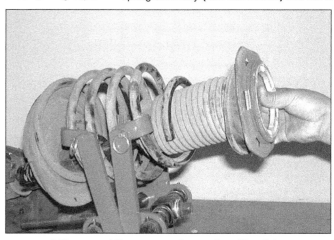

3.5b . . . and the upper spring seat and insulator from the damper shaft

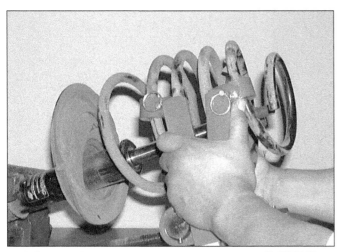

3.6 Remove the compressed spring assembly - be EXTREMELY CAREFUL when handling the spring!

3.7a Install the upper spring seat with the flat (arrow) facing the steering knuckle flange

in a safe place. **Warning:** *Don't position your head near the end of the spring!*

7   Assemble the new strut beginning with the lower spring insulator and spring, then the upper spring insulator **(see illustration)**, spring seat and bearing cap. Position the spring seat and bearing cap with the flats facing the steering knuckle flange **(see illustration)**. **Note:** *If the strut assembly is replaced with a new part, be sure to paint the alignment marks onto the new strut in exactly the same position as the original strut assembly.*

8   Install the damper shaft nut and tighten it to the torque listed in this Chapter's Specifications.

9   Install the strut and spring assembly as outlined in Section 2.

---

4   Steering knuckle - removal and installation

Warning: *Dust created by the brake system may contain asbestos, which is harmful to your health. Never blow it out with compressed air and don't inhale any of it. Do not,* *under any circumstances, use petroleum-based solvents to clean brake parts. Use brake system cleaner only.*

Removal

Refer to illustrations 4.2 and 4.10

1   Remove the center cap. **Note:** *If the driveaxle/hub nut cannot be accessed through the hub cap, install the spare tire and wheel.*

2   Apply the parking brake and block the rear wheels to keep the vehicle from rolling. Use a breaker bar and socket and loosen the driveaxle/hub nut with the weight of the vehicle on the wheel **(see illustration)**.

3   Loosen the wheel lug nuts slightly, raise the front of the vehicle and support it securely on jackstands placed under the frame. Remove the wheel and the driveaxle/hub nut.

4   Remove the brake caliper and the brake disc, and disconnect the brake hose from the strut (see Chapter 9). Suspend the brake caliper using a piece of wire or rope.

5   If the vehicle is equipped with ABS, disconnect and remove the wheel speed sensor (see Chapter 9).

6   Mark the position of the two strut-to-

knuckle nuts and remove them **(see illustration 2.4a)**. Don't drive out the bolts at this time.

7   Separate the tie-rod end from the steering knuckle arm (see Section 19) and the control arm balljoint from the steering knuckle (see Section 7).

8   Attach a puller to the hub flange and push the driveaxle out of the hub (see Chapter 8). Hang the driveaxle with a piece of wire to prevent damage to the inner CV joint.

9   Remove the hub and bearing assembly, if necessary (see Section 5).

10   Support the knuckle and drive out the two strut-to-knuckle bolts with a soft-face hammer. Remove the steering knuckle assembly from the strut **(see illustration)**.

Installation

11   Position the knuckle in the strut and insert the two bolts. Tap the bolts into place and install the nuts, but don't tighten them at this time.

12   Install the driveaxle in the hub.

13   Connect the control arm and tie-rod end to the steering knuckle and tighten the nuts

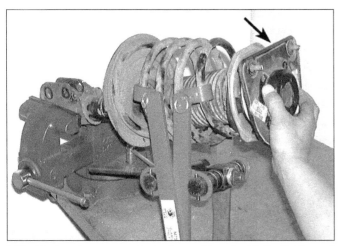

3.7b The bearing cap must also be positioned with the flat (arrow) facing the knuckle flange

4.2 Use a breaker bar and socket to loosen the driveaxle/hub nut

4.10 Separate the front steering knuckle from the driveaxle

5.6 Using the appropriate tools, remove the hub bolts

5.7 Pull the hub and bearing assembly and the brake disc shield off the steering knuckle

to the torque listed in this Chapter's Specifications. Install new cotter pins.
14   Align the strut-to-knuckle nuts with the previously applied marks and tighten them to the torque listed in this Chapter's Specifications.
15   Install the brake disc and caliper.
16   Tighten the driveaxle/hub nut securely.
17   Install the wheel, lower the vehicle and tighten the lug nuts to the torque listed in Chapter 1 Specifications.
18   Tighten the driveaxle/hub nut to the torque specified in Chapter 8.
19   Have the front end alignment checked and, if necessary, adjusted.

---

5   Hub and wheel bearing assembly (front) - removal and installation

Note: *The front hub and wheel bearing assembly is sealed-for-life and must be replaced as a unit.*

Removal

Refer to illustrations 5.6, 5.7 and 5.8
1   Apply the parking brake and block the rear wheels to keep the vehicle from rolling. Remove the hubcap or wheel cover and loosen the driveaxle/hub nut.

2   Loosen the wheel lug nuts, raise the front of the vehicle and support it securely on jackstands. Remove the wheel.
3   Remove the caliper from the steering knuckle and hang it out of the way with a piece of wire (see Chapter 9).
4   Remove the driveaxle/hub nut. Insert a prybar through the caliper inspection window and into the cooling vanes of the brake disc to hold the hub stationary while removing the nut. Pull the brake disc off the hub.
5   Attach a puller to the hub flange and break the driveaxle loose from the hub (see Chapter 8).
6   Remove the hub retaining bolts through the opening in the flange **(see illustration)**.
7   Wiggle the hub and bearing assembly back-and-forth and pull it out of the steering knuckle, along with the brake disc shield **(see illustration)**.
8   If the hub and bearing assembly is being replaced with a new one, it's a good idea to replace the dust seal in the back of the steering knuckle. To replace the seal, disconnect the stabilizer bar and the balljoint from the control arm and pull the steering knuckle and strut assembly off the driveaxle stub shaft. If you can't pull the knuckle out far enough to clear the stub shaft, remove the driveaxle. Pry it out of the knuckle with a screwdriver **(see illustration)**.

Installation

Refer to illustrations 5.9 and 5.10
9   Drive the new dust seal into the knuckle with a large socket or a seal driver and a hammer **(see illustration)**. Try not to cock the seal in the bore.
10   Install a new O-ring around the rear of the bearing and push it up against the bearing flange **(see illustration)**.
11   Clean the mating surfaces on the steering knuckle, bearing flange and knuckle bore. Lubricate the outside diameter of the bearing and the seal lips with high-temperature grease and insert the hub and bearing into the steering knuckle. Position the brake disc shield and install the three bolts, tightening them to the torque listed in this Chapter's Specifications.
12   Install the driveaxle (see Chapter 8).
13   Attach the control arm to the steering knuckle (see Section 7).
14   Reconnect the stabilizer bar to the control arm (see Section 6).
15   Install the brake disc and caliper (see Chapter 9).
16   Install the driveaxle/hub nut and tighten it securely.
17   Install the wheel, lower the vehicle and tighten the lug nuts to the torque listed in the

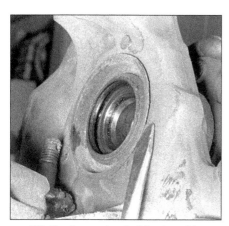

5.8 Pry the seal out of the knuckle with a screwdriver

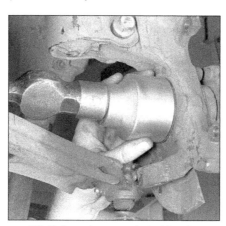

5.9 Using a large socket, drive the new seal into place

5.10 Install a new O-ring (arrow) on the wheel bearing assembly

6.2 Remove the stabilizer bar link nut/bolt (arrow) - note the arrangement of washers, bushings and spacer

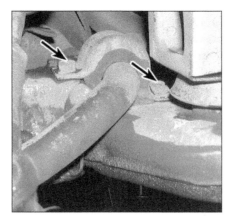

6.3 Remove the two bolts from each stabilizer bar bushing clamp (arrows)

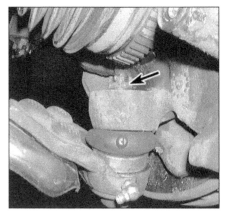

7.3 Remove the cotter pin and castellated nut (arrow) from the balljoint stud

Chapter 1 Specifications.

18   Tighten the driveaxle/hub nut to the torque listed in the Chapter 8 Specifications.

6   Stabilizer bar (front) - removal and installation

Removal

Refer to illustrations 6.2 and 6.3

1   Loosen the lug nuts on both front wheels, raise the front of the vehicle and support it securely on jackstands. Apply the parking brake and block the rear wheels to keep the vehicle from rolling off the jackstands. Remove the front wheels.

2   Remove the stabilizer bar link nut/bolt from each end of the stabilizer bar. Note how the link bushings, spacers and washers are arranged **(see illustration)**. **Note:** *The stabilizer link on Deville/Fleetwood models and 1989 and earlier Seville/Eldorado, Riviera and Toronado models is connected to the control arm; on 1990 and later Seville/Eldorado, Riviera and Toronado models, the stabilizer link is attached to the strut housing.*

3   Remove the stabilizer bar bushing

clamp bolts **(see illustration)**.

4   Detach the tie-rods from the steering knuckles on models where they would cause interference (see Section 19).

5   Detach the exhaust pipe from the rear manifold (see Chapter 2A or 2B).

6   Turn the right steering knuckle to the right.

7   Slide the bar over the right steering knuckle, then pull down until the bar clears the frame.

8   Inspect the bushings for wear and damage and replace them if necessary.

Installation

9   Guide the bar through the wheel well, over the suspension supports and into position.

10   Loosely install the bushings and clamps.

11   Center the bar in the vehicle and install the stabilizer bar-to-control arm bolts, spacers, bushings and washers or attach the stabilizer link to the strut housing as required. Tighten all of the fasteners to the torque listed in this Chapter's Specifications.

12   Install the wheels and lower the vehicle. Tighten the lug nuts to the torque specified in Chapter 1.

7   Control arm (front) - removal and installation

Removal

Refer to illustrations 7.3, 7.4 and 7.5

1   Loosen the wheel lug nuts, raise the front of the vehicle and support it securely on jackstands. Apply the parking brake and block the rear wheels to keep the vehicle from rolling off the jackstands. Remove the wheel.

2   If only one control arm is being removed, disconnect only that end of the stabilizer bar. If both control arms are being removed, disconnect both ends (see Section 6).

3   Remove the balljoint stud-to-steering knuckle castellated nut and cotter pin **(see illustration)**.

4   Using a balljoint separator (available at most auto parts stores) positioned between the control arm and steering knuckle, separate the balljoint from the knuckle **(see illustration)**. If the proper balljoint separator is not available, try striking the steering knuckle boss with a hammer to jar the balljoint stud loose. If that doesn't work, use a "picklefork" type balljoint separator, but note that the use of this tool will almost surely damage the

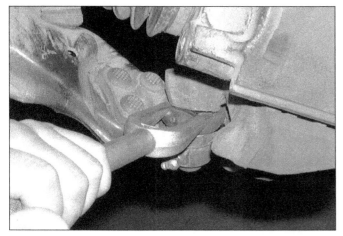

7.4 Separate the balljoint from the steering knuckle

7.5 Remove the control arm pivot bolts (arrows) - Deville/Fleetwood models

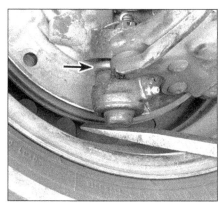

8.3a Check for movement between the balljoint and steering knuckle (arrow) when prying up

8.3b With the prybar positioned between the steering knuckle boss and the balljoint, pry down and check for play in the balljoint - if there's any play, replace the balljoint

balljoint boot. **Caution:** *When removing the balljoint from the knuckle, be careful not to overextend the inner CV joint or it may be damaged.*
5   On Deville/Fleetwood models, remove the two control arm pivot bolts and detach the control arm **(see illustration)**. On Seville/Eldorado, Riviera and Toronado models, remove the pivot bolt from the rear of the control arm and the brake reaction rod bushing and nut from the front.
6   The control arm bushings are replaceable, but special tools and expertise are necessary to do the job. Carefully inspect the bushings for hardening, excessive wear and cracks. If they appear to be worn or deteriorated, take the control arm to a dealer service department or suspension repair shop.

Installation

7   Position the control arm in the suspension support and install the pivot bolts. Do not tighten them completely at this time.
8   Insert the balljoint stud into the steering knuckle boss, install the castellated nut and tighten it to the torque listed in this Chapter's Specifications. If necessary, tighten the nut a little more if the cotter pin hole doesn't line up with an opening on the nut. Install a new cotter pin.
9   Install the stabilizer bar-to-control arm

8.8 Centerpunch the rivets and then drill out the rivet heads – use progressively larger drill bits until the rivet heads are removed

bolt, spacer, bushings and washers and tighten the nut to the torque listed in this Chapter's Specifications.
10   Using a floor jack, raise the outer end of the control arm to simulate normal ride height, then tighten the control arm pivot bolts and brake reaction rod bushing nut to the torque listed in this Chapter's Specifications. **Caution:** *If the bolts aren't tightened with the control arm raised to simulate normal ride height, control arm bushing damage may occur.*
11   Install the wheel and lower the vehicle. Tighten the lug nuts to the torque listed in the Chapter 1 Specifications.

---

8   Balljoints – check and replacement

Check

Refer to illustrations 8.3a and 8.3b
1   Apply the parking brake and block the rear wheels to keep the vehicle from rolling. Raise the front of the vehicle and support it securely on jackstands.
2   Visually inspect the rubber seal for damage, deterioration and leaking grease. If any of these conditions are noticed, the balljoint should be replaced.

3   Place a large prybar under the balljoint and attempt to push the balljoint up. Next, position the prybar between the steering knuckle and control arm and pry down **(see illustrations)**. If any movement is seen or felt during either of these checks, a worn out balljoint is indicated.
4   Have an assistant grasp the tire at the top and bottom and move the top of the tire in-and-out. Touch the balljoint stud castellated nut. If any looseness is felt, suspect a worn out balljoint stud or a widened hole in the steering knuckle boss. If the latter problem exists, the steering knuckle should be replaced as well as the balljoint.
5   Separate the control arm from the steering knuckle (see Section 4). Using your fingers (don't use pliers), try to twist the stud in the socket. If the stud turns, replace the balljoint.

Replacement
Front
Refer to illustrations 8.8 and 8.11
6   Apply the parking brake and block the rear wheels to keep the vehicle from rolling. Loosen the wheel lug nuts, raise the front of the vehicle and support it securely on jackstands. Remove the wheel.
7   Separate the balljoint from the steering knuckle (see Section 7). Temporarily insert the balljoint stud back into the steering knuckle (loosely). This will ease balljoint removal after Step 9 has been performed, as well as hold the assembly stationary while drilling out the rivets.
8   Using a 1/8-inch drill bit, drill a pilot hole into the center of each balljoint-to-control arm rivet **(see illustration)**. Be careful not to damage the CV joint boot in the process.
9   Using a 1/2-inch drill bit, drill the head off each rivet. Work slowly and carefully to avoid deforming the holes in the control arm.
10   Loosen (but don't remove) the stabilizer bar-to-control arm nut. Pull the control arm and balljoint down to remove the balljoint stud from the steering knuckle, then dislodge the balljoint from the control arm.
11   Position the new balljoint on the control arm and install the bolts (supplied in the balljoint kit) from the top of the control arm **(see illustration)**. Tighten the bolts to the

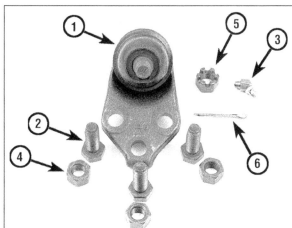

8.11 Replacement balljoint details (typical) - be sure to tighten the bolts to the torque specified on the instruction sheet

1   Replacement balljoint
2   Bolt
3   Grease fitting
4   Nut
5   Castle nut
6   Cotter pin

8.16 Remove the cotter pin and castellated nut (arrow) from the balljoint stud (rear balljoint shown)

8.17 Separate the balljoint from the knuckle - a pickle fork tool is shown in use here

9.7 If equipped with Electronic Level Control (ELC), disconnect the air line fitting (arrow) from the strut

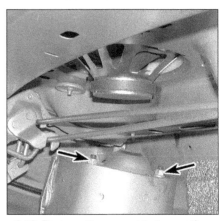

9.10 Remove the upper strut mounting nuts and washers

torque specified in the new balljoint instruction sheet.

12   Insert the balljoint into the steering knuckle, install the castellated nut, tighten it to the torque listed in this Chapter's Specifications and install a new cotter pin. It may be necessary to tighten the nut further to align the cotter pin hole with an opening in the nut, which is acceptable. Never loosen the castellated nut to allow cotter pin insertion.

13   Tighten the stabilizer bar-to-control arm nut to the torque listed in this Chapter's Specifications.

14   Install the wheel, lower the vehicle and tighten the lug nuts to the torque listed in the Chapter 1 Specifications.

Rear (Deville/Fleetwood models only)

Refer to illustrations 8.16 and 8.17
Note: *Replacement of the rear balljoint requires a number of special tools. If the special tools are not available, remove the control arm and take it to a dealer service department or other qualified automotive repair facility for balljoint replacement.*

15   Apply the parking brake and block the front wheels to keep the vehicle from rolling. Loosen the wheel lug nuts, raise the rear of the vehicle and support it securely on jackstands placed under the control arms. Remove the wheel. **Warning:** *The control arm must be supported by a jack, or jackstands placed under the control arm, when the balljoint is separated from the knuckle or the control arm will be forced down by spring pressure, possibly causing serious injury or damage. The jack or jackstands must remain in place under the control arm during the entire procedure.*

16   Separate the adjustment link from the knuckle (see Section 16). Loosen (but don't remove) the balljoint-to-knuckle nut **(see illustration)**.

17   Using a balljoint separator (available at most auto parts stores) positioned between the control arm and the knuckle, separate the balljoint from the knuckle **(see illustration)**. If the proper balljoint separator is not available, try striking the knuckle boss with a hammer

to jar the balljoint stud loose. If that doesn't work, use a "picklefork" type balljoint separator, but note that the use of this tool will almost surely damage the balljoint boot.

18   The balljoint must be replaced using a C-clamp type balljoint press and the proper adapters. If the proper tools are not available, remove the control arm (see Section 14) and have the balljoint installed by a qualified automotive suspension shop.

19   Insert the balljoint into the knuckle, install the castellated nut, tighten it to the torque listed in this Chapter's Specifications and install a new cotter pin. It may be necessary to tighten the nut further to align the cotter pin hole with an opening in the nut, which is acceptable. Never loosen the castellated nut to allow cotter pin insertion. Install the adjustment link to the steering knuckle (see Section 16).

20   Install the wheel, lower the vehicle and tighten the lug nuts to the torque listed in the Chapter 1 Specifications.

---

9   Strut assembly (rear) – removal and installation

Deville/Fleetwood models

Refer to illustrations 9.7, 9.10 and 9.11
1   Remove the rear speaker assembly.
2   Remove the trunk side cover.
3   Block the front wheels.
4   Loosen the rear wheel lug nuts.
5   Raise the rear of the vehicle and support it securely on jackstands placed under the control arms. **Warning:** *The control arm must be supported by a jack, or jackstands placed under the control arm, before removing the strut mounting bolts or the control arm will be forced down by spring pressure, possibly causing serious injury or damage. The jack or jackstands must remain in place under the control arms during the entire procedure.*
6   Remove the rear wheels.
7   If the vehicle is equipped with Electronic Level Control (ELC), disconnect the air tube from the strut assembly **(see illustration)**.
8   If the vehicle is equipped with electronic

level control (ELC), detach the ELC height sensor link from the control arm **(see illustration 11.6)**. **Note:** *The height sensor link is located on the left side control arm on ELC systems or on the right side control arm on CCR systems. Refer to Section 17 for additional information.*

9   If the vehicle is equipped with the Computer Command Ride (CCR), disconnect the CCR electrical connector from the front strut. Also on CCR systems, disconnect the air tube from the strut.

10   Remove the strut upper mounting nuts from inside the trunk **(see illustration)**.

11   Remove the nuts, bolts and washers from the knuckle **(see illustration)**.

12   Remove the strut.

13   Installation is the reverse of removal. Tighten the fasteners to the torque listed in this Chapter's Specifications.

14   When reconnecting the air tubes to the strut assemblies, be sure to use a slight amount of petroleum jelly on the fittings before installing the air tubes. Also, on ELC systems, pressurize the suspension system by grounding the compressor test lead in the engine compartment for several seconds to allow the struts to compress slightly. Do this before lowering the vehicle to the ground.

15   Have the rear end alignment checked and, if necessary, adjusted.

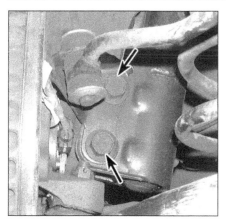

9.11 Remove the two strut-to-knuckle bolts (arrows)

9.24 Upper strut mounting nut - Seville/Eldorado, Riviera and Toronado models

10.7 A typical aftermarket internal-type spring compressor tool: the hooked arms grip the upper coils of the spring, the plate is inserted below the lower coils, and, when the nut on the threaded rod is turned, the spring is compressed

Seville/Eldorado, Riviera and Toronado models

Refer to illustration 9.24

16   Loosen the rear wheel lug nuts.

17   Block the front wheels. Raise the rear of the vehicle and support it securely on jackstands placed under the frame.

18   Remove the rear wheels.

19   Remove the rear brake caliper and suspend it with a piece of wire or rope (see Chapter 9).

20   Install two of the wheel lug nuts onto the hub to retain the brake disc.

21   If equipped with a stabilizer bar, disconnect the mounting bolts at the strut (see Section 15).

22   If the vehicle is equipped with electronic level control (ELC), detach the ELC height sensor link from the control arm **(see illustration 11.6)**. **Note:** *The height sensor link is located on the left side control arm on ELC systems or on the right side control arm on CCR systems. Refer to Section 17 for additional information.* If the vehicle is equipped with the Computer Command Ride (CCR), disconnect the CCR electrical connector from the top of the strut. Also, on both the ELC and CCR systems, disconnect the air tube from the strut **(see illustration 9.7)**.

23   Place a floor jack under the control arm and raise the jack slightly to support the control arm. **Warning:** *The control arm must be supported by the floor jack before removing any strut fasteners or the control arm will be forced down by spring pressure, possibly causing serious injury or damage. The floorjack must remain in place under the control arm until the strut in reinstalled.*

24   Remove the strut upper mounting nut(s) and insulator **(see illustration)**.

25   Loosen the knuckle-to-control arm pivot bolt. Slowly lower the control arm slightly with the floor jack and pull the strut upper mounting stud out of the suspension support.

26   Pivot the strut and knuckle outward and remove the strut lower pinch bolt. Observe the exact location of the strut in relation to the tang on the strut in the knuckle slot. Paint an alignment mark if

necessary and remove the strut.

27   Installation is the reverse of removal. Tighten the fasteners to the torque values listed in this Chapter's Specifications. Refer to Chapter 9 for the torque specifications for the brake caliper mounting bolts.

28   When reconnecting the air tubes to the strut assemblies, be sure to use a slight amount of petroleum jelly on the fittings before installing the air tubes. Also, on ELC systems, pressurize the suspension system by grounding the compressor test lead in the engine compartment for several seconds to allow the struts to compress slightly. Do this before setting the vehicle down onto the floor.

29   Have the rear end alignment checked and, if necessary, adjusted.

10   Coil spring (rear) (Deville/Fleetwood models) – removal and installation

Removal

Refer to illustration 10.7

1   Loosen the rear wheel lug nuts.

2   Block the front wheels.

3   Raise the rear of the vehicle and support it securely on jackstands placed under the frame.

4   Remove the rear wheels.

5   Detach the parking brake cable retaining clip from the left control arm.

6   If the vehicle is equipped with electronic level control (ELC), detach the ELC height sensor link from the control arm **(see illustration 11.6)**. **Note:** *The height sensor link is located on the left side control arm on ELC systems or on the right side control arm on CCR systems. Refer to Section 17 for additional information.*

7   Install a suitable interior style spring compressor onto the coil spring in accordance with the tool manufacturer's instructions **(see illustration)**. Compress the spring enough to relieve all pressure from the spring seats. Stop at this point and be sure not to compress the spring anymore than neces-

sary. When you can wiggle the spring, it is compressed enough. This tool is available at local auto parts stores and rental yards.

8   Detach the rear stabilizer bar (see Section 15).

9   Detach the suspension adjustment link (see Section 16).

10   Place a floor jack under the spring pocket of the control arm. Raise the jack just enough to support the control arm.

11   Paint alignment marks to indicate the exact position of the coil spring pigtail in the insulator slot. The coil spring must be positioned in the exact same position when it is reassembled.

12   Remove the control arm pivot bolts and detach the control arm from the frame crossmember (see Section 14).

13   Slowly lower the jack and remove the spring and insulators. Carefully remove the spring compressor from the spring.

Installation

14   Replace the insulators if damaged or worn. Be sure to position each insulator in the correct position in the control arm and the upper body in alignment with the end of the coil spring pigtail.

15   Installation is the reverse of removal. Tighten the fasteners to the torque listed in this Chapter's Specifications.

11   Leaf spring (rear) (Seville/Eldorado, Riviera and Toronado models) – removal and installation

Removal

Refer to illustrations 11.6, 11.13 and 11.14

1   Loosen the rear wheel lug nuts.

2   Block the front wheels.

3   Raise the rear of the vehicle and support it securely on jackstands placed under the frame.

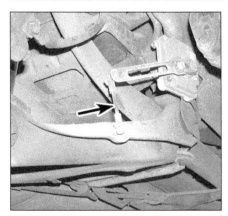

11.6 Disconnect the ELC height sensor link (arrow)

11.13 Position a jackstand under the leaf spring end and slowly lower the vehicle to relieve tension from the lower leaf spring retainer

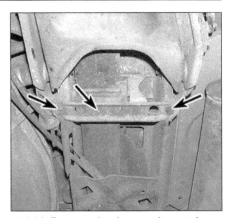

11.14 Remove the three spring retainer bolts (arrows)

4   Remove the rear wheels.
5   Detach the parking brake cable retaining clip from the left control arm.
6   If the vehicle is equipped with electronic level control (ELC), detach the ELC height sensor link from the control arm **(see illustration)**. **Note:** *The height sensor link is located on the left side control arm on ELC systems or on the right side control arm on CCR systems. Refer to Section 17 for additional information.*
7   Detach the rear stabilizer bar (see Section 15).
8   Remove the rear brake caliper (see Chapter 9) and suspend it from the body using wire or a piece of rope. Do NOT disconnect the brake hose or allow it to pinch while the caliper is suspended. Also, it will be necessary to disconnect the parking brake cable from the caliper (see Chapter 9).
9   Install two lug nuts on the rear brake disc to retain it to the hub assembly.
10   Loosen, but do not remove the knuckle pivot bolt on the outer end of the control arm. Support the control arm using a floorjack positioned toward the outer end of the control arm.
11   If the vehicle is equipped with electronic level control (ELC), disconnect the strut electrical connector from the strut. Remove the upper strut mounting nut and insulator. Slowly lower the floorjack and control arm until all spring pressure is relieved. Remove the knuckle pivot bolt and remove the strut, knuckle and brake disc as an assembly (see Section 9).
12   Remove the control arm (see Section 14).
13   Position a third jackstand under the end of the leaf spring **(see illustration)**.
14   Lower the vehicle so that the weight of the vehicle loads the spring downward onto the jackstand. Remove the three spring retainer bolts **(see illustration)**. **Warning:** *Do not remove the jackstands from under the vehicle. Change the jackstand adjustments down to allow the vehicle to drop several inches. The jackstands must be in position as a safety requirement when the vehicle is being lowered by the floorjack.*
15   Slowly raise the vehicle until the spring

does not exert any force on the jackstand and remove the jackstand. Secure the vehicle on jackstands.
16   Remove the spring retainer bolts, retainer and lower insulator from the other side of the vehicle. Check the condition of the upper insulator for cracks, tears or damage.
17   Remove the leaf spring from the rear suspension support on the disassembled side of the vehicle.

Installation

18   If the upper leaf spring insulators are damaged and must be replaced, access the retainer bolts from above the suspension support. When installing the upper insulators, the molded arrows must face toward the center of the vehicle. Torque the upper insulator mounting nuts to the torque listed in this Chapter's Specifications.
19   Slide the leaf spring into place from the disassembled side of the vehicle. **Caution:** *Make sure the outboard insulator bands are centered on the spring insulators. Failure to position the spring correctly can reduce vehicle handling.*
20   Position a third jackstand under the end of the leaf spring.
21   Slowly lower the vehicle to apply pressure to the leaf spring and position the spring into the suspension support. **Warning:** *Do not remove the jackstands from under the vehicle. Change the jackstand adjustments down to allow the vehicle to drop several inches. The jackstands must be in position as a safety requirement when the vehicle is being lowered by the floorjack.*
22   Install the lower insulator and spring retainer onto the leaf spring and suspension support. Torque the lower insulator mounting nuts to the torque listed in this Chapter's Specifications.
23   Install the control arm but do not tighten the pivot bolts fully at this time (see Section 14).
24   Install the strut, knuckle and hub assembly (see Section 9). Tighten the upper strut nut, knuckle pivot bolt and control arm pivot bolts to the torque listed in this Chapter's

Specifications.
25   Install the stabilizer bar (see Section 15).
26   Install the brake caliper (see Chapter 9).
27   Connect the electronic level control (ELC) height sensor link.
28   Have the rear end alignment checked and, if necessary, adjusted.

12   Hub and wheel bearing assembly (rear) - removal and installation

Refer to illustration 12.6
Note: *The rear hub and bearing assembly is sealed for life and must be replaced as a unit.*
1   Loosen the rear wheel lug nuts.
2   Block the front wheels.
3   Raise the rear of the vehicle and support it securely on jackstands placed under the frame.
4   Remove the rear wheel.
5   On disc brake models, remove the rear brake caliper (see Chapter 9) and suspend it from the body using wire or a piece of rope. Remove the brake disc or drum (see Chapter 9).
6   Rotate the hub to align one of the holes with each of the mounting bolts and remove the bolts **(see illustration)**. Remove the hub and bearing assembly. On drum brake models, support the backing plate assembly as the hub is removed and reinstall two of the bolts finger tight. This will prevent the brake line from being strained under the weight of the backing plate.
7   Installation is the reverse of removal. Be sure tighten the bolts to the torque listed in this Chapter's Specifications.

13   Knuckle (rear) – removal and installation

1   Loosen the rear wheel lug nuts.
2   Block the front wheels.
3   Raise the rear of the vehicle and support it securely on jackstands placed under the control arms. **Warning:** *The control arm must be supported by a jack, or jackstands placed*

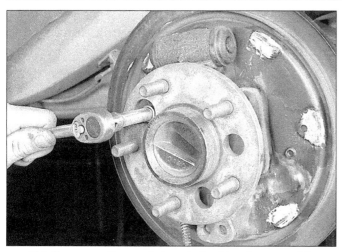

12.6 To remove the rear hub and wheel bearing assembly, rotate the hub until one of the holes in the flange is aligned with a mounting bolt

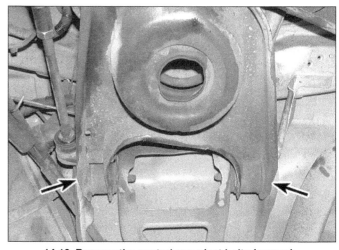

14.12 Remove the control arm pivot bolts (arrows) - Deville/Fleetwood models

under the control arm or control arm will be forced down by spring pressure, possibly causing serious injury or damage. The jack or jackstands must remain in place under the control arm during the entire procedure.

4   Remove the rear wheels.

5   If the vehicle is equipped with electronic level control (ELC), detach the ELC height sensor link from the control arm **(see illustration 11.6)**. **Note:** *The height sensor link is located on the left side control arm on ELC systems or on the right side control arm on CCR systems. Refer to Section 17 for additional information.*

6   Detach the rear stabilizer bar (see Section 15).

7   Remove the rear brake caliper (see Chapter 9) and suspend it from the body using wire or a piece of rope. Do NOT disconnect the brake hose or allow it to pinch while the caliper is suspended. Also, it will be necessary to disconnect the parking brake cable from the caliper (see Chapter 9).

8   Remove the rear brake disc and the hub assembly (see Section 12). On drum brake models, position the baking plate and brake assembly aside and suspend it from the body using wire or a piece of rope.

9   On Deville/Fleetwood models, remove the strut-to-knuckle bolts (see Section 9).

10   On Seville/Eldorado, Riviera and Toronado models, loosen, but do not remove the knuckle pivot bolt on the outer end of the control arm.

11   On Seville/Eldorado, Riviera and Toronado models, disconnect the Electronic Level Control (ELC) electrical connector from the strut (if equipped). Remove the upper strut mounting nut and insulator. Slowly lower the control arm and remove the strut (see Section 9).

12   Remove the control arm-to-knuckle pivot bolt (Seville/Eldorado, Riviera and Toronado models) or the balljoint-to-knuckle castle nut (Deville/Fleetwood models) (see Section 8).

13   Separate the knuckle from the control arm.

14   Installation is the reverse of removal. Be sure tighten the bolts to the torque listed in this Chapter's Specifications.

14   Control arm (rear) - removal and installation

Deville/Fleetwood models

Refer to illustration 14.12

Note: *Replacement of the bushings in the rear control arms require a number of special tools. If a bushing must be replaced, take the control arm to a dealer service department or other properly equipped repair facility.*

1   Loosen the rear wheel lug nuts.

2   Block the front wheels.

3   Raise the rear of the vehicle and support it securely on jackstands placed under the frame.

4   If the vehicle is equipped with electronic level control (ELC), detach the ELC height sensor link from the control arm **(see illustration 11.6)**. **Note:** *The height sensor link is located on the left side control arm on ELC systems or on the right side control arm on CCR systems. Refer to Section 17 for additional information.*

5   Remove the parking brake cable retaining clip from the left control arm.

6   Disconnect the stabilizer bar from the knuckle bracket (see Section 15).

7   Remove the suspension adjustment link from the control arm (see Section 16).

8   Refer to Section 10, install a coil spring compressor and compress the coil spring.

9   Remove the balljoint-to-knuckle castle nut (see Section 8).

10   Using a balljoint separator, separate the balljoint from the knuckle (see Section 8).

11   Place a floorjack under the control arm.

12   Remove the control arm pivot bolts **(see illustration)**.

13   Remove the control arm and coil spring from the vehicle.

14   Remove the spring compressor from the spring or place the compressed spring in a safe location.

15   Install the control arm and coil spring. Refer to Section 10 and make sure the spring is positioned correctly in the control arm and the insulators are in place. Install the pivot bolts, but do not tighten them fully at this time.

16   Connect the balljoint to the knuckle, install the castellated nut and a new cotter pin (see Section 8).

17   Using a floor jack, raise the outer end of the control arm to simulate normal ride height, then tighten the control arm pivot bolts to the torque listed in this Chapter's Specifications. **Caution:** *If the bolts aren't tightened with the control arm raised to simulate normal ride height, control arm bushing damage may occur.*

18   The remainder of installation is the reverse of removal. Tighten the fasteners to the torque listed in this Chapter's Specifications.

19   Install the wheels and torque the lug nuts to the torque Specifications in Chapter 1.

20   Remove the jackstands and lower the vehicle.

Seville/Eldorado, Riviera and Toronado models

Refer to illustration 14.33

Note: *Replacement of the bushings in the rear control arms require a number of special tools. If a bushing must be replaced, take the control arm to a dealer service department or other properly equipped repair facility.*

21   Loosen the rear wheel lug nuts.

22   Block the front wheels.

23   Raise the rear of the vehicle and support it securely on jackstands placed under the frame.

24   If the vehicle is equipped with electronic level control (ELC), detach the ELC height sensor link from the control arm **(see illustration 11.6)**. **Note:** *The height sensor link is*

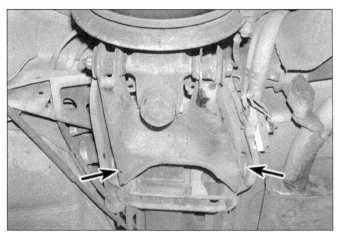

14.33 Remove the control arm pivot bolts (arrows) - Seville/Eldorado, Riviera and Toronado models

15.5 To detach the rear stabilizer bar link bolt assembly from the strut bracket, remove the nut on the upper end (arrow) and pull the link bolt out - note the arrangement of the bushings and washers

located on the left side control arm on ELC systems or on the right side control arm on CCR systems. Refer to Section 17 for additional information.

25 Disconnect the stabilizer bar from the strut (see Section 15).

26 Remove the rear brake caliper (see Chapter 9) and suspend it from the body using wire or a piece of rope. Do NOT disconnect the brake hose or allow it to pinch while the caliper is suspended. Also, it will be necessary to disconnect the parking brake cable from the caliper (see Chapter 9).

27 Install two wheel stud nuts to retain the rear brake disc on the hub.

28 If equipped with ABS, disconnect the rear wheel speed sensors (see Chapter 9).

29 Loosen, but do not remove the knuckle pivot bolt on the outer end of the control arm. Support the control arm using a floorjack positioned toward the outer end of the control arm.

30 Remove the upper strut mounting nut and insulator. Slowly lower the control arm and detach the strut upper mounting stud from the suspension support.

31 Support the knuckle and remove the control arm-to-knuckle pivot bolt. Remove the strut, knuckle and hub as an assembly.

32 Lower the jack to relieve spring pressure.

33 Remove the control arm pivot bolts **(see illustration).**

34 Remove the control arm from the vehicle.

35 Install the control arm and strut, knuckle and hub assembly, but do not fully tighten the fasteners at this time.

36 Using a floor jack, raise the outer end of the control arm to simulate normal ride height, then tighten the strut upper mounting bolt, knuckle pivot bolt and control arm bolts to the torque listed in this Chapter's Specifications. **Caution:** *If the bolts aren't tightened with the control arm raised to simulate normal ride height, control arm bushing damage may occur.*

37 The remainder of installation is the reverse of removal.

38 Install the wheels and tires and torque the wheel lug nuts to the torque Specifications in Chapter 1.

39 Remove the jackstands and lower the vehicle.

---

15 Stabilizer bar (rear) - removal and installation

Deville/Fleetwood models

Refer to illustrations 15.5, 15.6 and 15.7

1 Loosen the rear wheel lug nuts.

2 Block the front wheels.

3 Raise the rear of the vehicle and place it securely on jackstands placed under the control arms. **Note:** *Supporting the vehicle by the control arms will relieve tension on the stabilizer bar, making removal and installation easier.*

4 Remove the rear wheels.

5 Detach the stabilizer bar bolt/bushing assembly from the strut bracket on each side **(see illustration).**

6 If you're replacing the stabilizer bar and/or the lower bushing from the hanger

clamp only, remove the pinch bolt from each hanger clamp **(see illustration),** bend the end of the clamp open and remove the stabilizer bar and bushings.

7 If you plan to reinstall the same bar or replace both bushings, remove the upper bolt from the hanger clamp **(see illustration)** and leave the hanger attached to the stabilizer bar.

8 Installation is the reverse of removal. Tighten the fasteners to the torque listed in this Chapter's Specifications.

Seville/Eldorado, Riviera and Toronado models

9 Loosen the rear wheel lug nuts.

10 Block the front wheels.

11 Raise the rear of the vehicle and place it securely on jackstands placed under the control arms. **Note:** *Supporting the vehicle by the control arms will relieve tension on the stabilizer bar, making removal and installation easier.*

12 Remove the rear wheels.

13 Remove the stabilizer bar nut and bolt at the strut.

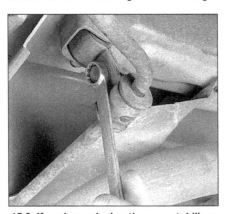

15.6 If you're replacing the rear stabilizer bar and/or lower bushing, remove this bolt on each hanger clamp, bend the clamp open and remove the bar

15.7 If you're going to reinstall the same stabilizer bar or replace both bushings, simply remove the upper hanger clamp bolt and leave the clamp attached to the bar during removal

14   If you plan to reinstall the same stabilizer bar or replace both end bushings, simply remove the upper bolt from the hanger clamp at the suspension support and leave the hanger attached to the stabilizer bar.

15   If you're replacing the stabilizer bar and/or the lower bushing from the hanger clamp only, remove the pinch bolt from each hanger clamp, bend the end of the clamp open and remove the stabilizer bar and bushings.

16   Installation is the reverse of removal. Tighten the fasteners to the torque values listed in this Chapter's Specifications.

16   Rear suspension adjustment link (Deville/Fleetwood models) - removal and installation

Refer to illustrations 16.5 and 16.7

1   Loosen the rear wheel lug nuts.
2   Block the front wheels.
3   Raise the vehicle and place it securely on jackstands.
4   Remove the wheel.
5   Remove the cotter pin and nut **(see illustration)**.
6   Separate the outer suspension adjustment link from the knuckle with a puller **(see illustration 19.2b)**.
7   To detach the inner end of the link from the control arm, remove the retaining nut, washer and spacer **(see illustration)**.
8   Remove the suspension adjustment link from the control arm.
9   Installation is the reverse of removal. Tighten the fasteners to the torque listed in this Chapter's Specifications, and use a new cotter pin.

17   Suspension level control systems – general information

Electronic Level Control (ELC) system

Some models are equipped with an optional Electronic Level Control (ELC) system which consists of an air compressor, a compressor relay, an air dryer, an exhaust solenoid, a height sensor, a pair of air adjustable rear shock absorbers and the air lines and fittings connecting the compressor to the shocks. The ELC automatically adjusts the rear trim height according to the loaded weight of the vehicle. The air compressor is activated when the ignition key is ON and extra weight is detected by the height sensor. The exhaust solenoid allows the ELC system to depressurize when the added weight is removed. The exhaust solenoid receives battery voltage at all times to allow suspension adjustments when the vehicle is sitting (not running).

The air compressor is mounted on top of the rear suspension support at the right side of the vehicle. The ELC compressor

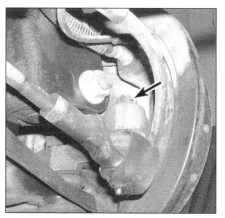

16.5 Remove the cotter pin and nut (arrow) from the suspension adjustment link stud

relay is located in the dash relay center (see Chapter 12). The exhaust solenoid is mounted in the compressor head assembly. The height sensor is an electronic device that detects ride height specifications. The height sensor is a mounted on the back of the rear suspension support with the sensor actuator arm attached to the left rear control arm. Each wheel is equipped with an air adjustable strut. Make sure the fittings at the shocks are tight and the link rod between the control arm and the height sensor is connected.

Repair procedures for the ELC system should be performed by a dealer service department or other qualified auto repair facility.

Computer Control Ride

Some models are equipped with an optional Computer Command Ride (CCR) system. The CCR system controls the firmness of the vehicle's ride by automatically controlling an actuator in each of the automatic struts at each wheel. Ride firmness increases with vehicle speed to gain more handling characteristics. The three modes of operation are – Comfort, Normal and Sport. Mode selection is controlled by the computer in response to driving conditions – winding roads, bumpy roads, straight pavement, etc.

18.4a Remove the airbag module screws from behind the steering wheel

16.7 To detach the inner end of the suspension adjustment link from the control arm, remove this nut (arrow) - note the arrangement of the washer and spacer

The CCR control module (computer) is located on a bracket under the driver's seat. The module receives input information from the height sensor, the accelerator, etc to adjust the ride for optimum handling. If the module receives an incorrect feedback signal, the computer will warn the driver by flashing the SERVICE CCR light on the dash. Some models are equipped with a speed version with a light marked SERVICE SSS. This speed sensitive suspension is an option on sport models. The CCR diagnostic system stores trouble codes that can be accessed at the Test Connector under the driver's side dash. Have a dealer service department or other qualified automotive repair facility extract the codes.

18   Steering wheel - removal and installation

Warning: *On models equipped with a Supplemental Inflatable Restraint (SIR) system (more commonly known as airbags), always disable the airbag system before working in the vicinity of any airbag system component to avoid the possibility of accidental deployment of the airbag, which could cause personal injury (see Chapter 12).*

Steering wheel removal and installation

Refer to illustrations 18.4a, 18.4b, 18.5, 18.6, 18.7, 18.8 and 18.9

1   Park the vehicle with the wheels pointing straight ahead.
2   On airbag-equipped models, disable the airbag. Refer to Chapter 12, Section 30 for the disabling procedure.
3   On models without an airbag, remove the two screws from the backside of the steering wheel and detach the horn pad from the steering wheel.
4   On airbag-equipped models, remove the screws that secure the airbag module to the steering wheel **(see illustrations)**.

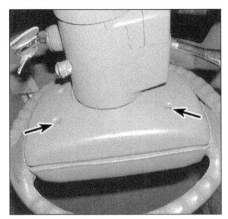

18.4b Locations of the airbag module screws (arrows)

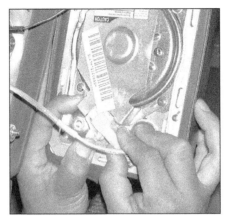

18.5 To detach the airbag module from the steering wheel, disconnect the yellow SIR harness connector

18.6 Push down and turn the horn contact lead counterclockwise to disengage the horn contact lead from the cam

5   On airbag-equipped models, lift the airbag module carefully away from the steering wheel and disconnect the yellow airbag electrical connector **(see illustration)**.

6   On all models, disconnect the horn contact lead from the steering wheel **(see illustration)**. Push down and twist the horn contact lead counterclockwise to disengage it from the cam. Remove the airbag module. **Warning:** *When carrying the airbag module, keep the driver's side of it away from your body, and when you set it down (in an isolated area), have the driver's side facing up.*

7   Remove the steering wheel nut retainer (if equipped) and the steering wheel nut **(see illustration)**.

8   Mark the relationship of the steering wheel to the shaft **(see illustration)**.

9   Install a steering wheel puller (available at most auto parts stores) and turn the center bolt until the wheel is free **(see illustration)**.

10   Remove the puller and disconnect the horn and ground connector. Remove the steering wheel.

11   Installation is the reverse of removal. On airbag-equipped models, if the clockspring was removed, make sure the clockspring is centered before installing the steering wheel

(see Steps 12 through 20). Be sure to enable the airbag system (see Chapter 12, Section 30).

Clockspring replacement and centering procedure (airbag models)

Refer to illustrations 18.14, 18.15, 18.17 and 18.20

Warning: **T***he airbag clockspring assembly is a ribbon-like mechanism which allows electrical current to flow to the airbag module regardless of steering wheel position. If the clockspring becomes uncentered, it may break when the vehicle is returned to service.*

Note: *It is not necessary to remove the clockspring unless repairs to the steering column are required. The clockspring does not have to be centered unless the center hub of the clockspring was moved while the clockspring was removed from the steering column. Follow the centering procedure if the clockspring was removed and the alignment was changed.*

12   Disconnect the negative battery cable.

13   Make sure the front wheels are pointed straight ahead.

18.7 Remove the steering wheel nut with a deep socket

14   Remove the clockspring assembly snap-ring **(see illustration)**.

15   Remove the clockspring from the steering column **(see illustration)**.

16   Carefully withdraw the clockspring harness connector through the steering column. Working on the harness connector at the base of the steering column (see Chapter 12), attach a piece of wire or string to the connec-

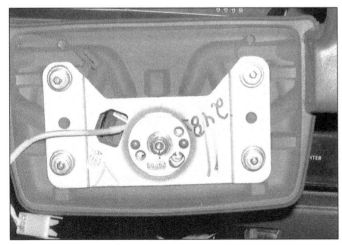

18.8 Mark the relationship of the steering wheel to the steering shaft

18.9 Use a steering wheel puller to remove the steering wheel

18.14 To remove the clockspring from the steering shaft, remove this snap-ring . . .

18.15 . . . and pull the clockspring off the steering shaft

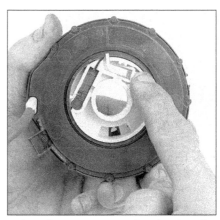

18.17 To center the clockspring, hold it with its underside facing UP, depress the spring lock and rotate the hub in the direction of the arrow on the coil assembly until it stops, then turn it in the opposite direction 2-1/2 turns and release the spring lock

tor and gently withdraw the assembly. Allow the wire to be drawn up through the column and out the top (face). Disconnect the wire from the harness connector and allow the wire to remain in position until the clockspring must be replaced. This extra step will make the installation of the clockspring harness through the steering column quicker and easier.

17   Holding the clockspring assembly with the bottom side facing up, depress the spring lock and rotate the hub in the direction of the arrow until it stops **(see illustration)**. The arrow is stamped onto the back of the clockspring assembly. The clockspring should wind up tight against the center hub. **Note:** *There are two different styles of clocksprings on these models. One rotates clockwise and the other rotates counterclockwise.*

18   Rotate the hub in the opposite direction 2-1/2 turns and then release the spring lock to secure the locking tab.

19   Before installing the clockspring onto the steering column, make sure the block tooth on the steering shaft is at the 12 o'clock position, the wheels straight ahead and the ignition in LOCK.

20   The tab should be at the top of the clockspring and the arrows on the face of the clockspring should be aligned **(see illustration)**.

21   Pull the clockspring harness through the steering column using the wire or string.

22   Install the steering wheel and the airbag (see Steps 1 through 10).

23   Enable the airbag system (see Chapter 12, Section 30).

19   Tie-rod ends - removal and installation

Removal

Refer to illustrations 19.2a, 19.2b and 19.3

1   Apply the parking brake and block the rear wheels to keep the vehicle from rolling. Loosen the wheel lug nuts, raise the front of the vehicle and support it securely on jackstands. Remove the wheel.

2   Remove the cotter pin and castellated nut from the tie-rod, then disconnect the tie-rod from the steering knuckle arm with a puller **(see illustrations)**.

3   Mark the relationship of the tie-rod end to the threaded adjuster **(see illustration)**. This will ensure the toe-in setting is restored when reassembled.

4   Unscrew the tie-rod end from the tie-rod.

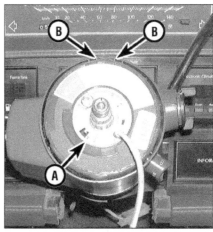

18.20 When properly aligned, the airbag coil will be centered with the marks aligned (A) and the tab fitted between the projections on the top of the steering column (B)

19.2a Remove the cotter pin and castellated nut from the tie-rod end stud . . .

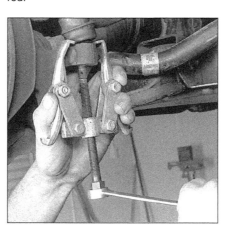

19.2b . . . then separate the tie-rod end from the steering knuckle arm with a two-jaw puller - DO NOT use a hammer!

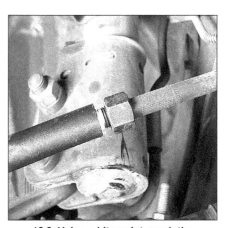

19.3 Using white paint, mark the relationship of the tie-rod end to the tie-rod

Installation

5   Thread the tie-rod end onto the tie-rod to the marked position and connect the tie-rod end to the steering arm. Install the castellated nut and tighten it to the torque listed in this Chapter's Specifications. Tighten the nut further, if necessary, to align a slot in the nut to the hole - DO NOT loosen the nut. Install a new cotter pin.

6   Install the wheel. Lower the vehicle and tighten the lug nuts to the torque listed in the Chapter 1 Specifications.

7   Have the front end alignment checked and, if necessary, adjusted.

20   Steering gear boots - replacement

Refer to illustrations 20.3a and 20.3b

1   Detach the tie-rod ends from the steering knuckle (see Section 19).

2   Remove the tie-rod ends from the steering gear tie-rods (see Section 19).

3   Cut off the inner boot clamp and discard it. Using pliers, squeeze the tangs and slide the outer boot clamp off the boot **(see illustrations)**.

4   Detach the boot from the vent cross tube and slide the boots off the tie-rod.

5   Before installing the new boots, wrap the threads on the ends of the steering rods with a layer of tape so the small ends of the new boots aren't damaged during installation.

6   Loosely install the inner boot clamp on each boot.

7   Slide the new boots onto the tie rod and secure them on the steering gear. Connect the vent cross tube.

8   Using the appropriate boot clamp pliers, tighten the inner boot clamp. Slide the outer clamp onto the boot.

9   Install the tie-rod ends and connect them to the steering knuckle (see Section 19).

10   Have the front end alignment checked and, if necessary, adjusted.

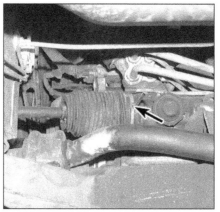

20.3a Cut the inner boot clamp and remove it (arrow)

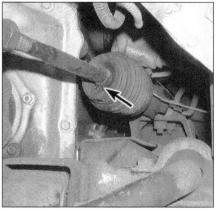

20.3b Remove the outer steering gear clamp (arrow)

21   Steering gear - removal and installation

Refer to illustrations 21.5, 21.7, 21.11, 21.12a and 21.12b

Warning 1: *On models equipped with a Supplemental Inflatable Restraint (SIR) system (more commonly known as airbags), always disable the airbag system before working in the vicinity of any airbag system component to avoid the possibility of accidental deployment of the airbag, which could cause personal injury (see Chapter 12).*

Warning 2: *On models equipped with an airbag, make sure the steering shaft is not turned while the steering gear is removed or the airbag clockspring could become damaged. To prevent the shaft from turning, place the ignition key in the LOCK position or thread the seat belt through the steering wheel and clip it into place.*

Note: *Some models are equipped with quick-connect fittings on the power steering lines to the steering gear. To separate the lines from the steering gear, special tools will be required. Purchase the proper tools at an automotive parts store or tool supplier.*

1   Disconnect the cable from the negative terminal of the battery.

2   Loosen the front wheel lug nuts.

3   Apply the parking brake and block the rear wheels to keep the vehicle from rolling. Raise the front of the vehicle and place it securely on jackstands.

4   Remove the front wheels.

5   Remove the steering column intermediate shaft pinch bolt at the steering gear **(see illustration)**.

6   Disconnect the compressor cutout switch electrical connector.

7   Detach the power steering line fittings from the steering gear **(see illustration)**.

8   Detach the tie-rod ends from the steering knuckles (see Section 19).

9   Support the engine/transaxle assembly from above with a 3-bar engine support fixture.

10   Loosen, but do not remove, the two front subframe mounting bolts (see Chapter 7).

11   Remove the four, rear subframe-to-body bolts **(see illustration)**.

12   Remove the two mounting bolts from

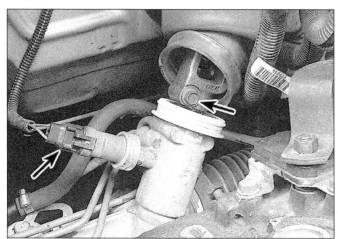

21.5 Slide the boot up and remove the pinch bolt from the steering column intermediate shaft universal joint (right arrow) - disconnect the compressor cutout switch electrical connector (left arrow)

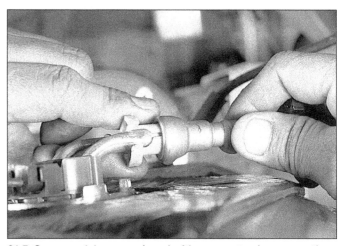

21.7 Some models are equipped with power steering connections that require quick-connect tools to separate power steering hoses from the metal lines - these tools can be purchased at most auto parts stores

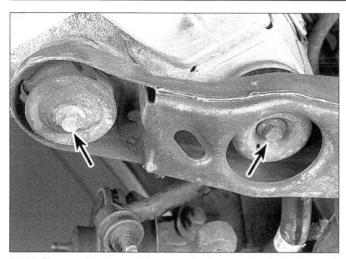

21.11 Remove the two subframe-to-body mounting bolts (arrows) at each rear corner of the subframe

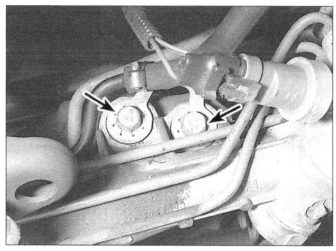

21.12a Remove the mounting bolts (arrows) from the left end of the rack-and-pinion assembly

the left side of the steering gear assembly and the bolts from the two clamps on the right **(see illustrations)**.

13 Lower the sub-frame about three inches.

14 Slide the steering gear assembly out the left wheel well to remove it.

15 Installation is the reverse of removal. Tighten the fasteners to the torque listed in this Chapter's Specifications and in Chapter 7 Specifications.

16 Have the front end alignment checked and, if necessary, adjusted.

22 Tilt/telescopic steering column - repair

Warning: *On models equipped with a Supplemental Inflatable Restraint (SIR) system (more commonly known as airbags), always disable the airbag system before working in the vicinity of any airbag system component to avoid the possibility of accidental deployment of the airbag, which could cause personal injury (see Chapter 12).*

21.12b Remove the bolts from each of the clamps (arrows) on the right end of the rack-and-pinion assembly

Steering column repair

Refer to illustrations 22.4a through 22.4p
Note: *Performing this procedure will make a "sloppy" steering column solid and tight again.*
1 Disconnect the negative battery cable.
2 Remove the steering wheel (see Section 18).

22.4a Remove the turn signal switch lever. On some models this is accomplished by pulling straight out; on other models push the lever in, rotate it 1/4 turn clockwise and pull it straight out

3 Remove the turn signal switch and the key lock cylinder (see Chapter 12).
4 Follow the accompanying photo sequence **(see illustrations)**. Stay in order and read each caption carefully.

22.4b Remove the turn signal switch and the ignition lock cylinder (see Chapter 12), then remove the three screws (arrows) which retain the switch cover assembly

22.4c Remove the switch cover assembly and set it aside

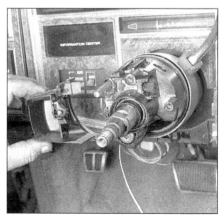

22.4d Remove the high-beam/turn signal switch actuator (if equipped)

22.4e To release the tilt mechanism retainer, push it in and turn it counterclockwise

22.4f The tilt pivot pins (arrow) are located on each side of the housing

22.4g A small slide hammer and an 8-32 machine screw . . .

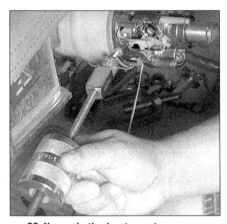

22.4h . . . is the best way to remove the pivot pins

5   Installation is the reverse of removal.

Adjusting the shift indicator

Refer to illustrations 22.8a and 22.8b

6   The shift indicator needle is linked to a guide clip on the steering column "shift bowl". The shift bowl is the ring encircling the steering column which is rotated as the shift

22.4i If you don't have a small slide hammer, screw an 8-32 machine screw into each pin, clamp a pair of locking pliers onto the screw and use a large screwdriver to pry out the pins as shown

lever is moved from one gear to another. As the shift bowl is rotated, it moves the indicator needle to the corresponding indicated gear on the gear position indicator panel just below the instrument cluster. These models use a wire cable between the clip and the

22.4j Before you remove the housing, pay very close attention to the relationship between the rack (A) and the sector (B); when the housing is installed again, this is how these two parts must fit together

indicator needle. Also, later cable-type units use a set screw threaded through the clip while other units have no screw. The shift indicator pointer on the instrument cluster must point to the correct gear – PARK, NEU-TRAL, DRIVE, etc. If the pointer for the shift

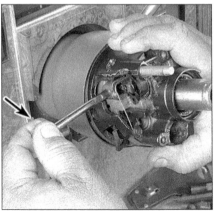

22.4k Temporarily install the tilt lever and pull toward you to release the tilt shoes from the shaft . . .

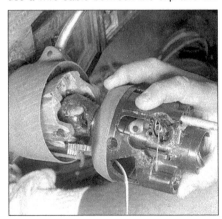

22.4l . . . and remove the housing (you will need to use this same procedure to install the housing)

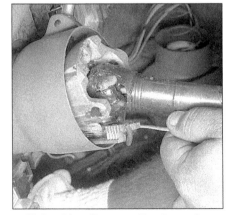

22.4m After disconnecting the cruise control lead down at the base of the column, carefully fish it out of the column and set the housing on the floor

22.4n Remove the rack; note how the rack is oriented - it must go back in exactly this way

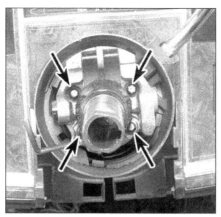

22.4o Remove these four screws (arrows), apply some Loctite to the threads, install the screws and tighten them securely; this will eliminate the "sloppiness" in the tilt mechanism

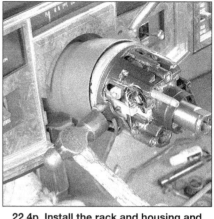

22.4p Install the rack and housing and pass the cruise control harness through the column (remember to pull back on the tilt lever when installing the housing). To install the pivot pins, carefully tap them into place with a hammer. The remainder of installation is the reverse of removal

lever indicates a gear position other than the gear the transaxle is in, the indicator must be adjusted. The adjustment procedure is similar for all models.

7   Remove the steering column cover and/or dashboard lower trim panel, as necessary (see Chapter 11).

8   Put the shift lever in the NEUTRAL position, remove the guide clip set screw, if equipped, pry the clip off the shift bowl **(see illustrations)**, move it until the indicator needle is pointing at the "N" on the shift indicator panel, then push the clip back onto the shift bowl. If the clip has a set screw, make sure the screw is snug (don't overtighten it or you'll strip the threads). Make sure the link or wire between the guide clip and the indicator needle is tight.

23   Power steering pump - removal and installation

Removal

Refer to illustrations 23.4 and 23.5

1   Disconnect the cable from the negative battery terminal.

2   Remove the pump drivebelt (see Chapter 1).

3   Remove the external reservoir, if equipped. **Note:** *Some models are equipped with an external reservoir that must be removed to gain access to the power steering pump. A one-piece adapter is used to link the pump with the reservoir. The adapter is equipped with internal O-rings. Be sure to install new O-rings into the adapter before installing the adapter onto the power steering pump. Some adapters are equipped with a nylon seal protector for the pump inlet tube. This seal protector covers the external seal on the 5/8 inch inlet tube. Do not attempt to install a seal protector onto the inlet tube unless it was designed as original equipment.*

4   Remove the pump mounting bolts and detach the pump from the engine **(see illustration)**.

5   Detach the return line from the pump **(see illustration)**, and drain the fluid in the line.

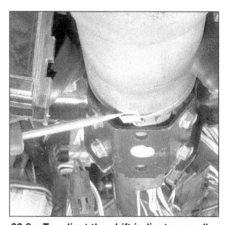

22.8a To adjust the shift indicator needle, pry the guide clip from the shift bowl, put the shift lever in the Neutral position (or as close to "N" as possible)

22.8b Move the clip until the pointer is pointing at the middle of the "N" on the shift indicator panel, make sure the wire between the guide slip and the indicator is tight and push the clip back onto the shift bowl

23.4 The power steering mounting bolts can be accessed through the holes in the pulley

23.5 Detach the power steering pump return line (1), drain the contents into a container, then disconnect the pressure line (2)

25.3 Use a press tool to push the stud out of the flange

| 1 | Hub flange | 2 | Lug nut on stud | 3 | Press tool |

25.4 Install four washers and a lug nut on the stud, then tighten the nut to draw the stud into place

| 1 | Hub flange | 2 | Spacer/washers |

6   Disconnect the pressure line from the pump.

Installation

7   Installation is the reverse of removal. **Note:** *It's a good idea to attach (but not tighten) the hoses to the pump before installing it in its bracket.*

8   Fill the reservoir with the recommended fluid and bleed the system, following the procedure described in the next Section.

24   Power steering system - bleeding

1   Following any operation in which the power steering fluid lines have been discon-nected, the power steering system must be bled to remove air and obtain proper steering performance.

2   With the front wheels turned all the way to the left, check the power steering fluid level and, if low, add fluid until it reaches the Cold mark on the dipstick.

3   Start the engine and allow it to run at fast idle. Recheck the fluid level and add more if necessary to reach the Cold mark on the dipstick.

4   Bleed the system by turning the wheels from side-to-side, without hitting the stops. This will work the air out of the system. Don't allow the reservoir to run out of fluid.

5   When the air is worked out of the sys-tem, return the wheels to the straight ahead position and leave the engine running for sev-eral minutes before shutting it off. Recheck the fluid level.

6   Road test the vehicle to be sure the steering system is functioning normally with no noise.

7   Recheck the fluid level to be sure it's up to the Hot mark on the dipstick while the engine is at normal operating temperature. Add fluid if necessary.

25   Wheel studs - replacement

Refer to illustrations 25.3 and 25.4
Note: *This procedure applies to both the front and rear wheel studs.*

1   Remove the hub and wheel bearing assembly (see Section 7 or 12).

2   Install a lug nut part way onto the stud being replaced.

3   Push the stud out of the hub flange with a press tool **(see illustration)**.

4   Insert the new stud into the hub flange from the back side and install four flat wash-ers and a lug nut on the stud **(see illustra-tion)**.

5   Tighten the lug nut until the stud is seated in the flange.

6   Reinstall the hub and wheel bearing assembly.

26   Wheels and tires - general information

Refer to illustration 26.1

All vehicles covered by this manual are equipped with metric-size steel belted radial tires **(see illustration)**. The use of other size or type tires may affect the ride and handling of the vehicle. Don't mix different types of tires, such as radials and bias belted, on the same vehicle, since handling may be seri-ously affected. Tires should be replaced in

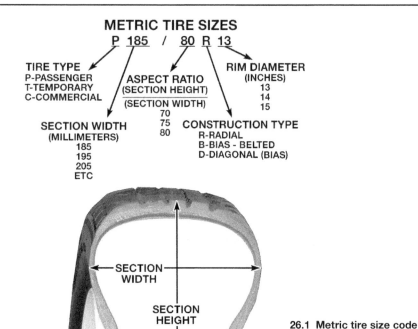

METRIC TIRE SIZES

P 185 / 80 R 13

TIRE TYPE
P-PASSENGER
T-TEMPORARY
C-COMMERCIAL

ASPECT RATIO
(SECTION HEIGHT)
―――――――
(SECTION WIDTH)
70
75
80

RIM DIAMETER
(INCHES)
13
14
15

SECTION WIDTH
(MILLIMETERS)
185
195
205
ETC

CONSTRUCTION TYPE
R-RADIAL
B-BIAS - BELTED
D-DIAGONAL (BIAS)

SECTION WIDTH

SECTION HEIGHT

26.1 Metric tire size code

pairs on the same axle, but if only one tire is being replaced, be sure it's the same size, structure and tread design as the other.

Because tire pressure affects handling and wear, the tire pressures should be checked at least once a month or before any extended trips (see Chapter 1).

Wheels must be replaced if they're bent, dented, leak air, have elongated bolt holes, are heavily rusted, out of vertical symmetry or if the lug nuts won't stay tight. Wheel repairs by welding or peening aren't recommended.

Tire and wheel balance is important to the overall handling, braking and performance of the vehicle. Unbalanced wheels can adversely affect handling and ride characteristics as well as tire life. Whenever a tire is installed on a wheel, the tire and wheel should be balanced by a shop with the proper equipment.

27 Wheel alignment - general information

Refer to illustration 27.1

Wheel alignment refers to the adjustments made to the wheels so they're in proper angular relationship to the suspension and the ground **(see illustration)**. Wheels that are out of proper alignment not only affect steering control, but also increase tire wear.

Getting the proper wheel alignment is a very exacting process, one in which complicated and expensive machines are necessary to perform the job properly. Because of this, you should have a technician with the proper equipment perform these tasks. We will, however, attempt to give you a basic idea of what's involved with front end alignment so you can better understand the process and deal intelligently with the shop that does the work.

Toe is the relationship between the front-to-rear alignment of the wheels and the centerline of the vehicle. The purpose of a toe specification is to ensure parallel rolling of the wheels. In a vehicle with zero toe, the distance between the front edges of the wheels will be the same as the distance between the rear edges of the wheels. A small amount of initial toe-in is normally desired because centrifugal force is attempting to push the wheels out as the vehicle is moving. Incorrect toe-in will cause the tires to wear improperly by making them scrub against the road surface.

On the front wheels, toe-in adjustment is controlled by the length of the tie-rod. The

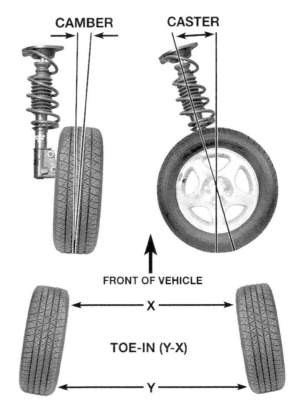

27.1 Wheel alignment details

tie-rod end may be threaded in-or-out on the inner tie-rod to lengthen or shorten the overall length of the tie-rod assembly. Rear wheel toe adjustments differ according to suspension design. Deville/Fleetwood models uses an adjustment link (similar to a front tie-rod) mounted to the front of the rear knuckle. The toe adjustment is accomplished by rotating the adjustment link to lengthen or shorten the overall length of the adjustment link. This in turn, pulls the front of the knuckle in-or-out. On Seville/Eldorado, Riviera and Toronado models, the toe adjustment is made by loosening the control arm pivot bolts and changing the position of the lower control arm. By prying between the rear inner bolt and the rear support assembly the toe specification can be adjusted in or out.

Camber is the tilting of the wheels from the vertical when viewed from the front or rear of the vehicle. When the wheels tilt out at the top, the camber is said to be positive (+). When the wheels tilt in at the top, the camber is negative (-). The amount of tilt is measured in degrees from the vertical and this measurement is called the camber angle. This

angle affects the amount of tire tread which contacts the road and compensates for changes in the suspension geometry when the vehicle is cornering or traveling over an undulating surface.

Camber, on the front suspension, is adjusted by loosening the strut-to-knuckle bolts/nuts and altering the relationship between the strut and knuckle. On Deville/Fleetwood models, the camber on the rear suspension may adjusted in the same manner. Rear camber is not adjustable on Seville/Eldorado, Riviera and Toronado models.

Caster is the tilting of the top of the front steering axis from the vertical. A tilt toward the rear is positive caster and a tilt toward the front is negative caster. Caster affects the straight-line steerability of the vehicle. Incorrect caster settings may cause the vehicle to pull to one side or the other. Caster is adjustable on the front end by elongating the holes for the strut upper mounting studs and moving the top of the strut towards the front or rear of the vehicle, as necessary. Caster is not adjustable on the rear of these vehicles.

Notes

Chapter 11   Body

Contents

1   General information

Warning: *Some models covered by this manual are equipped with Supplemental Restraint Systems (SRS), more commonly known as airbags. Always disable the airbag system before working in the vicinity of any airbag system components to avoid the possibility of accidental deployment of the airbag(s) which could cause personal injury (see Chapter 12).*

These models feature a "unibody" construction, using a floor pan with front and rear frame side rails which support the body components, front and rear suspension systems and other mechanical components.

Certain body components are particularly vulnerable to accident damage and can be unbolted and repaired or replaced. Among these parts are the body moldings, front fenders, doors, bumpers, the hood, trunk lid and all glass.

Only general body maintenance practices and body panel repair procedures within the scope of the do-it-yourselfer are included in this Chapter.

2   Body - maintenance

1   The condition of your vehicle's body is very important, because the resale value depends a great deal on it. It's much more difficult to repair a neglected or damaged body than it is to repair mechanical components. The hidden areas of the body, such as the wheel wells, the frame and the engine compartment, are equally important, although they don't require as frequent attention as the rest of the body.

2   Once a year, or every 12,000 miles, it's a good idea to have the underside of the body steam cleaned. All traces of dirt and oil will be removed and the area can then be inspected carefully for rust, damaged brake lines, frayed electrical wires, damaged cables and other problems. The front suspension components should be greased after completion of this job.

3   At the same time, clean the engine and the engine compartment with a steam cleaner or water-soluble degreaser.

4   The wheel wells should be given close attention, since undercoating can peel away and stones and dirt thrown up by the tires can cause the paint to chip and flake, allowing rust to set in. If rust is found, clean down to the bare metal and apply an anti-rust paint.

5   The body should be washed about once a week. Wet the vehicle thoroughly to soften the dirt, then wash it down with a soft sponge and plenty of clean soapy water. If the surplus dirt is not washed off very carefully, it can wear down the paint.

6   Spots of tar or asphalt thrown up from the road should be removed with a cloth soaked in solvent.

7   Once every six months, wax the body and chrome trim. If a chrome cleaner is used to remove rust from any of the vehicle's plated parts, remember that the cleaner also removes part of the chrome, so use it sparingly.

3   Vinyl trim - maintenance

Don't clean vinyl trim with detergents, caustic soap or petroleum-based cleaners.

Plain soap and water works just fine, with a soft brush to clean dirt that may be ingrained. Wash the vinyl as frequently as the rest of the vehicle. After cleaning, application of a high-quality rubber and vinyl protectant will help prevent oxidation and cracks. The protectant can also be applied to weatherstripping, vacuum lines and rubber hoses, which often fail as a result of chemical degradation, and to the tires.

4 Upholstery and carpets - maintenance

1 Every three months, remove the floor-mats and clean the interior of the vehicle (more frequently if necessary). Use a stiff whisk broom to brush the carpeting and loosen dirt and dust, then vacuum the upholstery and carpets thoroughly, especially along seams and crevices.

2 Dirt and stains can be removed from carpeting with basic household or automotive carpet shampoos available in spray cans. Follow the directions and vacuum again, then use a stiff brush to bring back the "nap" of the carpet.

3 Most interiors have cloth or vinyl upholstery, either of which can be cleaned and maintained with a number of material-specific cleaners or shampoos available in auto supply stores. Follow the directions on the product for usage, and always spot-test any upholstery cleaner on an inconspicuous area (bottom edge of a back seat cushion) to ensure that it doesn't cause a color shift in the material.

4 After cleaning, vinyl upholstery should be treated with a protectant. **Note:** *Make sure the protectant container indicates the product can be used on seats - some products may make a seat too slippery.* **Caution:** *Do not use protectant on vinyl-covered steering wheels.*

5 Leather upholstery requires special care. It should be cleaned regularly with saddlesoap or leather cleaner. Never use alcohol, gasoline, nail polish remover or thinner to clean leather upholstery.

6 After cleaning, regularly treat leather upholstery with a leather conditioner, rubbed in with a soft cotton cloth. Never use car wax on leather upholstery.

7 In areas where the interior of the vehicle is subject to bright sunlight, cover leather seating areas of the seats with a sheet if the vehicle is to be left out for any length of time.

5 Body repair - minor damage

Plastic body panels

The following repair procedures are for minor scratches and gouges. Repair of more serious damage should be left to a dealer service department or qualified auto body shop. Below is a list of the equipment and materials necessary to perform the following repair procedures on plastic body panels. Although a specific brand of material may be mentioned, it should be noted that equivalent products from other manufacturers may be used instead.

> *Wax, grease and silicone removing solvent*
> *Cloth-backed body tape*
> *Sanding discs*
> *Drill motor with three-inch disc holder*
> *Hand sanding block*
> *Rubber squeegees*
> *Sandpaper*
> *Non-porous mixing palette*
> *Wood paddle or putty knife*
> *Curved tooth body file*
> *Plastic body panel repair compound and material*

Flexible panels (front and rear bumper fascia)

1 Remove the damaged panel, if necessary or desirable. In most cases, repairs can be carried out with the panel installed.

2 Clean the area(s) to be repaired with a wax, grease and silicone removing solvent applied with a water-dampened cloth.

3 If the damage is structural, that is, if it extends through the panel, clean the backside of the panel area to be repaired as well. Wipe dry.

4 Sand the rear surface about 1-1/2 beyond the break.

5 Cut two pieces of fiberglass cloth large enough to overlap the break by about 1-1/2 in. Cut only to the required length.

6 Mix the adhesive from the repair kit according to the instructions included with the kit, and apply a layer of the mixture approximately 1/8-inch thick on the backside of the panel. Overlap the break by at least 1-1/2 inches.

7 Apply one piece of fiberglass cloth to the adhesive and cover the cloth with additional adhesive. Apply a second piece of fiberglass cloth to the adhesive and immediately cover the cloth with additional adhesive insufficient quantity to fill the weave.

8 Allow the repair to cure for 20 to 30 minutes at 60-degrees to 80-degrees F.

9 If necessary, trim the excess repair material at the edge.

10 Remove all of the paint film over and around the area(s) to be repaired. The repair material should not overlap the painted surface.

11 With a drill motor and a sanding disc (or a rotary file), cut a "V" along the break line approximately 1/2-inch wide. Remove all dust and loose particles from the repair area.

12 Mix and apply the repair material. Apply a light coat first over the damaged area; then continue applying material until it reaches a level slightly higher than the surrounding finish.

13 Cure the mixture for 20 to 30 minutes at 60-degrees to 80-degrees F.

14 Roughly establish the contour of the area being repaired with a body file. If low areas or pits remain, mix and apply additional adhesive.

15 Block sand the damaged area with sandpaper to establish the actual contour of the surrounding surface.

16 If desired, the repaired area can be temporarily protected with several light coats of primer. Because of the special paints and techniques required for flexible body panels, it is recommended that the vehicle be taken to a paint shop for completion of the body repair.

Steel body panels

See photo sequence

Repair of minor scratches

17 If the scratch is superficial and does not penetrate to the metal of the body, repair is very simple. Lightly rub the scratched area with a fine rubbing compound to remove loose paint and built-up wax. Rinse the area with clean water.

18 Apply touch-up paint to the scratch, using a small brush. Continue to apply thin layers of paint until the surface of the paint in the scratch is level with the surrounding paint. Allow the new paint at least two weeks to harden, then blend it into the surrounding paint by rubbing with a very fine rubbing compound. Finally, apply a coat of wax to the scratch area.

19 If the scratch has penetrated the paint and exposed the metal of the body, causing the metal to rust, a different repair technique is required. Remove all loose rust from the bottom of the scratch with a pocket knife, then apply rust inhibiting paint to prevent the formation of rust in the future. Using a rubber or nylon applicator, coat the scratched area with glaze-type filler. If required, the filler can be mixed with thinner to provide a very thin paste, which is ideal for filling narrow scratches. Before the glaze filler in the scratch hardens, wrap a piece of smooth cotton cloth around the tip of a finger. Dip the cloth in thinner and then quickly wipe it along the surface of the scratch. This will ensure that the surface of the filler is slightly hollow. The scratch can now be painted over as described earlier in this Section.

Repair of dents

20 When repairing dents, the first job is to pull the dent out until the affected area is as close as possible to its original shape. There is no point in trying to restore the original shape completely as the metal in the damaged area will have stretched on impact and cannot be restored to its original contours. It is better to bring the level of the dent up to a point which is about 1/8-inch below the level of the surrounding metal. In cases where the dent is very shallow, it is not worth trying to pull it out at all.

21 If the back side of the dent is accessible, it can be hammered out gently from behind using a soft-face hammer. While

doing this, hold a block of wood firmly against the opposite side of the metal to absorb the hammer blows and prevent the metal from being stretched.

22 If the dent is in a section of the body which has double layers, or some other factor makes it inaccessible from behind, a different technique is required. Drill several small holes through the metal inside the damaged area, particularly in the deeper sections. Screw long, self-tapping screws into the holes just enough for them to get a good grip in the metal. Now the dent can be pulled out by pulling on the protruding heads of the screws with locking pliers.

23 The next stage of repair is the removal of paint from the damaged area and from an inch or so of the surrounding metal. This is done with a wire brush or sanding disk in a drill motor, although it can be done just as effectively by hand with sandpaper. To complete the preparation for filling, score the surface of the bare metal with a screwdriver or the tang of a file, or drill small holes in the affected area. This will provide a good grip for the filler material. To complete the repair, see the subsection on filling and painting later in this Section.

Repair of rust holes or gashes

24 Remove all paint from the affected area and from an inch or so of the surrounding metal using a sanding disk or wire brush mounted in a drill motor. If these are not available, a few sheets of sandpaper will do the job just as effectively.

25 With the paint removed, you will be able to determine the severity of the corrosion and decide whether to replace the whole panel, if possible, or repair the affected area. New body panels are not as expensive as most people think and it is often quicker to install a new panel than to repair large areas of rust.

26 Remove all trim pieces from the affected area except those which will act as a guide to the original shape of the damaged body, such as headlight shells, etc. Using metal snips or a hacksaw blade, remove all loose metal and any other metal that is badly affected by rust. Hammer the edges of the hole in to create a slight depression for the filler material.

27 Wire brush the affected area to remove the powdery rust from the surface of the metal. If the back of the rusted area is accessible, treat it with rust inhibiting paint.

28 Before filling is done, block the hole in some way. This can be done with sheet metal riveted or screwed into place, or by stuffing the hole with wire mesh.

29 Once the hole is blocked off, the affected area can be filled and painted. See the following subsection on filling and painting.

Filling and painting

30 Many types of body fillers are available, but generally speaking, body repair kits which contain filler paste and a tube of resin hardener are best for this type of repair work.

A wide, flexible plastic or nylon applicator will be necessary for imparting a smooth and contoured finish to the surface of the filler material. Mix up a small amount of filler on a clean piece of wood or cardboard (use the hardener sparingly). Follow the manufacturer's instructions on the package, otherwise the filler will set incorrectly.

31 Using the applicator, apply the filler paste to the prepared area. Draw the applicator across the surface of the filler to achieve the desired contour and to level the filler surface. As soon as a contour that approximates the original one is achieved, stop working the paste. If you continue, the paste will begin to stick to the applicator. Continue to add thin layers of paste at 20-minute intervals until the level of the filler is just above the surrounding metal.

32 Once the filler has hardened, the excess can be removed with a body file. From then on, progressively finer grades of sandpaper should be used, starting with a 180-grit paper and finishing with 600-grit wet-or-dry paper. Always wrap the sandpaper around a flat rubber or wooden block, otherwise the surface of the filler will not be completely flat. During the sanding of the filler surface, the wet-or-dry paper should be periodically rinsed in water. This will ensure that a very smooth finish is produced in the final stage.

33 At this point, the repair area should be surrounded by a ring of bare metal, which in turn should be encircled by the finely feathered edge of good paint. Rinse the repair area with clean water until all of the dust produced by the sanding operation is gone.

34 Spray the entire area with a light coat of primer. This will reveal any imperfections in the surface of the filler. Repair the imperfections with fresh filler paste or glaze filler and once more smooth the surface with sandpaper. Repeat this spray-and-repair procedure until you are satisfied that the surface of the filler and the feathered edge of the paint are perfect. Rinse the area with clean water and allow it to dry completely.

35 The repair area is now ready for painting. Spray painting must be carried out in a warm, dry, windless and dust free atmosphere. These conditions can be created if you have access to a large indoor work area, but if you are forced to work in the open, you will have to pick the day very carefully. If you are working indoors, dousing the floor in the work area with water will help settle the dust which would otherwise be in the air. If the repair area is confined to one body panel, mask off the surrounding panels. This will help minimize the effects of a slight mismatch in paint color. Trim pieces such as chrome strips, door handles, etc., will also need to be masked off or removed. Use masking tape and several thickness of newspaper for the masking operations.

36 Before spraying, shake the paint can thoroughly, then spray a test area until the spray painting technique is mastered. Cover the repair area with a thick coat of primer. The thickness should be built up using sev-

eral thin layers of primer rather than one thick one. Using 600-grit wet-or-dry sandpaper, rub down the surface of the primer until it is very smooth. While doing this, the work area should be thoroughly rinsed with water and the wet-or-dry sandpaper periodically rinsed as well. Allow the primer to dry before spraying additional coats.

37 Spray on the top coat, again building up the thickness by using several thin layers of paint. Begin spraying in the center of the repair area and then, using a circular motion, work out until the whole repair area and about two inches of the surrounding original paint is covered. Remove all masking material 10 to 15 minutes after spraying on the final coat of paint. Allow the new paint at least two weeks to harden, then use a very fine rubbing compound to blend the edges of the new paint into the existing paint. Finally, apply a coat of wax.

6 Body repair - major damage

1 Major damage must be repaired by an auto body/frame repair shop with the necessary welding and hydraulic straightening equipment.

2 If the damage has been serious, it is vital that the structure be checked for proper alignment or the vehicle's handling characteristics may be adversely affected. Other problems, such as excessive tire wear and wear in the driveline and steering may occur.

3 Due to the fact that all of the major body components (hood, fenders, etc.) are separate and replaceable units, any seriously damaged components should be replaced rather than repaired. Sometimes these components can be found in a wrecking yard that specializes in used vehicle components, often at considerable savings over the cost of new parts.

7 Hinges and locks - maintenance

Once every 3000 miles, or every three months, the hinges and latch assemblies on the doors, hood and trunk should be given a few drops of light oil or lock lubricant. The door latch strikers should also be lubricated with a thin coat of grease to reduce wear and ensure free movement. Lubricate the door and trunk locks with spray-on graphite lubricant.

8 Windshield and fixed glass - replacement

Replacement of the windshield and fixed glass requires the use of special fast setting adhesive/caulk materials. These operations should be left to a dealer or a shop specializing in glass work.

These photos illustrate a method of repairing simple dents. They are intended to supplement *Body repair - minor damage* in this Chapter and should not be used as the sole instructions for body repair on these vehicles.

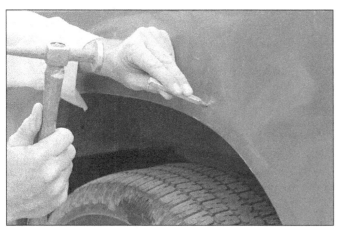

1 If you can't access the backside of the body panel to hammer out the dent, pull it out with a slide-hammer-type dent puller. In the deepest portion of the dent or along the crease line, drill or punch hole(s) at least one inch apart . . .

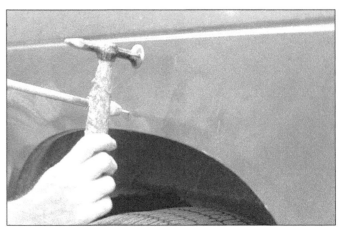

2 . . . then screw the slide-hammer into the hole and operate it. Tap with a hammer near the edge of the dent to help 'pop' the metal back to its original shape. When you're finished, the dent area should be close to its original contour and about 1/8-inch below the surface of the surrounding metal

3 Using coarse-grit sandpaper, remove the paint down to the bare metal. Hand sanding works fine, but the disc sander shown here makes the job faster. Use finer (about 320-grit) sandpaper to feather-edge the paint at least one inch around the dent area

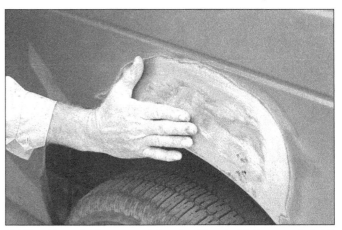

4 When the paint is removed, touch will probably be more helpful than sight for telling if the metal is straight. Hammer down the high spots or raise the low spots as necessary. Clean the repair area with wax/silicone remover

5 Following label instructions, mix up a batch of plastic filler and hardener. The ratio of filler to hardener is critical, and, if you mix it incorrectly, it will either not cure properly or cure too quickly (you won't have time to file and sand it into shape)

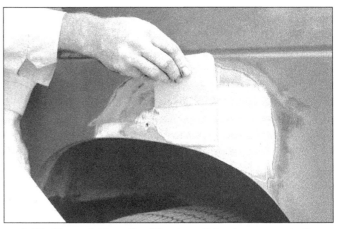

6 Working quickly so the filler doesn't harden, use a plastic applicator to press the body filler firmly into the metal, assuring it bonds completely. Work the filler until it matches the original contour and is slightly above the surrounding metal

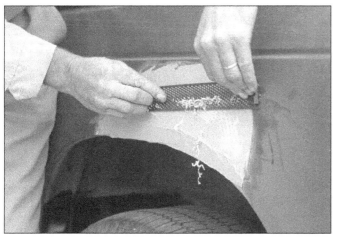

7 Let the filler harden until you can just dent it with your fingernail. Use a body file or Surform tool (shown here) to rough-shape the filler

8 Use coarse-grit sandpaper and a sanding board or block to work the filler down until it's smooth and even. Work down to finer grits of sandpaper - always using a board or block - ending up with 360 or 400 grit

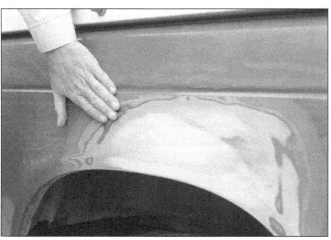

9 You shouldn't be able to feel any ridge at the transition from the filler to the bare metal or from the bare metal to the old paint. As soon as the repair is flat and uniform, remove the dust and mask off the adjacent panels or trim pieces

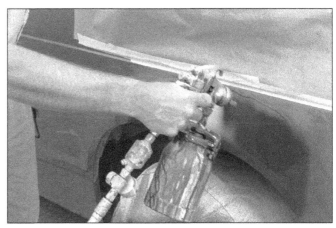

10 Apply several layers of primer to the area. Don't spray the primer on too heavy, so it sags or runs, and make sure each coat is dry before you spray on the next one. A professional-type spray gun is being used here, but aerosol spray primer is available inexpensively from auto parts stores

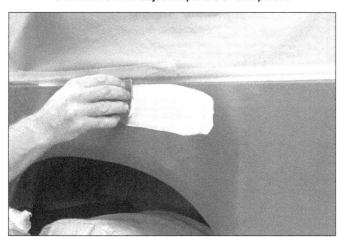

11 The primer will help reveal imperfections or scratches. Fill these with glazing compound. Follow the label instructions and sand it with 360 or 400-grit sandpaper until it's smooth. Repeat the glazing, sanding and respraying until the primer reveals a perfectly smooth surface

12 Finish sand the primer with very fine sandpaper (400 or 600-grit) to remove the primer overspray. Clean the area with water and allow it to dry. Use a tack rag to remove any dust, then apply the finish coat. Don't attempt to rub out or wax the repair area until the paint has dried completely (at least two weeks)

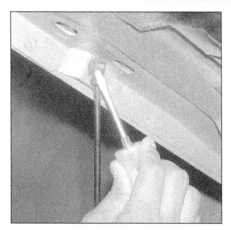

9.2a Pry the retaining clip off the end of the upper end . . .

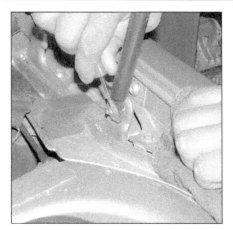

9.2b . . . and the lower end of the strut, then detach the strut from the mounting studs

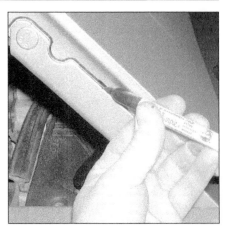

11.3 Before removing the hood, draw a mark around the hinge plate

9   Hood support struts - replacement

Refer to illustrations 9.2a and 9.2b

1   Open the hood and support it securely.
2   Detach the clips from the upper and lower ends of the strut with a screwdriver and remove the strut **(see illustrations)**.
3   Installation is the reverse of removal.

10   Radiator grille - removal and installation

Warning: *Some models covered by this manual are equipped with Supplemental Restraint Systems (SRS), more commonly known as airbags. Always disable the airbag system before working in the vicinity of any airbag system components to avoid the possibility of accidental deployment of the airbag(s) which could cause personal injury* (see Chapter 12).

1   Open the hood.
2   On models with hood-mounted grilles, remove the retaining bolts/nuts and detach the grille/bracket assembly as a unit.
3   Remove the screws/nuts and separate

the grille from the bracket.
4   On body-mounted grilles, remove the air deflector panel for access.
5   Remove the grille retaining screws or plastic retainers.
6   Rotate the top of the grille out and lift it from the vehicle.
7   Installation is the reverse of removal.

11   Hood - removal, installation and adjustment

Note: *The hood is somewhat awkward to remove and install, at least two people should perform this procedure.*

Removal and installation

Refer to illustrations 11.3 and 11.5

1   Open the hood, then place blankets or pads over the fenders and cowl area of the body. This will protect the body and paint as the hood is lifted off.
2   Disconnect any cables or wires that will interfere with removal.
3   Make marks or scribe a line around the hood hinge to ensure proper alignment dur-

ing installation **(see illustration)**.
4   Support the hood and detach the support struts (see Section 9).
5   Have an assistant support one side of the hood while you support the other, then remove the hinge-to-hood bolts and lift off the hood **(see illustration)**.
6   Installation is the reverse of removal. Align the hinge bolts with the marks made in step 3.

Adjustment

Refer to illustrations 11.10 and 11.11

7   Fore-and-aft and side-to-side adjustment of the hood is done by moving the hinge plate slot after loosening the bolts or nuts.
8   Scribe a line around the entire hinge plate so you can determine the amount of movement.
9   Loosen the bolts or nuts and move the hood into correct alignment. Move it only a little at a time. Tighten the hinge bolts and carefully lower the hood to check the position.
10   If necessary after installation, the entire hood latch assembly can be adjusted up-and-down as well as from side-to-side on the radiator support so the hood closes securely

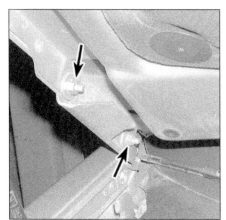

11.5 Remove the hood mounting nuts and bolts (arrows)

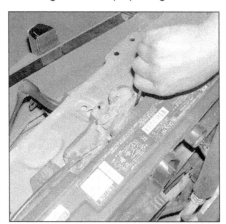

11.10 Loosen the bolts and move the hood latch to adjust the hood closed position

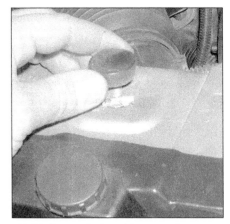

11.11 Screw the hood bumpers in or out to adjust the hood flush with the fenders

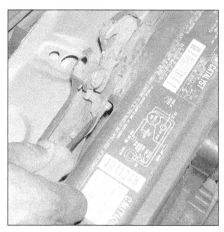

12.2 Pry out the cable retainer from the hood latch assembly, then disengage the cable

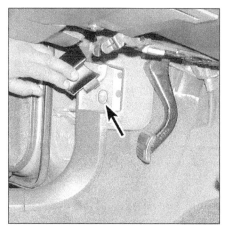

12.6a Remove the screw (arrow) and . . .

and flush with the fenders. To make the adjustment, first remove the air deflector covering the hood latch. Scribe a line or mark around the hood latch mounting bolts to provide a reference point, then loosen them and reposition the latch assembly, as necessary **(see illustration)**. Following adjustment, retighten the mounting bolts.

11   Finally, adjust the hood bumpers on the radiator support so the hood, when closed, is flush with the fenders **(see illustration)**.

12   The hood latch assembly, as well as the hinges, should be periodically lubricated with white, lithium-base grease to prevent binding and wear.

12 Hood latch and release cable - removal and installation

Warning: *Some models covered by this manual are equipped with Supplemental Restraint Systems (SRS), more commonly known as airbags. Always disable the airbag system before working in the vicinity of any airbag*

system components to avoid the possibility of accidental deployment of the airbag(s) which could cause personal injury (see Chapter 12).

Latch

Refer to illustration 12.2

1   Scribe a line around the latch to aid alignment when installing, then remove the retaining bolts securing the hood latch to the radiator support **(see illustration 11.10)**. Remove the latch.

2   Disconnect the hood release cable by disengaging the cable from the latch assembly **(see illustration)**.

3   Installation is the reverse of removal. **Note:** *Adjust the latch so the hood engages securely when closed and the hood bumpers are slightly compressed.*

Cable

Refer to illustrations 12.6a, 12.6b and 12.7

4   Disconnect the hood release cable from the latch assembly as described in step 2.

5   Attach a piece of thin wire or string to the end of the cable and unclip all remaining cable retaining clips. **Note:** *It may be necessary on some vehicles to remove the air*

cleaner assembly (see Chapter 4) to allow access for release cable removal.

6   Working in the passenger compartment, remove the driver's side kick panel **(see illustrations)**.

7   Detach the cable from the hood release lever **(see illustration)**.

8   Pull the cable and grommet rearward into the passenger compartment until you can see the wire or string. Ensure that the new cable has a grommet attached then remove the wire or string from the old cable and fasten it to the new cable.

9   With the new cable attached to the wire or string, pull the wire or string back through the firewall until the new cable reaches the latch assembly.

10   Working in the passenger compartment, reinstall the new cable into the hood release lever. Use pliers to crimp the retaining clip which secures the cable to the release lever.

11   The remainder of the installation is the reverse of removal. **Note:** *Push on the grommet with your fingers from the passenger compartment to seat the grommet in the firewall correctly.*

13 Bumpers - removal and installation

Warning: *Some models covered by this manual are equipped with Supplemental Restraint Systems (SRS), more commonly known as airbags. Always disable the airbag system before working in the vicinity of any airbag system components to avoid the possibility of accidental deployment of the airbag(s) which could cause personal injury (see Chapter 12).*

Front bumper

Refer to illustration 13.5

1   Open the hood and disconnect the negative battery cable.

2   Raise the front of the vehicle and support it securely on jackstands.

3   Working in the front wheel opening,

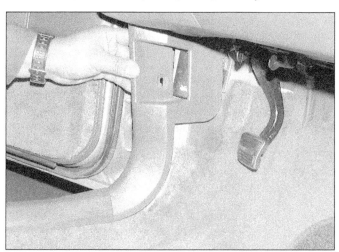

12.6b . . . detach the left side kick panel

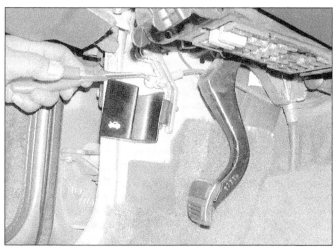

12.7 Detach the cable and pull it rearward into the passenger compartment

13.5 Remove the bumper-to-energy absorber nuts and bolts (arrows)

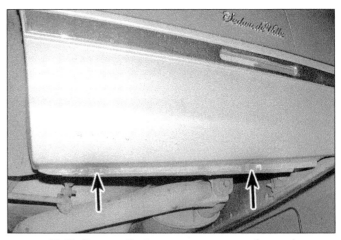

13.10   Remove the bumper cover screws (arrows)

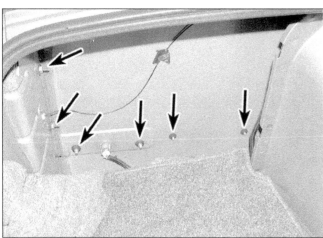

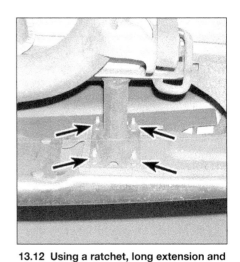

13.11 Working in the trunk compartment, remove the nuts (arrows) securing the bumper cover to the rear quarter panels

13.12 Using a ratchet, long extension and a socket, remove the rear bumper retaining nuts (arrows)

detach any retaining screws securing the inner fenderwell extension panels to the bumper.

4   Remove the retaining screws securing the bumper to each fender. Disconnect the turn signal and side marker lamp electrical connectors.

5   Working under the front of the vehicle, remove the retaining nuts and bolts securing the bumper to the bumper energy absorbers **(see illustration)**.

6   Separate the bumper cover (if equipped) from the fenders, then pull the bumper assembly straight out and away from the vehicle to remove it.

7   Installation is the reverse of removal.

Rear bumper

Refer to illustrations 13.10, 13.11 and 13.12

8   Apply the parking brake, raise the rear of the vehicle, support it securely on jackstands and remove the rear wheels.

9   Disconnect the rear side marker lights and the backup lamps (see Chapter 12).

10   Remove screws retaining the bumper cover to the rear quarter panels **(see illustration)**.

11   Working inside the trunk compartment, remove the trunk finishing panels. Peel back

the trunk finishing panels and remove the remaining nuts and bolts securing the bumper cover to rear quarter panels and the trunk opening **(see illustration)**.

12   Working under the vehicle, remove the retaining nuts securing the bumper to the bumper energy absorbers **(see illustration)**. Pull the bumper assembly straight out and away from the vehicle to remove it.

13   To remove the bumper cover (if equipped) from the bumper, detach the plastic clips securing the upper and lower edges of the bumper cover.

14   Installation is the reverse of removal.

14   Front fender - removal and installation

Refer to illustrations 14.3, 14.4, 14.6a, 14.6b, 14.6c, 14.6d, 14.7a and 14.7b

Warning: *Some models covered by this manual are equipped with Supplemental Restraint Systems (SRS), more commonly known as airbags. Always disable the airbag system before working in the vicinity of any airbag system components to avoid the possibility of accidental deployment of the airbag(s) which*

could cause personal injury (see Chapter 12).

1   Remove the hood (see Section 11).

2   Raise the vehicle, support it securely on jackstands and remove the front wheel.

3   Remove the rocker panel finish mould-

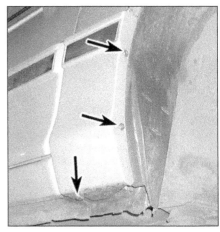

14.3 Remove the bolts (arrows) and detach the fender moulding

14.4 Remove the inner retainers and detach the inner fenderwell

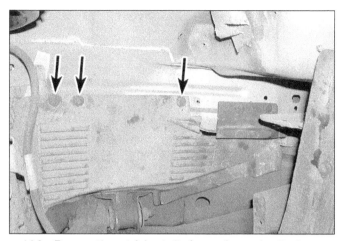

14.6a Remove the retaining bolts (arrows) securing the inner fender splash shield

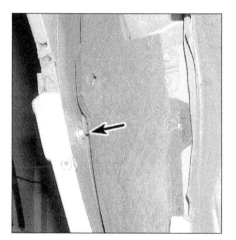

14.6b Remove the fender retaining bolt (arrow) located in the wheel opening

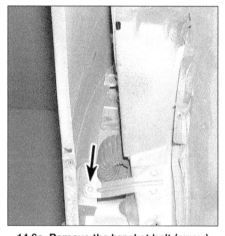

14.6c Remove the bracket bolt (arrow) securing the front . . .

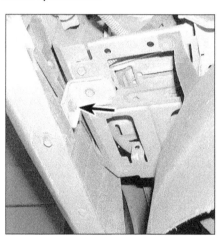

14.6d . . . and the rear (arrow) of the fender

ing (if equipped) from the side of the vehicle that the fender is to be removed **(see illustration)**.

4   Remove the inner fenderwell splash shield **(see illustration)**.

5   Remove the front bumper assembly (see Section 13).

6   Remove any remaining nuts and bolts attaching the fender to chassis and inner splash shields **(see illustrations)**.

7   Remove the fender mounting bolts **(see illustrations)**.

8   Detach the fender. It's a good idea to have an assistant support the fender while it's being moved away from the vehicle to prevent damage to the surrounding body panels.

9   Installation is the reverse of removal.

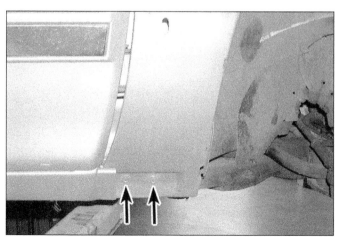

14.7a Remove the retaining bolts (arrows) securing the fender to the rocker panel

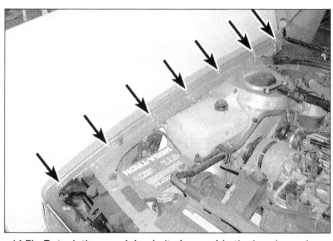

14.7b Detach the remaining bolts (arrows) in the hood opening, then remove the fender from the vehicle

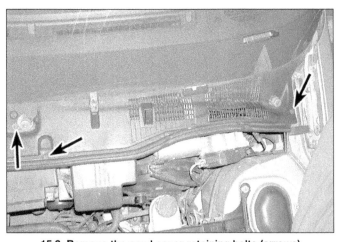

15.2 Remove the cowl cover retaining bolts (arrows)

16.1a Pull out the handle, remove the screw and pry out the inside door handle bezel

15 Cowl cover - removal and installation

Refer to illustration 15.2

1   Remove the windshield wiper arms (see Chapter 12).
2   Remove the bolts securing the cowl cover **(see illustration)**.
3   Lift the cover up, disconnect the windshield washer hoses, then remove the cover from the vehicle.
4   Installation is the reverse of removal.

16 Door trim panel - removal and installation

Removal

Refer to illustrations 16.1a, 16.1b, 16.2a, 16.2b, 16.2c, 16.3, 16.4a, 16.4b, 16.5 and 16.6

1   Disconnect the cable from the negative terminal of the battery. Remove the screw and detach the inside door handle trim bezel **(see illustrations)**.
2   Pry out the armrest switch control plate and disconnect the electrical connections.

16.1b Use a small screwdriver to disconnect the electrical connector

Working through the armrest opening, remove the retaining screws **(see illustrations)**.
3   Lift up the door pull handle for access to the retaining screws **(see illustration)**.
4   Remove the remaining door panel retaining screws **(see illustrations)**.
5   Insert a wide putty knife or a special trim

16.2a Pry out the armrest control switch . . .

panel removal tool between the trim panel and the head of the retaining clip to disengage the door panel retaining clips **(see illustration)**. **Note:** *Door trim panel retaining clips are approximately six to ten inches apart. Pry at the clip location only. Prying in between clips will result in distorted or damaged door trim panels.*

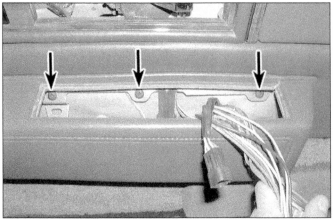

16.2b . . . and remove the screws (arrows) from the side . . .

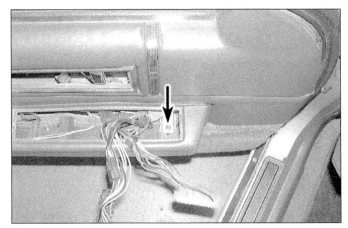

16.2c . . . and the front (arrow) of the armrest opening

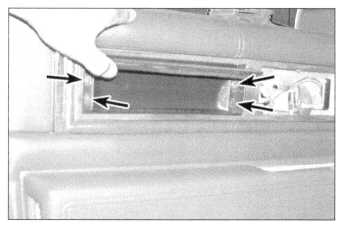

16.3 Remove the screws (arrows) under the door pull handle

16.4a Pry out the interior lamps and remove the screw in the end (arrow) . . .

16.4b . . . and below (arrow) the armrest

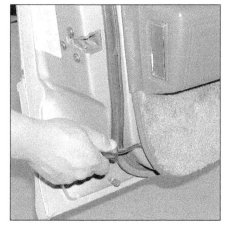

16.5 Insert a trim removal tool or a putty knife between the door and the trim panel to disengage the clips along the outer edge of the door trim panel

16.6 Use a small screwdriver or similar tool to detach the door trim panel electrical connectors

6   Once all of the clips and screws are disengaged, detach the trim panel and remove the trim panel from the vehicle by gently pulling it up and out **(see illustration)**.

7   For access to the inner door, peel back the watershield, taking care not to tear it.

8   To install the door panel, work through the window opening and engage the hooks on the back of the trim panel onto the door and push down until they are seated, then press the door panel retaining clips into place. **Note:** *When installing door trim panel retaining clips, make sure the clips are lined up with their mating holes first, then gently tap inward with the palm of your hand.*

9   The remainder of the installation is the reverse of removal.

17 Door - removal, installation and adjustment

Removal and installation

Refer to illustrations 17.7a, 17.7b and 17.7c

1   Raise the window completely and disconnect the negative cable from the battery.

2   Open the door all the way and support it on jacks or blocks covered with rags to prevent damaging the paint.

3   Remove the door trim panel and water-

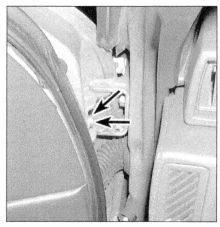

17.7a Front door retaining bolts (arrows)

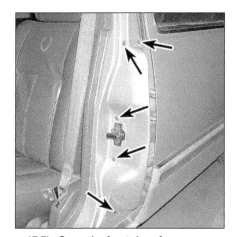

17.7b Open the front door for access, remove the rear door retaining cover screws (arrows) . . .

shield as described in Section 16.

4   Unplug all electrical connections, ground wires and harness retaining clips from the door. **Note:** *It is a good idea to label all connections to aid the reassembly process.*

5   Working on the door side, detach the rubber conduit between the body and the

door. Pull the wiring harness through the conduit hole and remove the wiring from the door.

6   Mark around the door hinges with a pen or a scribe to facilitate realignment during reassembly.

7   Have an assistant hold the door, remove

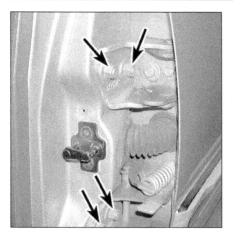

17.7c . . . and remove the rear door retaining nuts (arrows)

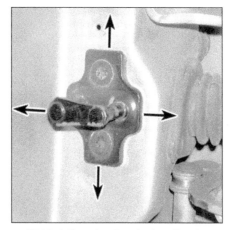

17.12 Adjust the door lock striker by loosening the mounting screws and gently tapping the striker in the desired direction

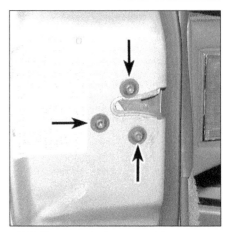

18.2 Remove the latch retaining screws (arrows) from the end of the door, then detach the actuating rods and pull the latch assembly through the access hole

the hinge-to-door bolts and lift the door off **(see illustrations)**.

8   Installation is the reverse of removal.

Adjustment

Refer to illustration 17.12

9   Having proper door-to-body alignment is a critical part of a well functioning door assembly. First check the door hinge pins and bushings for excessive play. **Note:** *If the door can be lifted (1/16-inch or more) without the car body lifting with it the hinge pins and bushings should be replaced.*

10   Door-to-body alignment adjustments are made by loosening the hinge-to-body bolts or hinge-to-door bolts and moving the door. Proper body alignment is achieved when the top of the doors are parallel with the roof section, the front door is flush with the fender, the rear door is flush with the rear quarter panel and the bottom of the doors are aligned with the lower rocker panel. If these goals can't be reached by adjusting the hinge-to-body or hinge-to-door bolts, body alignment shims may have to be purchased and inserted behind the hinges to achieve correct alignment.

11   To adjust the door closed position, first check that the door latch is contacting the center of the latch striker bolt. If not, remove the striker bolt and add or subtract washers to achieve correct alignment.

12   Finally, adjust the latch striker bolt as necessary (up-and-down or sideways) to provide positive engagement with the latch mechanism **(see illustration)** and the door panel is flush with the center pillar or rear quarter panel.

18   Door latch, lock cylinder and handles - removal and installation

Door latch

Refer to illustration 18.2

1   Raise the window then remove the door trim panel and watershield as described in Section 16.

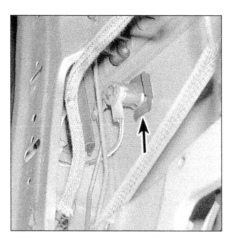

18.9 Remove the lock cylinder retaining clip (arrow)

2   Remove the screws securing the latch to the door **(see illustration)**.

3   Working through the large access hole, position the latch as necessary to disengage the outside door handle and outside lock cylinder to latch rods and the inside handle to latch rod.

4   All door locking rods are attached by plastic or pressed-metal clips. The plastic clips can be removed by unsnapping the portion engaging the connecting rod and then by pulling the rod out of its locating hole. Metal clips are removed by disengaging the small locking tab from the hole, then sliding the clip off the rod.

5   Position the latch as necessary to disengage the door lock actuator rod. Then remove the latch assembly from the door.

6   Installation is the reverse of removal.

Door lock cylinder and outside handle

Refer to illustrations 18.9 and 18.11

7   To remove the lock cylinder, raise the window and remove the door trim panel and watershield as described in Section 16.

8   Working through the large access hole,

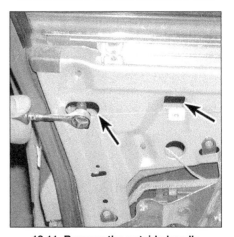

18.11 Remove the outside handle retaining nuts or bolts with a socket and extension working through the access holes in the door (arrows)

disengage the clip that secures the lock cylinder to the latch rod.

9   Using a pair of pliers, slide the lock cylinder retaining clip out of engagement and remove the lock cylinder from the door **(see illustration)**.

10   To remove the outside handle, work through the access hole and disengage the plastic clip that secures the outside handle-to-latch rod.

11   Working through the access holes in the door, use a socket and extension to remove the outside door handle retaining nuts or bolts **(see illustration)**.

12   Remove the outside handle. Installation is the reverse of removal.

Inside handle

Refer to illustration 18.15

13   Remove the door trim panel as described in Section 16 and peel away the watershield.

14   Detach the actuating rods from the handle.

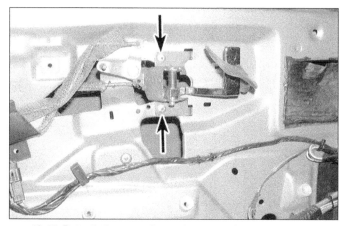

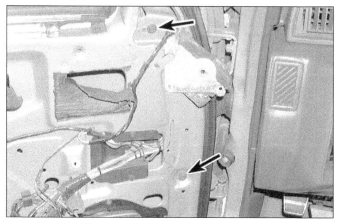

18.15 Detach the actuating rods, then drill out the handle retaining rivets (arrows) and rotate the handle out

19.3 Remove the two front window glass run channel bolts (arrows)

15   Drill out the rivets securing the inside handle **(see illustration)**. Pull forward on the handle to disengage it from the inner door panel.

16   Installation is the reverse of removal.

19   Door window glass - removal and installation

1   Remove the door trim panel and water shield (Section 16).

Eldorado/Seville/Riviera/Toronado models

Refer to illustration 19.3

2   Remove the trim panel retainer and sealer strip.

Front door

3   Remove the bolts and detach the window glass front run channel **(see illustration)**.

4   On Eldorado, Riviera and Toronado models, lower the glass to the bottom of the door, slide the window glass forward to detach it from the rollers and glass guides. Tilt the glass and lift it inward, and then out of the door frame to remove it.

5   On Seville models, first slide the window glass forward, then rearward to detach it from the rollers. Tilt the glass and lift it outward and out of the door frame to remove it.

6   To install, insert the glass into the door and engage the regulator guide block to the sash channel. Engage the rear guide clip on the glass to the rear run channel weatherstrip.

7   Raise the glass to the half way up position and install the front run channel bolts.

8   The remainder of installation is the reverse of removal.

Rear door

9   Remove the sash screws while supporting the glass.

10   Disengage the glass and lift it inward and out of the door frame.

11   Remove the vent glass assembly.

12   Lift the glass from the door.

13   To install, insert the glass into the door and connect the channel, making sure to engage the glass guide securely to the channel.

14   The remainder of installation is the reverse of removal.

Deville/Fleetwood models

Front door

15   Remove the trim panel retainer and sealer strip.

16   Remove the rubber stop bumper, screws and front run channel.

17   Lower the glass halfway, slide the window regulator guide block off the sash channel and tilt the glass outboard of the door frame to remove it.

18   To install, insert the glass into the door and engage the regulator guide block to the sash channel. Engage the rear guide clip on the glass to the rear run channel weatherstrip.

19   Raise the glass to the half way up position and install the front run channel bolts finger tight.

20   Engage the front clip on the glass in the front run channel and tighten the bolts securely.

21   Install the rubber stop bumper.

22   The remainder of installation is the reverse of removal.

Rear door

23   Remove the sash screws while supporting the glass.

24   Disengage the glass and lower it to the bottom of the door.

25   Remove the vent glass assembly.

26   Lift the glass from the door.

27   To install, insert the glass into the door and connect the channel, making sure to engage the glass guide securely to the channel.

28   The remainder of installation is the reverse of removal.

20   Door window glass regulator - removal and installation

Refer to illustrations 20.5a and 20.5b

Removal

1   Disconnect the negative cable at the battery.

2   With the window glass in the full up position, remove the door trim panel and water shield (Section 16).

4   Secure the window glass in the up position with strong adhesive tape fastened to the glass and wrapped over the door frame.

5   Punch out the center pins of the rivets that secure the window regulator and drill the

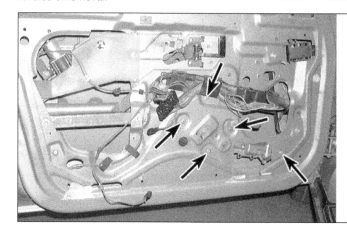

20.5a Drill out the regulator retaining rivets (arrows) - E/K model (Eldorado, Seville, Riviera and Toronado) shown

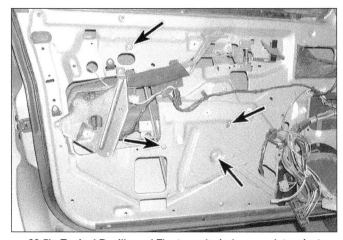

20.5b   Typical Deville and Fleetwood window regulator rivet locations (arrows)

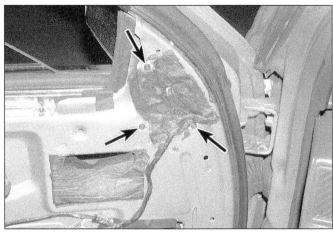

21.3 Remove the screws (arrows) and detach the mirror cover

heads of the rivets off with a 1/4-inch drill bit **(see illustrations)**. **Caution:** *Be careful not to enlarge the holes in the door sheetmetal.*
6   Unplug the electrical connector.
7   Remove the retaining rivets and move the regulator until it is disengaged from the sash channel. Lift the regulator from the door.
8   The window motor can be replaced if necessary. This involves drilling out the rivets securing the motor to the regulator frame. The new motor should come with screws and nuts for attaching the motor to the regulator. **Warning:** *On Eldorado, Seville, Riviera and Toronado models the regulator arms are under extreme spring tension and can cause serious injury if the motor is removed without locking the sector gear to the regulator frame. This can be done by inserting a bolt through one of the holes in the regulator frame and sector gear, then fastening it in place with a nut (if no holes are provided, drill through the sector gear and regulator frame).*

Installation

9   Place the regulator in position in the door and engage it in the sash channel.
10   Secure the regulator to the door using new rivets and a riveting tool. Or, secure the regulator to the door with screws, washers and nuts.
11   Install the bolts and tighten them securely.
12   Plug in the electrical connector (if equipped).
13   Install the water shield, door trim panel and window regulator handle. Connect the negative battery cable.

---

21   Outside mirrors - removal and installation

Refer to illustrations 21.3 and 21.5

1   Disconnect the negative cable from the battery.
2   Remove the door trim panel and the plastic watershield (see Section 16 15).
3   Remove the screws and detach the mir-

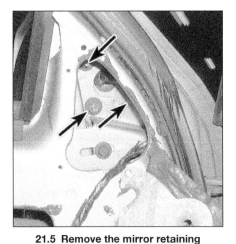

21.5 Remove the mirror retaining nuts (arrows)

ror cover from the door window frame **(see illustration)**.
4   Disconnect the electrical connector from the mirror.
5   Remove the three mirror retaining nuts or bolts and detach the mirror from the vehicle **(see illustration)**.
6   Installation is the reverse of removal.

---

22   Trunk lid - removal, installation and adjustment

Note: *The trunk lid is heavy and somewhat awkward to remove and install - at least two people should perform this procedure.*

Removal and installation

Refer to illustration 22.3

1   Open the trunk lid and cover the edges of the trunk compartment with pads or cloths to protect the painted surfaces when the lid is removed.
2   Disconnect any cables or wire harness connectors attached to the trunk lid that would interfere with removal.
3   Make alignment marks around the trunk

22.3 Before removing the trunk lid, draw marks around the bolt heads to aid in the reinstallation process

lid hinge bolts **(see illustration)**.
4   While an assistant helps you support the lid, remove the hinge bolts from both sides and lift the trunk lid off the vehicle.
5   Installation is the reverse of removal.
Note: *When reinstalling the trunk lid, align the hinge bolts with the marks made during removal.*

Adjustment

Refer to illustrations 22.9 and 22.10

6   Fore-and-aft and side-to-side adjustment of the trunk lid is done by moving the hood in relation to the hinge plate after loosening the bolts or nuts.
7   Scribe a line around the hinge bolt heads as described earlier in this Section so you can judge the amount of movement.
8   Loosen the bolts or nuts and move the trunk lid into correct alignment. Move it only a little at a time. Tighten the hinge bolts or nuts and carefully lower the trunk lid to check the alignment.
9   If necessary after installation, the entire trunk lid latch assembly can be adjusted up and down as well as from side to side on the trunk lid so the lid closes securely and is flush

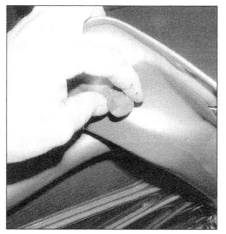

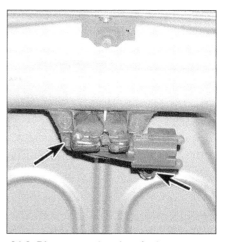

22.9 Loosen the bolts (arrows) and move the latch assembly as necessary to adjust the trunk lid flush with the quarter panels in the closed position

22.10 Adjust the bumpers so they're slightly compressed when the trunk lid is in the closed position

24.2 Disconnect the electrical connector and remove the bolts (arrows), then detach the release actuator

with the rear quarter panels. To do this, scribe a line around the trunk lid latch mounting bolts to provide a reference point. Then loosen the bolts and reposition the latch assembly as necessary **(see illustration)**. Following adjustment, retighten the mounting bolts.

10 Adjust the bumpers on the trunk lid, so that the trunk lid is flush with the rear quarter panels when closed **(see illustration)**.

11 The trunk lid latch assembly, as well as the hinges, should be periodically lubricated with white lithium-base grease to prevent sticking and wear.

23 Trunk lid latch, striker and lock cylinder - removal and installation

Latch and latch striker

1 Unplug any electrical connectors and detach any cables from the latch assembly.

2 The trunk lid latch is retained by bolts which can readily be removed with a wrench

(see illustration 22.9). For adjustment procedures, see Section 22.

3 To access the striker, remove the plastic retaining nuts and clips securing the rear trunk finishing panel.

4 Detach the striker retaining nuts and remove the striker from the vehicle.

5 Installation is the reverse of removal.

Lock cylinder

6 Remove the trunk lid rear trim panel.

7 Using a small drill bit, drill out the rivet securing the lock cylinder retaining clip.

8 Detach the retaining clip, then pull the lock cylinder out of the trunk lid and remove it from the vehicle.

9 Installation is the reverse of removal.

24 Trunk lid release actuator and pull-down unit - removal and installation

Release actuator

Refer to illustration 24.2

1 Open the trunk lid.

2 Disconnect the electrical connector, remove the retaining screws and detach the actuator from the latch **(see illustration)**.

3 Installation is the reverse of removal.

Pull down unit

Refer to illustration 24.5

4 Remove the screws and nut, unplug the electrical connector and detach the pull-down unit from the striker **(see illustration)**.

5 Installation is the reverse of removal.

25 Fuel door solenoid - removal and installation

1 Disconnect the negative cable from the battery.

2 Open the trunk lid and remove the left side trim panel.

3 Open the fuel door, remove the latch assembly screws and disconnect actuating cable.

4 Remove the two retaining bolts, then detach the cable grommet and remove the solenoid and cable assembly from the vehicle.

5 Installation is the reverse of removal.

26 Center console - removal and installation

Warning: *Some models covered by this manual are equipped with Supplemental Restraint Systems (SRS), more commonly known as airbags. Always disable the airbag system before working in the vicinity of any airbag system components to avoid the possibility of accidental deployment of the airbag(s) which could cause personal injury (see Chapter 12).*

1 Apply the parking brake lever and place the gear selector into the neutral position.

2 Pry out the shift lever knob retaining clip and remove the knob.

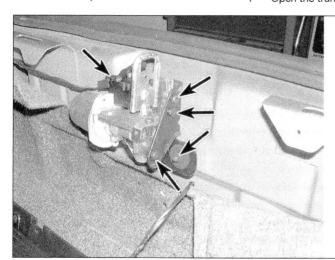

24.5 Remove the pull-down unit screws and nut (right arrows), then disconnect the electrical connector (left arrow) and detach the unit

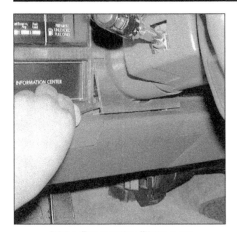

27.5a Pry out the filler panel (some models) . . .

27.5b . . . and remove the screws (arrows) from the top . . .

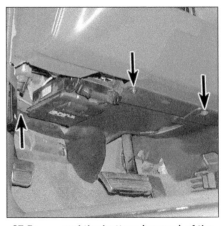

27.5c . . . and the bottom (arrows) of the trim cover

3   Pry out the gear selector trim bezel by disengaging the clips on the front edge. Then disconnect the electrical connectors and remove the bezel from the console assembly.

4   Remove the retaining screws securing the front half of the console.

5   Working in the console glove box, detach the retaining screws and nuts securing the rear half of the console.

6   Lift the console up and over the shift lever. Disconnect any electrical connections and remove the console from the vehicle.

7   Installation is the reverse of removal.

---

27   Dashboard trim panels - removal and installation

Warning: *Some models covered by this manual are equipped with Supplemental Restraint Systems (SRS), more commonly known as airbags. Always disable the airbag system before working in the vicinity of any airbag system components to avoid the possibility of accidental deployment of the airbag(s) which could cause personal injury (see Chapter 12).*

Instrument cluster bezel

1   Tilt the steering wheel down to the lowest position.

2   Remove the screws, grasp the bezel securely and pull back sharply to detach any clips from the instrument panel. On Deville/Fleetwood models the instrument cluster and bezel are combined into a single unit (see Chapter 12). It will be necessary to remove the instrument panel pad (see Section 28) for access to the instrument cluster.

3   Unplug any electrical connectors that interfere with removal.

4   Installation is the reverse of removal.

Lower steering column trim cover

Refer to illustrations 27.5a, 27.5b and 27.5c

5   Remove the steering column filler panel (some models) then remove the steering column cover retaining screws **(see illustrations)**.

6   Unplug any electrical connectors, then lower the trim panel from the instrument panel.

7   Installation is the reverse of removal.

Knee bolster

Refer to illustration 27.9

8   Remove the lower steering column trim cover as described above.

9   Detach the knee bolster retaining screws and remove it from the vehicle **(see illustration)**. On some models, the bolster is a press fit and can be removed by pulling it straight back.

10   Installation is the reverse of removal.

Lower insulator panels

Refer to illustration 27.11

11   Remove the bolts, detach the panels and lower them from the dash **(see illustration)**.

12   Installation is the reverse of removal.

Glove box

Refer to illustration 27.13

13   To remove the glove box, simply remove the screws and detach the glove box from the instrument panel **(see illustration)**.

14   Installation is the reverse of the removal procedure.

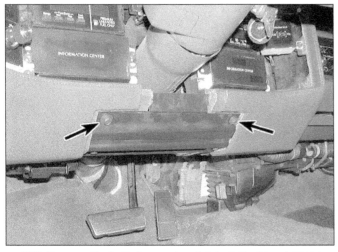

27.9 Remove the bolts (arrows) and detach the knee bolster

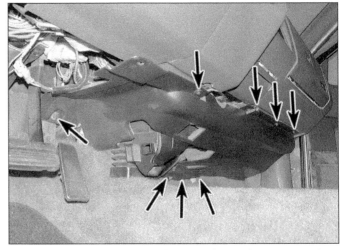

27.11 Remove the bolts and fasteners (arrows) and detach the insulator panels

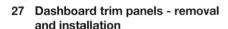

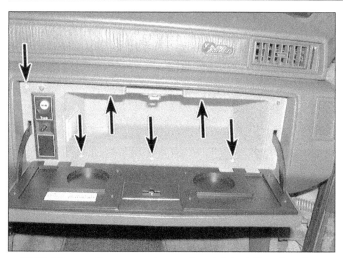

27.13 Remove the screws (arrows) and detach the glove box

28.2 Remove the retaining screws (arrows) along the top of the trim pad

28 Instrument panel upper trim pad (Deville/Fleetwood models) - removal and installation

Refer to illustrations 28.2, 28.3a, 28.3b, 28.4 and 28.5

Warning: *The models covered by this manual are equipped with Supplemental Restraint Systems (SRS), more commonly known as airbags. Always disable the airbag system before working in the vicinity of any airbag system components to avoid the possibility of accidental deployment of the airbag(s) which could cause personal injury* (see Chapter 12).

1   Disconnect the negative cable from the battery.
2   Remove the screws along the top of the trim pad **(see illustration)**.
3   Detach the air vent grilles and remove the instrument panel trim pad screws **(see illustrations)**.
4   Remove the glove box (see Section 27)

28.3a Use needle-nose pliers to detach the air vent grilles

and remove the instrument panel-to-pad screws **(see illustration)**.
5   Detach the trim pad, disconnect the

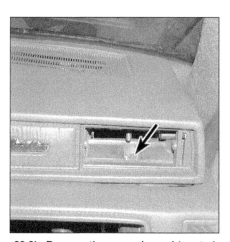

28.3b Remove the screw (arrow) located in each air vent grille opening

electrical connectors and lift the pad off **(see illustration)**.
6   Installation is the reverse of removal.

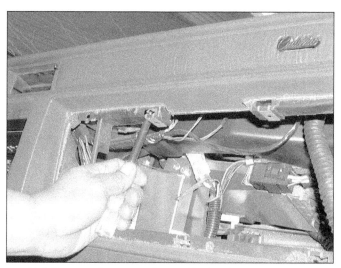

28.4 Remove the screws located in the glove box opening

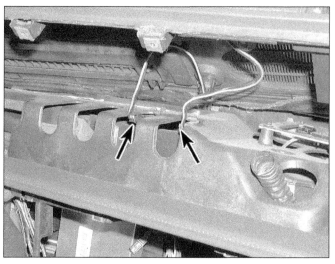

28.5 Lift the pad up and disconnect the electrical connectors (arrows)

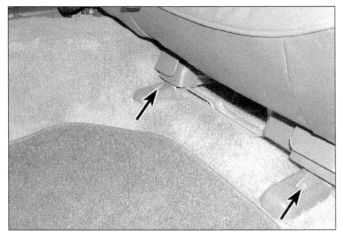

29.2a Remove the screws (arrows) at the front . . .

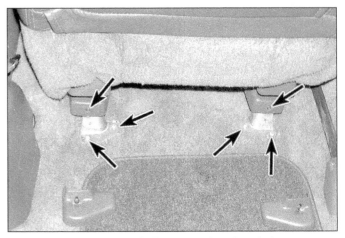

29.2b . . . and rear (arrows) of the seat, then detach the covers . . .

29 Seats - removal and installation

29.2c . . . and remove the retaining nuts (arrows)

Front seat

Refer to illustration 29.2a, 29.2b and 29.2c

1   Position the seat all the way forward or all the way to the rear to access the front seat retaining bolts.

2   Detach the trim covers and remove the retaining nuts **(see illustrations)**.

3   Tilt the seat upward to access the underside, then disconnect any electrical connectors and lift the seat from the vehicle.

4   Installation is the reverse of removal.

Rear seat

5   Push the bottom front of the seat to the rear while pushing down slightly. Then lift up on the front edge and remove the cushion from the vehicle.

6   Detach the retaining bolts at the lower edge of the seat back, then lift up on the lower edge of the seat back and remove it from the vehicle.

7   Installation is the reverse of removal.

Chapter 12
Chassis electrical system

Contents

1   General information

The electrical system is a 12-volt, negative ground type. Power for the lights and all electrical accessories is supplied by a lead/acid-type battery, which is charged by the alternator.

This Chapter covers repair and service procedures for the various electrical components not associated with the engine. Information on the battery, alternator, distributor and starter motor can be found in Chapter 5. It should be noted that when portions of the electrical system are serviced, the negative battery cable should be disconnected from the battery to prevent electrical shorts and/or fires.

Warning: *On models equipped with a Supplemental Inflatable Restraint (SIR) system (more commonly known as airbags), always disable the airbag system before working in the vicinity of any airbag system component to avoid the possibility of accidental deployment of the airbag, which could cause personal injury (see Section 30).*

2   Electrical troubleshooting

Refer to illustration 2.15

A typical electrical circuit consists of an electrical component, any switches, relays, motors, fuses, fusible links or circuit breakers related to that component and the wiring and connectors that link the component to both the battery and the chassis. To help you pinpoint an electrical circuit problem, wiring diagrams are included at the end of this Chapter.

Before tackling any troublesome electrical circuit, first study the appropriate wiring diagrams to get a complete understanding of what makes up that individual circuit. Trouble spots, for instance, can often be narrowed down by noting if other components related to the circuit are operating properly. If several components or circuits fail at one time, chances are the problem is in a fuse or ground connection, because several circuits are often routed through the same fuse and ground connections.

Electrical problems usually stem from simple causes, such as loose or corroded connections, a blown fuse, a melted fusible link or a failed relay. Visually inspect the condition of all fuses, wires and connections in a problem circuit before troubleshooting the circuit.

If test equipment and instruments are going to be utilized, use the diagrams to plan ahead of time where you will make the necessary connections in order to accurately pinpoint the trouble spot.

The basic tools needed for electrical troubleshooting include a circuit tester or voltmeter (a 12-volt bulb with a set of test leads can also be used), a continuity tester, which includes a bulb, battery and set of test leads, and a jumper wire, preferably with a circuit breaker incorporated, which can be used to bypass electrical components. Before attempting to locate a problem with test instruments, use the wiring diagram(s) to decide where to make the connections.

Voltage checks

Voltage checks should be performed if a circuit is not functioning properly. Connect one lead of a circuit tester to either the negative battery terminal or a known good ground. Connect the other lead to a connector in the circuit being tested, preferably nearest to the battery or fuse. If the bulb of the tester lights, voltage is present, which means that the part of the circuit between the connector and the battery is problem free. Continue checking the rest of the circuit in the same fashion. When you reach a point at which no voltage is present, the problem lies between that point and the last test point with voltage. Most of the time the problem can be traced to a loose connection. **Note:** *Keep in mind that some circuits receive voltage only when the ignition key is in the Accessory or Run position.*

Finding a short

One method of finding shorts in a circuit is to remove the fuse and connect a test light or voltmeter in place of the fuse terminals. There should be no voltage present in the circuit. Move the wiring harness from side-to-side while watching the test light. If the bulb goes on, there is a short to ground somewhere in that area, probably where the insulation has rubbed through. The same test can be performed on each component in the circuit, even a switch.

Ground check

Perform a ground test to check whether a component is properly grounded. Disconnect the battery and connect one lead of a self-powered test light, known as a continuity tester, to a known good ground. Connect the other lead to the wire or ground connection being tested. If the bulb goes on, the ground is good. If the bulb does not go on, the ground is not good.

Continuity check

A continuity check is done to determine if there are any breaks in a circuit - if it is passing electricity properly. With the circuit off (no power in the circuit), a self-powered continuity tester can be used to check the circuit. Connect the test leads to both ends of the circuit (or to the "power" end and a good ground), and if the test light comes on the circuit is passing current properly. If the light doesn't come on, there is a break somewhere in the circuit. The same procedure can be used to test a switch, by connecting the continuity tester to the switch terminals. With the switch turned On, the test light should come on.

Finding an open circuit

When diagnosing for possible open circuits, it is often difficult to locate them by sight because oxidation or terminal misalignment are hidden by the connectors. Merely wiggling a connector on a sensor or in the wiring harness may correct the open circuit condition. Remember this when an open circuit is indicated when troubleshooting a circuit. Intermittent problems may also be caused by oxidized or loose connections.

Electrical troubleshooting is simple if you keep in mind that all electrical circuits are basically electricity running from the battery, through the wires, switches, relays, fuses and fusible links to each electrical component (light bulb, motor, etc.) and to ground, from which it is passed back to the battery. Any electrical problem is an interruption in the flow of electricity to and from the battery.

Connectors

Most electrical connections on these vehicles are made with multiwire plastic connectors. The mating halves of many connectors are secured with locking clips molded into the plastic connector shells. The mating halves of large connectors, such as some of those under the instrument panel, are held together by a bolt through the center of the connector.

To separate a connector with locking clips, use a small screwdriver to pry the clips apart carefully, then separate the connector halves. Pull only on the shell, never pull on the wiring harness as you may damage the individual wires and terminals inside the connectors. Look at the connector closely before trying to separate the halves. Often the locking clips are engaged in a way that is not immediately clear. Additionally, many connectors have more than one set of clips.

Each pair of connector terminals has a male half and a female half. When you look at the end view of a connector in a diagram, be sure to understand whether the view shows the harness side or the component side of the connector. Connector halves are mirror images of each other, and a terminal shown on the right side end view of one half will be on the left side end view of the other half.

It is often necessary to take circuit voltage measurements with a connector connected. Whenever possible, carefully insert the test probes of your meter into the rear of the connector shell to contact the terminal inside. This kind of connection is called "backprobing" **(see illustration)**. When inserting a test probe into a male terminal, be careful not to distort the terminal opening. Doing so can lead to a poor connection and corrosion at that terminal later.

Troubleshooting a typical circuit

Most of your troubleshooting will center on discovering why something on the vehicle doesn't work, for example, a power window. A typical circuit will encompass a motor (or other device at the end of the circuit such as a light), a switch, a relay and a fuse.

Begin by checking for power and ground at the device end of the circuit, i.e. at the motor or light. Refer to the wiring diagrams at the end of this Chapter for the circuit diagram, terminals and wire colors. Disconnect the connector at the device and, on the harness side of the connector, clip a 12-volt test light to the ground side of the connector and probe the power side of the con-

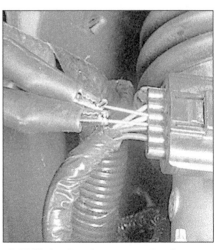

2.15 To backprobe a connector, insert a small, sharp probe (such as a T-pin) into the back of the connector alongside the desired wire until it contacts the metal terminal inside; connect your meter leads to the probes - this allows you to test a functioning circuit

nector with the switch turned On. If the test light glows brightly, both power and ground circuits to the device are good.

If there is power at the connector on the device, but the device doesn't operate when the switch is turned On, disconnect the electrical connector at the device and ground the ground side terminal of the device with a jumper wire to a known-good chassis ground. Attach a fused jumper wire (the fuse should have a rating the same as the fuse that normally feeds that device) to the power supply terminal on the device. If it's doesn't operate, that device, motor or light is faulty.

If the device is OK, but the test light did not light at the connector, trace the circuit according to the wiring diagram and check the fuse that is the power source for the circuit. If the fuse is good, check for power at the relay that sends power to the device. If there is power to the relay, refer to Section 6 and test the relay. If the relay is good, check the continuity of wiring from the relay to the switch, and from the relay to the device.

In a process of elimination, you will find an open or short in the circuit, or a faulty switch.

---

3 Fuses - general information

Refer to illustrations 3.1a, 3.1b, 3.1c, 3.1d, 3.1e and 3.3

1   The electrical circuits of the vehicle are protected by a combination of fuses and circuit breakers. Most models covered by this manual have a fuse/relay panel in the engine compartment, plus a fuse/relay panel (or a separate fuse panel and a relay panel) in the passenger compartment **(see illustrations)**.

2   Each of the fuses is designed to protect a specific circuit (or circuits), and the various

3.1a On most models, the underhood fuse/relay panel is located in the left-front of the engine compartment

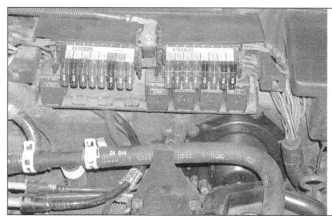

3.1b On Deville and Fleetwood models, a large underhood fuse/relay panel is located at the center of the firewall (shown with plastic cover removed)

3.1c On Deville and Fleetwood models, the interior fuse panel is located under a plastic cover at the left end of the dashboard (cover shown removed) - pull the latch under the dash and the panel drops down for access - pull the latch again when pushing the fuse panel back in place

3.1d On Seville, Riviera and Toronado models, the interior fuse/relay panel is located under the right side of the dash, below the glove box (shown with lower dash sound panel removed) - on some models, this fuse/relay panel is located in the right side of the console

circuits are identified on the fuse panel cover or the fuse panel itself.

3   Miniaturized fuses are employed in the fuse block. These compact fuses, with blade terminal design, allow fingertip removal and replacement. If an electrical component fails, always check the fuse first. The easiest way to check fuses is with a test light. Check for power at the exposed terminal tips of each fuse. If power is present on one side of the fuse but not the other, the fuse is blown. When removed, a blown fuse is easily identified through the clear plastic body. Visually inspect the element for evidence of damage **(see illustration)**.

4   Be sure to replace blown fuses with the correct type. Fuses of different ratings are physically interchangeable, but only fuses of the proper rating should be used. Replacing a fuse with one of a higher or lower value than specified is not recommended. Each electrical circuit needs a specific amount of protection. The amper-

age value of each fuse is molded into the fuse body.

5   If the replacement fuse immediately fails, don't replace it again until the cause of

the problem is isolated and corrected. In most cases, the cause will be a short circuit in the wiring caused by a broken or deteriorated wire.

3.1e On early Seville/Eldorado models, an interior fuse/circuit breaker panel is located just under the glove compartment lid, under a cover

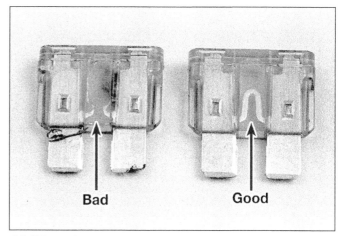

3.3 When a fuse blows, the element between the terminals melts - the fuse on the left is blown, the fuse on the right is good

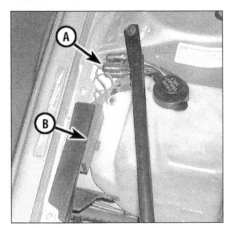

4.1 Fusible links (A) on most models are located next to the isolated power block (B), usually on the right fenderwell or next to the fuse relay panel

4   Fusible links - general information

Refer to illustration 4.1

Some circuits are protected by fusible links. Fusible links are circuit protection devices that are part of the wiring harness itself, that are designed to melt and open the circuit when a short causes excessive current flow. Fusible links on these models are on the right (passenger's side) of the engine compartment, on the inner wheelwell, or next to the fuse/relay panel **(see illustration)**. Fusible links are used in circuits that are not ordinarily fused, such as the ignition circuit.

Although the fusible links appear to be a heavier gauge than the wire they are protecting, the appearance is due to the thick insulation. All fusible links are several wire gauges smaller than the wire they are designed to protect.

Fusible links cannot be repaired, but a new link of the same size wire can be put in its place. The procedure is as follows:

a) *Disconnect the negative cable from the battery.*
b) *Disconnect the fusible link from the wiring harness.*
c) *Cut the damaged fusible link out of the wiring just behind the electrical connector.*
d) *Strip the insulation back approximately 1/2-inch.*
e) *Position the electrical connector on the new fusible link and crimp it into place.*
f) *Use rosin-core solder at each end of the new link to obtain a good solder joint.*
g) *Use plenty of electrical tape around the soldered joint. No wires should be exposed.*
h) *Connect the battery ground cable. Test the circuit for proper operation.*

5   Circuit breakers - general information

Circuit breakers protect components such as sunroof motors, power seat motors, power window motors and airbag inflator resistors.

On some models the circuit breaker resets itself automatically, so an electrical overload in a circuit-breaker-protected system will cause the circuit to fail momentarily, then come back on. If the circuit does not come back on, check it immediately. Once the condition is corrected, the circuit breaker will resume its normal function. Some circuit breakers must be reset manually. On the covered models, most circuit breakers are located in the underhood or interior fuse/relay panels (see Section 3).

6   Relays - general information and testing

General information

1   Several electrical components, such as the fuel pump, engine cooling fans and air conditioning compressor clutch use relays to transmit the electrical power to the component. Relays use a low-current circuit (the control circuit) to open and close a high-current circuit (the power circuit). If the relay is defective, that component will not operate properly. The various relays are mounted in engine compartment and interior fuse/relay panels **(see Section 3)** and several locations throughout the vehicle. If a faulty relay is suspected, it can be removed and tested using the procedure below or by a dealer service department or a repair shop. Defective relays must be replaced as a unit. Consult the wiring diagrams at the end of this Chapter for relay usage.

Testing

Refer to illustrations 6.4a and 6.4b

2   It's best to refer to the wiring diagram for the circuit to determine the proper hook-ups for the relay you're testing. However, if you're not able to determine the correct hook-up from the wiring diagrams, you may be able to determine the test hook-ups from the information that follows.

3   On most relays, two of the terminals are the relay's control circuit (they connect to the relay coil which, when energized, closes the large contacts to complete the circuit). The other terminals are the power circuit (they are connected together when the relay when the control-circuit coil is energized).

4   Most relays are marked as an aid to help you determine which terminals are the control circuit and which are the power circuit **(see**

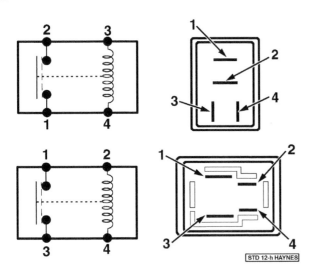

6.4a Typical four-terminal relay designs, terminal numbering and internal circuitry

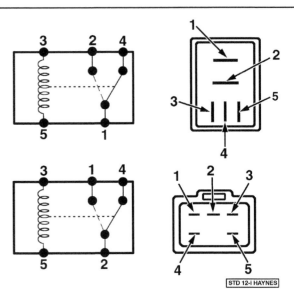

6.4b Typical five-terminal relay designs, terminal numbering and internal circuitry

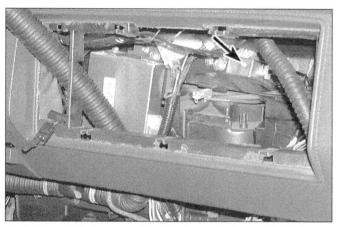

7.3a The Body Computer Module (arrow) is located behind the glove box - Deville model shown

7.3b On some models, such as this Seville model, the BCM (arrow) is under the center of the dash - remove the center underdash insulator panel for access

illustrations). Most the relays have a symbol on top, such as a fan, light, horn or other identification as to the circuit it controls, in addition to the identification on the schematic on the inside of the fuse-relay panel cover.

5   Connect a fused jumper wire between one of the two control circuit terminals and the positive battery terminal. Connect another jumper wire between the other control circuit terminal and ground. When the connections are made, the relay should click. On some relays, polarity may be critical, so, if the relay doesn't click, try swapping the jumper wires on the control circuit terminals.

6   With the jumper wires connected, check for continuity between the power circuit terminals as indicated by the markings on the relay.

7   If the relay fails the above tests, replace it.

7   Body Computer Module (BCM) - general information and replacement

Warning: *On models equipped with a Supplemental Inflatable Restraint (SIR) system (more commonly known as airbags), always disable the airbag system before working in the vicinity of any airbag system component to avoid the possibility of accidental deployment of the airbag, which could cause personal injury (see Section 30).*

General information

1   The vehicles covered by this manual have a complex electrical system. There are several "computers", modules and electrical subsystems, but the Body Computer Module (BCM) is the central communication device for all of the various components. The BCM receives input from the ECM (which controls the powertrain) and other sources, and communicates to the ECM, the air conditioning/heating system programmer, the instrument panel and the airbag system, among other devices.

2   The BCM is also the source for in-depth troubleshooting information about many systems on the vehicle, especially the climate control components. Diagnostic codes can be retrieved through the display on the electronic climate control panel. Refer to Chapter 6 for the code listings and the code access procedures.

Replacement

Refer to illustrations 7.3a and 7.3b

3   The BCM is located behind the glove compartment, at the right end of the instrument panel, on all models except the 1986 and 1987 Riviera models **(see illustrations)**. On 1986 and 1987 Riviera models, it is located in the floor console.

4   Disconnect the negative battery cable.

5   Refer to Chapter 11 and remove the glove box door and glove box liner.

6   Disconnect the two large electrical connectors at the BCM.

7   Remove the screw securing the BCM to the mounting bracket and remove the BCM. **Note:** *On some models, an accessory relay bracket may have to be set aside for BCM removal.*

8   Installation is the reverse of the removal procedure. **Note:** *The BCM contains two PROMs that must be transferred.*

PROM transfer

9   The PROMs in the BCM contain information specific to you vehicle. When a new BCM is purchased, it will come without PROMs, although it will come with a special PROM-removal tool so you can transfer the PROMs from your old BCM to the new one. **Caution:** *The factory tool must be used to remove or install a PROM.*

10   The PROMs are located under a cover panel. Remove the screws and the cover.

11   Using the tool, rock the PROM carrier by pushing on the top bar of the tool. The PROM pins are very delicate, so work carefully.

12   When installing the new PROM, make sure it is oriented the same way as the old one. If it is installed backward, the PROM will be ruined the first time the ignition is turned on.

13   Some models are equipped with an

EEPROM, which contains information such as the mileage of the vehicle. An EEPROM is not removable; you must take the vehicle to a GM dealer and have the EEPROM calibrated to your vehicle with special equipment.

14   Installation is the reverse of the removal procedure. Be sure to reinstall the PROM cover and screws before installing the BCM.

8   Turn signal/hazard flasher - check and replacement

Refer to illustrations 8.1a and 8.1b
Warning: *On models equipped with a Supplemental Inflatable Restraint (SIR) system (more commonly known as airbags), always disable the airbag system before working in the vicinity of any airbag system component to avoid the possibility of accidental deployment of the airbag, which could cause personal injury (see Section 30).*

1   The turn signal and hazard flashers are controlled either from a turn signal/hazard module (all Seville/Eldorado models and 1986 through 1989 Riviera models) or from two electronic flasher units (Deville/Fleetwood models, Toronado models, and 1990

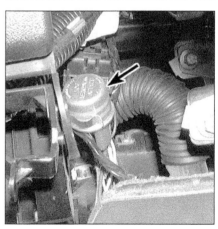

8.1a Turn signal flasher location (arrow), next to the fuse panel at the lower left of the steering column . . .

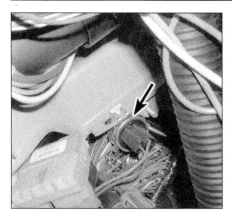

8.1b ... while the hazard flasher (arrow)
is located to the right of the
steering column

9.3 Disconnect the turn signal connector
on the firewall (arrow) or at the base of
the steering column (depending on model)
to test the switch circuit

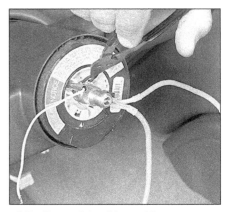

9.7a On models with an airbag, remove
the snap-ring and lift the airbag coil off
the steering column shaft

and later Riviera models). The turn signal/hazard module is located under the center console. The turn signal/hazard flashers are mounted near the steering column **(see illustrations)**. The turn signal/hazard module is a solid-state device and cannot be checked using conventional methods, have the turn signal/hazard module checked at a dealer service department or other properly equipped repair facility.

2   When the flasher unit is functioning properly, an audible click can be heard during its operation. If the turn signals fail on one side or the other and the flasher unit does not make its characteristic clicking sound, a faulty turn signal bulb is indicated.

3   If both turn signals fail to blink, the problem may be due to a blown fuse, a faulty flasher unit, a broken switch or a loose or open connection. If a quick check of the fuse box indicates that the turn signal fuse has blown, check the wiring for a short before installing a new fuse.

4   To replace the flasher, simply unplug it from its electrical connector.

5   Make sure that the replacement unit is identical to the original. Compare the old one to the new one before installing it.

6   Installation is the reverse of removal.

---

9   Steering column switches - check and replacement

Warning: *On models equipped with a Supplemental Inflatable Restraint (SIR) system (more commonly known as airbags), always disable the airbag system before working in the vicinity of any airbag system component to avoid the possibility of accidental deployment of the airbag, which could cause personal injury (see Section 30).*

Check

Refer to illustration 9.3

1   Within the span of covered models, two types of switch arrangements have been used. On some models, the turn signal lever controls only the turn signals, while on other

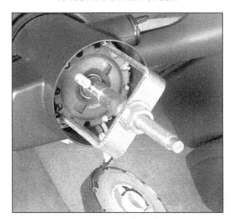

9.7b Use a special tool (available at most
auto parts stores) to depress the lock
plate, then pry out the retaining ring

models, the cruise control and wiper functions are controlled by the same lever. Some models use small flasher units in the turn signal and hazard circuits, and others use a turn signal/hazard module (see Section 8). The procedure for testing the turn signal switch is similar for all models.

2   If one or both turn signals fail to operate when the key is On and the turn signal lever is actuated, check the bulbs first (see Section 20). Refer to the wiring diagrams at the end of this Chapter and examine the turn signal circuit. Check the fuse(s).

3   To check the turn signal switch itself, refer to Chapter 11 and remove the driver's knee bolster for access to the bottom of the steering column. The turn signal switch has a wiring harness attached to it that goes though the steering column and ends at a connector near the base of the steering column **(see illustration)**.

4   Disconnect the connector and use an ohmmeter to check the continuity between terminals on the turn signal switch side of that connector. The circuits make a loop, from the vehicle electrical system to the connector, up to the turn signal switch, and back to the system through different terminals of

9.7c Lift the lockplate out

the same connector. Thus, to check the turn signal switch, check the wiring diagrams and look for continuity (at the connector) between the wires from each side of the switch. There should be continuity between one pair with the lever in the Left turn position, and continuity between another pair in the Right turn position. If there isn't, replace the switch. On models with cruise control, headlight dimming and wiper functions on the turn signal lever, refer to the wiring diagrams for those circuits and test for continuity with the switch in each position at the connector.

Turn signal switch

Removal

Refer to illustrations 9.7a, 9.7b, 9.7c, 9.8, 9.9a, 9.9b, 9.11 and 9.14

5   Detach the cable from the negative battery terminal (see Chapter 1) and disable the airbag system (see Section 30).

6   Remove the steering wheel (see Chapter 10).

7   On models equipped with a driver's airbag, remove the airbag coil retaining snap-ring and remove the coil assembly; let the coil hang by the wiring harness **(see illustration)**. Remove the wave washer. Using a lock plate removal tool, depress the lock plate for access to the retaining ring **(see illustra-**

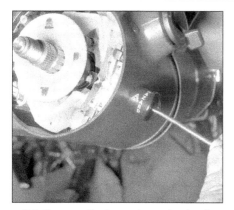

9.8 Remove the hazard warning knob screw (arrow) and the knob

9.9a Remove the cancel cam . . .

9.9b . . . and the spring

9.11 Remove the screw and the turn signal lever actuating arm (shown here with screwdriver), then remove the switch mounting screws (arrows) - to get at the upper screw in this photo, the top of the switch must be clicked down

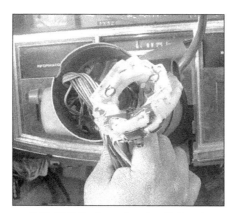

9.14 Pull the turn signal switch out, carefully drawing the wiring harness through the steering column

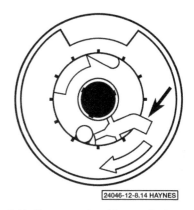

9.19 To center the airbag coil, depress the spring lock (arrow); rotate the hub in the direction of the arrow until it stops; back the hub off 2-1/2 turns and release the spring lock

9.20 When properly installed, the airbag coil will be centered with the marks aligned (circle) and the tab fitted between the projections on the top of steering column (arrow)

tions). Use a small screwdriver to pry the retaining ring out of the groove in the steering column and remove the lock plate.

8   Use a small Phillips head screwdriver to remove the hazard warning knob **(see illustration)**.

9   Remove the turn signal cancel cam and the spring **(see illustrations)**.

10   Remove the turn signal lever actuating arm screw and remove the actuating arm.

11   Remove the turn signal switch mounting screws **(see illustration)**.

12   Remove the knee bolster panel below the steering column (see Chapter 11).

13   Locate the turn signal switch electrical connector and unplug it **(see illustration 9.3)**. Remove the wiring protector. **Note:** *On certain models it may be necessary to remove the terminals for the key reminder buzzer from the turn signal switch connector. To do this, insert a narrow pin into the electrical connector, from the side opposite the wire, depress the tang on the terminal and pull the wire out* **(see illustration 10.6b for technique)**. *Be sure the wires get installed into their proper positions upon reassembly.*

14   Attach a length of mechanic's wire to the electrical connector (this will aid in the installation of the new switch). Pull the wiring harness

and electrical connector up through the steering column and remove the switch assembly **(see illustration)**. **Note:** *On some models it may be necessary to remove the steering column mounting bracket to provide clearance for the electrical connector.*

Installation

Refer to illustrations 9.19 and 9.20

15   Attach the mechanic's wire to the electrical connector of the new switch. Slowly pull the wiring harness through the steering column. **Note:** *If the column mounting bracket was removed, install it and tighten the bracket-to-column bolts to 22 ft-lbs. Tighten the bracket-to-instrument panel bolts or nuts to 20 ft-lbs.*

16   Seat the turn signal switch on the column and install the turn signal switch mounting screws and lever actuating arm. Install the hazard knob.

17   Connect the switch electrical connector and reinstall the wiring protector.

18   Install the spring, cancel cam and the lock plate. Depress the lock plate and install the retaining ring.

19   If necessary on airbag-equipped models, center the airbag coil as follows (it will only become uncentered if the spring lock is depressed and the hub rotated with the coil off the column) **(see illustration)**: **Note:** *Make sure the front wheels are pointing straight*

ahead before installing the airbag coil.

a) Turn the coil over and depress the spring lock.

b) Rotate the hub in the direction of the arrow until it stops.

c) Rotate the hub in the opposite direction 2-1/2 turns and release the spring lock.

20   Install the wave washer and the airbag coil **(see illustration)**. Pull the slack out of the

9.24 Gently pry off the cover behind the multifunction lever

9.25 Disconnect the cruise-control electrical connector (arrow)

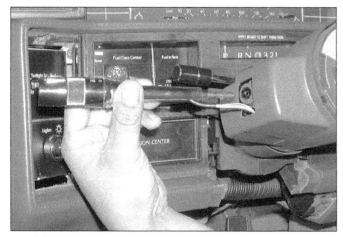

9.26 To remove the multi-function lever, grasp it firmly and pull straight out

9.28 Remove the three Torx bolts (arrows) and pull the cover off the steering column

airbag coil lower wiring harness to keep it tight through the steering column, or it may be cut when the steering wheel is turned. Install the airbag coil retaining snap-ring.

21 The remainder of installation is the reverse of removal.

Wiper/washer switch (column-mounted switch)

Removal

Refer to illustrations 9.24, 9.25, 9.26, 9.28 and 9.29

22 The wiper/washer functions are controlled by a switch mounted on the steering column under the column cover assembly. The multi-function lever controls the wiper switch. Only the cruise control switch is contained within the multi-function lever.

23 Refer to the procedure above and remove the turn signal switch. It isn't necessary to pull the switch wiring harness and connector out of the column, just let the switch hang from the column by the wires. Refer to Section 10 and remove the ignition key lock cylinder.

24 Remove the plastic cover behind the multi-function lever at the left side of the

steering column **(see illustration)**.

25 Disconnect the cruise control wiring connector at the column, just below the tilt lever **(see illustration)**. If necessary, remove the tilt lever by turning it counterclockwise.

26 Grasp the multi-function lever and pull it straight out of the wiper switch/pivot assembly, fishing the cruise control wires out with it **(see illustration)**.

27 Disconnect the wiper switch wire connector at the base of the steering column near the turn signal switch connector **(see illustration 9.3)**. Attach a length of mechanics wire to the electrical connector (this will aid in the installation of the new switch).

28 Remove the three bolts and pull the steering column cover away from the column **(see illustration)**.

29 Remove the pivot pin from the cover assembly **(see illustration)**. Remove the wiper switch/pivot assembly and pull the wiring harness and connector through the lower steering column shroud.

Installation

30 Attach the mechanic's wire to the electrical connector of the new switch. Slowly pull the wiring harness through the steering col-

umn and install the switch/pivot assembly into the cover.

31 Insert the pivot pin through the switch and into the cover.

32 Connect the switch wire connector at

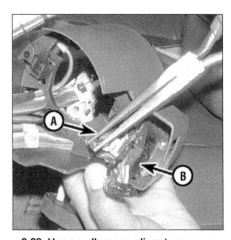

9.29 Use needle-nose pliers to remove the pivot pin (A), then remove the wiper switch/pivot assembly (B) from the steering column cover

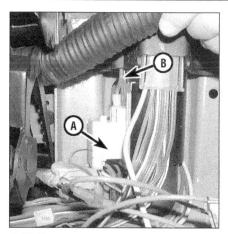

9.35 Remove the nut and screw securing the dimmer switch (A) to the steering column, and withdraw the switch from the actuator rod (B)

10.4 Pull the buzzer switch out of the column with needle-nose pliers

10.5 Remove the lock cylinder retaining screw

the base of the steering column.

33   The remainder of the installation is the reverse of the removal procedure. Make sure the front wheels are pointing straight ahead and the airbag coil is centered (as described in the turn signal switch installation procedure) before installing it.

Headlight dimmer switch

Refer to illustration 9.35

34   The dimmer switch is located at the base of the steering column, just below the ignition switch. If continuity tests reveal a problem with the dimmer switch, disconnect the negative battery cable and remove the trim panel and knee bolster from below the steering column.

35   Remove the nut from the stud, remove the screw and pull out the dimmer switch, disengaging it from the dimmer switch actuator rod **(see illustration)**.

36   Follow the dimmer switch wiring harness to the bulkhead connector at the base of the column.

37   Withdraw the dimmer switch and wiring harness.

38   Install the new dimmer switch leaving the nut/screw loose and adjust it as follows:
 a) *Insert a 3/32-inch drill bit into the adjustment hole provided on the dimmer switch.*
 b) *Slide the dimmer switch toward the actuating rod, removing all clearance between the switch and rod.*
 c) *Tighten the nut/screw and remove the drill bit.*

39   The remainder of installation is the reverse of removal.

---

10   Ignition switch and key lock cylinder - check and replacement

Warning: *On models equipped with a Supplemental Inflatable Restraint (SIR) system (more commonly known as airbags), always disable the airbag system before working in*

10.6a Pull the lock cylinder straight out of the steering column

the vicinity of any airbag system component to avoid the possibility of accidental deployment of the airbag, which could cause personal injury (see Section 30).

Lock cylinder

Refer to illustrations 10.4, 10.5, 10.6a and 10.6b

1   The lock cylinder is located on the upper right-hand side of the steering column.

2   Disconnect the negative battery cable (see Chapter 1). Refer to Section 30 and disable the airbag system (if equipped).

3   Remove the steering wheel (see Chapter 10) and turn signal switch (see Section 9). **Note:** *The turn signal switch need not be fully removed provided that it can be slipped over the end of the shaft. Do not pull the harness out of the column.*

4   Insert the key and place the lock cylinder in the Run position. Using needle-nose pliers, remove the buzzer switch **(see illustration)**.

5   Remove the lock cylinder retaining screw **(see illustration)**.

6   Remove the lock cylinder by pulling the assembly straight out of the column **(see illustration)**. On models equipped with a

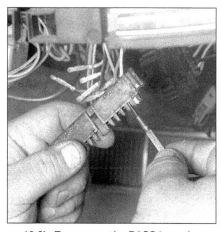

10.6b To remove the PASS key wire harness terminals from the connector, insert a small screwdriver or pin into the front of the connector, depress the retaining tab (part of the terminal) and pull the wire and terminal out from the backside

PASS key security system, remove the PASS key wiring harness as follows:
 a) *Note exactly how the wires are routed through the column and remove the wiring harness retaining clips and wire protector.*
 b) *Follow the PASS key wires at the base of the steering column to the turn signal electrical connector, disconnect the electrical connector and remove the PASS key wire terminals from the connector by inserting a small screwdriver into the connector to depress the terminal retaining tab **(see illustration)**.*
 c) *Tape the wire terminals to a section of mechanics wire and pull the wire harness and mechanics wire up through the steering column. Remove the terminals from the mechanics wire but leave the mechanics wire in place.*

7   Insert the ignition key into the new lock cylinder and install the lock cylinder, aligning the tab on the cylinder with the keyway in the steering column housing.

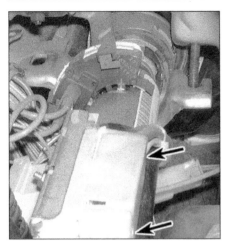

10.13 After the steering column has been lowered, remove the ignition switch wiring cover screws (arrows) - with the cover off remove the switch-to-column screws

8   Push the lock all the way in and install the retaining screw. If equipped with a PASS key system, tape the wire terminals to the mechanics wire used in Step 6 and pull the wiring harness through the steering column. Route the wiring harness exactly as originally installed, insert the wiring terminals into the electrical connector at the base of the column and connect the connector. Install the retaining clips and wire protector. **Note:** *Make sure the PASS key wiring harness terminals are properly installed in the connector. If they are not properly installed, the vehicle will not start on completion of the procedure.*
9   Install the remaining components, referring to the appropriate Sections.

Ignition switch

Refer to illustration 10.13
10   The ignition switch is located on the lower portion of the steering column and is activated by a rod connected to the ignition lock cylinder. Disconnect the negative cable at the battery (see Chapter 1). Refer to Sec-

tion 30 and disable the airbag system, if equipped.
11   Place the ignition switch in the Lock position. If the key lock cylinder has been removed, pull the actuating rod up until a definite stop can be felt, then move it down one detent.
12   Remove the trim panel under the steering column and the knee bolster. Remove the steering column bracket-to-instrument panel bolts or nuts and lower the steering column for access to the ignition switch.
13   Remove the ignition switch retaining screws and remove the ignition switch **(see illustration)**. Disconnect the ignition switch harness connector from the ignition switch.
14   Prior to installation, make sure the ignition switch is in the Lock position.
15   Connect the actuating rod to the switch.
16   Connect the electrical connector to the new switch.
17   Press the switch into position and install the screws.
18   Raise the steering column into position and install and tighten the bolts or nuts to 20 ft-lbs. Make sure the switch is actuated when the ignition key is turned to the Start position. If it doesn't, loosen the switch screws and adjust the position of the switch on the steering column.

11 Instrument panel switches - check and replacement

Warning: *On models equipped with a Supplemental Inflatable Restraint (SIR) system (more commonly known as airbags), always disable the airbag system before working in the vicinity of any airbag system component to avoid the possibility of accidental deployment of the airbag, which could cause personal injury (see Section 30).*

Check

Refer to illustrations 11.3a and 11.3b
1   The instrument panel switches are

located next to the instrument cluster. Most models have a switch pod on either side of the cluster, later models may have only one pod on the left. Each pod contains several switches, with the headlight, interior light dimmer and Twilight Sentinel (where equipped) switches generally in the left pod. The Twilight Sentinel option uses a sensor to detect daylight, and when activated, the system will turn on the headlights when it gets dark, and turn them off during daylight. The Sentinel switch turns the system on and off and controls the delay function.
2   To test any of the instrument panel switches, the switch must first be removed from the instrument panel (see *Replacement* below).
3   By referring to the wiring diagrams at the end of this Chapter and examining the wire colors attached to each switch connector, you can determine which pins are responsible for the power in, power out and ground **(see illustrations)**. Test most any typical ON/OFF switch, with an ohmmeter or self-powered test light, for continuity between the power in and various power out terminals with the switch in each position. Other switches, such as the Twilight Sentinel switch, are solid-state and can only be checked by applying battery power with a jumper wire to the power-in terminal and checking for power out at the proper terminal with a test light connected to a good chassis ground.

Replacement

Refer to illustrations 11.6a, 11.6b and 11.7
4   Disconnect the negative battery cable.
5   Refer to Chapter 11 and remove the instrument panel bezel.
6   Remove the screws and pull the switch pod out from the instrument panel **(see illustrations)**.
7   Pull the switch or switch pod out far enough to disconnect the electrical connector on the back **(see illustration)**.
8   Installation is the reverse of the removal procedure.

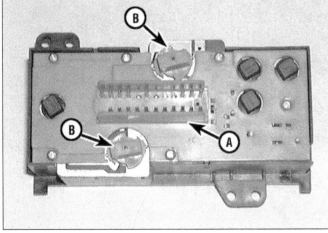

11.3a Connector pins (A) on the left switch pod - (B) are the twist-out bulb holders

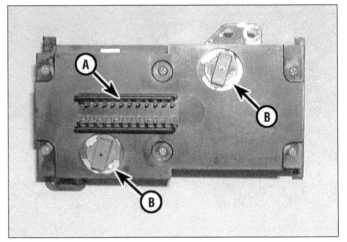

11.3b Connector pins (A) on the right switch pod - (B) are the twist-out bulb holders

12-11

11.6a Left switch pods screws (arrows) - 1991 Deville model

11.6b Right switch pods screws (arrows) - 1987 Seville model

11.7 Disconnect the electrical connector (arrow) at the back of the switch - window lock switch shown

13.3a Remove the instrument cluster mounting screws (arrows) - on Deville/Fleetwood models (shown) the dash pad must first be removed as shown here

13.3b Disconnect the electrical connectors at the back of the cluster

12 Instrument panel gauges - check

Warning: *On models equipped with a Supplemental Inflatable Restraint (SIR) system (more commonly known as airbags), always disable the airbag system before working in the vicinity of any airbag system component to avoid the possibility of accidental deployment of the airbag, which could cause personal injury (see Section 30).*

1 All of the covered models have digital instrument clusters with warning lights. Only a few models have gauges, and these are digital as well. The ECM supplies information to the instrument cluster on all models. Refer to Chapter 6 for procedures to extract trouble codes from the ECM and BCM that may relate to the instrument cluster, or gauge sending units.

2 The Coolant Temperature Sensor(CTS) supplies engine temperature input to the ECM and BCM. The BCM then sends a coded signal to the instrument cluster's temperature warning light or gauge. For testing of the CTS itself, see Chapter 6.

3 Input about engine oil pressure is supplied by the oil pressure switch. In most models, the oil pressure switch is open above 4 psi of pressure, and should trigger the oil warning light if pressure is below 4 psi. To

check the warning light, ground the connector from the oil pressure switch (see Chapter 2, Part C for location of the oil pressure switch) and the warning light should come on. If not, check for trouble codes (see Chapter 6).

4 Fuel level input to the instrument cluster comes from the sending unit in the fuel tank. If your instrument panel isn't showing the proper fuel level, and you have tested the sending unit (See Chapter 4), then look at the wiring diagrams at the end of this Chapter and check the circuit. If the circuit to the BCM is good, check for ECM and BCM trouble codes (see Chapter 6).

13 Instrument cluster - removal and installation

Refer to illustrations 13.3a and 13.3b
Warning: *On models equipped with a Supplemental Inflatable Restraint (SIR) system (more commonly known as airbags), always disable the airbag system before working in the vicinity of any airbag system component to avoid the possibility of accidental deployment of the airbag, which could cause personal injury (see Section 30).*

1 Disconnect the negative cable from the battery.
2 Remove the instrument cluster bezel

(see Chapter 11). **Note:** *On some models, the lens over the face of the instrument cluster must be removed to access the screws for the cluster. On Deville/Fleetwood models, the dash pad must be removed before the cluster can be removed (see Chapter 11).*

3 Remove the instrument cluster screws, pull out the cluster and unplug the electrical connectors **(see illustrations)**.
4 Installation is the reverse of removal.

14 Radio and speakers - removal and installation

Warning: *On models equipped with a Supplemental Inflatable Restraint (SIR) system (more commonly known as airbags), always disable the airbag system before working in the vicinity of any airbag system component to avoid the possibility of accidental deployment of the airbag, which could cause personal injury (see Section 30).*

Radio

Refer to illustrations 14.2, 14.3a, 14.3b and 14.3c
Note: *The following procedure applies to models without the CRT display. On models with the CRT display the radio receiver is located at the front of the center console; see*

14.2 On Deville/Fleetwood models, feel with your finger under the top-center of the bezel until you find the notch (arrow) - pull there and the bezel can be removed

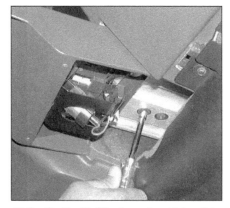

14.3a On Deville/Fleetwood models, slide the ashtray out as far as it will go, then remove the two radio mounting screws from below

14.3b Remove the radio mounting screws (arrows) from the dash - Deville shown, others similar

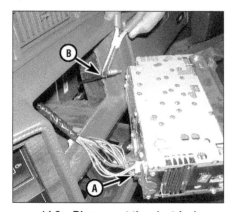

14.3c Disconnect the electrical connectors (A) and the antenna cable (B) at the back of the radio

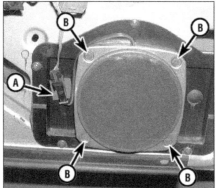

14.6 Disconnect the electrical connector (A) at the front speaker, then remove the mounting screws (B)

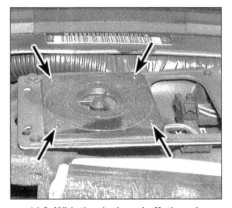

14.9 With the dash pad off, there is access to the four tweeter screws (arrows)

Chapter 11 for center console removal.
1   Disconnect the negative battery cable.
2   On Riviera, Toronado and Eldorado/Seville models, remove the instrument cluster trim panel (see Chapter 11). On Deville/Fleetwood models, the radio is in the center of the dash, with its own small bezel **(see illustration)**.
3   Remove the retaining screws and pull

14.12 With the package shelf removed, you can remove the rear speaker mounting screws (arrows)

the radio outward to access the backside, then disconnect the electrical connectors and the antenna lead and lift the radio out of the vehicle **(see illustrations)**. **Note:** *On Riviera and Toronado models, the radio and climate control panel are attached to each other with brackets. Remove the two nuts to separate the radio from the climate control panel.*
4   Installation is the reverse of removal.

Front speakers

Refer to illustration 14.6
5   Remove the front door trim panel (see Chapter 11).
6   Remove the speaker retaining bolts. Disconnect the electrical connector and remove the speaker from the vehicle **(see illustration)**.
7   Installation is the reverse of removal.

Front tweeters

Refer to illustration 14.9
8   On models equipped with front tweeters in the top of the dash, remove the dash pad (see Chapter 11).
9   Remove the tweeter retaining bolts. Disconnect the electrical connector and remove the tweeter from the vehicle **(see illustration)**.
10   Installation is the reverse of removal.

Rear speakers

Refer to illustration 14.12
11   Refer to Chapter 11 and remove the rear seat, then remove the package shelf.
12   Remove the speaker retaining screws, leaving the plastic enclosure in place **(see illustration)**. Disconnect the electrical connector and remove the speaker from the vehicle.
13   Installation is the reverse of removal.

---

15   Antenna - general information and check

Antenna lead

1   Detach the radio and disconnect the antenna lead from the backside of the radio (see Section 14). Attach a piece of mechanic's wire to the end of the antenna lead, then remove any retaining clips under the instrument panel securing the antenna lead.
2   Fish the lead out from under the instrument panel and follow it. It may go under the carpet or console. Release any clips securing the lead to the body until you reach the back seat.

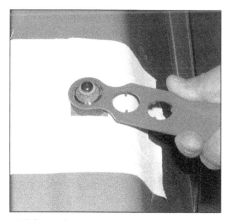

15.6 Apply tape to the body around the antenna to ensure the paint isn't scratched and remove the antenna base nut

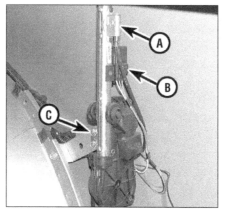

15.8 Remove the antenna cable (A), disconnect the electrical connector at the antenna relay (B), and remove the bolts (C) securing the antenna assembly to the body

16.10 Remove the two thumbscrews (arrows) retaining the headlight housing

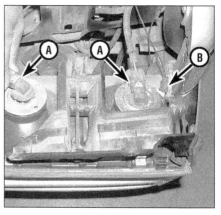

16.12 Rotate the bulb holders (A) counterclockwise and remove them from the housing - (B) is the fiber-optic cable

3   Working in the trunk, pry out the plastic clips securing the passenger side trunk finishing panels to allow access to the antenna motor.
4   Release the lead from the antenna motor assembly and fish the lead from the trunk through to the interior at the back seat **(see illustration 15.8)**.
5   Installation is the reverse of the removal procedure. Attach the mechanic's wire to the lead and fish the new lead through the body in the same route as the original.

Power antenna assembly

Refer to illustrations 15.6 and 15.8
6   Using a pair of snap-ring pliers or an antenna tool, remove the antenna base retaining nut. An antenna removal tool is available at most auto parts stores **(see illustration)**, and won't slip off and possibly scratch the body, like pliers or other tools.
7   Working in the trunk, pry out the plastic clips securing the passenger side trunk finishing panels to allow access to the antenna motor.
8   Detach the motor/base assembly retaining bolt(s). Disconnect the antenna lead and the electrical connector (if equipped) and remove the antenna motor/base assembly from the vehicle **(see illustration)**.
9   Installation is the reverse of removal.

---

16 Headlight bulb - replacement

Warning: *Halogen gas filled bulbs are under pressure and may shatter if the surface is scratched or the bulb is dropped. Wear eye protection and handle the bulbs carefully, grasping only the base whenever possible. Do not touch the surface of the bulb with your fingers because the oil from your skin could cause it to overheat and fail prematurely. If you do touch the bulb surface, clean it with rubbing alcohol.*

Buick Riviera

1   Open the hood.

2   Remove the plastic rivet from the inner end of the trim strip under the headlight assembly. To remove the rivet, pop the center up with a sharp tool, then pull out the rivet body. Unsnap the trim strip from its retaining clips.
3   Detach the plastic shield from above the headlight housing.
4   Remove the bolts from the top at each end of the headlight housing.
5   Remove the two nuts from the back side of the housing, then pull the housing forward and disconnect the bulb holder(s).
6   Remove the bulb from the holder by pulling it straight out.
7   Without touching the glass with your bare fingers, insert the new bulb into the socket and then into the headlight housing, install the holder.
8   Reinstall the headlight housing by reversing the removal procedure.
9   Test headlight operation, then close the hood.

Cadillac

Deville/Fleetwood

Refer to illustrations 16.10 and 16.12
10   Open the hood.
11   Remove the two thumbscrews securing the headlight assembly **(see illustration)**. Lean the headlight assembly forward for access to the bulb holders.
12   Rotate the headlight bulb holders counterclockwise **(see illustration)**.
13   Withdraw the bulb holders from the headlight housing.
14   Remove the bulb from the holder by pulling it straight out.
15   Without touching the glass with your bare fingers, insert the new bulb into the socket and then into the headlight housing, install and tighten the holder.
16   Test headlight operation, then close the hood.

Eldorado/Seville

17   Open the hood.

18   Remove the Torx screw from the inner end of the trim strip under the headlight assembly. Unsnap the trim strip from its retaining clips.
19   Remove the plastic cover over the radiator, then remove the grille (see Chapter 11).
20   Remove the headlight retaining screws and pull the headlight housing forward.
21   Remove the bulb holder(s).
22   Remove the bulb from the holder.
23   Without touching the glass with your bare fingers, insert the new bulb into the socket and then into the headlight housing, install the holder.
24   Reinstall the headlight housing by reversing the removal procedure.
25   Test headlight operation, then close the hood.

Oldsmobile Toronado

Refer to illustration 16.27
26   Turn the headlights On to open the headlight doors, then remove the no. 2 fuse from the underhood fuse/relay panel. Now turn the headlights off.
27   Remove the four screws from the headlight retainer **(see illustration)**. **Note:** *Don't disturb the headlight aiming screws. Pull the bulbs forward to access the electrical connectors.*

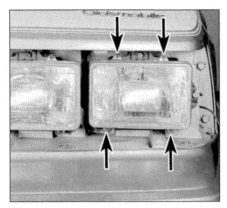

16.27 Remove the four screws (arrows), then remove the retainer and the headlight (typical)

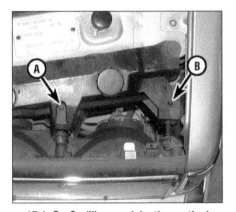

17.1 On Cadillac models, the vertical headlight adjusting screw (A) is in the center, and the horizontal adjusting screw (B) is on the side

28   Unplug the electrical connector(s) and detach the bulb(s).
29   Installation is the reverse of removal, but be sure the headlight is installed in the upright position. Reinstall the no. 2 fuse in the underhood fuse/relay panel.

17 Headlights - adjustment

Refer to illustrations 17.1 and 17.3
Caution: *The headlights must be aimed correctly. If adjusted incorrectly they could blind the driver of an oncoming vehicle and cause a serious accident or seriously reduce your ability to see the road. The headlights should be checked for proper aim every 12 months and any time a new headlight is installed or front end body work is performed. It should be emphasized that the following procedure is only an interim step, which will provide temporary adjustment until the headlights can be adjusted by a properly equipped shop.*
1   Headlights have two spring loaded adjusting screws, one on the top controlling up-and-down movement and one on the side controlling left-and-right movement **(see illustration)**. **Note:** *On Toronado models, the headlight adjustment screws are located in*

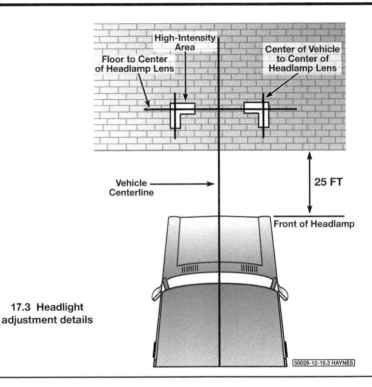

17.3 Headlight adjustment details

front, at the headlight housing retaining bezel. The screws going through the bezel just retain the bezel, while the adjuster screws are the ones in the cutouts. The bezel does not have to be removed to aim the headlights. On Riviera models, the adjuster screws are on the back of the headlight housing, but a plastic shield must be removed first for access.
2   There are several methods of adjusting the headlights. The simplest method requires a large level area with a blank wall.
3   Position masking tape vertically on the wall in reference to the vehicle centerline and the centerlines of both headlights **(see illustration)**.
4   Position a horizontal tape line in reference to the centerline of all the headlights. **Note:** *It may be easier to position the tape on the wall with the vehicle parked only a few inches away.*
5   Adjustment should be made with the vehicle parked 25 feet from the wall, sitting level, the gas tank half-full and no unusually heavy load in the vehicle.
6   Starting with the low beam adjustment, position the high intensity zone so it is two inches below the horizontal line and two inches to the side of the headlight vertical line, away from oncoming traffic. Adjustment is made by turning the top adjusting screw clockwise to raise the beam and counterclockwise to lower the beam. The adjusting screw on the side should be used in the same manner to move the beam left or right.
7   With the high beams on, the high intensity zone should be vertically centered with the exact center just below the horizontal line. **Note:** *It may not be possible to position the headlight aim exactly for both high and low beams. If a compromise must be made,*

keep in mind that the low beams are the most used and have the greatest effect on safety.
8   Have the headlights adjusted by a dealer service department or service station at the earliest opportunity.

18 Headlight door actuator (Toronado models) - check and replacement

1   On Toronado models, doors cover the headlights when the lights are Off. When the lights are turned On, the doors open.
2   When working on the headlights, either to replace the headlight bulbs or the housings, the door must be kept open. Turn the headlights On, then remove the #2 fuse in the underhood fuse/relay panel supplying the headlight door actuator motor.

Check

Refer to illustration 18.6
3   The headlight doors should swing out of the way when the headlight switch is turned On. If they don't, Refer to the label on the cover to the fuse/relay panel and check the fuse indicated as supplying this circuit. If the fuse is OK, check for power at the actuator with a 12-volt test light. Probe each terminal in the electrical connector to the motor while an assistant turns the headlight switch on and off.
4   The headlight doors can be manually operated, either for testing purposes or in an emergency when the doors fail to open. Disconnect the electrical connector to the motor, located between the grille and the radiator (see Chapter 11 for grille removal).

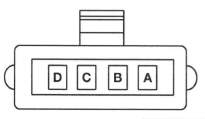

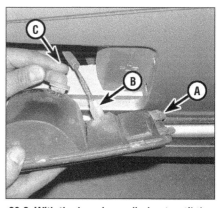

18.6 Toronado headlight door module connector terminal identification

20.1 Remove the screws (arrows) on the outside of the cornering lamp/side marker housing

20.2 With the housing pulled out until the tabs (A) clear the body, you can twist out and remove either the side marker (B) or the cornering lamp (C) bulb holder

5   Remove the plastic cover over the actuator's manual knob and turn the knob clockwise until the doors fully open. Turning the knob counterclockwise should close the doors. If the doors don't operate properly with the manual knob, the problem is binding of the actuator or the doors, possible from body damage.

6   If the doors operate smoothly with the manual knob, disconnect the electrical connector at the headlight door module, which is below the underhood fuse/relay panel. Ground the D terminal at the module with a jumper wire, then apply a fuse jumper wire with battery voltage to terminal C **(see illustration)**. The doors should open.

7   Check that power is supplied at the B terminal of the module connector, and that the test light lights when connected to terminals B and D. With the headlight switch On, there should be power at C and A. If the headlight door module fails these tests, replace the module.

8   If the module is OK but the doors don't operate, refer to the wiring diagram and check the wiring from the module to the headlight door actuator motor. If the wiring is OK, replace the actuator motor.

Replacement

9   To replace the actuator, open the headlight doors as described in Step 2.

10   Disconnect the electrical connector at the actuator motor. Refer to Chapter 11 if necessary and remove the grille for access.

11   Remove the screws on each side of the actuator that connects the actuator shaft to

the shafts at each headlight door.

12   Remove the bolts and nuts securing the actuator assembly to the radiator center support brace and remove the actuator from the vehicle.

13   Installation is the reverse of the removal procedure. Check for proper operation of the actuator when connected.

19   Headlight housing - replacement

1   Remove the headlight bulb(s) (see Section 16).

2   Remove the two thumbscrews and pull the headlight housing forward **(see illustration 16.10)**.

3   With the headlight bulbs and holders removed, carefully disconnect the fiber-optic cable (on Twilight Sentinel-equipped models) from the housing.

4   The front bezel over the headlight housing is removed by releasing the clips. Transfer the bezel and the mounting bracket on the back to the new headlight housing. **Note:** *On Riviera models, remove the plastic cover over the top of the headlight housing, the filler strip below the headlight housing (it's retained by one plastic rivet), and the nuts retaining the housing to the body.*

5   The remainder of the installation is the reverse of removal.

20   Bulb replacement

Warning: *Bulbs may remain hot for up to twenty minutes after they're turned off. Be sure bulbs are off and cool before you touch them.*

Front cornering lamp/side marker light

Refer to illustrations 20.1 and 20.2

1   Remove the two screws securing the outer bezel of the cornering lamp/side marker assembly **(see illustration)**.

2   Pull the housing out a little and forward until the tabs on the housing clear the body **(see illustration)**.

2   Turn the bulb holder counterclockwise and pull it out of the housing **(see illustration 20.2)**.

3   Pull the bulb out of the holder.

4   Installation is the reverse of removal.

Front park/turn lights

Refer to illustrations 20.5 and 20.6

5   From below the car, look up at the back of the front bumper, near the left or right end. Through an opening in the reinforcement panel behind the bumper, the outer bulb holder can be removed with the assembly in place **(see illustration)**. To access the inner bulb, remove the outer bulb holder and remove one nut holding the park/turn assembly. Through another bumper access hole at the inner end of the light housing, use a screwdriver to push up the clip, then pull the light housing out of the front of the bumper.

6   Twist and pull out either bulb holder **(see illustration)**. Remove the bulb from the holder. Take care not to damage the fiber-optic cable attached to the outer corner of the light housing (on Twilight Sentinel-equipped models).

7   Installation is the reverse of removal.

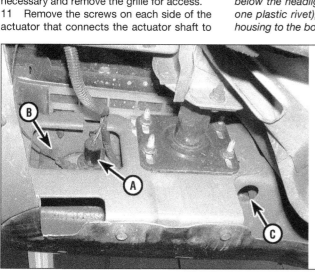

20.5 From below, looking at the back of the bumper, remove the outer bulb holder (A), the nut (B), then depress the park-turn housing clip through the access hole (C)

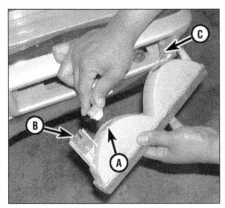

20.6 With the housing pulled out through the front of the bumper, there is access to the inner bulb holder (A) - (B) is the clip that holds the inner end of the housing to the bumper, (C) is the fiber-optic cable

20.8 Remove the taillight mounting screws (arrows) - Deville model shown, Eldorado/Seville models have only one screw, at the bottom

20.11 Rotate the stop/turn or tail light bulb holders from the taillight assembly to access the bulb(s) - Cadillac shown

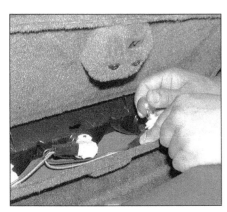

20.12 On Cadillac models, lift the package shelf cover for access to the high-mount brake light bulbs

20.16 Remove the two screws to pull down the license plate light

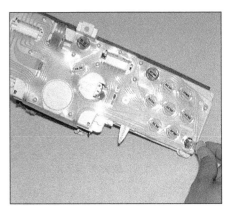

20.21 To remove an instrument cluster bulb, depress it and turn 90 degrees counterclockwise

Rear turn signal and stoplight/taillight bulbs

Refer to illustrations 20.8 and 20.11

8   On Cadillac models, remove the screw(s) at the taillight lens and remove the taillight body **(see illustration)**.

9   On Riviera models, open the trunk and remove the plastic caps and screws holding the rearmost section of trim carpet. Remove the two long screws at the top of the trunk opening and pull the light housing out.

10   On Toronado models, the light housing runs the full width of the vehicle. Open the trunk and remove the five white plastic wingnuts securing the taillight panel to the body. Pull the taillight panel out.

11   Rotate the bulb holder(s) 1/8-turn counterclockwise and pull it out of the housing, then pull the bulb out **(see illustration)**.

High-mounted brake light

Refer to illustration 20.12

12   Access to the high-mounted brake light is at the package shelf behind the rear seat. On Cadillac models, lift the cover in the center of the package shelf and twist the bulb holders out to replace the bulbs **(see illustration)**.

13   On Riviera models, remove the two screws at the side of the high-mount brake light housing on the package shelf, then lift the housing for access to the bulbs. On Toronado models, Pry off the plastic cover (from inside, facing the rear window) over the high-mount brake light assembly. Remove the two screws, then lift the assembly for bulb access.

14   On all models, twist the bulb holder counterclockwise to remove it, then pull the bulb straight out of the holder.

15   Installation is the reverse of removal.

License plate lights

Refer to illustration 20.16

16   Remove the two screws retaining the license plate light over the license plate **(see illustration)**.

17   Remove the bulb holder by releasing the clips on the lens housing.

18   Pull the bulb straight out of the holder.

19   Installation is the reverse of removal.

Instrument panel lights

Refer to illustration 20.21

20   To gain access to the instrument panel lights, the instrument cluster will have to be removed first (see Section 13).

21   Rotate the bulb counterclockwise and remove it from the instrument cluster **(see illustration)**.

22   Pull the bulb straight out of the holder.

23   Installation is the reverse of removal. **Note:** *Make sure you replace the bulb with one of the same wattage as the original bulb. The gauge bulbs have higher wattage than the indicator bulbs.*

Dome/door/vanity/trunk lights

Refer to illustrations 20.24 and 20.25

24   The courtesy lights in the front footwells are attached to the underdash sound panels. They are uncovered, so just unscrew the bulbs for replacement. The dome light, door light, trunk light and vanity (right sun visor) lights are replaced in the same manner. Carefully pry the lens off **(see illustration)**. In some models, the dome light is in an overhead panel with map lights. On these models, remove the screws and drop the light housing to access the bulbs.

25   Remove the bulb from the terminals. It may be necessary to pry the bulb out - if this is the case, pry on the ends of the glass (otherwise the glass may shatter) **(see illustration)**.

26   Installation is the reverse of removal.

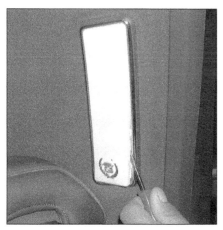

20.24 Most interior bulbs are replaced by prying off the plastic lens - typical example is this Deville rear courtesy light

20.25 Some bulbs fit into spring-loaded contacts at each end - when installing the new bulb, hold the bulb with gloves or a cloth, as fingerprints can shorten the bulb life

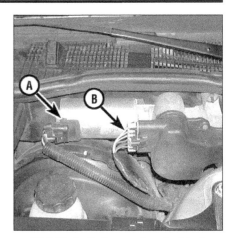

21.2 Typical wiper motor electrical connectors - three-pin connector (A) and six-pin connector (B)

21 Wiper motor - check and replacement

Wiper motor circuit check

Refer to illustration 21.2

Note: *Refer to the wiring diagrams at the end of this Chapter for wire colors and locations in the following checks. When checking for voltage, probe a grounded 12-volt test light to each terminal at a connector until it lights; this verifies voltage (power) at the terminal. If the following checks fail to locate the problem, have the system diagnosed by a dealer service department or other properly equipped repair facility.*

1   If the wipers work slowly, make sure the battery is in good condition and has a strong charge (see Chapter 1). If the battery is in good condition, remove the wiper motor (see below) and operate the wiper arms by hand. Check for binding linkage and pivots. Lubricate or repair the linkage or pivots as necessary. Reinstall the wiper motor. If the wipers still operate slowly, check for loose or corroded connections, especially the ground connection. If all connections look OK, replace the motor.

2   If the wipers fail to operate when activated, check the fuse. If the fuse is OK, connect a jumper wire between the wiper motor and ground, then retest. If the motor works now, repair the ground connection. If the motor still doesn't work, turn the wiper switch to the HI position and check for voltage at the motor **(see illustration)**. If there's voltage at the connector, remove the motor and check it off the vehicle with fused jumper wires from the battery. If the motor now works, check for binding linkage (see Step 1 above). If the motor still doesn't work, replace it. If there's no voltage to the motor, check for voltage at the wiper control relays. If there's voltage at the wiper control relays and no voltage at the wiper motor, check the switch for continuity (see Section 11 for instrument panel-

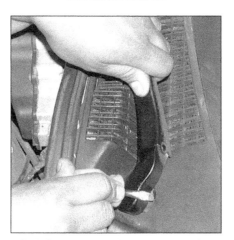

21.7 Remove the wiper arms by prying out the retainer, then pull the arm off the shaft

mounted switches or Section 9 for steering column-mounted switches). If the switch is OK, there may be a problem in the circuits. Refer to the wiring diagrams at the end of this Chapter.

3   If the interval (delay) function is inoperative, check the continuity of all the wiring between the switch and wiper control module. If the wiring is OK, check the resistance of the delay control, the resistance should vary as the control is turned.

4   If the wipers stop at the position they're in when the switch is turned off (fail to park), check for voltage at the park feed wire of the wiper motor connector when the wiper switch is OFF but the ignition is ON. If no voltage is present, check for an open circuit between the wiper motor and the fuse panel.

5   If the wipers won't shut off unless the ignition is OFF, disconnect the wiring from the wiper control switch. If the wipers stop, replace the switch. If the wipers keep running, there's a defective limit switch in the motor; replace the motor.

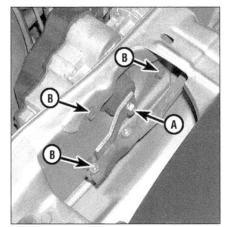

21.9 Remove the nut (A) holding the linkage to the back of the motor, then remove the three motor mounting bolts (B)

6   If the wipers won't retract below the hood line, check for mechanical obstructions in the wiper linkage or on the vehicle's body that would prevent the wipers from parking. If there are no obstructions, check the wiring between the switch and motor for continuity. If the wiring is OK, replace the wiper motor.

Wiper motor replacement

Refer to illustrations 21.7 and 21.9

7   Remove the windshield wiper arms **(see illustration)**.

8   Remove the cowl cover (see Chapter 11).

9   Detach the wiper motor/linkage assembly from the cowl **(see illustration)**.

10   Remove the wiper motor retaining bolts **(see illustration 21.9)**.

11   Remove the wiper motor from the wiper linkage assembly.

12   Disconnect the electrical connector from the wiper motor.

13   Installation is the reverse of removal.

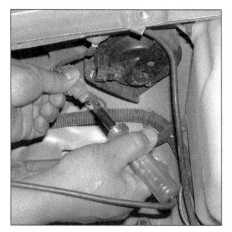

22.3 Check for voltage at the horn while an assistant pushes the horn button

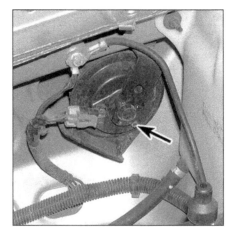

22.8 Use the current adjusting screw (arrow) to achieve a draw of 4.5 to 5.5 amps for proper horn operation

24.5 When measuring the voltage at the rear window defogger grid, wrap a piece of aluminum foil around the positive probe of the voltmeter and press the foil against the wire with your finger

22   Horn - check and replacement

Check

Refer to illustrations 22.3 and 22.8

Note: *Check the fuses before beginning electrical diagnosis.*

1   Disconnect the electrical connector from the horns.

2   To test the horns, connect battery voltage and ground to the horn terminals with a pair of jumper wires. If either horn doesn't sound, replace it.

3   If the horn does sound, check for voltage at the horn connector when the horn switch is depressed **(see illustration)**. If there's voltage at the connector, check for a bad ground at the horn.

4   If there's no voltage at the horn, check the relay (see Section 6).

5   If the relay is OK, refer to the wiring diagrams at the end of this Chapter and check for voltage to the relay power and control circuits. If either of the circuits is not receiving voltage, inspect the wiring between the relay and the fuse panel.

6   If both relay circuits are receiving voltage, depress the horn switch and check the circuit from the relay to the horn switch for continuity to ground. If there's no continuity, check the circuit for an open. If there's no open circuit, replace the horn switch.

7   If there's continuity to ground through the horn switch, check for an open or short in the circuit from the relay to the switch.

8   If the horn works, but the tone is off, either weak or strained, check the current draw of the horn, using an ammeter during operation. Hook the ammeter between the horn connector and the terminal on the horn. A correctly operating horn should draw 4.5 to 5.5 amps. Use the adjusting screw on the horn to achieve this draw **(see illustration)**.

Replacement

9   The horn on Cadillac models is located

in the right fenderwell, in front of the coolant recovery tank. On Toronado and Riviera models, the horn is located in the left wheelwell, ahead of the wheel. Turn the wheels to the left for better access.

10   Disconnect the electrical connectors and remove the bracket bolts **(see illustration 22.8)**.

11   Installation is the reverse of removal. Make sure the area of the body is clean where the horn bracket attaches. This is the ground point for the horn.

23   Daytime Running Lights (DRL) - general information

The Daytime Running Lights (DRL) system used on Canadian models illuminates the headlights whenever the engine is running. The only exception is with the engine running and the parking brake engaged. Once the parking brake is released, the lights will remain on as long as the ignition switch is on, even if the parking brake is later applied.

The DRL system supplies reduced power to the headlights so they won't be too bright for daytime use, while prolonging headlight life.

24   Rear window defogger and heated windshield - check and repair

Rear window defogger

1   The rear window defogger consists of a number of horizontal heating elements baked onto the inside surface of the glass. Power is supplied through a large fuse from the power distribution box in the engine compartment. The heater is controlled by the instrument panel switch.

2   Small breaks in the element can be repaired without removing the rear window.

Check

Refer to illustrations 24.5, 24.6 and 24.8

3   Turn the ignition switch and defogger switches to the ON position.

4   Using a voltmeter, place the positive probe against the defogger grid positive terminal and the negative probe against the ground terminal. If battery voltage is not indicated, check the fuse, defogger switch and related wiring. If voltage is indicated, but all or part of the defogger doesn't heat, proceed with the following tests.

5   When measuring voltage during the next two tests, wrap a piece of aluminum foil around the tip of the voltmeter positive probe and press the foil against the heating element with your finger **(see illustration)**. Place the negative probe on the defogger grid ground terminal.

6   Check the voltage at the center of each heating element **(see illustration)**. If the voltage is 5 to 6 volts, the element is okay (there is no break). If the voltage is 0 volts, the element is broken between the center of the element and the positive end. If the voltage is 10 to 12 volts the element is broken between the center of the element and the ground side. Check each heating element.

7   If none of the elements are broken, connect the negative probe to a good chassis ground. The voltage reading should stay the same, if it doesn't the ground connection is bad.

8   To find the break, place the voltmeter negative probe against the defogger ground terminal. Place the voltmeter positive probe with the foil strip against the heating element at the positive side and slide it toward the negative side. The point at which the voltmeter deflects from several volts to zero is the point where the heating element is broken **(see illustration)**.

Repair

Refer to illustration 24.14

9   Repair the break in the element using a

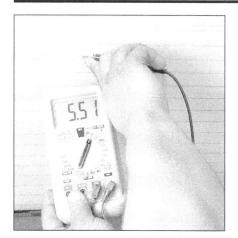

24.6 To determine if a wire has broken, check the voltage at the center of each wire. If the voltage is 5 to 6 volts, the wire is unbroken; if the voltage is 10 to 12 volts, the wire is broken between the center of the wire and the ground side; if the voltage is 0 volts, the wire is broken between the center of the wire and the power side

24.8 To find the break, place the voltmeter negative lead against the defogger ground terminal, place the voltmeter positive lead with the foil strip against the heat wire at the positive terminal end and slide it toward the negative terminal end - the point at which the voltmeter deflects from several volts to zero volts is the point at which the wire is broken

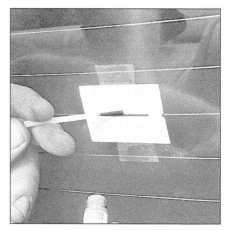

24.14 To use a defogger repair kit, apply masking to the inside of the window at the damaged area, then brush on the special conductive coating

repair kit specifically for this purpose (available at most auto parts stores). The kit includes conductive plastic epoxy.

10   Before repairing a break, turn off the system and allow it to cool for a few minutes.

11   Lightly buff the element area with fine steel wool; then clean it thoroughly with rubbing alcohol.

12   Use masking tape to mask off the area being repaired.

13   Thoroughly mix the epoxy, following the kit instructions.

14   Apply the epoxy material to the slit in the masking tape, overlapping the undamaged area about 3/4-inch on either end **(see illustration)**.

15   Allow the repair to cure for 24 hours before removing the tape and using the system.

Heated windshield

16   Some models have an optional heated windshield, which operates much like a standard rear window defogger.

Check

Refer to illustrations 24.19 and 24.22

17   Inspect the windshield and the power strip (usually located at the lower right on the inside of the windshield) for any signs of damage before making any tests. The heating element should not be operated on a damaged windshield, or hot spots could develop that could be dangerous. The heated windshield control module has a sensor that should detect a cracked windshield and keep the system from operating.

18   Disconnect the three-pin connect from the windshield.

19   Check the resistance between terminals at the windshield with an ohmmeter **(see illustration)**. There should be 10 ohms resistance between A and B, 4 ohms between B and C, and more than 10,000 ohms (no continuity) between any of the terminals and a body ground. If the windshield fails any of these tests, replace it.

20   If the windshield fails to heat after the switch is pushed, check the fuse first. Also make sure the battery is fully charged (at least 11.5 volts).

21   If the heated windshield switch is operating properly, with observed voltage coming into it and voltage coming out when the switch is on, then begin checking the heater windshield control module, located under the instrument panel, just to the right of the steering column.

22   Disconnect the electrical connector at the module and check for voltage and resistance at the terminals on the harness side of the connector with the engine running **(see illustration)**.

23   There should be battery voltage between terminal B6 and ground, between A6 and A8, and between B6 and A8. If not, check for an open or short in those wires.

24   Connect an ohmmeter to terminals A3 and ground. Continuity should be indicated when the heated windshield switch is depressed, and no continuity when the switch is released.

25   If the circuits are good (no shorts or opens) the heated windshield control module may be defective.

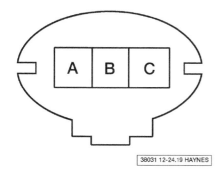

24.19 Heated windshield electrical connector terminal identification

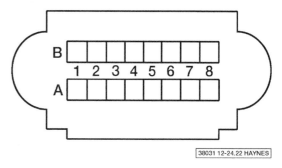

24.22 Heated windshield control module electrical connector terminal identification (harness side)

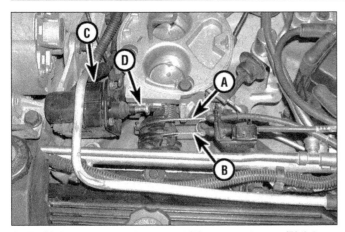

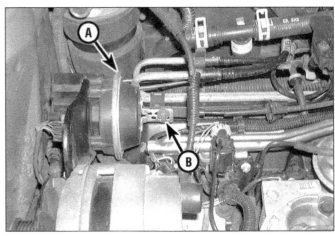

25.5a Make sure the cruise control (A) and accelerator (B) linkage mounted on the throttle body are not damaged and that they operate smoothly together when the throttle is opened - (C) is the idle speed control motor and its plunger (D)

25.5b Typical cruise control servo (A) - adjust the cruise cable by attaching it to the servo with the clip (B) in the hole that provides the minimum slack

25 Cruise control system - description, check and cable adjustment

Refer to illustrations 25.5a and 25.5b

1   The cruise control system maintains vehicle speed with a servo motor, located in the engine compartment, that is connected to the throttle linkage by a cable. On V8 models, the servo is located near the left end of the front valve cover on early models, and at the right end of the rear valve cover on later models. On V6 models, it is usually located near the engine/transaxle juncture, below the throttle body. The system consists of the servo motor, brake switch, control switches, speed sensors and relays. Some features of the system require special testers and diagnostic procedures that are beyond the scope of this manual. Listed below are some general procedures that may be used to locate common problems.

2   Locate and check the fuse (see Section 3).

3   The brake light switch deactivates the cruise control system. Have an assistant press the brake pedal while you check the brake light operation.

4   If the brake lights do not operate properly, correct the problem and retest the cruise control.

5   Check the control cable between the cruise control servo/amplifier and the throttle linkage and adjust/replace as necessary **(see illustrations)**. Refer to Chapter 6 for the procedure to retract the plunger of the idle speed control motor, so that it doesn't touch the throttle linkage while you are adjusting the cruise-control linkage.

6   The cruise control system uses a speed sensing device. The speed sensor is located in the transmission. To test the speed sensor, see Chapter 6.

7   Test drive the vehicle to determine if the cruise control is now working. Refer to Chapter 6 for the procedure to extract trouble

codes from the vehicle. There are several cruise-control-related trouble codes that may help you pinpoint the problem area. If there are no codes, or you can't locate the problem, don't use the cruise control until you can take it to a dealer service department or an automotive electrical specialist for further diagnosis.

26 Power window system - description and check

Refer to illustrations 26.11a and 26.11b
Caution: *The power window system will operate even when the ignition switch is Off. If the engine is shut off, the power windows can still operate for 10 minutes, or until one of the doors is opened. Do not operate the window switch when any part of your body is in the window opening, unless the battery has been disconnected.*

1   The power window system operates electric motors, mounted in the doors, which lower and raise the windows. The system consists of the control switches, the motors, regulators, glass mechanisms and associated wiring.

2   The power windows can be lowered and raised from the master control switch by the driver or by remote switches located at the individual windows. Each window has a separate motor that is reversible. The position of the control switch determines the polarity and therefore the direction of operation.

3   The circuit is protected by a fuse and a circuit breaker. Each motor is also equipped with an internal circuit breaker; this prevents one stuck window from disabling the whole system.

4   These procedures are general in nature, so if you can't find the problem using them, take the vehicle to a dealer service department or qualified repair facility.

5   If the power windows won't operate, always check the fuse and circuit breaker first. Refer to the wiring diagrams at the end

of this Chapter to study the circuit in detail.

6   Check the wiring between the switches and fuse panel for continuity. Repair the wiring, if necessary.

7   Some models are equipped with an "express-down" module mounted near the driver's door window regulator. When equipped with this feature, a quick touch (less than 0.3 seconds) on the driver's window down button will lower the window a little. Touching the button for a longer period will send the window down all the way without the driver having to keep his finger on the button while driving.

8   If one of the passenger door windows is inoperative from the master control switch, try the other control switch at the window.

9   If the same window works from one switch, but not the other, check the switch for continuity.

10   If the switch tests OK, check for a short or open in the circuit between the affected switch and the window motor.

11   If one window is inoperative from both switches, remove the trim panel from the affected door and check for voltage at the

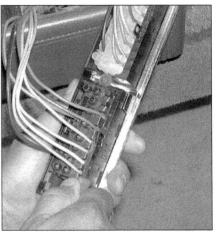

26.11a With the ignition On, check that power is coming in to the power window switch

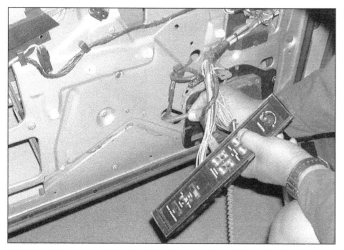

26.11b Disconnect the electrical connector at the window motor and check for power when the switch is pushed

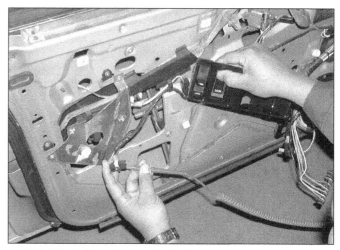

27.9 Check for voltage at the lock actuator connector while operating the switch

switch and at the motor while the switch is operated **(see illustrations)**.

12   If voltage is reaching the motor, disconnect the glass from the regulator (see Chapter 11). Move the window up and down by hand while checking for binding and damage. Also check for binding and damage to the regulator. If the regulator is not damaged and the window moves up and down smoothly, replace the motor. If there's binding or damage, lubricate, repair or replace parts, as necessary.

13   If voltage isn't reaching the motor, check the wiring in the circuit for continuity between the switches and motors. You'll need to consult the wiring diagram for the vehicle. If the circuit is equipped with a relay, check that the relay is grounded properly and receiving voltage.

14   Test the windows after you're done to confirm proper repairs.

27   Power door lock and keyless entry system - description and check

Power door lock system

Refer to illustration 27.9

1   The power door lock system operates the door lock actuators mounted in each door. The system consists of the switches, actuators, and associated wiring. Diagnosis can usually be limited to simple checks of the wiring connections and actuators for minor faults that can be easily repaired.

2   Power door lock systems are operated by bi-directional solenoids or motors located in the doors. The lock switches have two operating positions: Lock and Unlock. On later models with keyless entry and theft-deterrent options, the switches activate a module that in turn connects voltage to the door lock solenoids or motors. Depending on which way the switch is activated, it reverses polarity, allowing the two sides of the circuit

to be used alternately as the feed (positive) and ground side.

3   If you are unable to locate the trouble using the following general steps, consult a dealer service department or other qualified repair facility.

4   Always check the circuit protection first. Examine the wiring diagrams at the end of this Chapter.

5   Operate the door lock switches in both directions (Lock and Unlock) with the engine off. Listen for the faint click of the door lock relay operating.

6   If there's no click, check for voltage at the switches. If no voltage is present, check the wiring between the fuse block and the switches for shorts and opens.

7   If voltage is present but no click is heard, test the switch for continuity. Replace it if there's no continuity in both switch positions.

8   If the switch has continuity but the solenoid or motor doesn't click, check the wiring between the switch and solenoid for continuity. Repair the wiring if there's no continuity. Also check the wiring to and from the power door lock relay, usually located behind the driver's side kick panel.

9   If all but one lock solenoids operate, remove the trim panel from the affected door (see Chapter 11) and check for voltage at the solenoid while the lock switch is operated **(see illustration)**. One of the wires should have voltage in the lock position; the other should have voltage in the unlock position.

10   If the inoperative solenoid is receiving voltage but won't operate, replace the solenoid. **Note:** *It's common for wires to break in the portion of the harness between the body and door (opening and closing the door fatigues and eventually breaks the wires).*

Keyless entry system

11   The keyless entry system consists of a remote control transmitter that sends a coded infrared signal to a receiver, which

then operates the door, lock system.

12   Replace the transmitter batteries when the red LED light on the side of the case doesn't light when the button is pushed.

13   Use a small screwdriver to carefully separate the case halves.

14   Replace the two DL-2016 batteries.

15   Snap the case halves together.

28   Electric side view mirrors - description and check

Refer to illustration 28.6

1   Most electric rear view mirrors use two motors to move the glass; one for up-and-down adjustments and one for left-to-right adjustments. In addition, some models have electrically-heated glass defroster circuits, which are usually powered through the rear window defogger relay.

2   The control switch usually has a selector portion that sends voltage to the left or right side mirror. With the ignition ON but the engine OFF, roll down the windows and operate the mirror control switch through all functions (left-right and up-down) for both the left and right side mirrors.

3   Listen carefully for the sound of the electric motors running in the mirrors.

4   If the motors can be heard but the mirror glass doesn't move, there's probably a problem with the drive mechanism inside the mirror. Remove and disassemble the mirror to locate the problem.

5   If the mirrors don't operate and no sound comes from the mirrors, check the fuse (see Chapter 1).

6   If the fuse is OK, remove the mirror control switch from its mounting without disconnecting the wires attached to it. On models where the power mirror switch is in the instrument panel, remove the switch from the instrument panel to probe the wires behind it. On models where the switch is in the driver's armrest, just pull up the switch panel. Turn the ignition ON and check for voltage at the

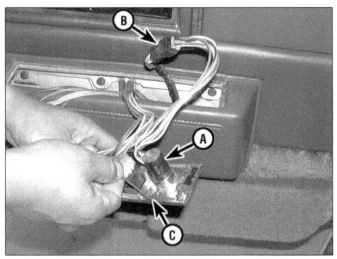

28.6 Check for voltage coming into the switch (A) and leaving the switch - if the switch must be replaced, disconnect the electrical connector (B) and remove the screw with an Allen wrench (C)

30.2 The DERM (arrow indicates its mounting bracket) is mounted under the glove compartment on Deville/Fleetwood models, and under the dash above the accelerator pedal on other models

switch **(see illustration)**. There should be voltage at one terminal. If there's no voltage at the switch, check for an open or short in the wiring between the fuse panel and the switch.

7   Locate the wire going from the switch to ground. Leaving the switch connected, connect a jumper wire between this wire and ground. If the mirror works normally with this wire in place, repair the faulty ground connection.

8   If the mirror still doesn't work, remove the cover and check the wires at the mirror for voltage with a test light. Check with ignition ON and the mirror selector switch on the appropriate side. Operate the mirror switch in all its positions. There should be voltage at one of the switch-to-mirror wires in each switch position (except the neutral "off" position).

9   If there's no voltage in any switch position, check the wiring between the mirror and control switch for opens and shorts.

10   If there's voltage, remove the mirror and test it off the vehicle with jumper wires. Replace the mirror if it fails this test (see Chapter 11).

---

29   Electric sunroof - description and check

1   The electric sunroof is powered by a single motor located in the roof above the center of the windshield. The power circuit is protected by a circuit breaker, and a sunroof control module located in the roof at the right-front of the sunroof opening signals the motor is response to the sunroof switch.

2   The control switch sends a ground signal to the sunroof motor when the switch is pressed. With the ignition On but the engine Off, operate the sunroof control switch through the Forward and Reverse functions.

3   Listen carefully for the sound of the sunroof motor running in the roof.

4   If the motors can be heard but the sunroof glass doesn't move, there's probably a problem with the drive mechanism or drive cables.

5   If the sunroof does not operate and no sound comes from the motor, check the circuit breaker (23) in the interior fuse/relay panel.

6   If the circuit breaker is OK, remove the control switch at the left-front of the overhead interior light panel. Disconnect the wires attached to it. Turn the ignition On and check for voltage at the switch. If there's no voltage at the switch, check for power and ground at the motor. If power and ground exist at the motor and there's still no voltage at the switch, replace the motor. If there's no voltage at the motor, check the control module or look for an open or short in the wiring between the module and the motor.

7   If there's voltage at the switch, disconnect it. Check the switch for continuity in all its operating positions. If the switch does not have continuity, replace it.

8   If the switch has continuity, re-connect the switch. Locate the wire going from the switch to ground. Leaving the switch connected, connect a jumper wire between this wire and ground. If the motor works normally with this wire in place, repair the faulty ground connection.

9   The sunroof can be closed manually by inserting a 4mm Allen wrench into the motor shaft and rotating it clockwise. **Note:** *To access the motor, pull back the trim edging around the front of the sunroof opening and release the front portion of the headliner.*

---

30   Airbag system - general information

Description

Refer to illustration 30.2

1   Some models are equipped with a Sup-

plemental Inflatable Restraint (SIR), more commonly known as an airbag. The SIR system is designed to protect the driver from serious injury in the event of a head-on or frontal collision. The horn pad on your vehicle should have an SIR designation facing the driver if the vehicle is equipped with an airbag.

2   The SIR system consists of an impact sensor mounted in front of the radiator and one under the dashboard, a resistor module, an airbag assembly in the center of the steering wheel, and a Diagnostic Energy Reserve Module (DERM). The heart of the system is the DERM, which determines when the airbag should be deployed and supplies reserve energy to the SIR system, even if the crash takes out the vehicle's battery **(see illustration)**. It also monitors the rest of the system, records trouble codes, and turns on the "Inflatable Restraints" warning light on the instrument panel if there is a problem with the system.

Operation

3   The sensor in the passenger compartment contains an impact sensor and an arming sensor. The arming sensor powers the airbag, while the impact sensors (front and interior) provide ground to the airbag causing it to deploy in the event of sufficient impact. If the battery is destroyed by the impact, or is too low to power the inflator, the DERM provides power.

Self-diagnosis system

4   A self-diagnosis circuit in the SIR unit displays a light when the ignition switch is turned to the On position. If the system is operating normally, the light should flash seven to nine times, then go out. If the light doesn't come on, or doesn't go out after six seconds, or if it comes on while you're driving the vehicle, there's a malfunction in the SIR system. Have it inspected and repaired as

30.7 Disconnect the yellow two-wire connector (arrow) near the bottom of the steering column

soon as possible. Do not attempt to troubleshoot or service the SIR system yourself. Even a small mistake could cause the SIR system to malfunction when you need it.

Servicing components near the SIR system

5   Nevertheless, there are times when you need to remove the steering wheel, radio or service other components on or near the dashboard. At these times, you'll be working around components and wire harnesses for the SIR system. The SIR wiring harnesses are easy to identify: They're all bright yellow. Except for the one connector mentioned in Step 7, do not unplug the connectors for these wires. And do not use electrical test equipment on yellow wires; it could cause the airbag(s) to deploy. *ALWAYS DISABLE THE SIR SYSTEM BEFORE WORKING NEAR THE SIR SYSTEM COMPONENTS OR RELATED WIRING.*

Disabling the SIR system

Refer to illustration 30.7

Warning: *Any time you are working in the vicinity of airbag wiring, the sensors or components, DISABLE THE SIR SYSTEM.*

6   With the ignition key OFF, refer to Section 3 and locate the airbag fuse in the interior fuse panel or underhood fuse/relay panel. Remove the fuse.

7   Remove the driver's knee bolster (see Chapter 11) and unplug the yellow two-pin connector located near the bottom of the steering column **(see illustration).**

Enabling the system

8   After you've disabled the airbag and performed the necessary service, reconnect the two-pin airbag connector near the bottom of the steering column **(see illustration 30.7).**

9   Reinstall the driver's knee bolster.
10   Reinstall the airbag fuse and turn the ignition key to ON. If the airbag warning light doesn't flash seven to nine times and go out, as it should, bring the vehicle to a dealer service department for diagnosis of the SIR system.

Removal and installation

11   Refer to Chapter 10 for removal and installation of the driver's side airbag.

31   Wiring diagrams - general information

Since it isn't possible to include all wiring diagrams for every year and model covered by this manual, the following diagrams are those that are typical and most commonly needed.

Prior to troubleshooting any circuits, check the fuse and circuit breakers (if equipped) to make sure they're in good condition. Make sure the battery is properly charged and check the cable connections (see Chapter 1).

When checking a circuit, make sure that all connectors are clean, with no broken or loose terminals. When unplugging a connector, do not pull on the wires. Pull only on the connector housings themselves.

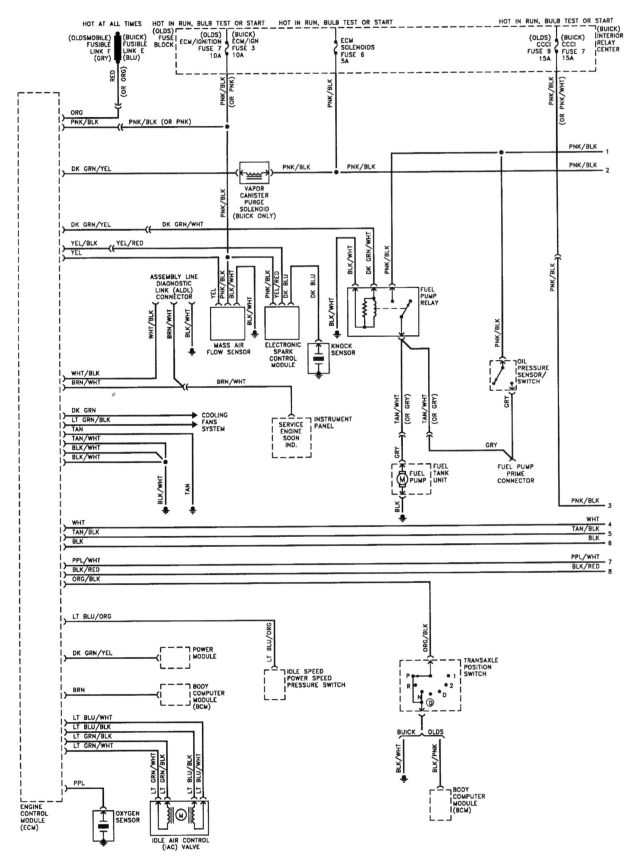

Engine control system - 1986 and 1987 Buick Riviera/Oldsmobile Toronado (1 of 2)

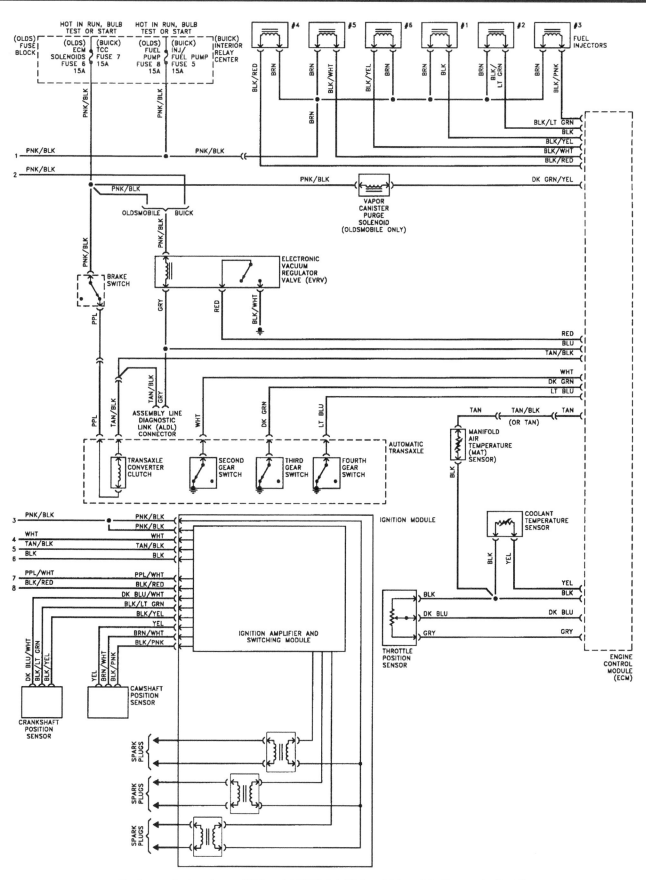

Engine control system - 1986 and 1987 Buick Riviera/Oldsmobile Toronado (2 of 2)

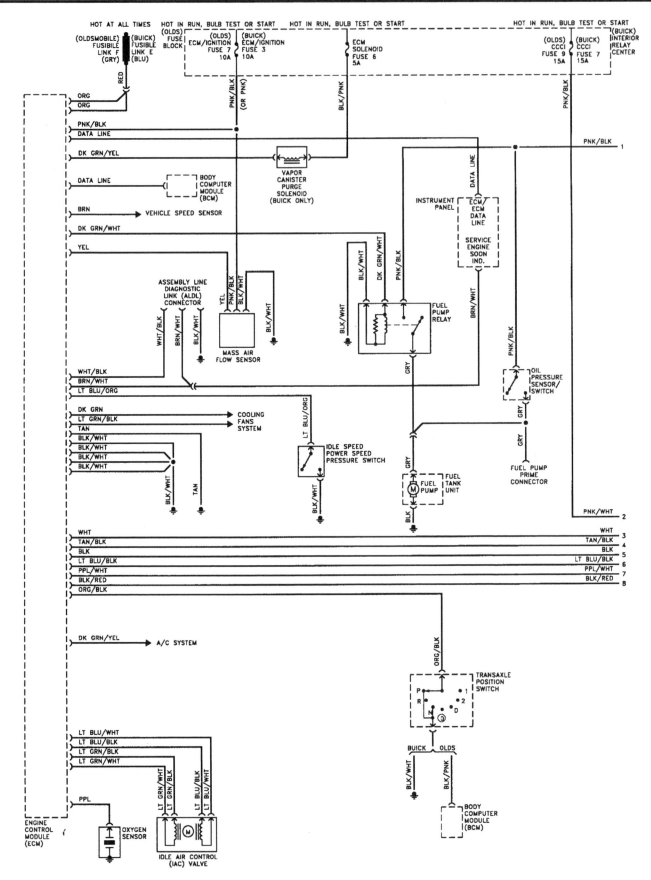

Engine control system - 1988 Buick Riviera/Oldsmobile Toronado (1 of 2)

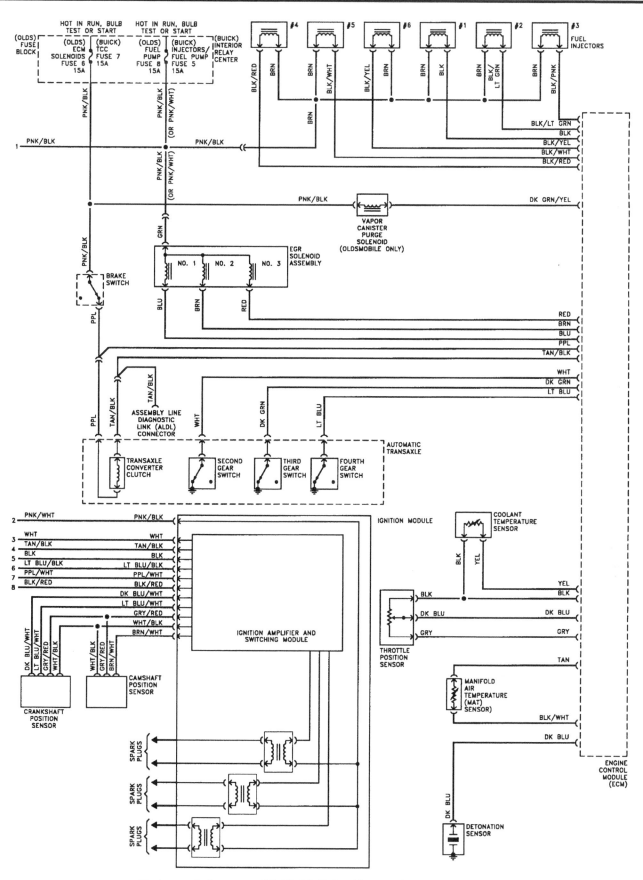

Engine control system - 1988 Buick Riviera/Oldsmobile Toronado (2 of 2)

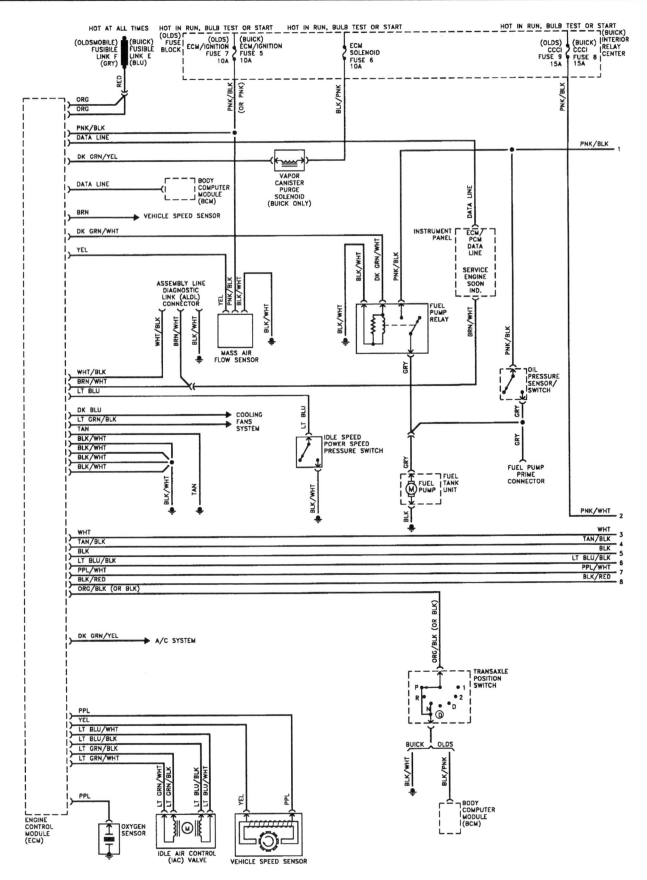

Engine control system - 1989 Buick Riviera/Oldsmobile Toronado (1 of 2)

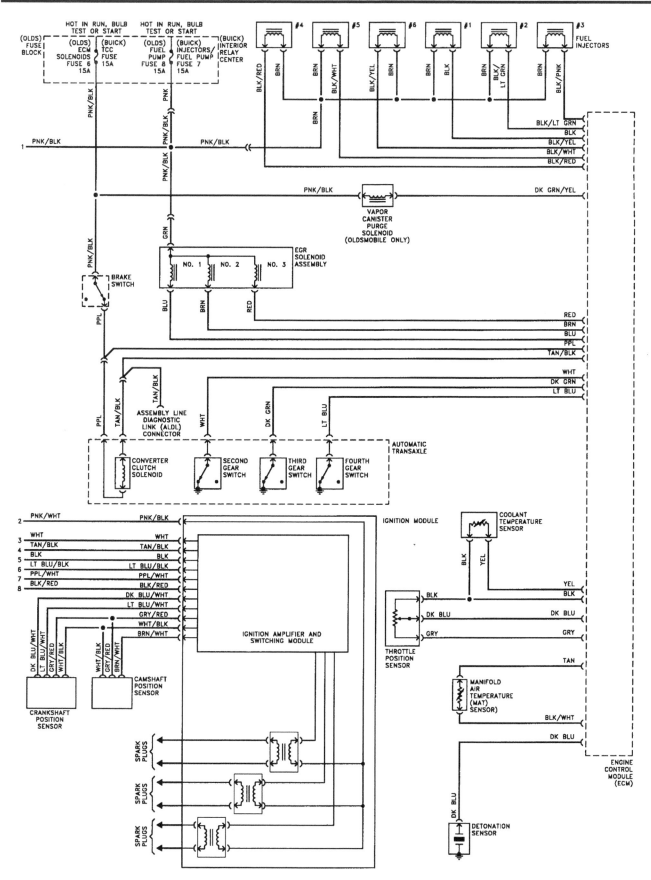

Engine control system - 1989 Buick Riviera/Oldsmobile Toronado (2 of 2)

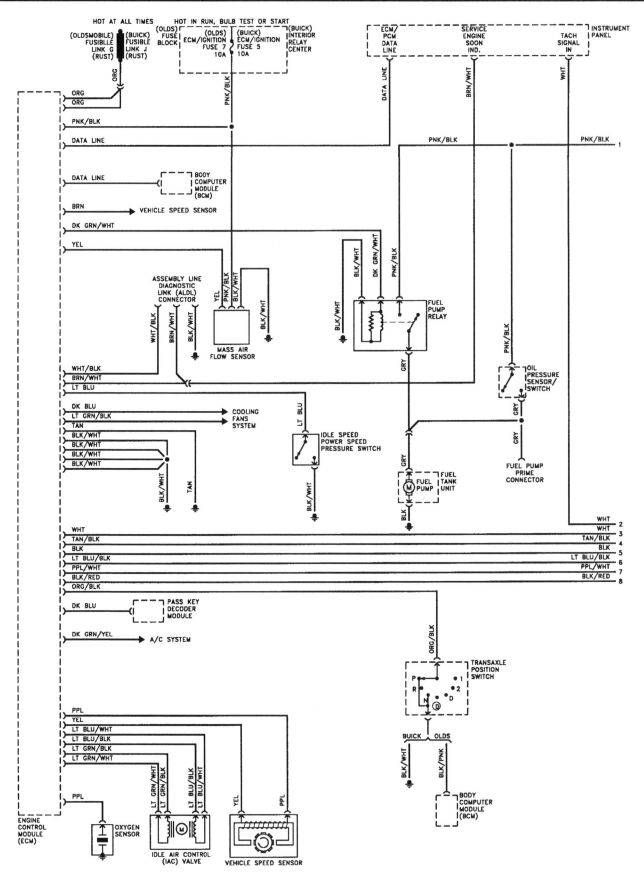

Engine control system - 1990 Buick Riviera/Oldsmobile Toronado (1 of 2)

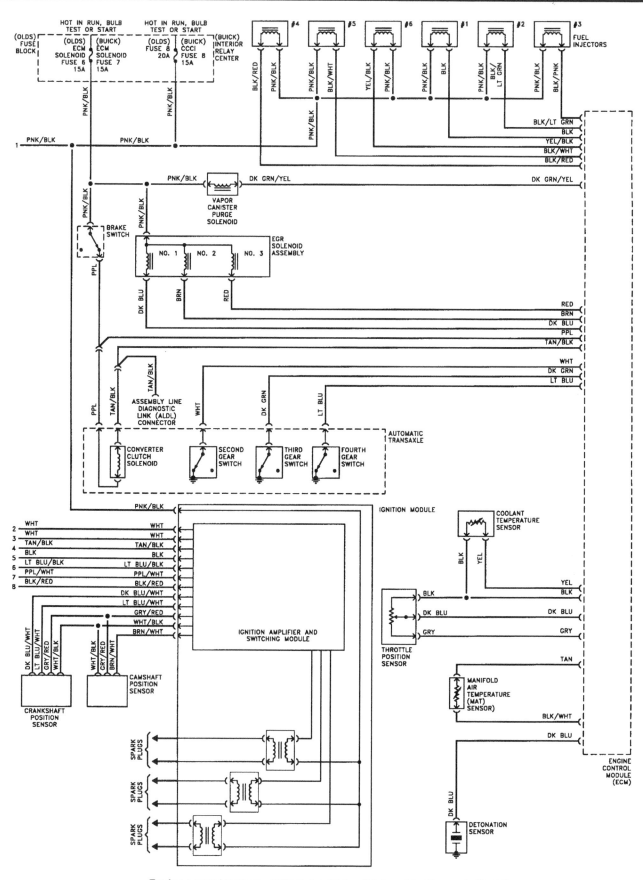

Engine control system - 1990 Buick Riviera/Oldsmobile Toronado (2 of 2)

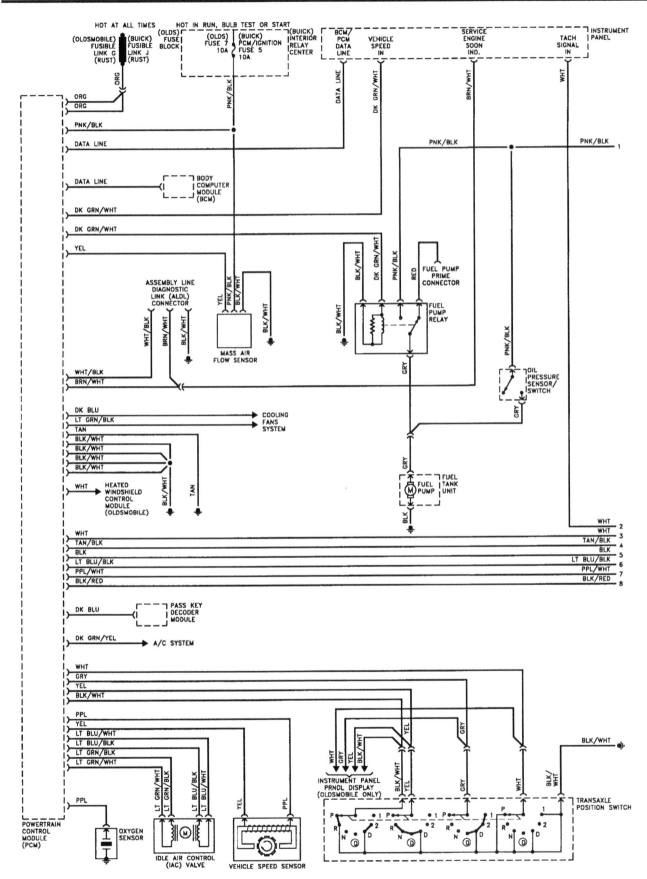

Engine control system - 1991 and later Buick Riviera/Oldsmobile Toronado (1 of 2)

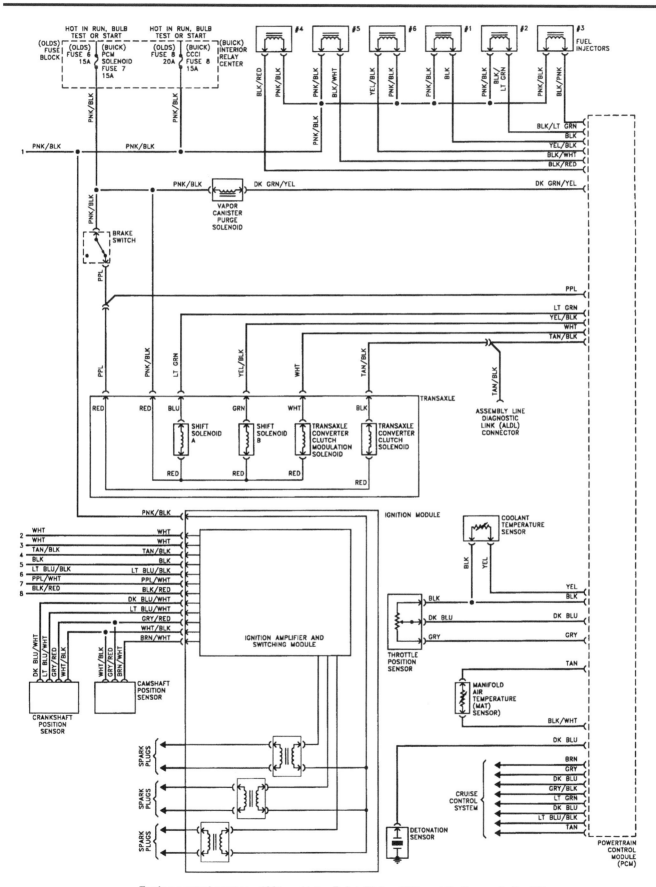

Engine control system - 1991 and later Buick Riviera/Oldsmobile Toronado (2 of 2)

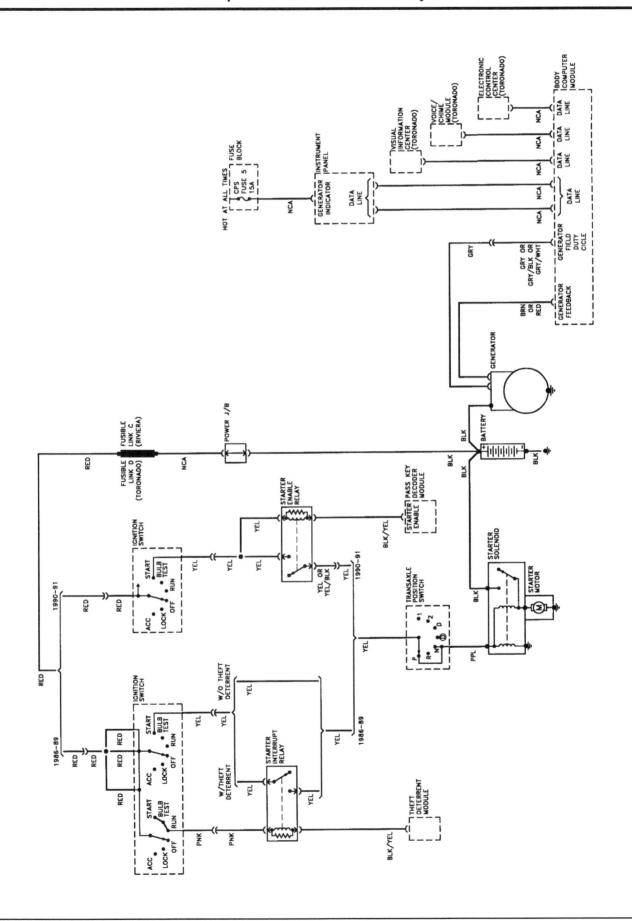

Starting and charging systems - Buick Riviera/Oldsmobile Toronado

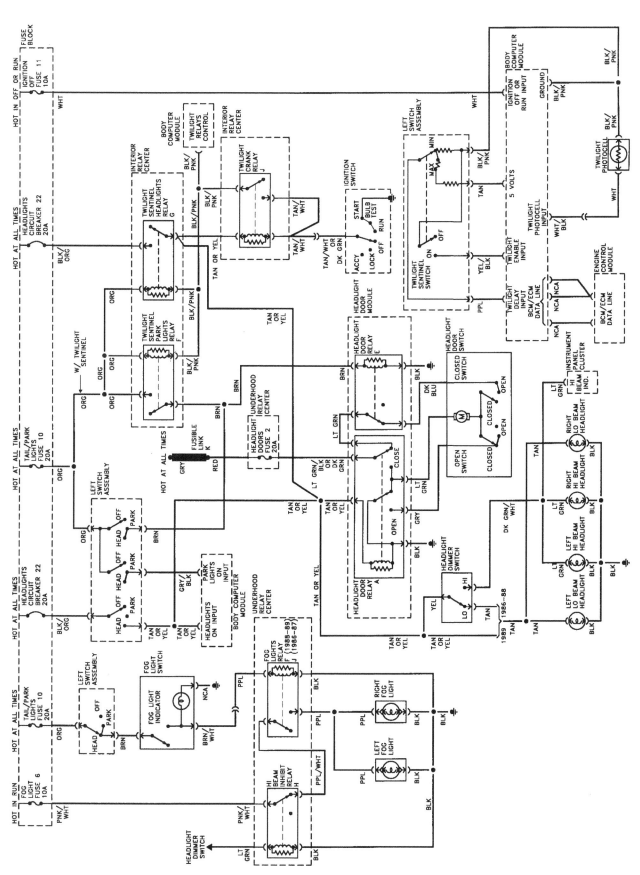

Headlight system - 1989 and earlier Oldsmobile Toronado

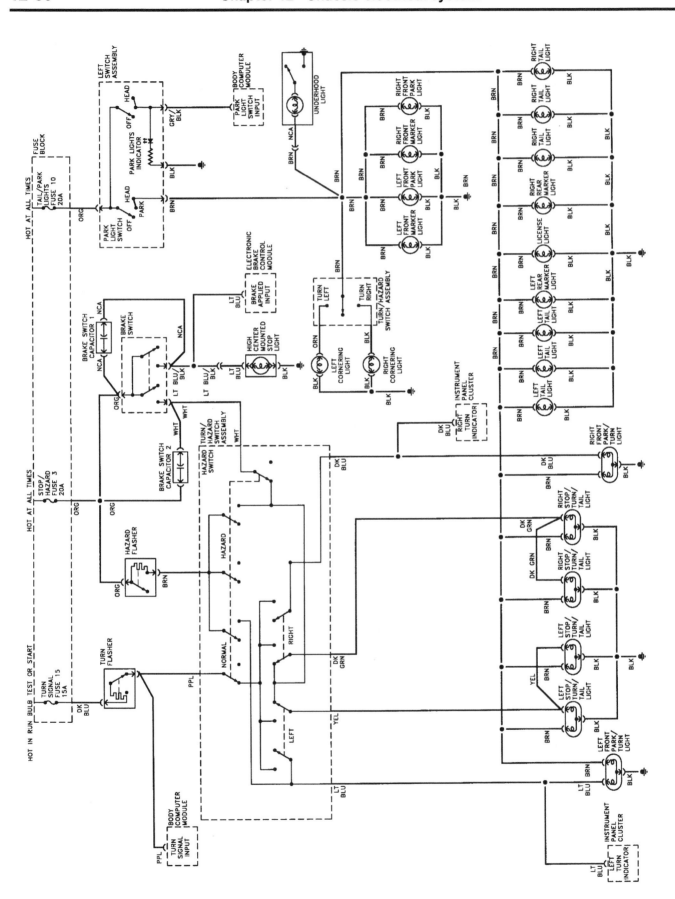

Exterior lighting system - 1989 and earlier Oldsmobile Toronado

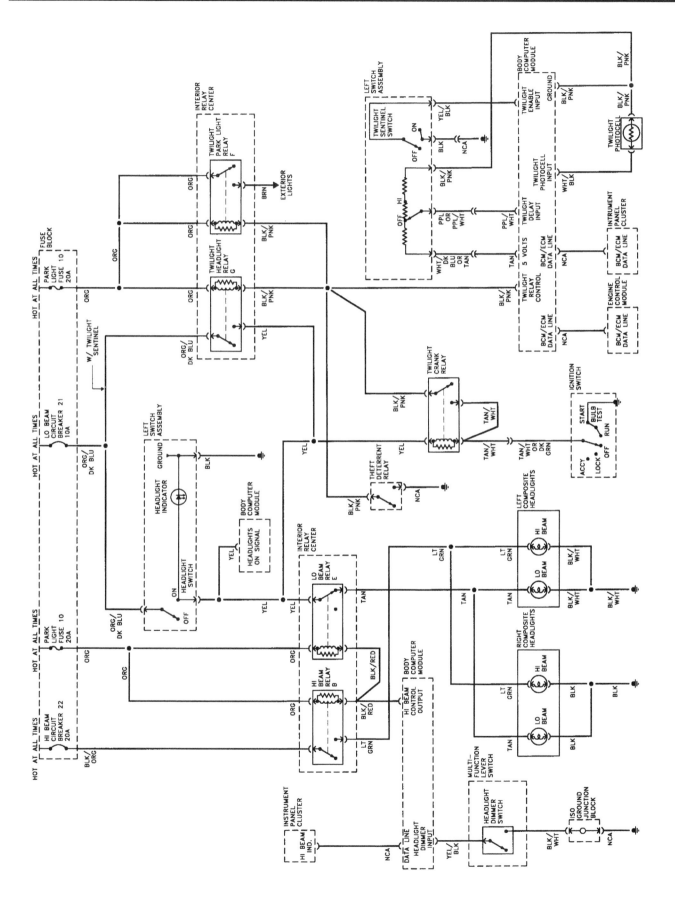

Headlight system - 1989 and earlier Buick Riviera

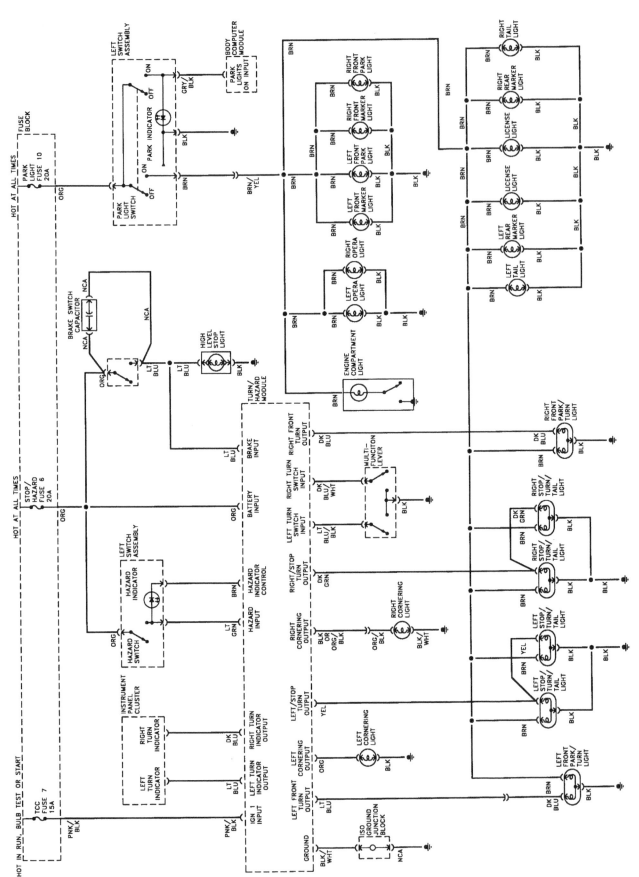
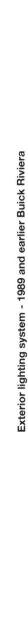

Exterior lighting system - 1989 and earlier Buick Riviera

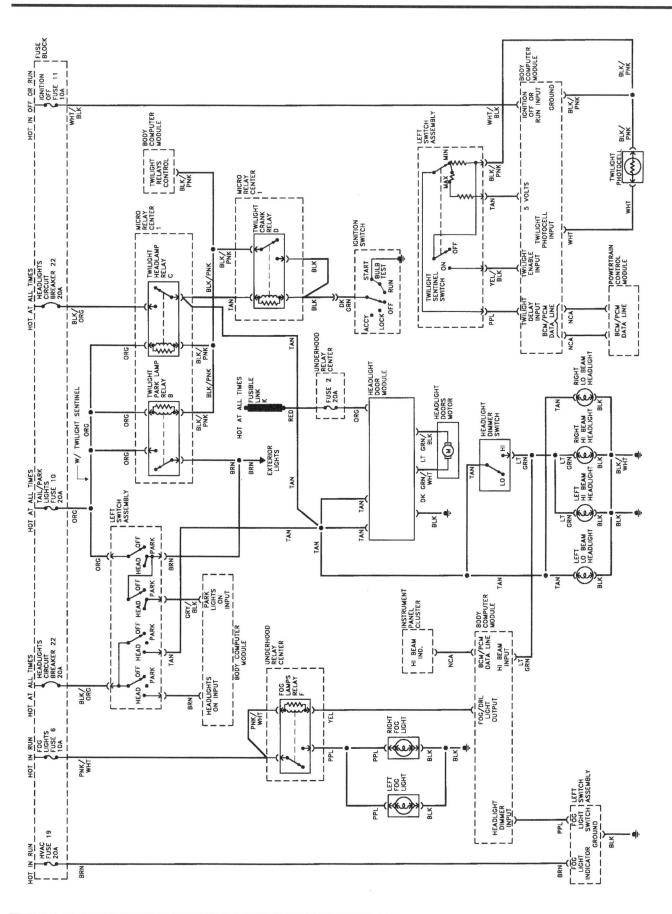

Headlight system - 1990 and later Oldsmobile Toronado

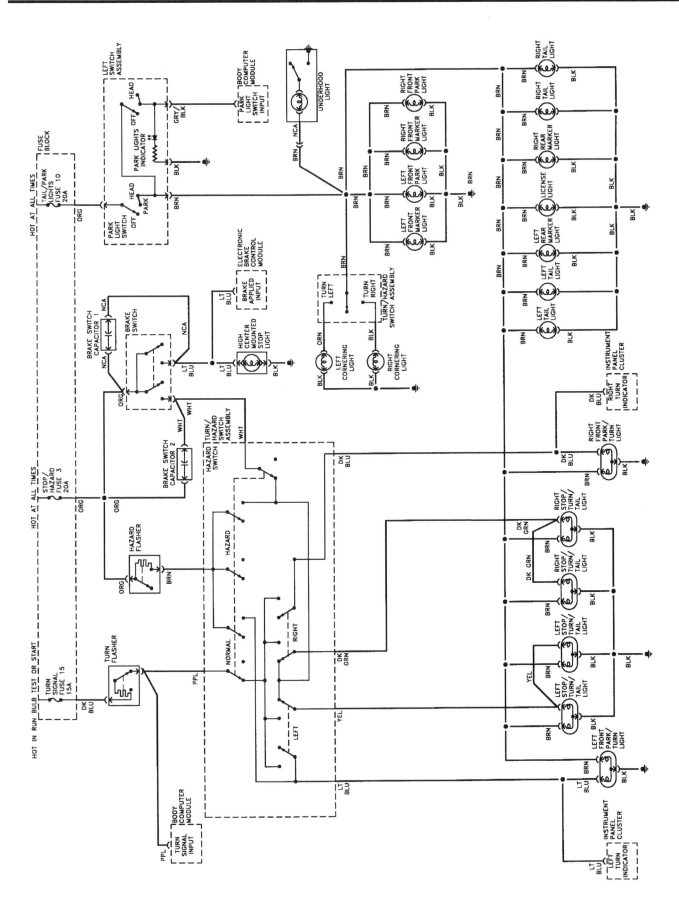

Exterior lighting system - 1990 later Oldsmobile Toronado

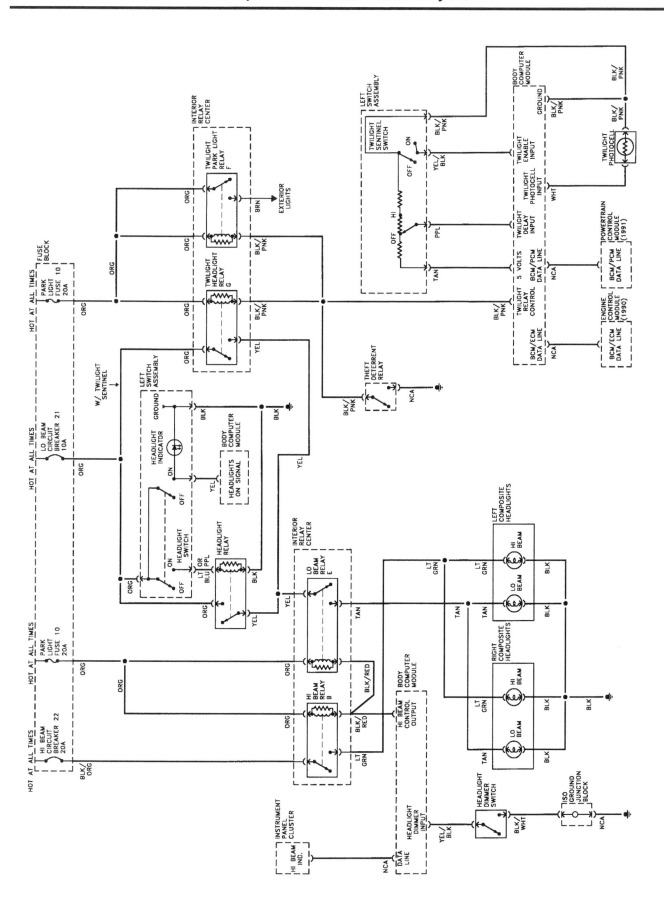

Headlight system - 1990 and later Buick Riviera

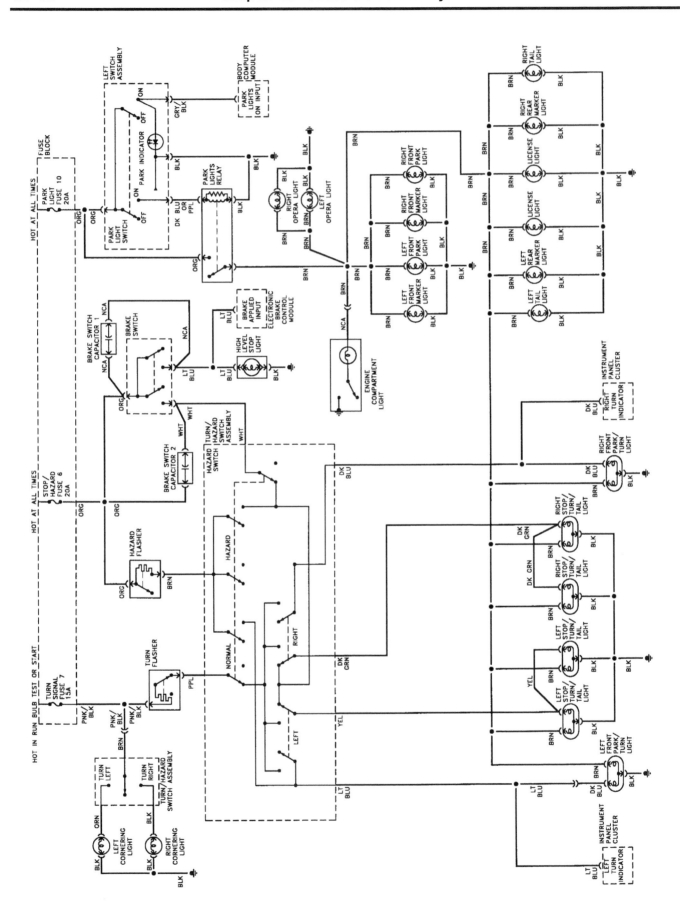

Exterior lighting system - 1990 and later Buick Riviera

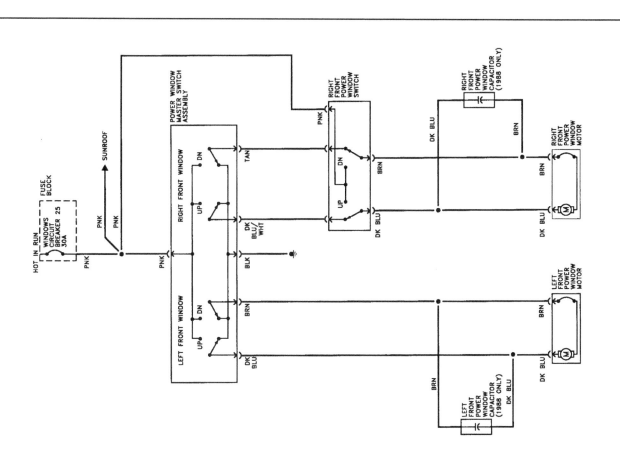

Power window system - 1990 and later Oldsmobile Toronado

Power window system - 1989 and earlier Oldsmobile Toronado

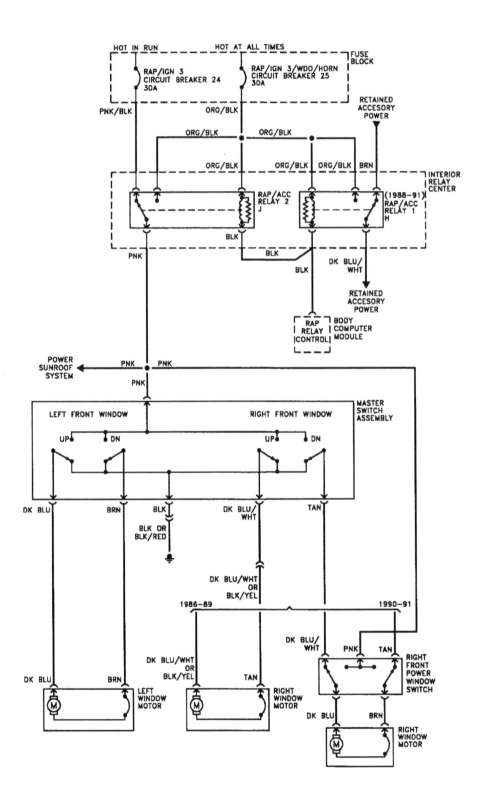

Power window system - Buick Riviera

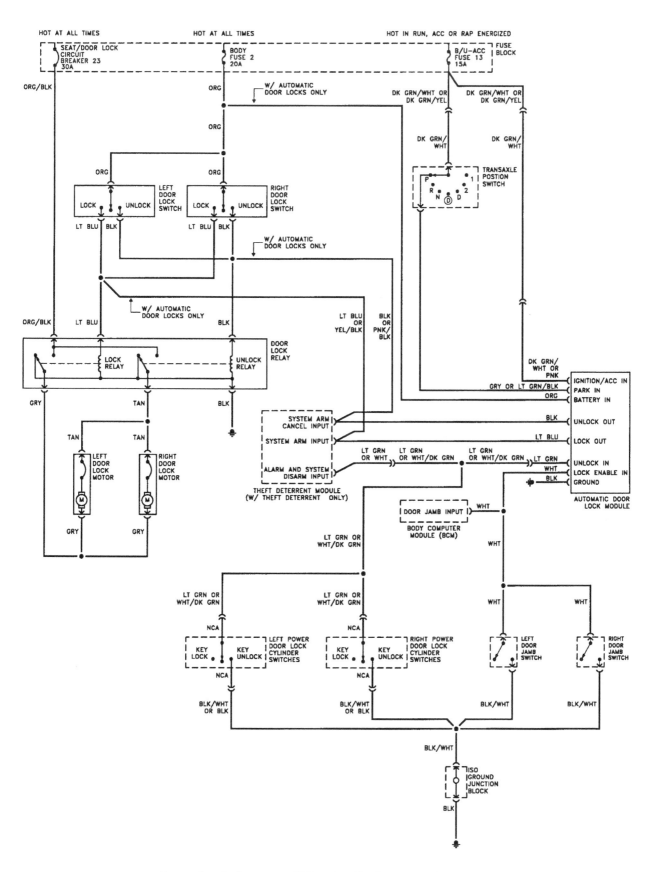

Power door lock system - 1989 and earlier Oldsmobile Toronado

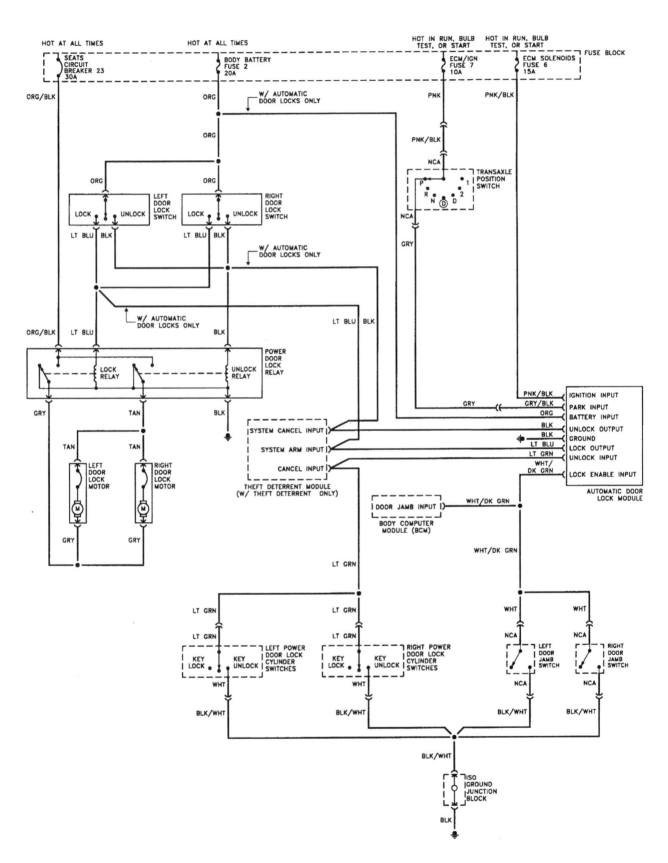

Power door lock and keyless entry system - 1988 and earlier Buick Riviera

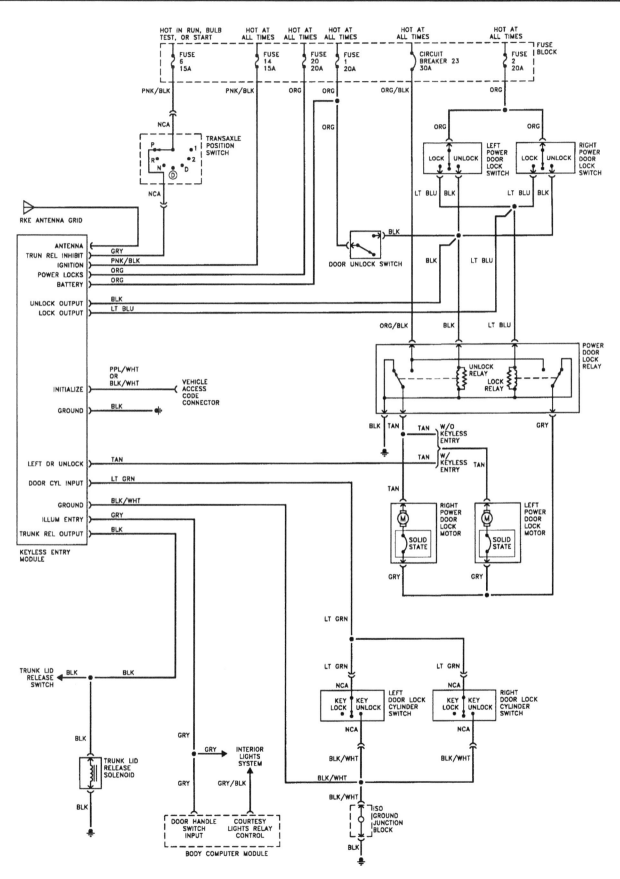

Power door lock and keyless entry system - 1990 and later Oldsmobile Toronado

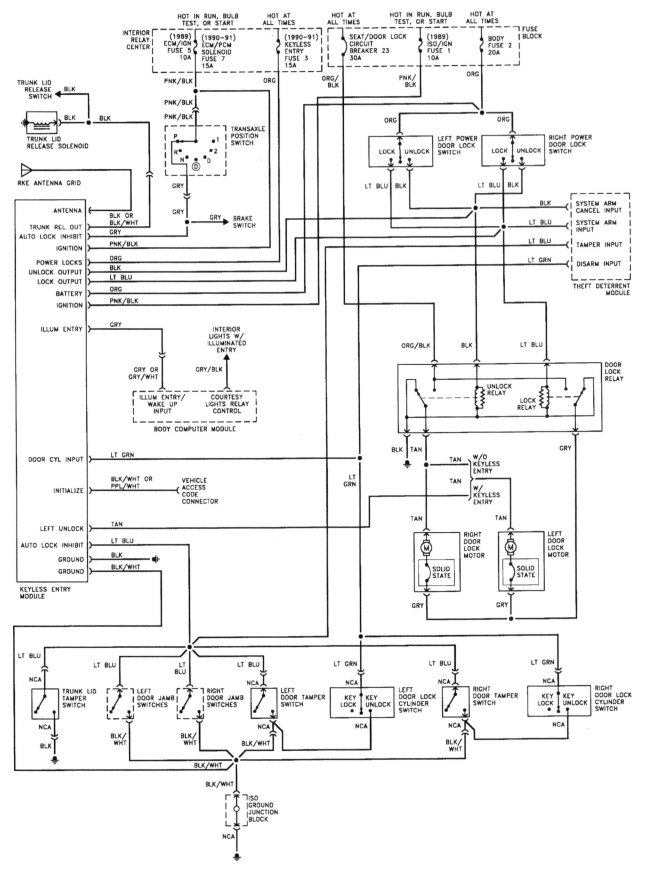

Power door lock and keyless entry system - 1989 and later Buick Riviera

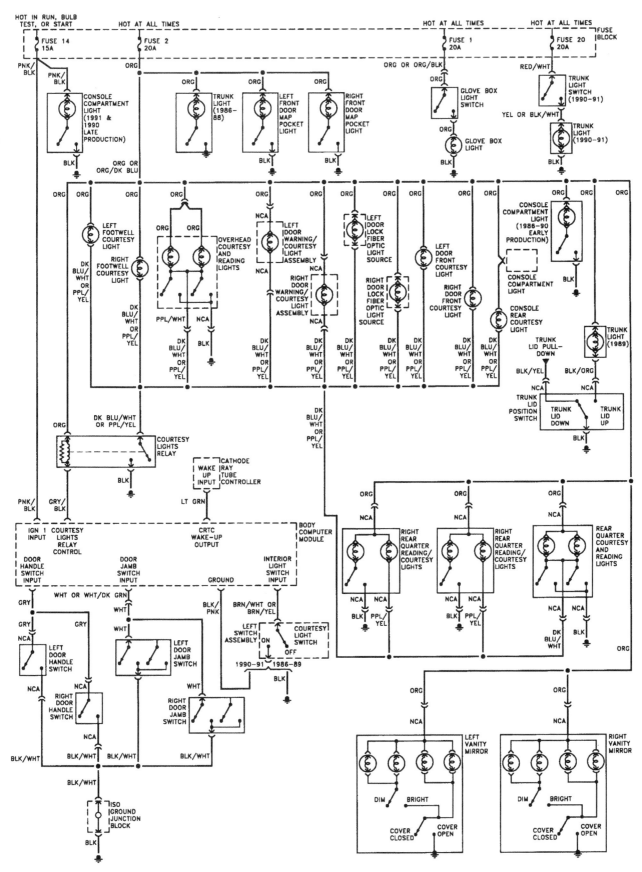

Courtesy lighting system - Oldsmobile Toronado

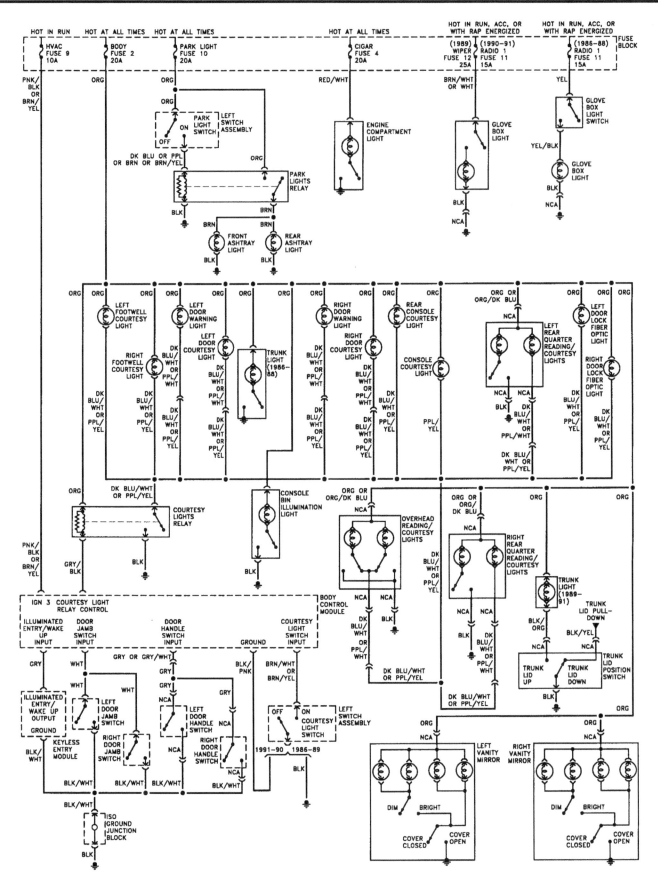

Courtesy lighting system - Buick Riviera

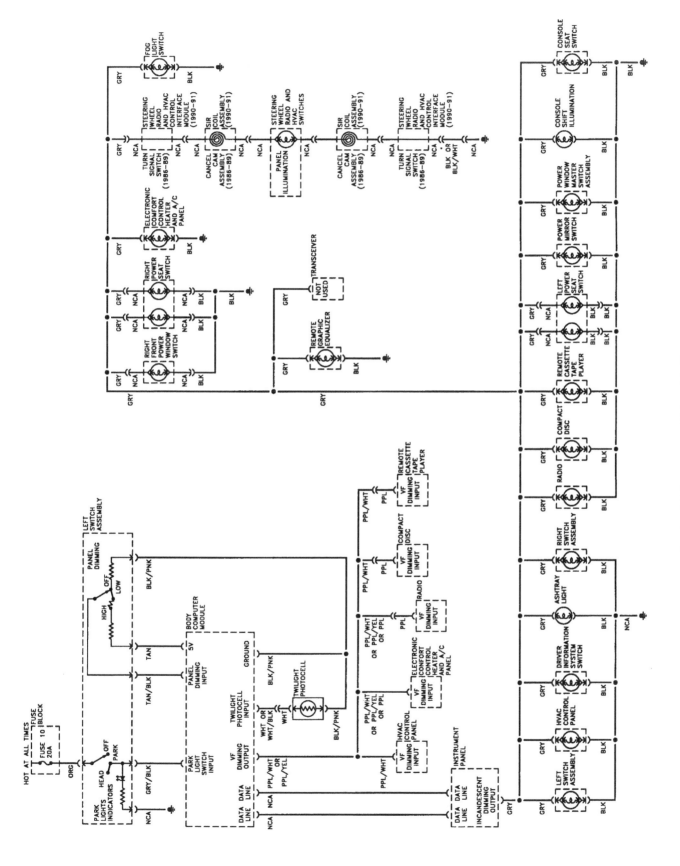

Interior lighting system - Oldsmobile Toronado

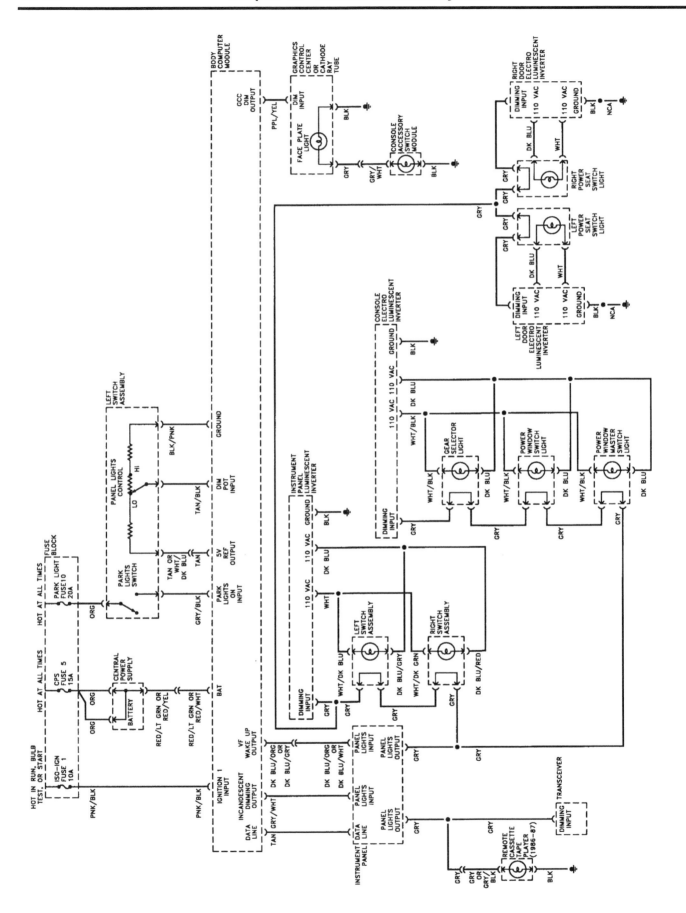

Interior lighting system - 1989 and earlier Buick Riviera

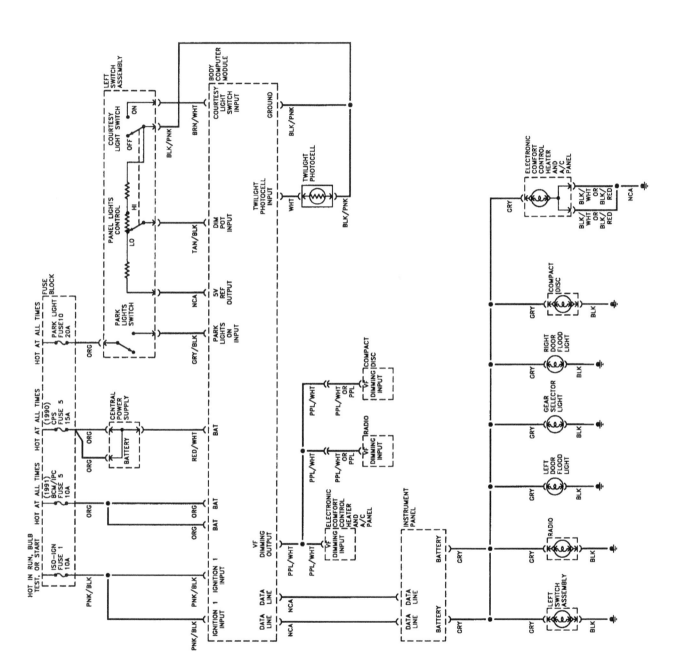

Interior lighting system – 1990 and later Buick Riviera

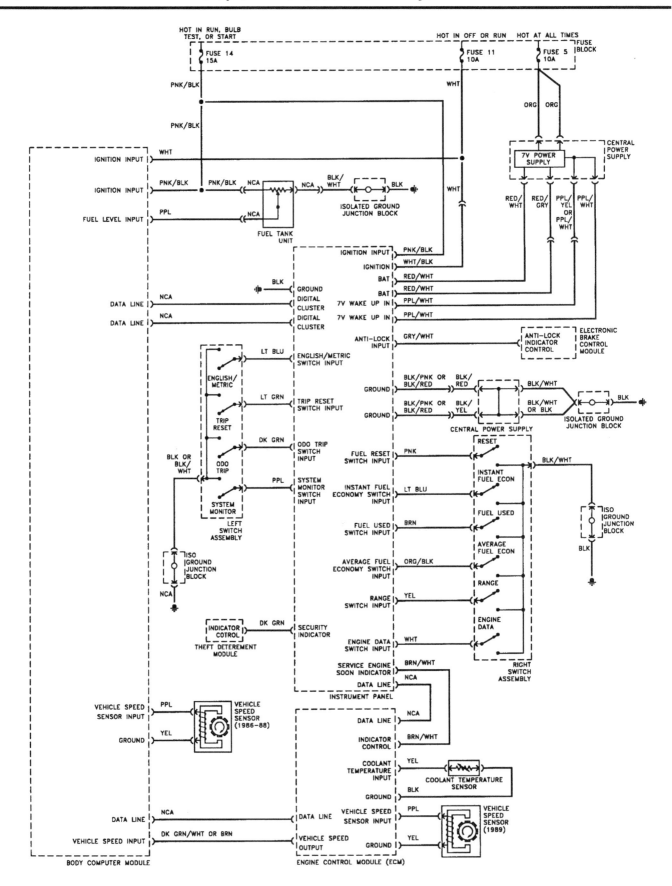

Instrument panel gauges and warning lights - 1989 and earlier Oldsmobile Toronado

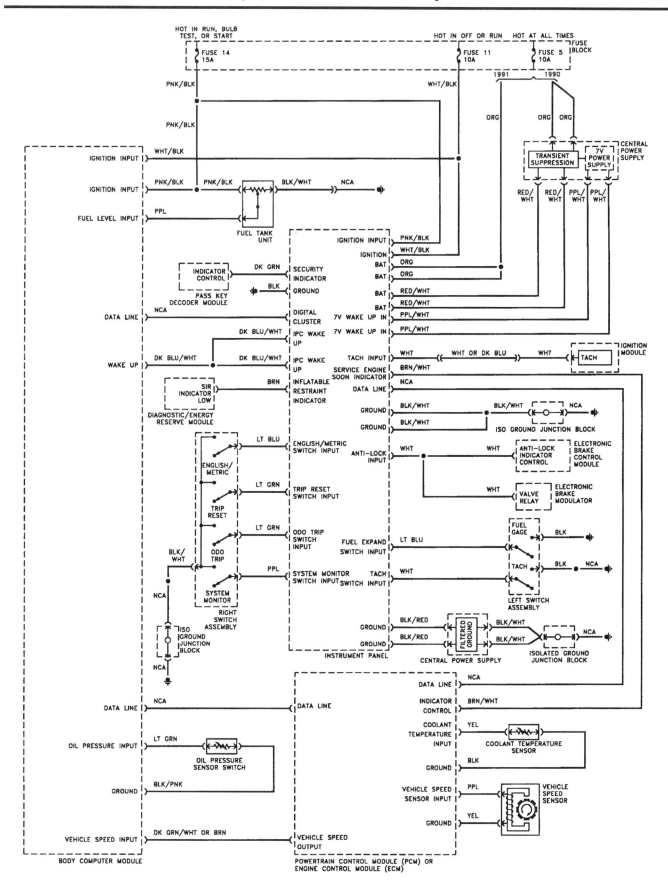

Instrument panel gauges and warning lights - 1990 and later Oldsmobile Toronado

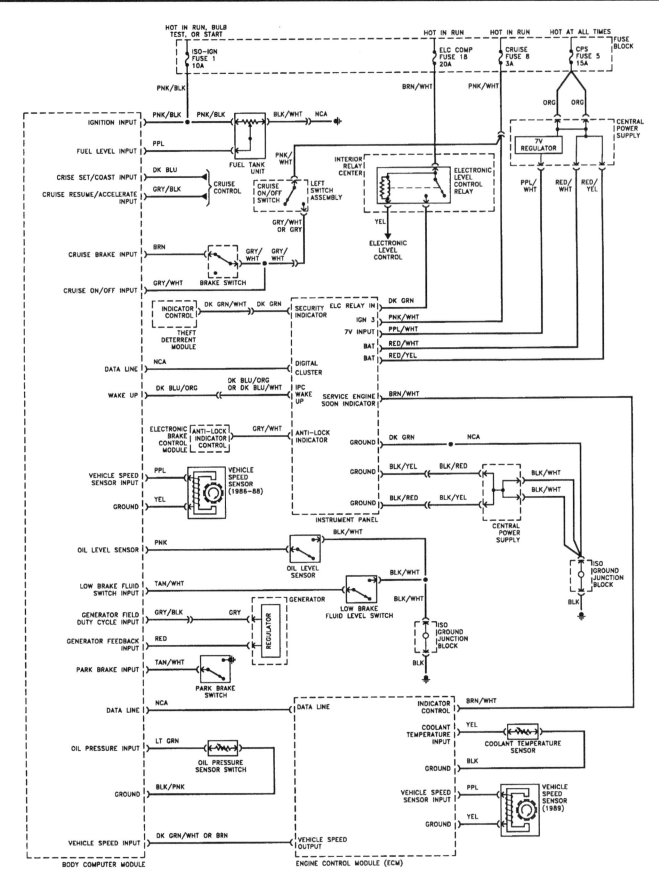

Instrument panel gauges and warning lights - 1989 and earlier Buick Riviera

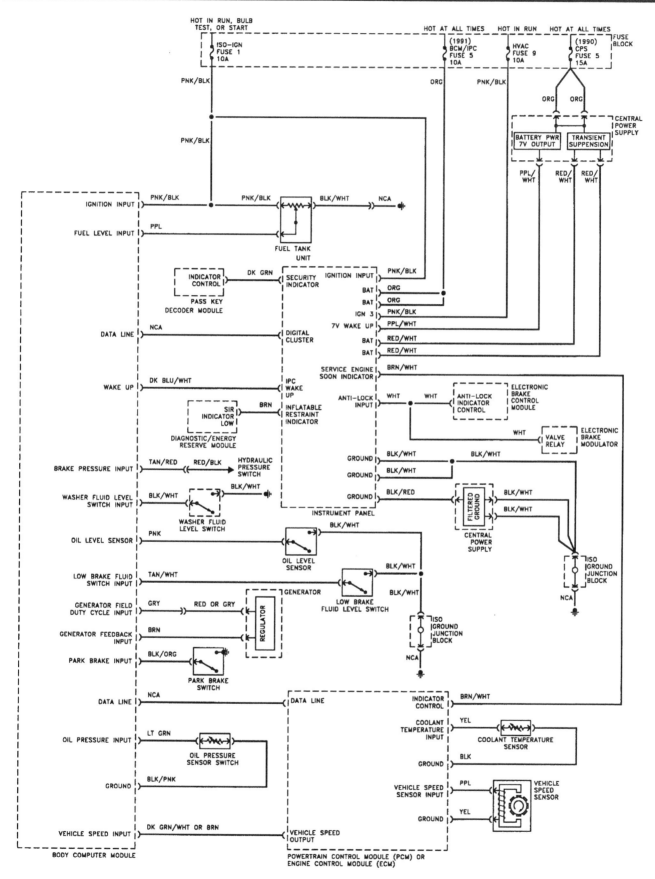

Instrument panel gauges and warning lights - 1990 and later Buick Riviera

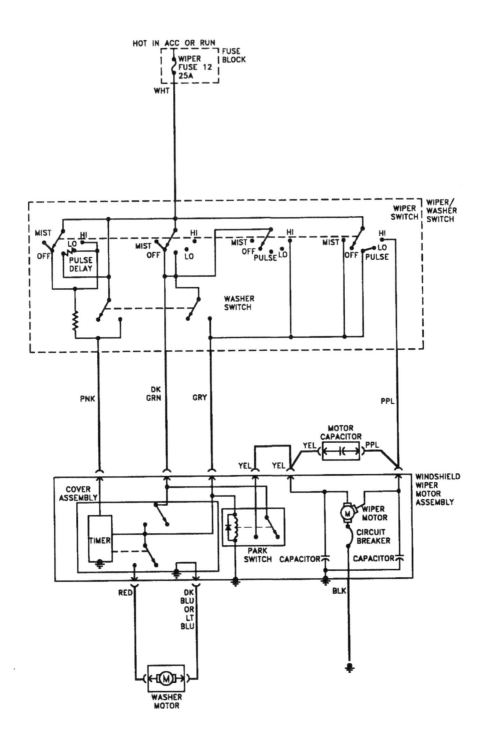

Windshield wiper and washer system - Buick Riviera/Oldsmobile Toronado

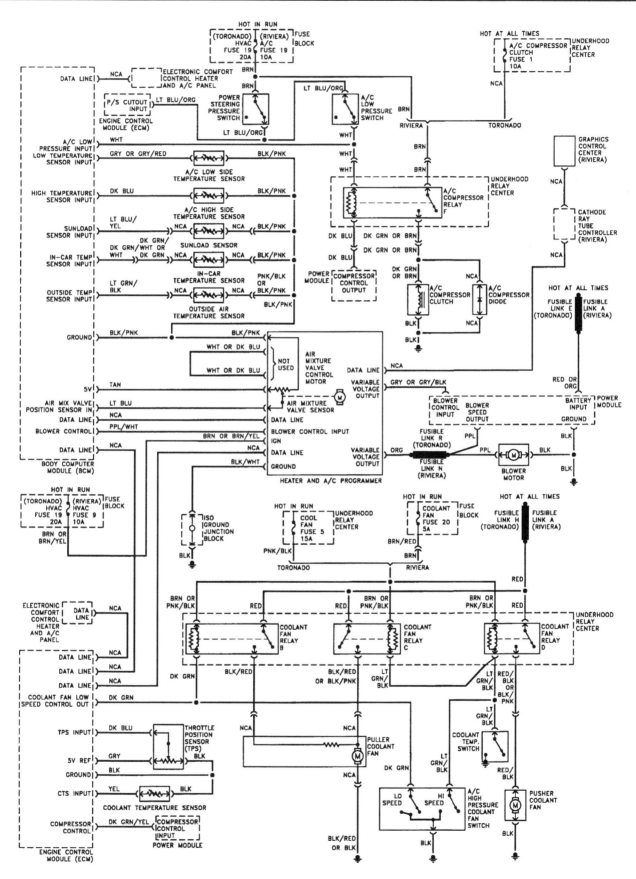

Heating and air conditioning system (including engine cooling fan system) - 1986 and 1987 Buick Riviera/Oldsmobile Toronado

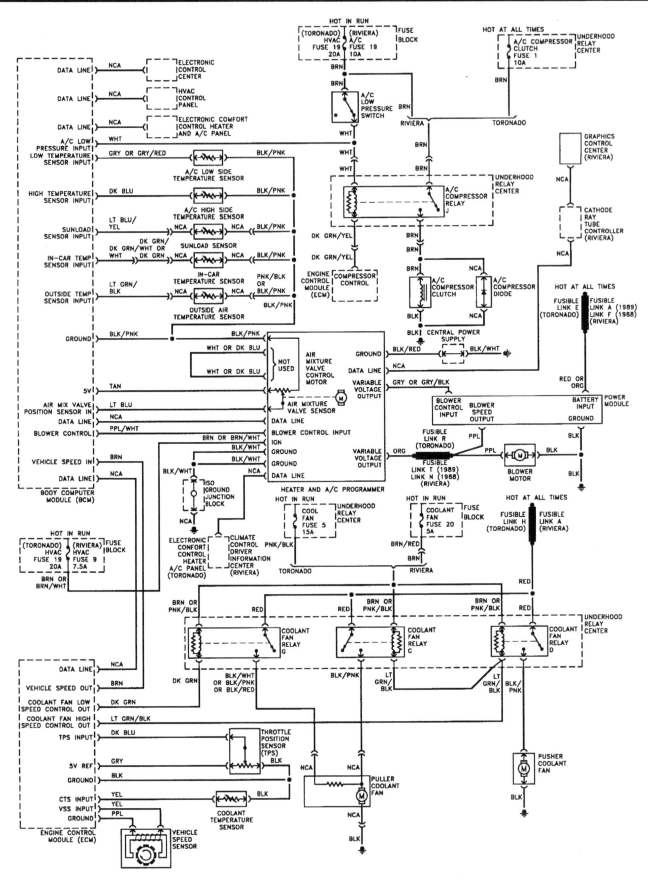

Heating and air conditioning system (including engine cooling fan system) - 1988 and 1989 Buick Riviera/Oldsmobile Toronado

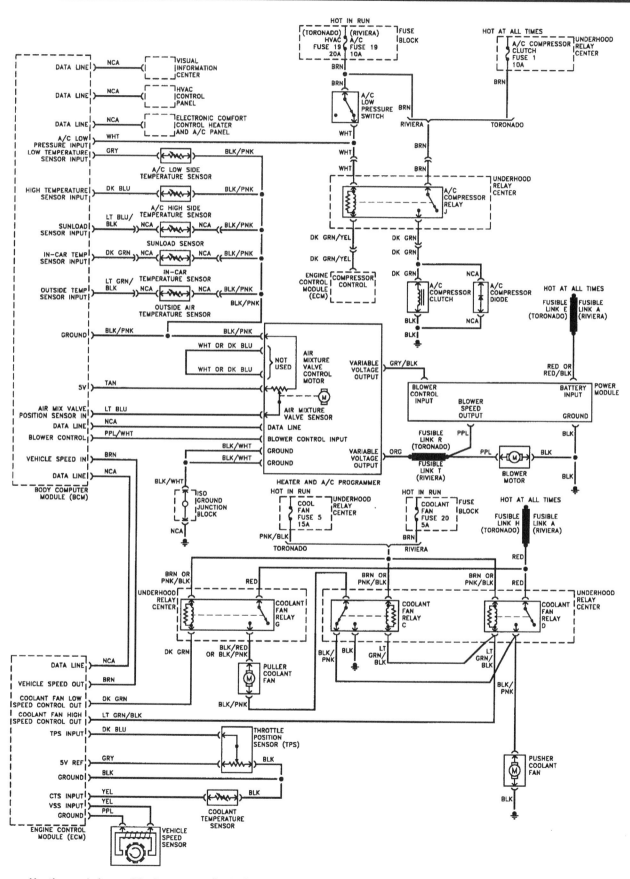

Heating and air conditioning system (including engine cooling fan system) - 1990 Buick Riviera/Oldsmobile Toronado

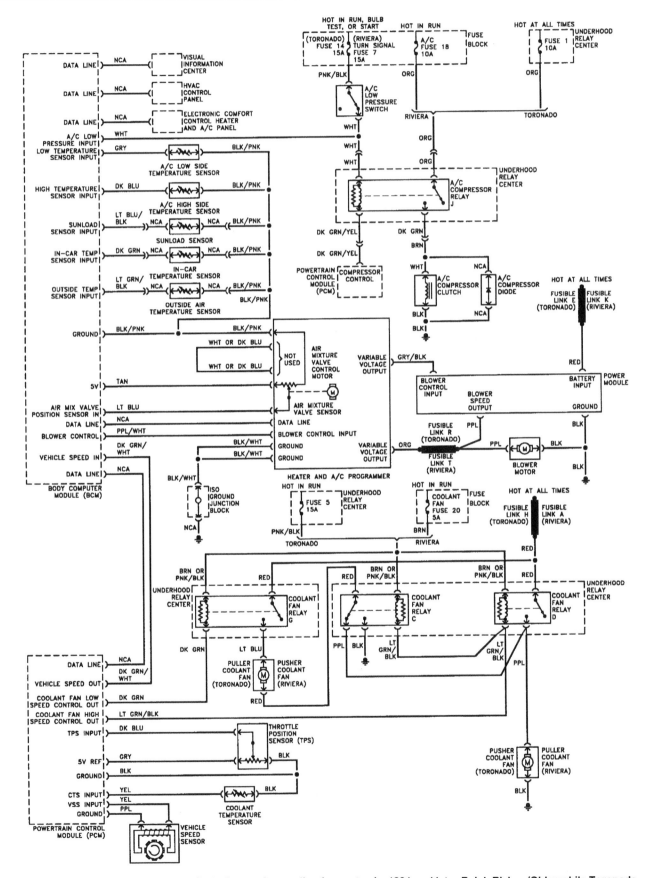

Heating and air conditioning system (including engine cooling fan system) - 1991 and later Buick Riviera/Oldsmobile Toronado

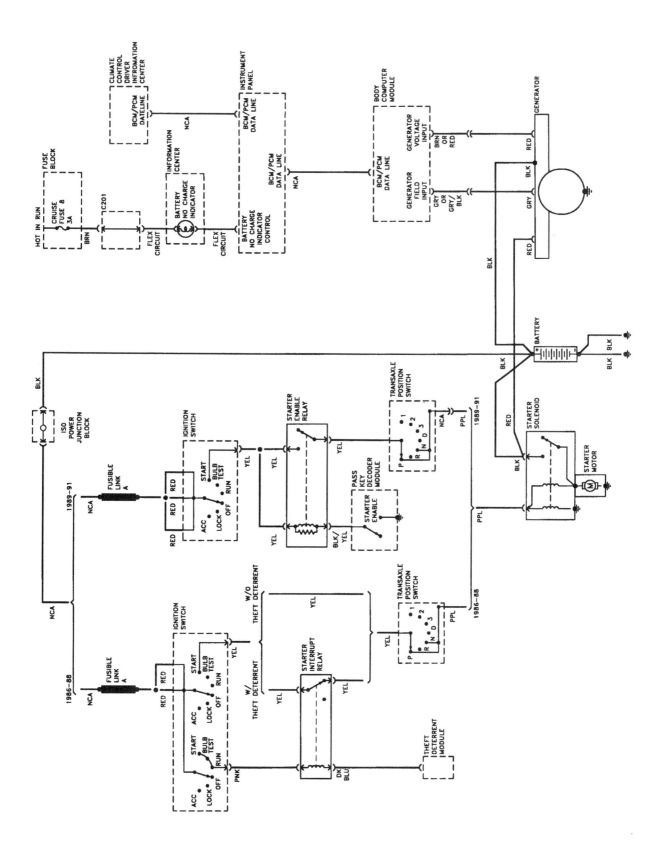

Starting and charging systems - Cadillac Seville/Eldorado

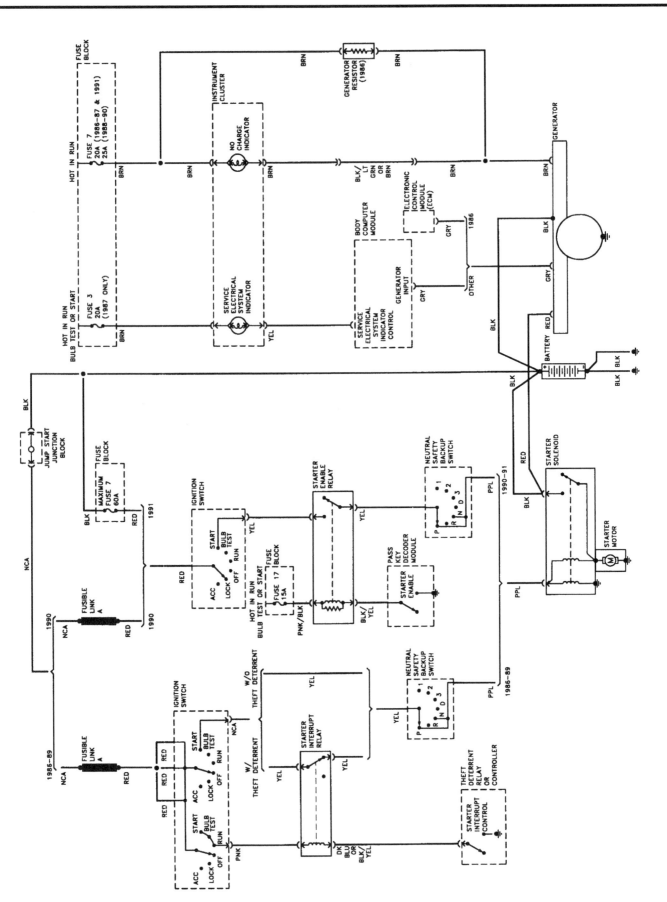
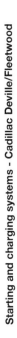

Starting and charging systems - Cadillac Deville/Fleetwood

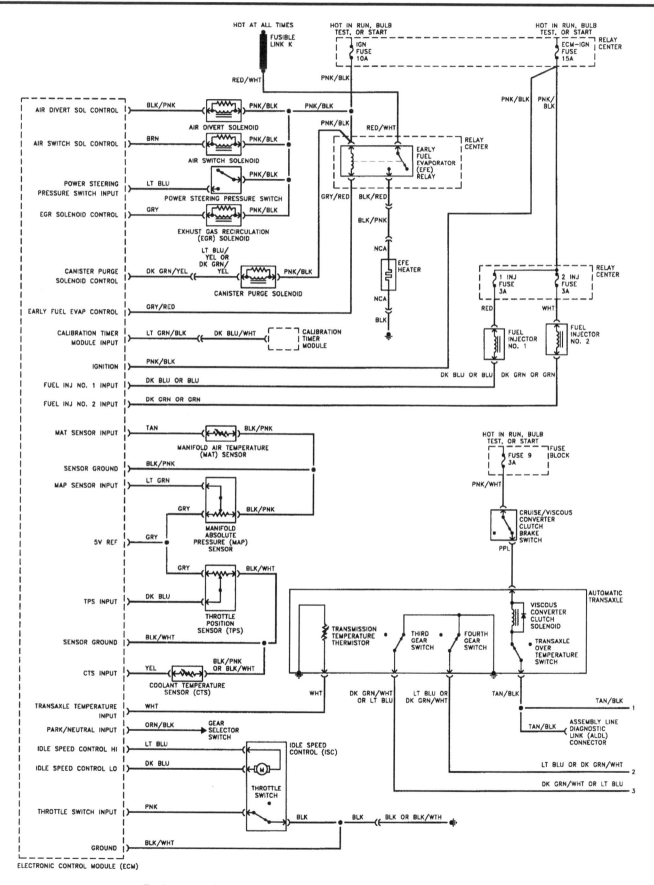

Engine control system - 1986 and 1987 Cadillac Deville/Fleetwood (1 of 2)

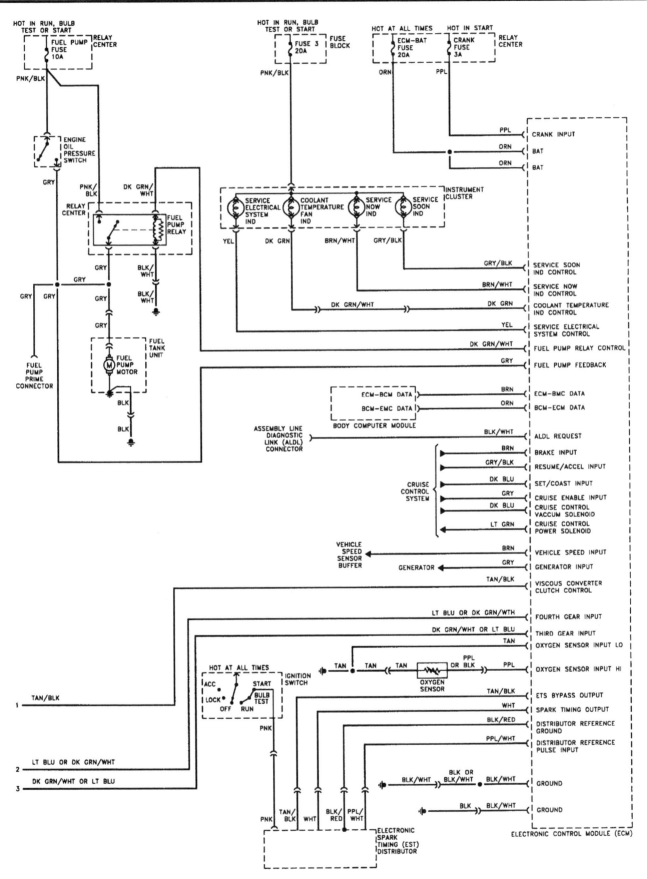

Engine control system - 1986 and 1987 Cadillac Deville/Fleetwood (2 of 2)

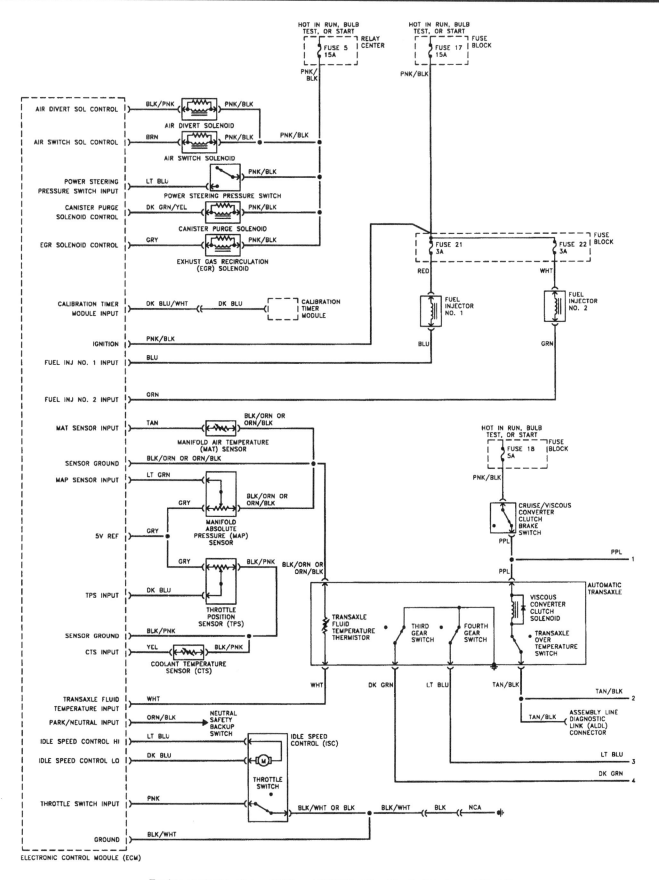

Engine control system - 1988 and 1989 Cadillac Deville/Fleetwood (1 of 2)

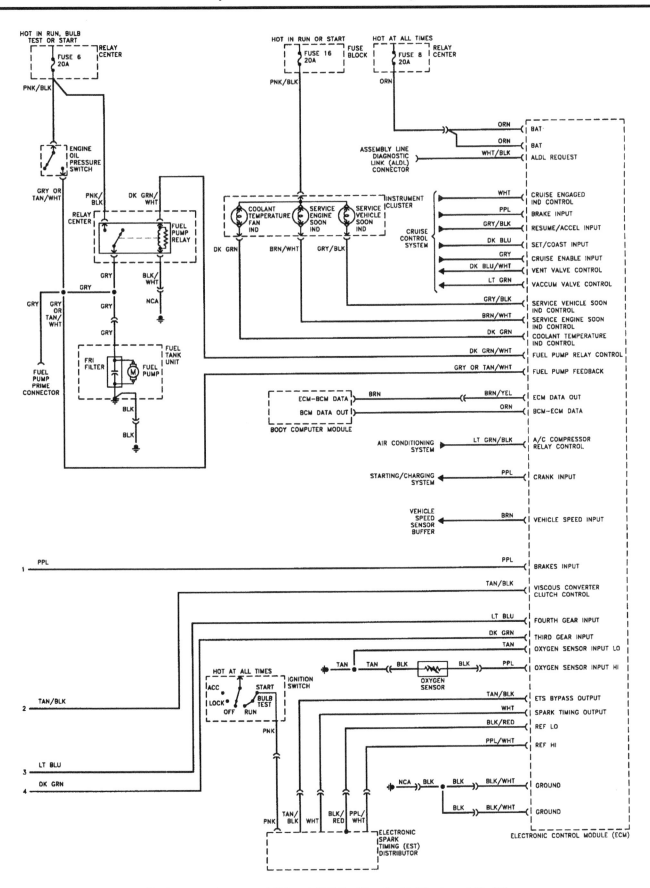

Engine control system - 1988 and 1989 Cadillac Deville/Fleetwood (2 of 2)

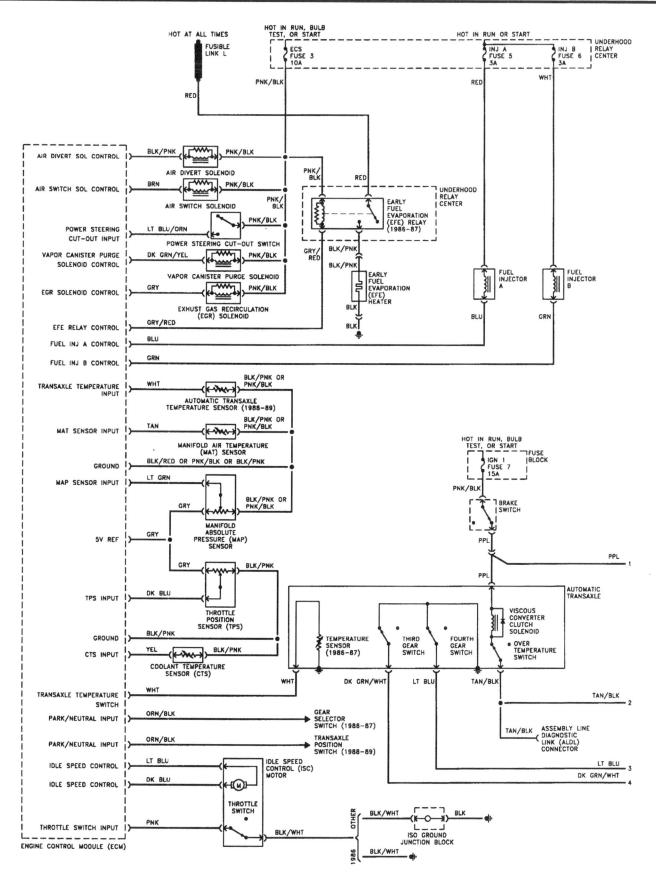

Engine control system - 1986 through 1989 Cadillac Seville/Eldorado (1 of 2)

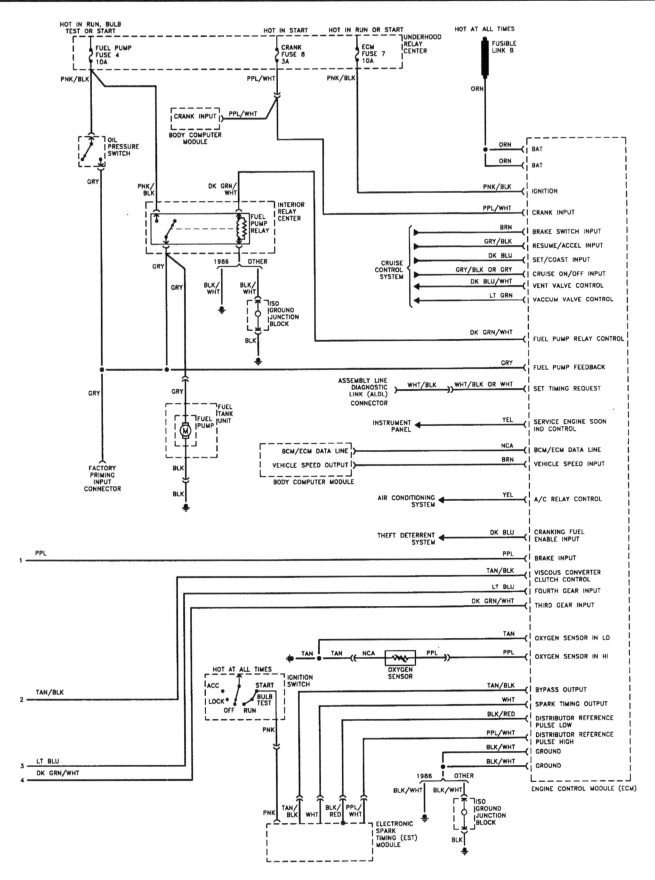

Engine control system - 1986 through 1989 Cadillac Seville/Eldorado (2 of 2)

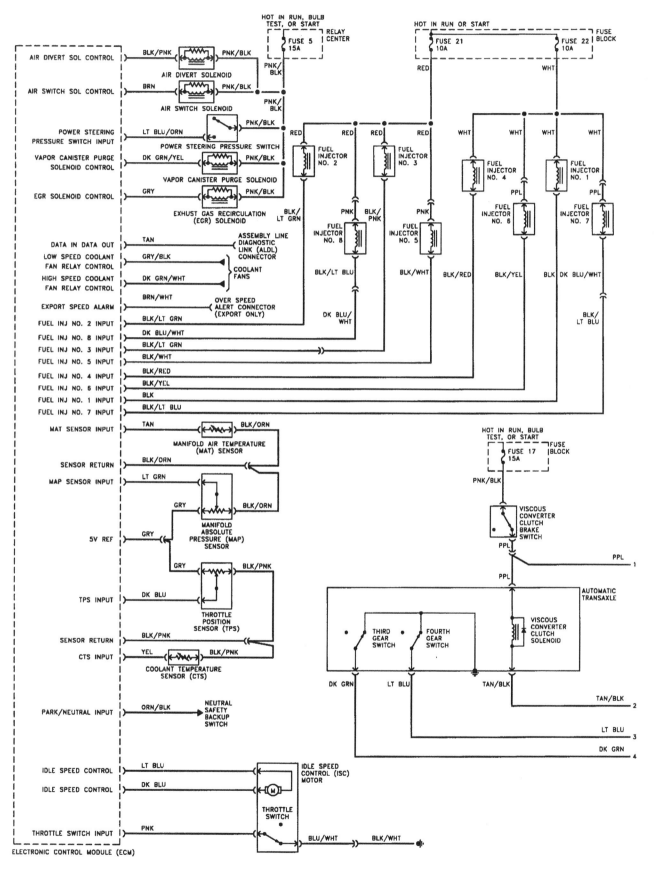

Engine control system - 1990 Cadillac Deville/Fleetwood (1 of 2)

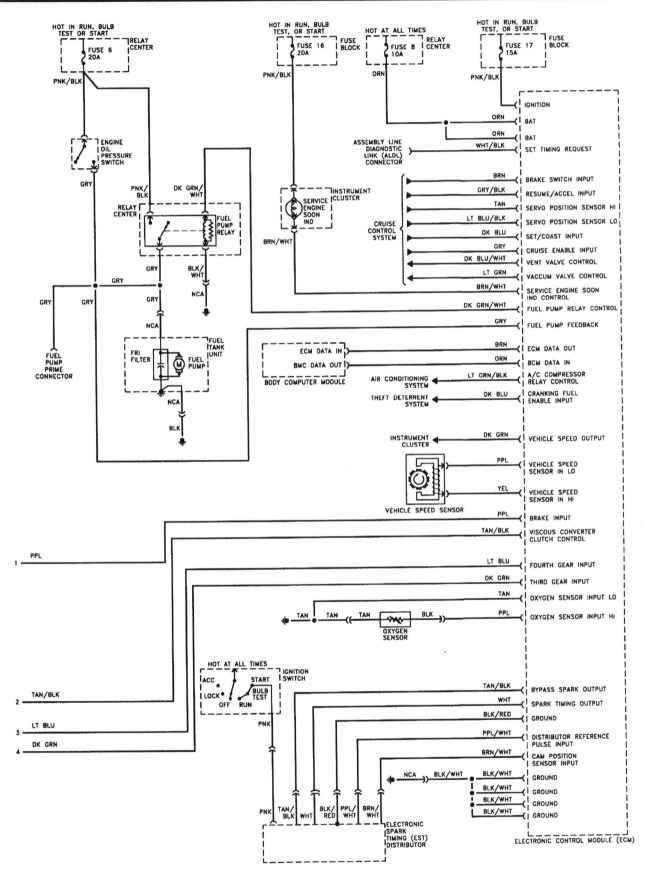

Engine control system - 1990 Cadillac Deville/Fleetwood (2 of 2)

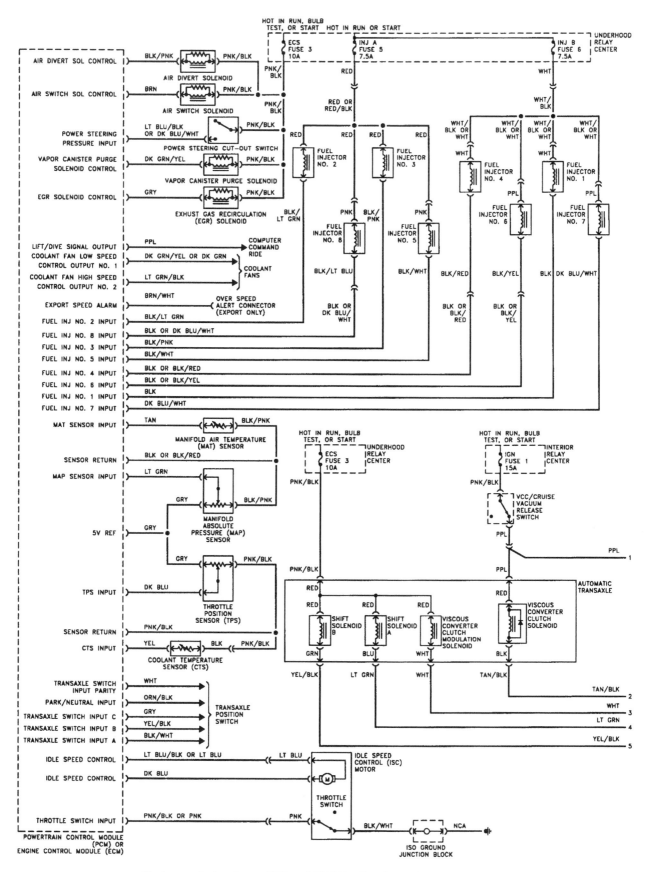

Engine control system - 1990 and 1991 Cadillac Seville/Eldorado (1 of 2)

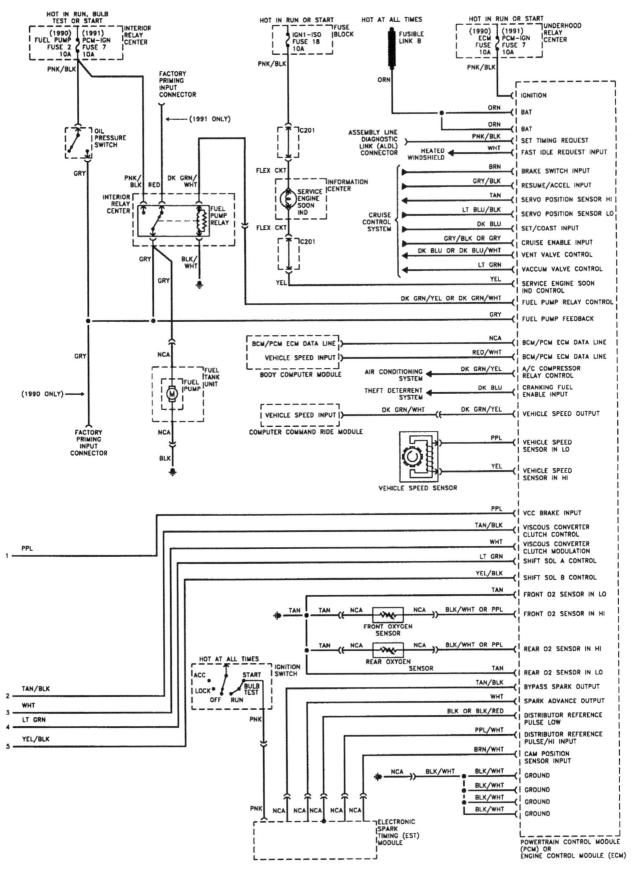

Engine control system - 1990 and 1991 Cadillac Seville/Eldorado (2 of 2)

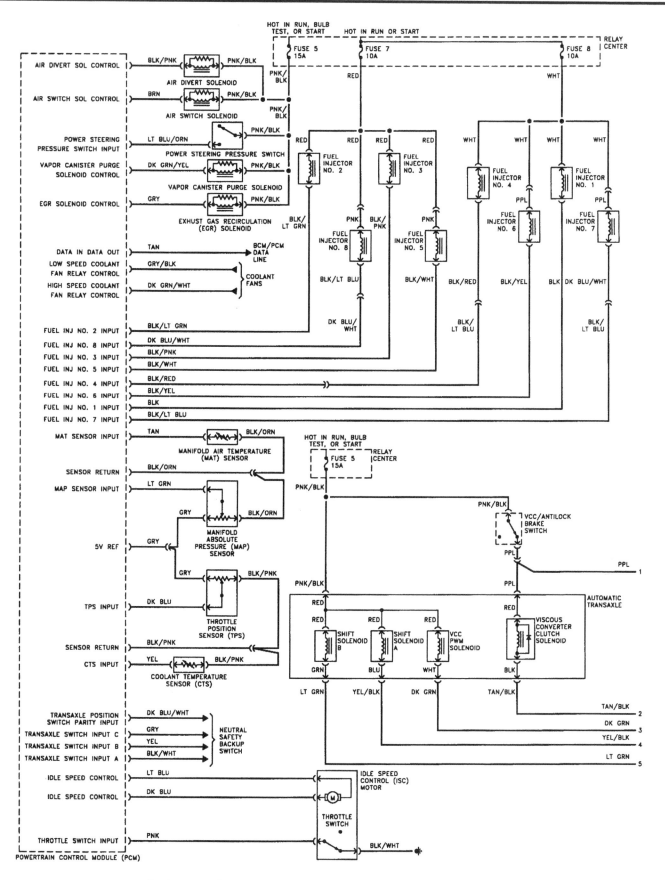

Engine control system - 1991 and later Cadillac Deville/Fleetwood (1 of 2)

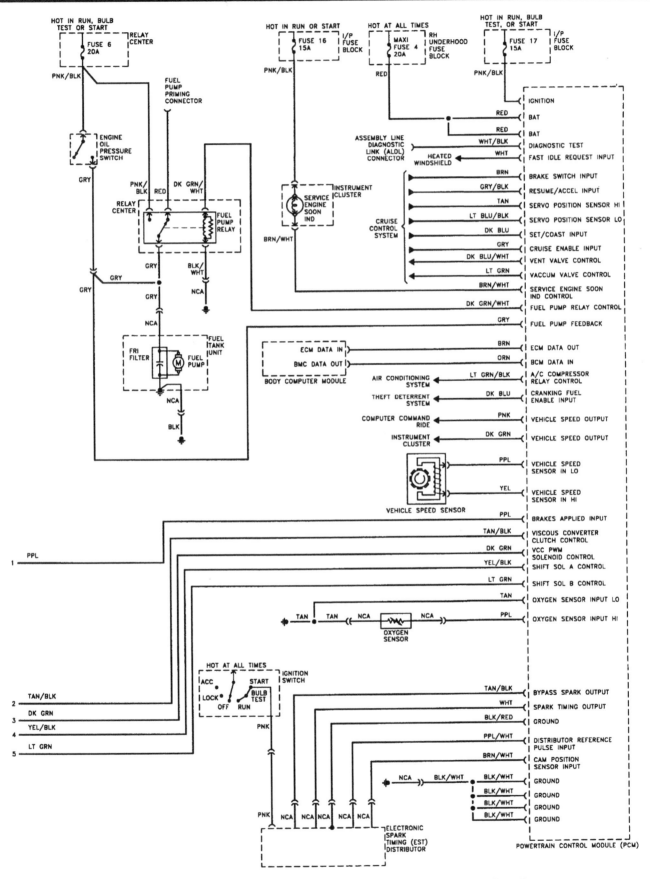

Engine control system - 1991 and later Cadillac Deville/Fleetwood (2 of 2)

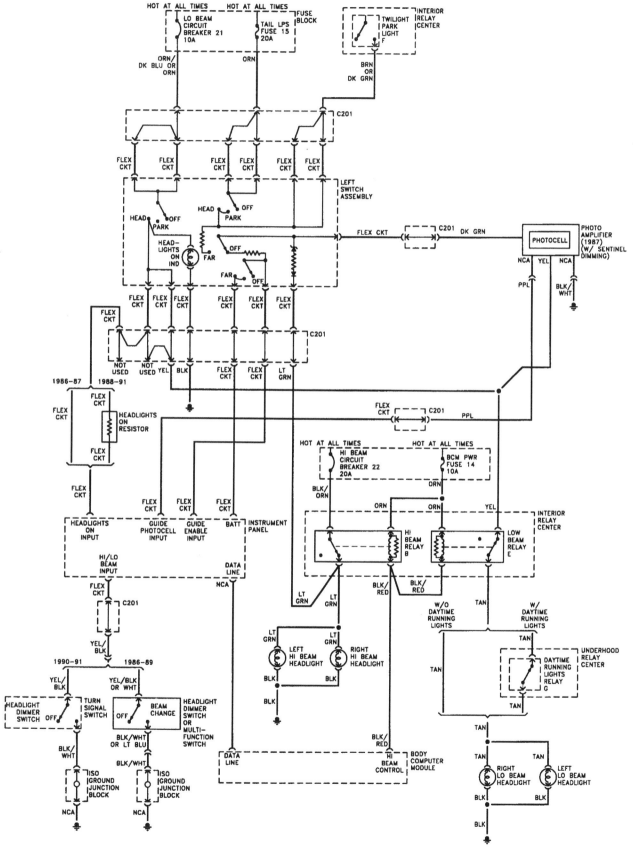

Headlight system - Cadillac Seville/Eldorado (1 of 2)

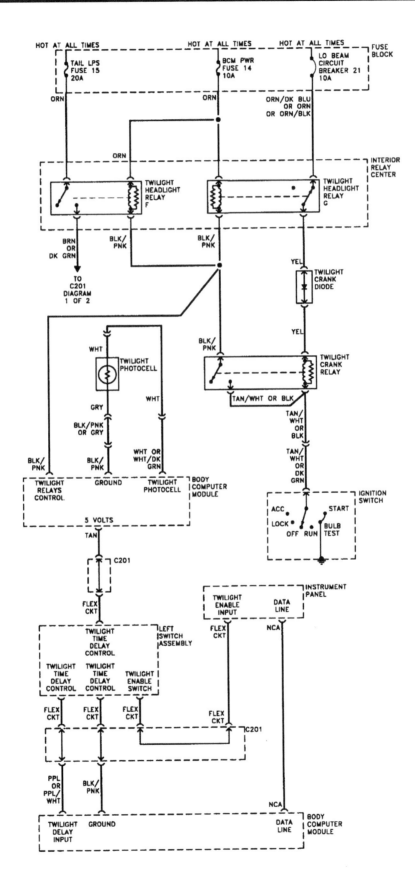

Headlight system - Cadillac Seville/Eldorado (2 of 2)

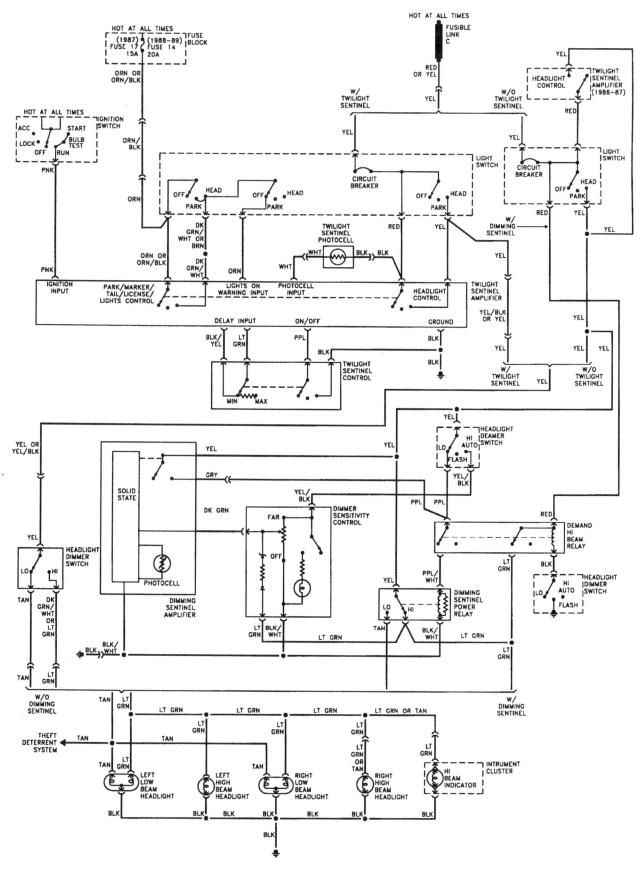

Headlight system - 1990 and earlier Cadillac Deville/Fleetwood

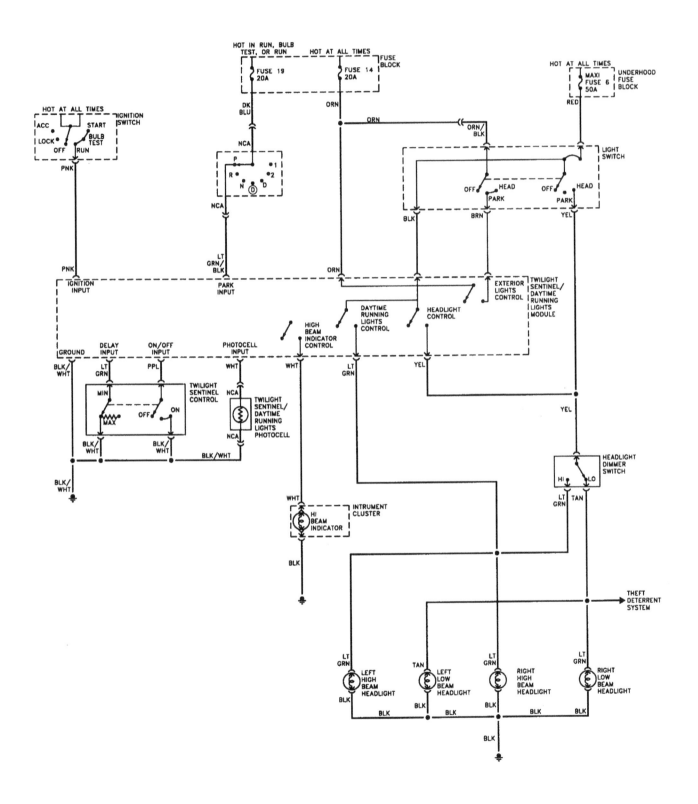

Headlight system - 1991 and later Cadillac Deville/Fleetwood

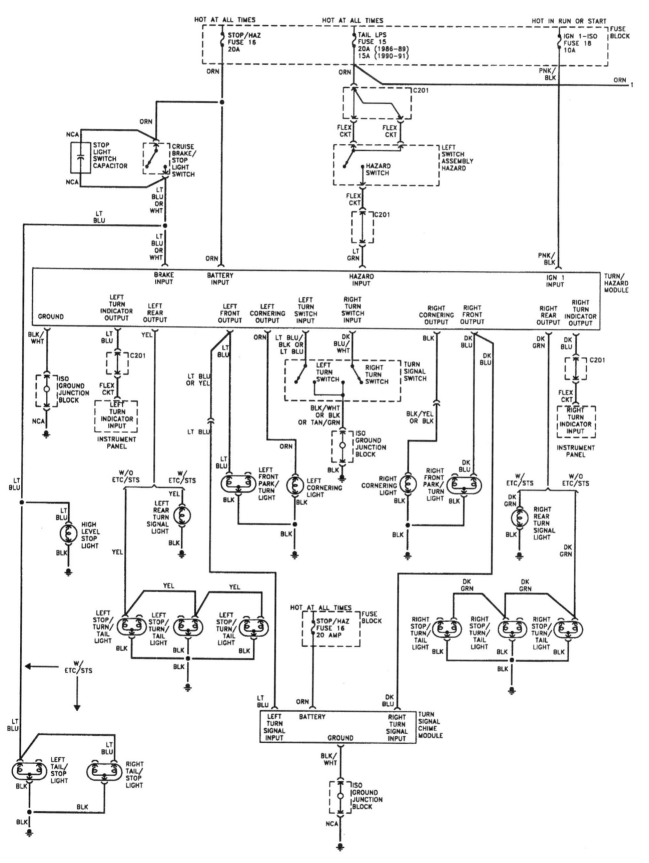

Exterior lighting system - Cadillac Seville/Eldorado (1 of 2)

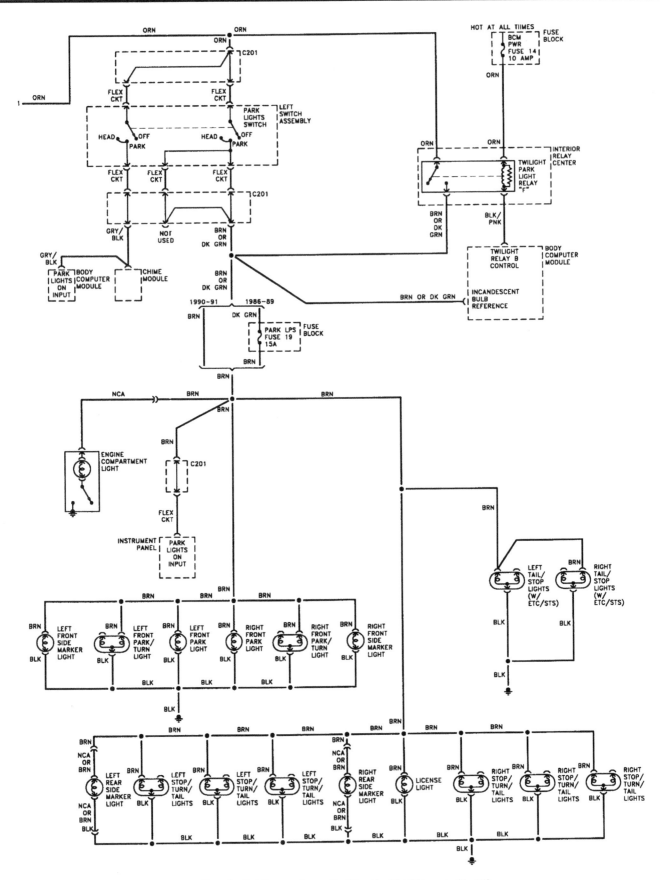

Exterior lighting system - Cadillac Seville/Eldorado (2 of 2)

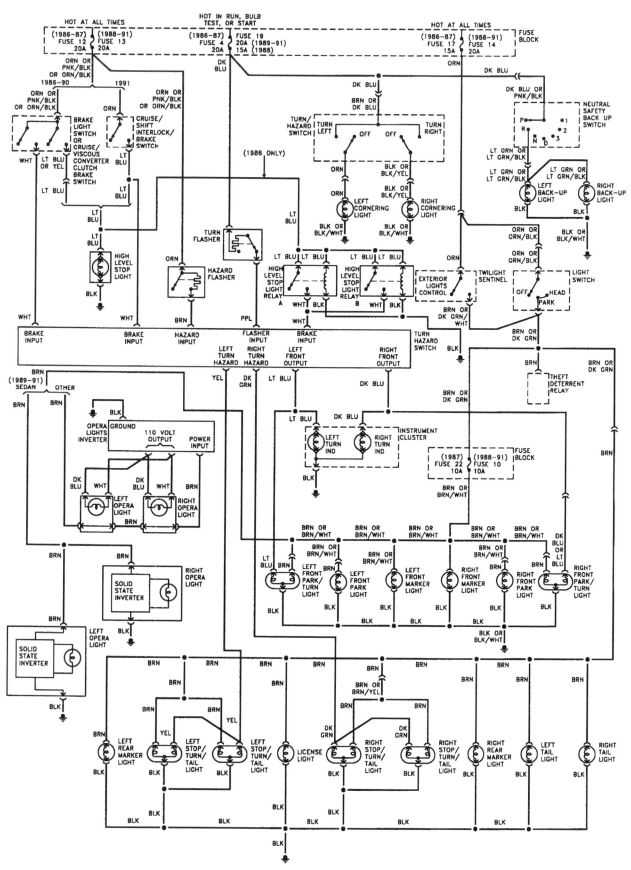

Exterior lighting system - Cadillac Deville/Fleetwood

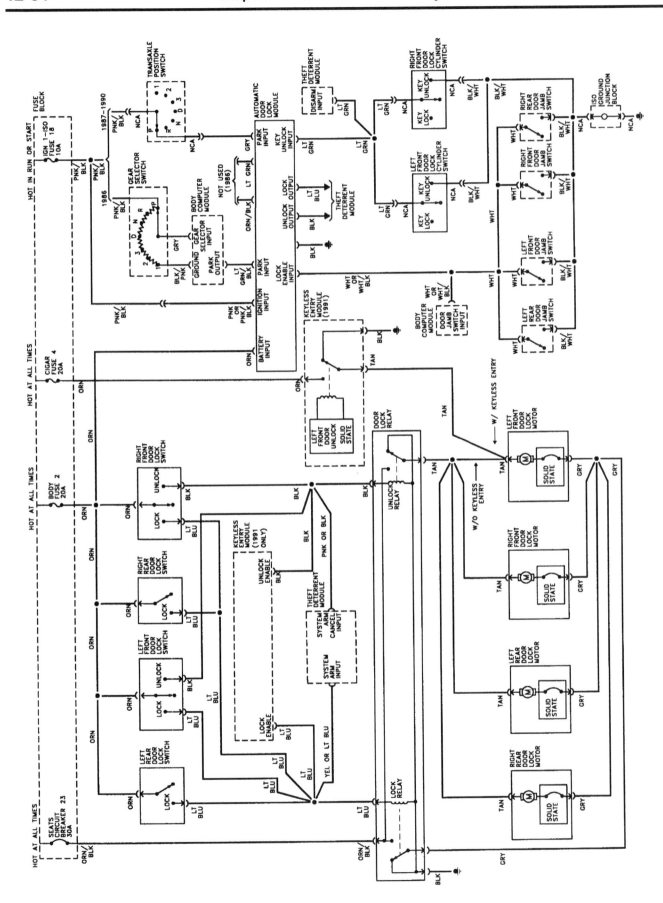

Power window system - Cadillac Seville/Eldorado

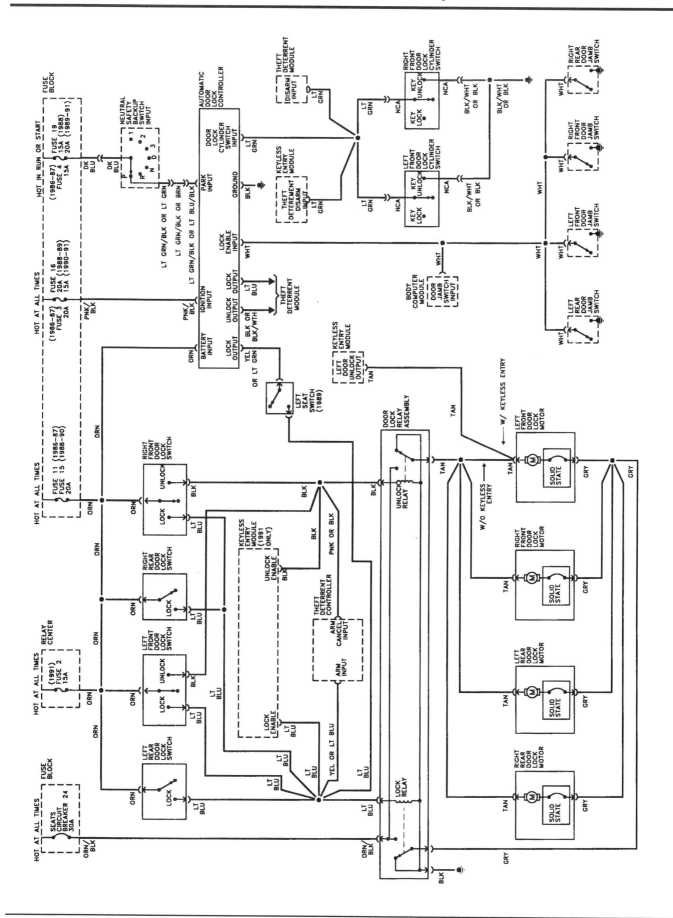

Power window system - Cadillac Deville/Fleetwood

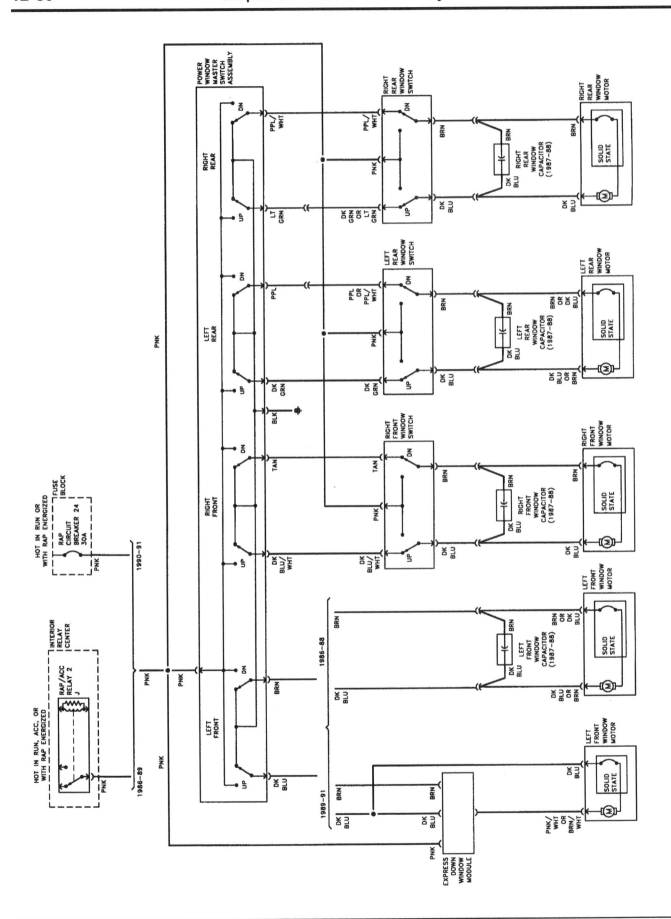

Power door lock system - Cadillac Seville/Eldorado

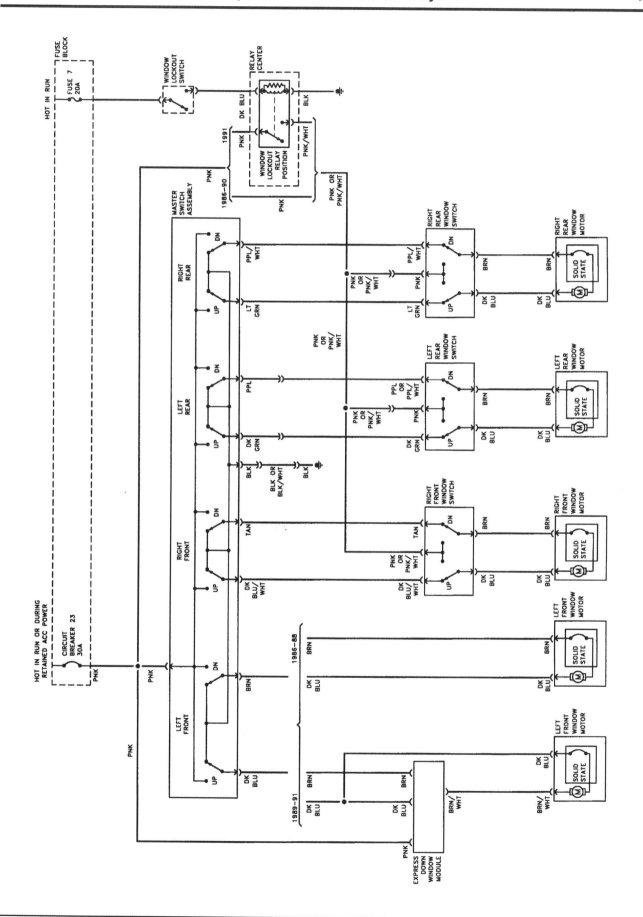

Power door lock system - Cadillac Deville/Fleetwood

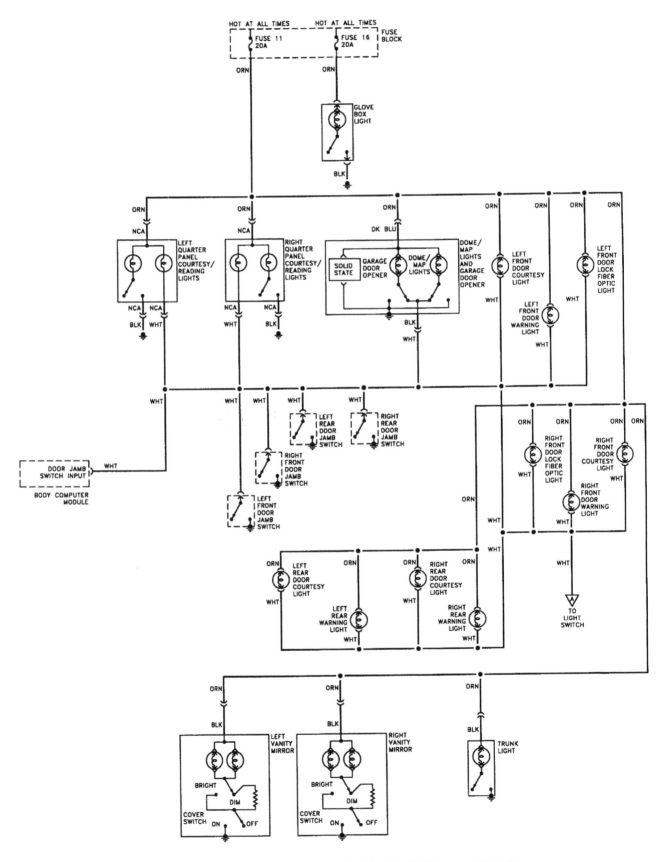

Interior lighting system - 1986 and 1987 Cadillac Deville/Fleetwood (1 of 2)

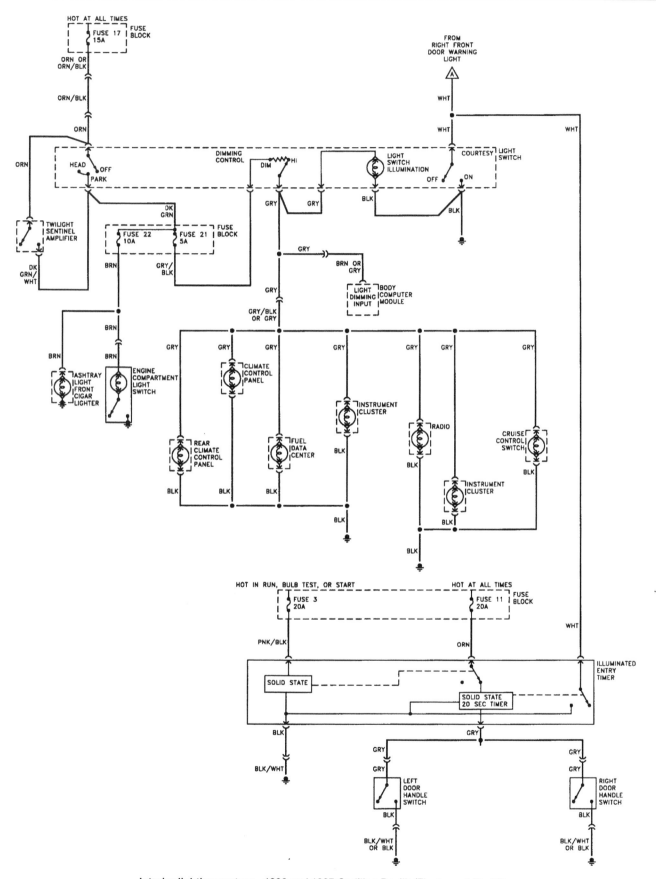

Interior lighting system - 1986 and 1987 Cadillac Deville/Fleetwood (2 of 2)

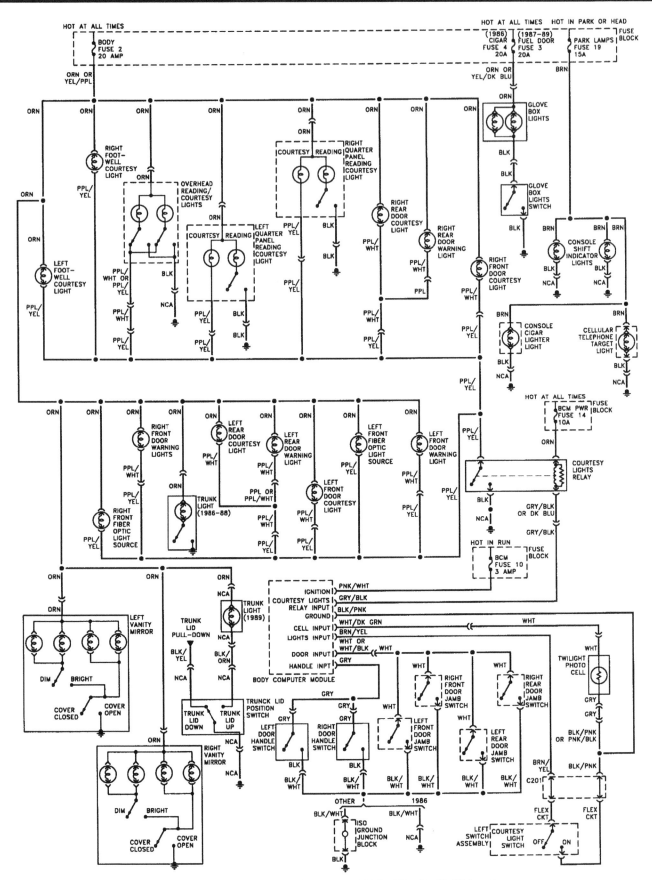

Interior lighting system - 1986 through 1989 Cadillac Seville/Eldorado

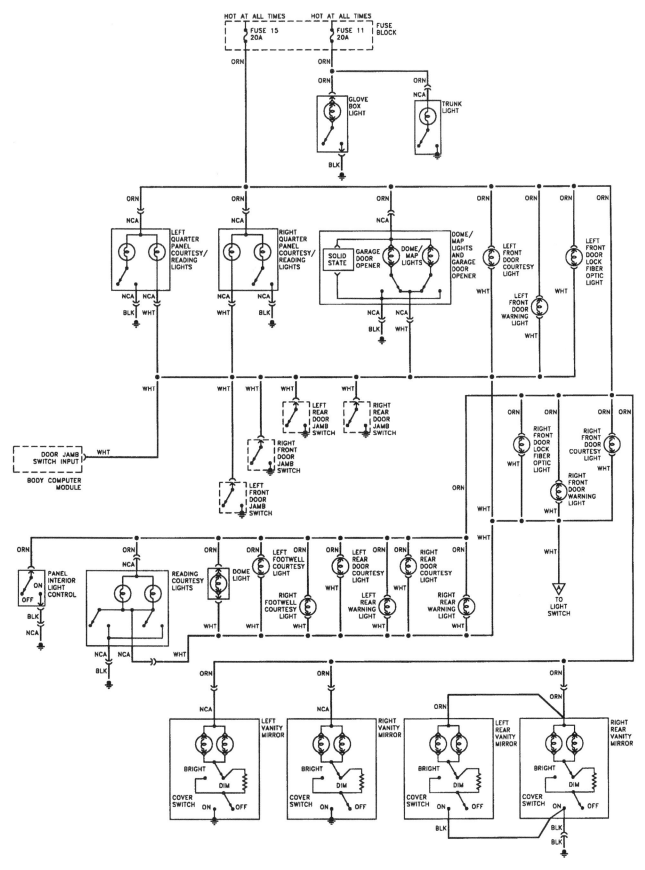

Interior lighting system - 1988 and 1989 Cadillac Deville/Fleetwood (1 of 2)

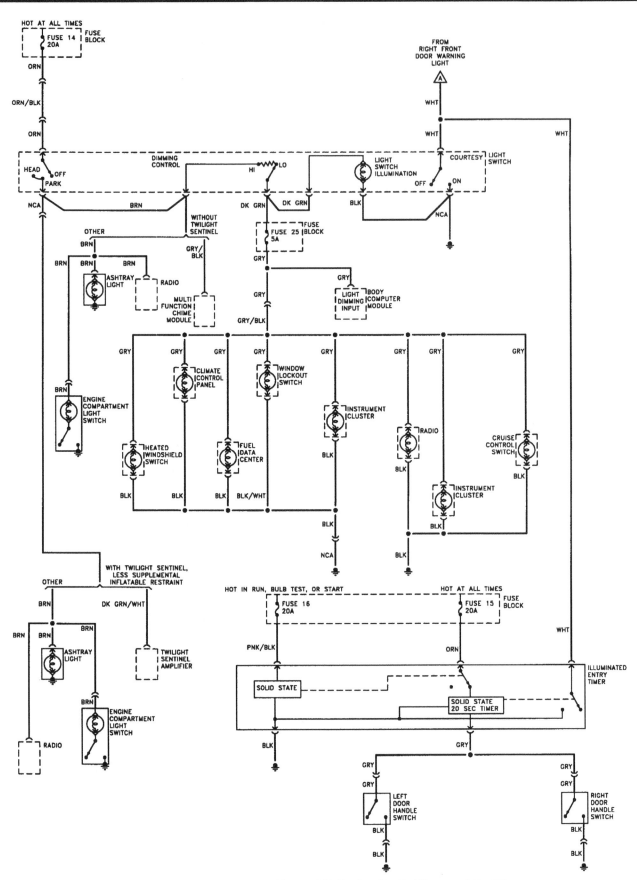

Interior lighting system - 1988 and 1989 Cadillac Deville/Fleetwood (2 of 2)

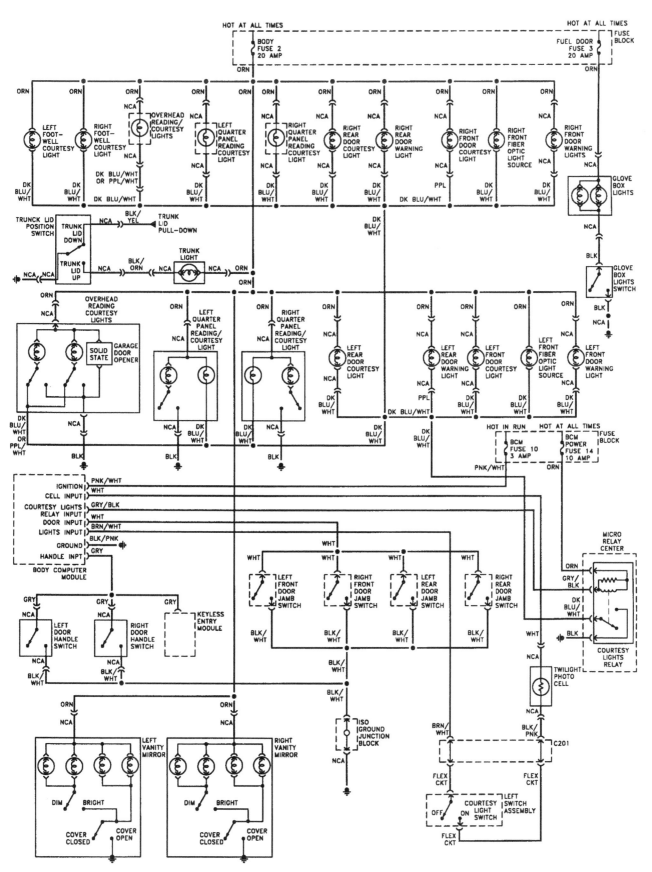

Interior lighting system - 1990 and 1991 Cadillac Seville/Eldorado

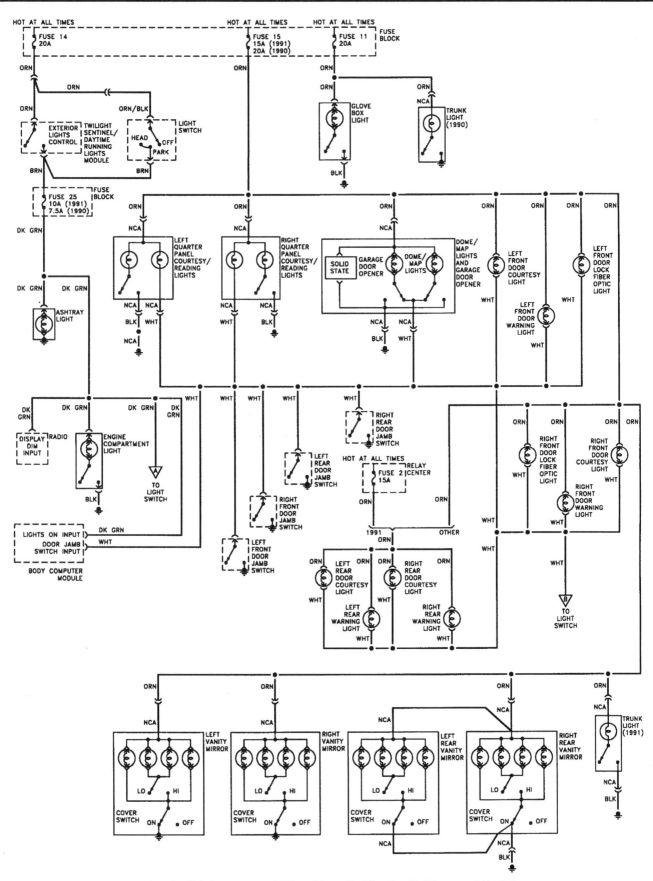

Interior lighting system - 1990 and later Cadillac Deville/Fleetwood (1 of 2)

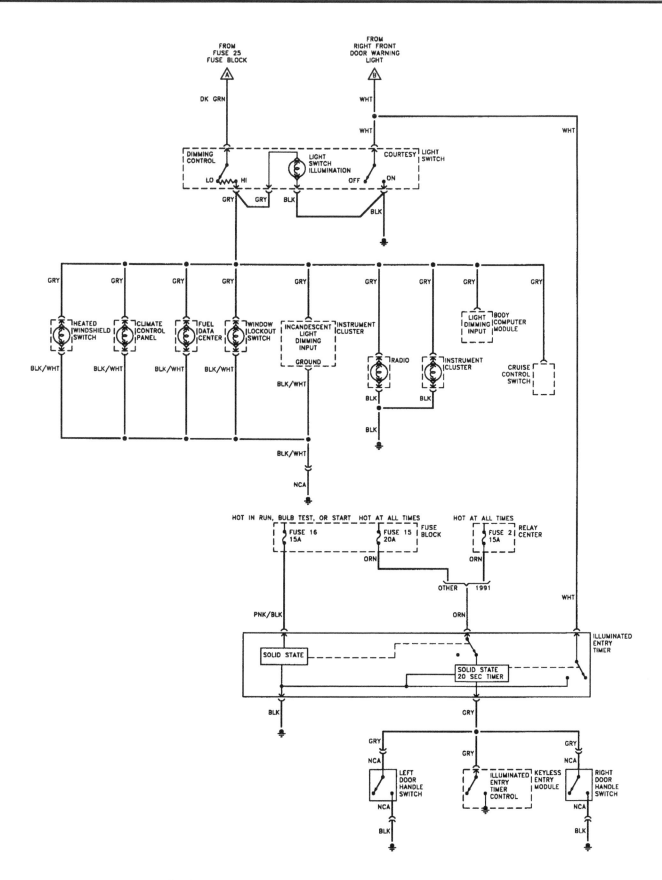

Interior lighting system - 1990 and later Cadillac Deville/Fleetwood (2 of 2)

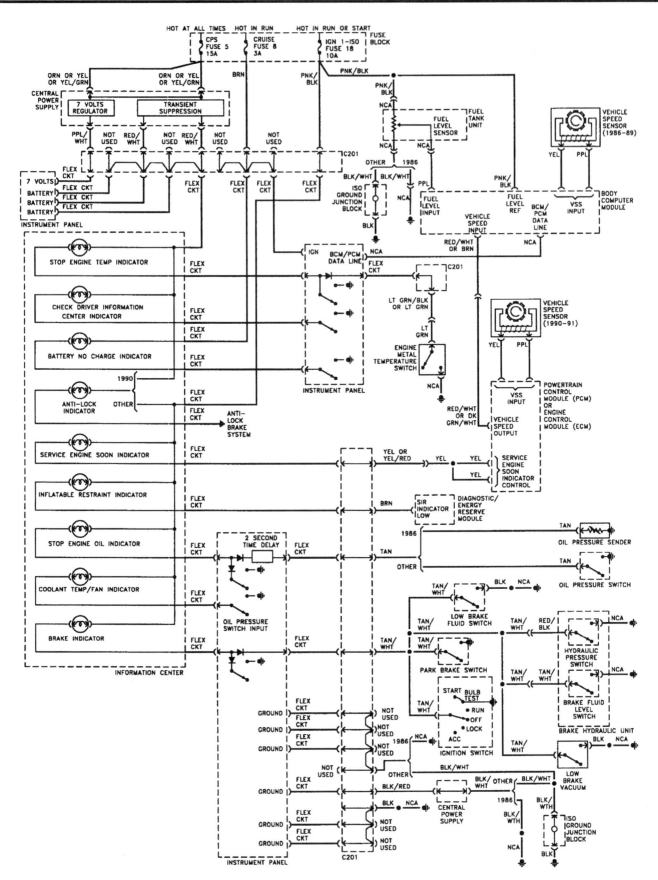

Instrument panel gauges and warning light system - Cadillac Seville/Eldorado

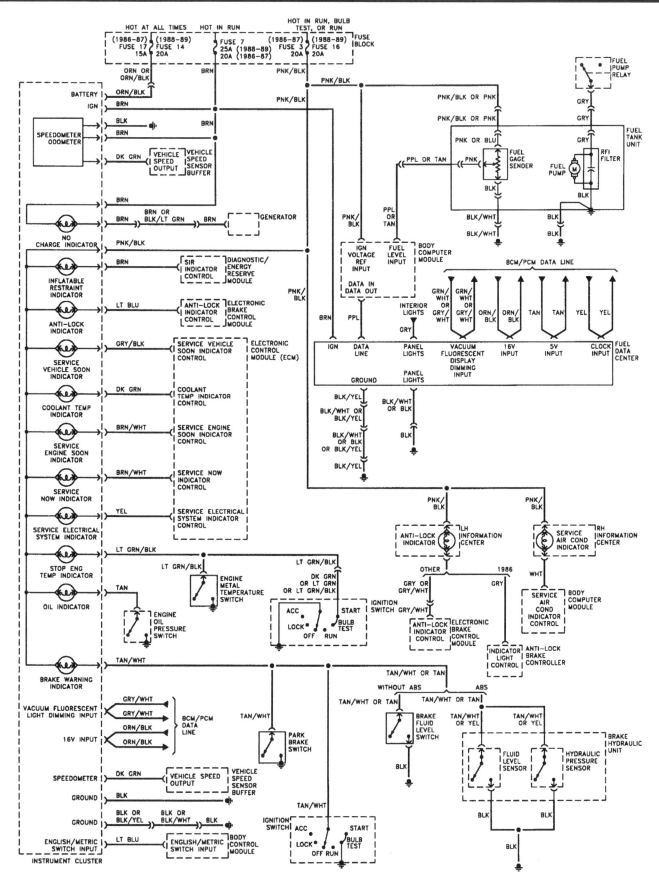

Instrument panel gauges and warning light system - 1989 and earlier Cadillac Deville/Fleetwood

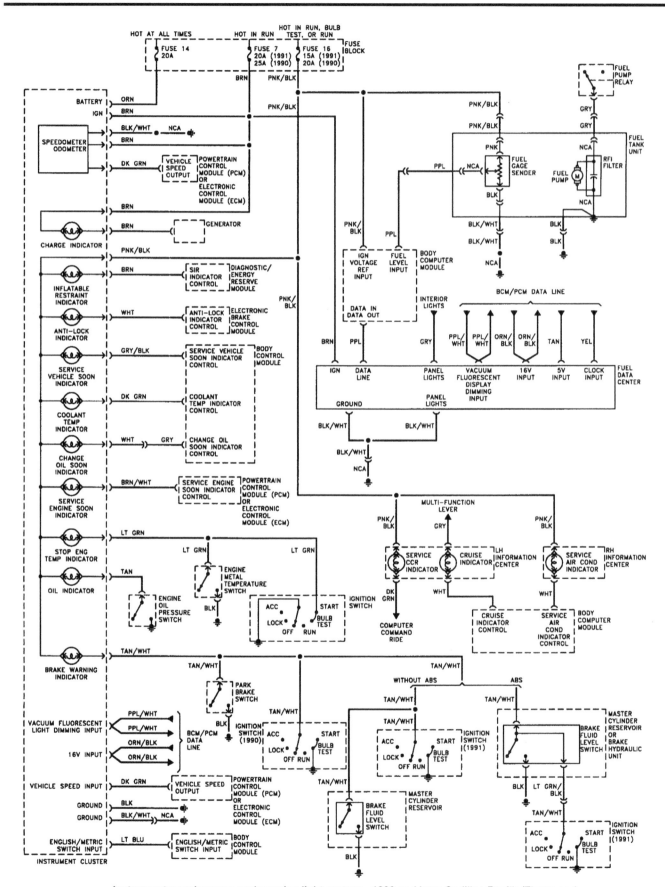

Instrument panel gauges and warning light system - 1990 and later Cadillac Deville/Fleetwood

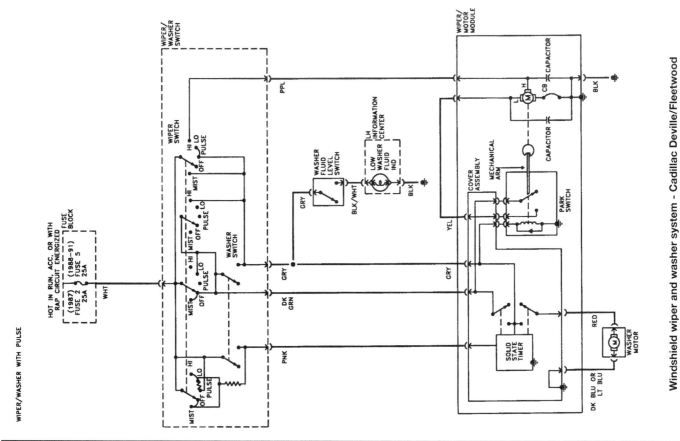

Windshield wiper and washer system - Cadillac Deville/Fleetwood

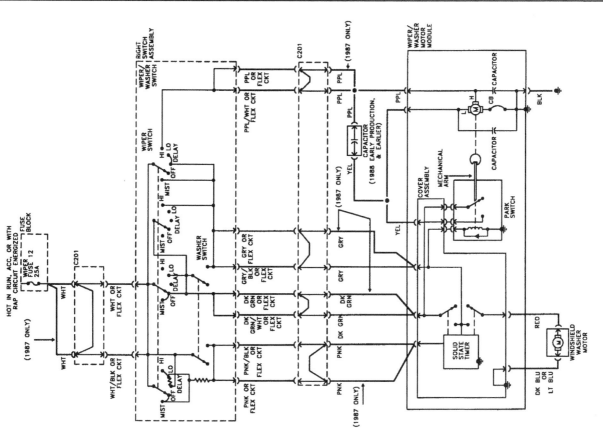

Windshield wiper and washer system - Cadillac Seville/Eldorado

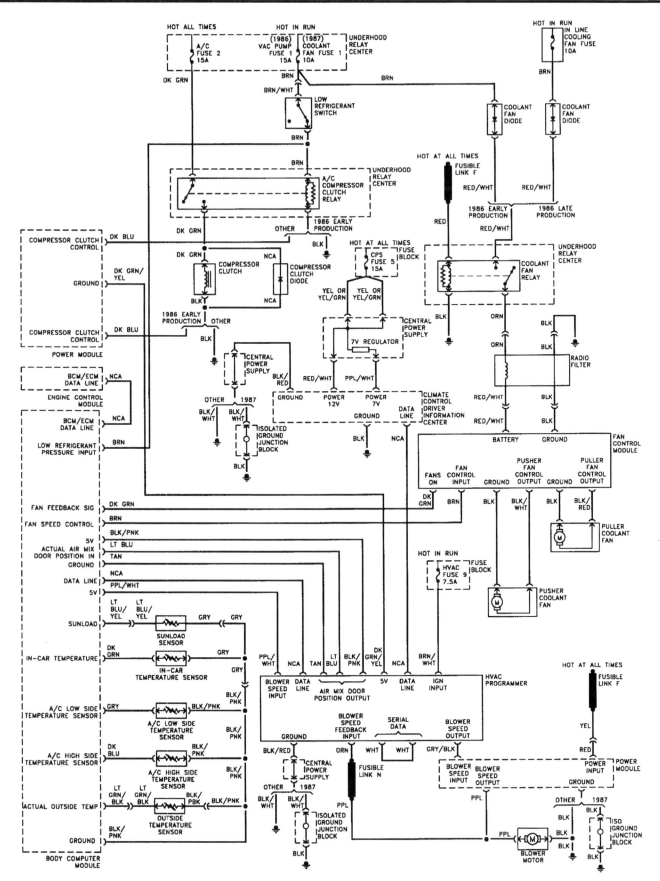

Heating and air conditioning system (including engine cooling fan system) - 1986 and 1987 Cadillac Seville/Eldorado

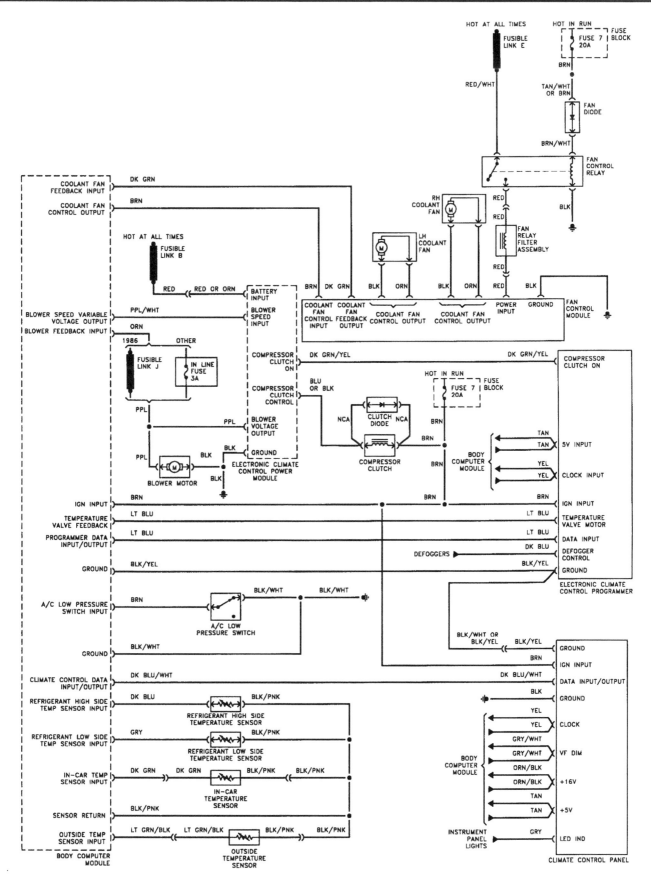

Heating and air conditioning system (including engine cooling fan system) - 1986 through 1988 Cadillac Deville/Fleetwood

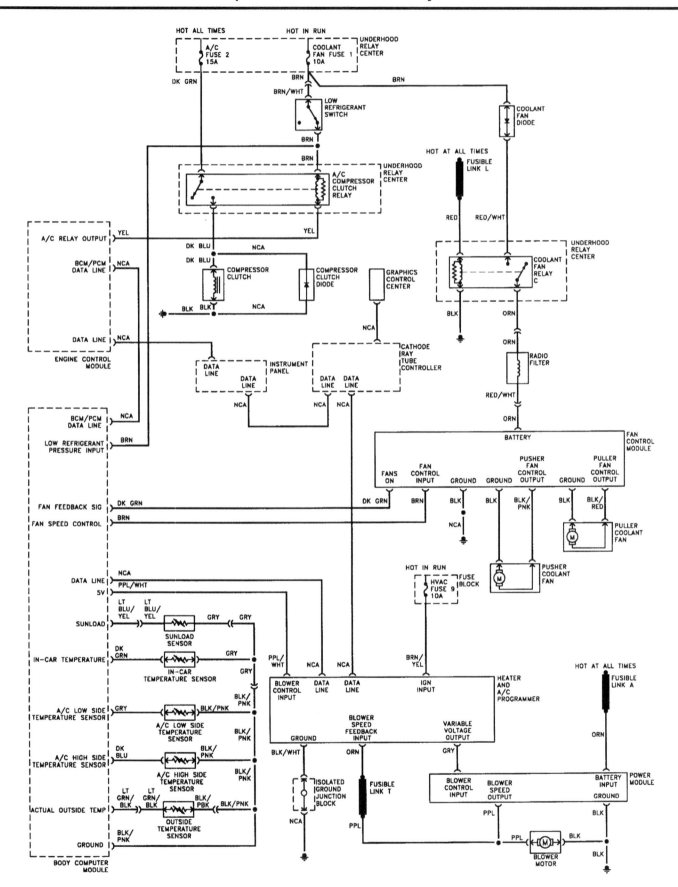

Heating and air conditioning system (including engine cooling fan system) - 1988 Cadillac Seville/Eldorado

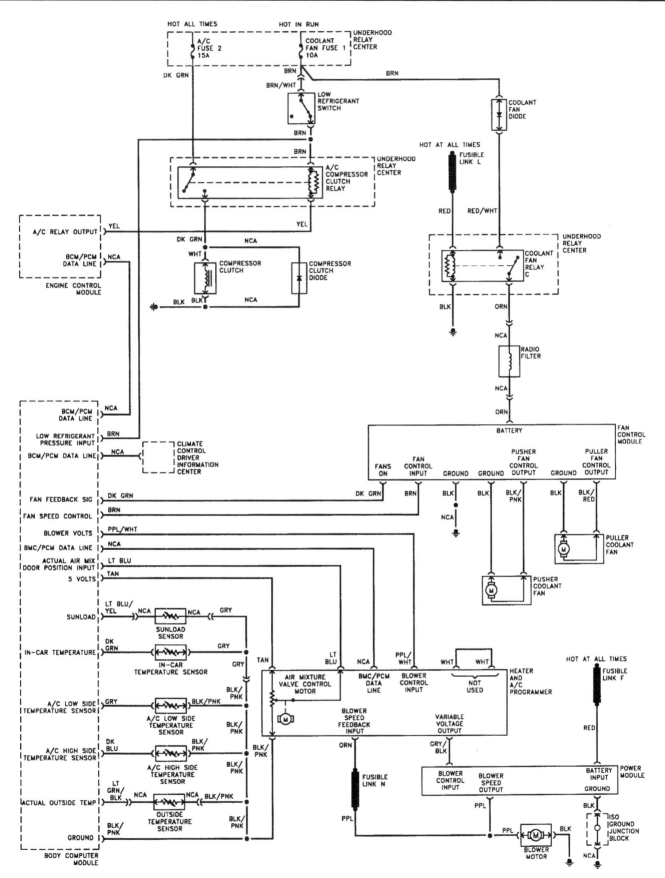

Heating and air conditioning system (including engine cooling fan system) - 1989 Cadillac Seville/Eldorado

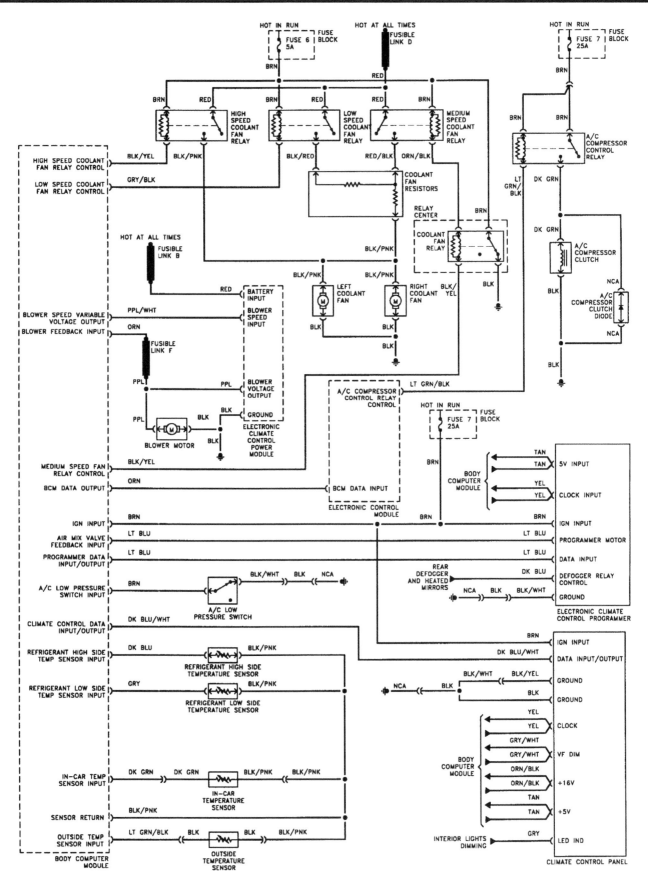

Heating and air conditioning system (including engine cooling fan system) - 1989 and 1990 Cadillac Deville/Fleetwood

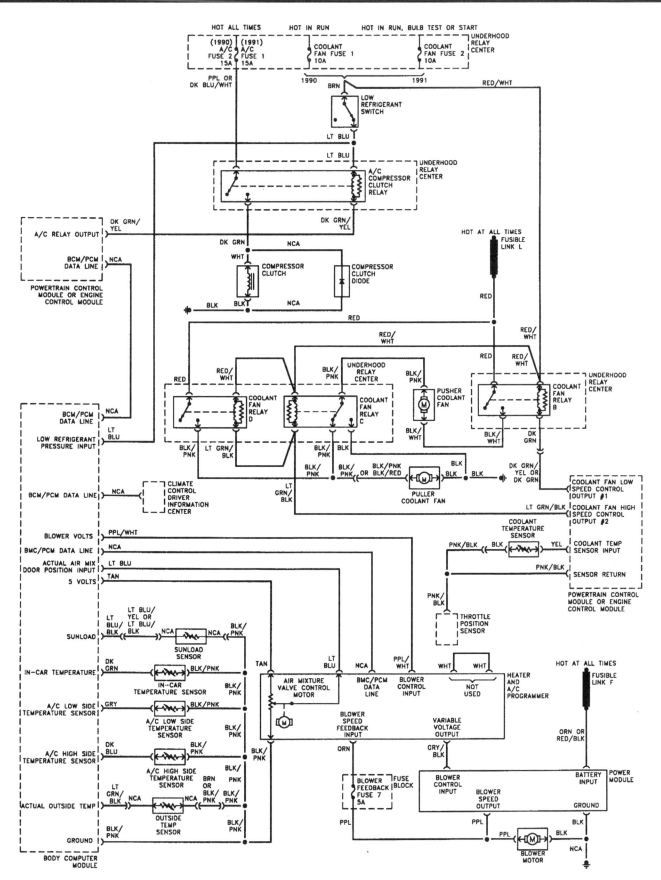

Heating and air conditioning system (including engine cooling fan system) - 1990 and 1991 Cadillac Seville/Eldorado

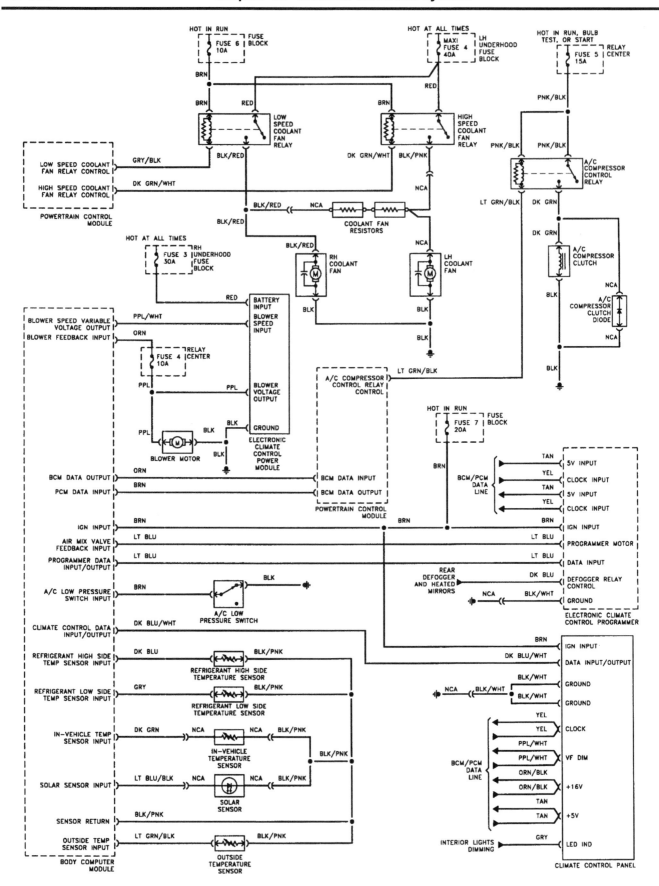

Heating and air conditioning system (including engine cooling fan system) - 1991 and later Cadillac Deville/Fleetwood

Index

Haynes Automotive Manuals

NOTE: If you do not see a listing for your vehicle, please visit haynes.com for the latest product information and check out our Online Manuals!

ACURA
12020 Integra '86 thru '89 & Legend '86 thru '90
12021 Integra '90 thru '93 & Legend '91 thru '95
Integra '94 thru '00 - see HONDA Civic (42025)
MDX '01 thru '07 - see HONDA Pilot (42037)
12050 Acura TL all models '99 thru '08

AMC
14020 Concord/Hornet/Gremlin/Spirit '70 thru '83
14025 (Renault) Alliance & Encore '83 thru '87

AUDI
15020 4000 all models '80 thru '87
15025 5000 all models '77 thru '83
15026 5000 all models '84 thru '88
Audi A4 '96 thru '01 - see VW Passat (96023)
15030 Audi A4 '02 thru '08

AUSTIN-HEALEY
Sprite - see MG Midget (66015)

BMW
18020 3/5 Series '82 thru '92
18021 3-Series including Z3 models '92 thru '98
18022 3-Series including Z4 models '99 thru '05
18023 3-Series '06 thru '14
18025 320i all 4-cylinder models '75 thru '83
18050 1500 thru 2002 except Turbo '59 thru '77

BUICK
19010 Buick Century '97 thru '05
Century (front-wheel drive) - see GM (38005)
19020 Buick, Oldsmobile & Pontiac Full-size (Front-wheel drive) '85 thru '05
19025 Buick, Oldsmobile & Pontiac Full-size (Rear-wheel drive) '70 thru '90
19027 Buick LaCrosse '05 thru '13
Regal - see GENERAL MOTORS (38010)
Skyhawk - see GM (38015)
Skylark - see GM (38020, 38025)
Somerset - see GENERAL MOTORS (38025)

CADILLAC
21015 CTS & CTS-V '03 thru '14
21030 Cadillac Rear Wheel Drive '70 thru '93
Cimarron, Eldorado & Seville - see GM (38015, 38030, 38031)

CHEVROLET
10305 Chevrolet Engine Overhaul Manual
24010 Astro & GMC Safari Mini-vans '85 thru '05
24015 Camaro V8 all models '70 thru '81
24016 Camaro all models '82 thru '92
Cavalier - see GM (38016)
Celebrity - see GM (38005)
24017 Camaro & Firebird '93 thru '02
24018 Camaro '10 thru '15
24020 Chevelle, Malibu, El Camino '69 thru '87
24024 Chevette & Pontiac T1000 '76 thru '87
Citation - see GENERAL MOTORS (38020)
24027 Colorado & GMC Canyon '04 thru '12
24032 Corsica & Beretta all models '87 thru '96
24040 Corvette all V8 models '68 thru '82
24041 Corvette all models '84 thru '96
24042 Corvette all models '97 thru '13
24044 Cruze '11 thru '19
24045 Full-size Sedans Caprice, Impala, Biscayne, Bel Air & Wagons '69 thru '90
24046 Impala SS & Caprice and Buick Roadmaster '91 thru '96
Impala '00 thru '05 - see LUMINA (24048)
24047 Impala & Monte Carlo all models '06 thru '11
Lumina '90 thru '94 - see GM (38010)
24048 Lumina & Monte Carlo '95 thru '05
Lumina APV - see GM (38035)
24050 Luv Pick-up all 2WD & 4WD '72 thru '82
24051 Malibu '13 thru '19
24055 Monte Carlo all models '70 thru '88
Monte Carlo '95 thru '01 - see LUMINA (24048)
24059 Nova all V8 models '69 thru '79
24060 Nova/Geo Prizm '85 thru '92
24064 Pick-ups '67 thru '87 - Chevrolet & GMC
24065 Pick-ups '88 thru '98 - Chevrolet & GMC
24066 Pick-ups '99 thru '06 - Chevrolet & GMC
24067 Chevy Silverado & GMC Sierra '07 thru '14
24068 Chevy Silverado & GMC Sierra '14 thru '19
24070 S-10 & S-15 Pick-ups '82 thru '93
24071 S-10 & Sonoma Pick-ups '94 thru '04
24072 Chevrolet TrailBlazer, GMC Envoy & Oldsmobile Bravada '02 thru '09
24075 Sprint '85 thru '88, Geo Metro '89 thru '01
24080 Vans - Chevrolet & GMC '68 thru '96
24081 Chevrolet Express & GMC Savana Full-size Vans '96 thru '19

CHRYSLER
10310 Chrysler Engine Overhaul Manual
25015 Chrysler Cirrus, Dodge Stratus, Plymouth Breeze, '95 thru '00
25020 Full-size Front-Wheel Drive '88 thru '93
K-Cars - see DODGE Aries (30008)
25025 Chrysler LHS, Concorde & New Yorker, Dodge Intrepid, Eagle Vision, '93 thru '97
25026 Chrysler LHS, Concorde, 300M, Dodge Intrepid '98 thru '04
25027 Chrysler 300 '05 thru '18, Dodge Charger '06 thru '18, Magnum '05 thru '08 & Challenger '08 thru '18
25030 Chrysler & Plymouth Mid-size '82 thru '95
Rear-wheel Drive - see DODGE (30050)
25035 PT Cruiser all models '01 thru '10
25040 Chrysler Sebring '95 thru '06, Dodge Stratus '01 thru '06 & Dodge Avenger '95 thru '00
25041 Chrysler Sebring '07 thru '10, 200 '11 thru '17 & Dodge Avenger '08 thru '14

DATSUN
28005 200SX all models '80 thru '83
28012 240Z, 260Z & 280Z Coupe '70 thru '78
28014 280ZX Coupe & 2+2 '79 thru '83
300ZX - see NISSAN (72010)
28018 510 & PL521 Pick-up '68 thru '73
28020 510 all models '78 thru '81
28022 620 Series Pick-up all models '73 thru '79
720 Series Pick-up - see NISSAN (72030)

DODGE
30008 Aries & Plymouth Reliant '81 thru '89
30010 Caravan & Plymouth Voyager '84 thru '95
30011 Caravan & Plymouth Voyager '96 thru '02
30012 Challenger/Plymouth Sapporo '78 thru '83
30013 Caravan, Chrysler Voyager & Town & Country '03 thru '07
30014 Grand Caravan & Chrysler Town & Country '08 thru '18

30020 Dakota Pick-ups all models '87 thru '96
30021 Durango '98 & '99 & Dakota '97 thru '99
30022 Durango '00 thru '03 & Dakota '00 thru '04
30023 Durango '04 thru '09 & Dakota '05 thru '11
30025 Dart, Challenger/Plymouth Barracuda & Valiant 6-cylinder models '67 thru '76
30030 Daytona & Chrysler Laser '84 thru '89
Intrepid - see Chrysler (25025, 25026)
30034 Neon all models '95 thru '99
30035 Omni & Plymouth Horizon '78 thru '90
30036 Dodge & Plymouth Neon '00 thru '05
30040 Pick-ups full-size '74 thru '93
30041 Pick-ups full-size '94 thru '01
30042 Pick-ups full-size '02 thru '08
30043 Pick-ups full-size '09 thru '18
30045 Ram 50/D50 Pick-ups & Raider and Plymouth Arrow Pick-ups '79 thru '93
30050 Dodge/Plymouth/Chrysler RWD '71 thru '89
30055 Shadow & Plymouth Sundance '87 thru '94
30060 Spirit & Plymouth Acclaim '89 thru '95
30065 Vans - Dodge & Plymouth '71 thru '03

EAGLE
Talon - see MITSUBISHI (68030, 68031)
Vision - see CHRYSLER (25025)

FIAT
34010 124 Sport Coupe & Spider '68 thru '78
34025 X1/9 all models '74 thru '80

FORD
10320 Ford Engine Overhaul Manual
10355 Ford Automatic Transmission Overhaul
11500 Mustang '64-1/2 thru '70 Restoration Guide
36004 Aerostar Mini-vans '86 thru '97
36006 Contour & Mercury Mystique '95 thru '00
36008 Courier Pick-up all models '72 thru '82
36012 Crown Victoria & Mercury Grand Marquis '88 thru '11
36016 Escort & Mercury Lynx '81 thru '90
36020 Escort & Mercury Tracer '91 thru '02
36022 Escape '01 thru '17, Mazda Tribute '01 thru '11 & Mercury Mariner '05 thru '11
36024 Explorer & Mazda Navajo '91 thru '01
36025 Explorer & Mercury Mountaineer '02 thru '10
36026 Explorer '11 thru '17
36028 Fairmont & Mercury Zephyr '78 thru '83
36030 Festiva & Aspire '88 thru '97
36032 Fiesta all models '77 thru '80
36034 Focus all models '00 thru '11
36035 Focus '12 thru '14
36045 Ford Fusion '06 thru '14 & Mercury Milan '06 thru '10
36048 Mustang V8 all models '64-1/2 thru '73
36049 Mustang II 4-cylinder, V6 & V8 '74 thru '78
36050 Mustang & Mercury Capri '79 thru '93
36051 Mustang all models '94 thru '04
36052 Mustang '05 thru '14
36054 Pick-ups and Bronco '73 thru '79
36058 Pick-ups and Bronco '80 thru '96
36059 F-150 '97 thru '03, Expedition '97 thru '17, F-250 '97 thru '99, F-150 Heritage '04 & Lincoln Navigator '98 thru '17
36060 Super Duty Pick-up & Excursion '99 thru '10
36061 F-150 full-size '04 thru '14
36062 Pinto & Mercury Bobcat '75 thru '80
36063 F-150 full-size '15 thru '17
36064 Super Duty Pick-ups '11 thru '16
36066 Probe all models '89 thru '92
36070 Ranger & Bronco II gas models '83 thru '92
36071 Ranger '93 thru '11 & Mazda Pick-ups '94 thru '09
36074 Taurus & Mercury Sable '86 thru '95
36075 Taurus & Mercury Sable '96 thru '07
36076 Taurus '08 thru '14, Five Hundred '05 thru '07, Mercury Montego '05 thru '07 & Sable '08 thru '09
36078 Tempo & Mercury Topaz '84 thru '94
36082 Thunderbird & Mercury Cougar '83 thru '88
36086 Thunderbird & Mercury Cougar '89 thru '97
36090 Vans all V8 Econoline models '69 thru '91
36094 Vans full size '92 thru '14
36097 Windstar '95 thru '03, Freestar & Mercury Monterey Mini-van '04 thru '07

GENERAL MOTORS
10360 GM Automatic Transmission Overhaul
38005 Buick Century, Chevrolet Celebrity, Olds Cutlass Ciera & Pontiac 6000 '82 thru '96
38010 Buick Regal, Chevrolet Lumina, Oldsmobile Cutlass Supreme & Pontiac Grand Prix front wheel drive '88 thru '07
38015 Buick Skyhawk, Cadillac Cimarron, Chevrolet Cavalier, Oldsmobile Firenza Pontiac J-2000 & Sunbird '82 thru '94
38016 Chevrolet Cavalier/Pontiac Sunfire '95 thru '05
38017 Chevrolet Cobalt & Pontiac G5 '05 thru '11
38020 Buick Skylark, Chevrolet Citation, Olds Omega, Pontiac Phoenix '80 thru '85
38025 Buick Skylark & Somerset, Olds Achieva, Calais & Pontiac Grand Am '85 thru '98
38026 Chevrolet Malibu, Olds Alero & Cutlass, Pontiac Grand Am '97 thru '03
38027 Chevrolet Malibu '04 thru '12
38030 Cadillac Eldorado, Seville, Oldsmobile Toronado & Buick Riviera '71 thru '85
38031 Cadillac Eldorado, Seville, DeVille, Fleetwood, Oldsmobile Toronado & Buick Riviera '86 thru '93
38032 Cadillac DeVille '94 thru '05, Seville '92 thru '04 & Cadillac DTS '06 thru '10
38035 Chevrolet Lumina APV, Olds Silhouette & Pontiac Trans Sport all models '90 thru '96
38036 Chevrolet Venture, Olds Silhouette, Pontiac Trans Sport & Montana '97 thru '05
38040 Chevrolet Equinox '05 thru '17, GMC Terrain '10 thru '17 & Pontiac Torrent '06 thru '09

GEO
Metro - see CHEVROLET Sprint (24075)
Prizm - '85 thru '92 see CHEVY (24060), '93 thru '02 see TOYOTA Corolla (92036)
40030 Storm all models '90 thru '93
Tracker - see SUZUKI Samurai (90010)

GMC
Vans & Pick-ups - see CHEVROLET

HONDA
42010 Accord CVCC all models '76 thru '83
42011 Accord all models '84 thru '89
42012 Accord all models '90 thru '93
42013 Accord all models '94 thru '97
42014 Accord all models '98 thru '02
42015 Accord '03 thru '12 & Crosstour '10 thru '14
42016 Accord '13 thru '17

42020 Civic 1200 all models '73 thru '79
42021 Civic 1300 & 1500 CVCC '80 thru '83
42022 Civic 1500 CVCC all models '75 thru '79
42023 Civic all models '84 thru '91
42024 Civic & del Sol '92 thru '95
42025 Civic '96 thru '00 & CR-V '97 thru '01 & Acura Integra '94 thru '00
42026 Civic '01 thru '11 & CR-V '02 thru '11
42027 Civic '12 thru '15 & CR-V '12 thru '16
42030 Fit '07 thru '13
42035 Odyssey models '99 thru '10
Passport - see ISUZU Rodeo (47017)
42037 Honda Pilot '03 thru '08, Ridgeline '06 thru '14 & Acura MDX '01 thru '07
42040 Prelude CVCC all models '79 thru '89

HYUNDAI
43010 Elantra all models '96 thru '19
43015 Excel & Accent all models '86 thru '13
43050 Santa Fe all models '01 thru '12
43055 Sonata all models '99 thru '14

INFINITI
G35 '03 thru '08 - see NISSAN 350Z (72011)

ISUZU
Hombre - see CHEVROLET S-10 (24071)
47017 Rodeo '91 thru '02, Amigo '89 thru '94 & '98 thru '02 & Honda Passport '95 thru '02
47020 Trooper '84 thru '91 & Pick-up '81 thru '93

JAGUAR
49010 XJ6 all 6-cylinder models '68 thru '86
49011 XJ6 all models '88 thru '94
49015 XJ12 & XJS all 12-cylinder models '72 thru '85

JEEP
50010 Cherokee, Comanche & Wagoneer Limited all models '84 thru '01
50011 Cherokee '14 thru '19
50020 CJ all models '49 thru '86
50025 Grand Cherokee all models '93 thru '04
50026 Grand Cherokee '05 thru '19 & Dodge Durango '11 thru '19
50029 Grand Wagoneer & Pick-up '72 thru '91
50030 Wrangler all models '87 thru '17
50035 Liberty '02 thru '12 & Dodge Nitro '07 thru '11
50050 Patriot & Compass '07 thru '17

KIA
54050 Optima '01 thru '10
54060 Sedona '02 thru '14
54070 Sephia '94 thru '01, Spectra '00 thru '09, Sportage '05 thru '20
54077 Sorento '03 thru '13

LEXUS
ES 300/330 - see TOYOTA Camry (92007, 92008)
RX 330/350/350 - see TOYOTA Highlander (92095)

LINCOLN
Navigator - see FORD Pick-up (36059)
59010 Rear-Wheel Drive Continental '70 thru '87, Mark Series '70 thru '92 & Town Car '81 thru '10

MAZDA
61010 GLC (rear wheel drive) '77 thru '83
61011 GLC (front wheel drive) '81 thru '85
61012 Mazda3 '04 thru '11
61015 323 & Protegé '90 thru '03
61016 MX-5 Miata '90 thru '14
61020 MPV all models '89 thru '98
61030 Pick-ups '72 thru '93
Pick-ups '94 thru '09 - see Ford (36071)
61035 RX-7 all models '79 thru '85
61036 RX-7 all models '86 thru '91
61040 626 (rear-wheel drive) all models '79 thru '82
61041 626 & MX-6 (front-wheel drive) '83 thru '92
61042 626 '93 thru '01 & MX-6/Ford Probe '93 thru '02
61043 Mazda6 '03 thru '13

MERCEDES-BENZ
63012 123 Series Diesel '76 thru '85
63015 190 Series 4-cylinder gas models '84 thru '88
63020 230, 250 & 280 6-cylinder SOHC '68 thru '72
63025 280 123 Series gas models '77 thru '81
63030 350 & 450 all models '71 thru '80
63040 C-Class: C230/C240/C280/C320/C350 '01 thru '07

MERCURY
64200 Villager & Nissan Quest '93 thru '01
All other titles, see FORD listing.

MG
66010 MGB Roadster & GT Coupe '62 thru '80
66015 MG Midget & Austin Healey Sprite Roadster '58 thru '80

MINI
67020 Mini '02 thru '13

MITSUBISHI
68020 Cordia, Tredia, Galant, Precis & Mirage '83 thru '93
68030 Eclipse, Eagle Talon & Plymouth Laser '90 thru '94
68031 Eclipse '95 thru '05 & Eagle Talon '95 thru '98
68035 Galant '94 thru '12
68040 Pick-up '83 thru '96 & Montero '83 thru '93

NISSAN
72010 300ZX all models incl. Turbo '84 thru '89
72011 350Z & Infiniti G35 all models '03 thru '08
72015 Altima all models '93 thru '06
72016 Altima '07 thru '12
72020 Maxima all models '85 thru '92
72021 Maxima all models '93 thru '08
72025 Murano '03 thru '14
72030 Pick-ups '80 thru '97 & Pathfinder '87 thru '95
72031 Frontier, Xterra & Pathfinder '96 thru '04
72032 Frontier & Xterra '05 thru '14
72037 Pathfinder '05 thru '14
72040 Pulsar all models '83 thru '86
72042 Rogue all models '08 thru '20
72050 Sentra all models '82 thru '94
72051 Sentra & 200SX all models '95 thru '06
72060 Stanza all models '82 thru '90
72070 Titan pick-ups '04 thru '10 & Armada '05 thru '10
72080 Versa all models '07 thru '19

OLDSMOBILE
73015 Cutlass V6 & V8 gas models '74 thru '88
For other OLDSMOBILE titles, see BUICK, CHEVROLET or GM listings.

PLYMOUTH
For PLYMOUTH titles, see DODGE.

PONTIAC
79008 Fiero all models '84 thru '88
79018 Firebird V8 models except Turbo '70 thru '81
79019 Firebird all models '82 thru '92
79025 G6 all models '05 thru '09
79040 Mid-size Rear-wheel Drive '70 thru '87
Vibe '03 thru '10 - see TOYOTA Corolla (92037)
For other PONTIAC titles, see BUICK, CHEVROLET or GM listings.

PORSCHE
80020 911 Coupe & Targa models '65 thru '89
80025 914 all 4-cylinder models '69 thru '76
80030 924 all models including Turbo '76 thru '82
80035 944 all models including Turbo '83 thru '89

RENAULT
Alliance & Encore - see AMC (14025)

SAAB
84010 900 all models including Turbo '79 thru '88

SATURN
87010 Saturn all S-series models '91 thru '02
Saturn Ion '03 thru '07- see GM (38017)
87020 Saturn L-series all models '00 thru '04
87040 Saturn VUE '02 thru '09

SUBARU
89002 1100, 1300, 1400 & 1600 '71 thru '79
89003 1600 & 1800 2WD & 4WD '80 thru '94
89080 Impreza '02 thru '11, WRX '02 thru '14 & WRX STI '04 thru '14
89100 Legacy all models '90 thru '99
89101 Legacy & Forester '00 thru '09
89102 Legacy '10 thru '16 & Forester '12 thru '16

SUZUKI
90010 Samurai/Sidekick & Geo Tracker '86 thru '01

TOYOTA
92005 Camry all models '83 thru '91
92006 Camry '92 thru '96 & Avalon '95 thru '96
92007 Camry, Avalon, Solara, Lexus ES 300 '97 thru '01
92008 Camry, Avalon, Lexus ES 300/330 '02 thru '06 & Solara '02 thru '08
92009 Camry, Avalon & Lexus ES 350 '07 thru '17
92015 Celica Rear-wheel Drive '71 thru '85
92020 Celica Front-wheel Drive '86 thru '99
92025 Celica Supra all models '79 thru '92
92030 Corolla all models '75 thru '79
92032 Corolla rear-wheel drive models '80 thru '87
92035 Corolla front-wheel drive models '84 thru '92
92036 Corolla & Geo/Chevrolet Prizm '93 thru '02
92037 Corolla '03 thru '13, Matrix '03 thru '14, & Pontiac Vibe '03 thru '10
92040 Corolla Tercel all models '80 thru '82
92045 Corona all models '74 thru '82
92050 Cressida all models '78 thru '82
92055 Land Cruiser FJ40/43/45/55 '68 thru '82
92056 Land Cruiser FJ60/62/80/FZJ80 '80 thru '96
92060 Matrix '03 thru '11 & Pontiac Vibe '03 thru '10
92065 MR2 all models '85 thru '87
92070 Pick-up all models '69 thru '78
92075 Pick-up all models '79 thru '95
92076 Tacoma '95 thru '04, 4Runner '96 thru '02 & T100 '93 thru '98
92077 Tacoma all models '05 thru '18
92078 Tundra '00 thru '06, Sequoia '01 thru '07
92079 4Runner all models '03 thru '09
92080 Previa all models '91 thru '95
92081 Prius all models '01 thru '12
92082 RAV4 all models '96 thru '12
92085 Tercel all models '87 thru '94
92090 Sienna all models '98 thru '10
92095 Highlander '01 thru '19 & Lexus RX330/330/350 '99 thru '19
92179 Tundra '07 thru '19 & Sequoia '08 thru '19

TRIUMPH
94007 Spitfire all models '62 thru '81
94010 TR7 all models '75 thru '81

VW
96008 Beetle & Karmann Ghia '54 thru '79
96009 New Beetle '98 thru '10
96016 Rabbit, Jetta, Scirocco & Pick-up (gas models '75 thru '92 & Convertible '80 thru '92
96017 Golf, GTI & Jetta '93 thru '98, Cabrio '95 thru '02
96018 Golf, GTI & Jetta '99 thru '05
96019 Jetta, Rabbit, GLI, GTI & Golf '05 thru '11
96020 Rabbit, Jetta, Pick-up diesel '77 thru '84
96021 Jetta '11 thru '18 & Golf '15 thru '19
96023 Passat '98 thru '05, Audi A4 '96 thru '01
96030 Transporter 1600 all models '68 thru '79
96035 Transporter 1700, 1800, 2000 '72 thru '79
96040 Type 3 1500 & 1600 '63 thru '73
96045 Vanagon Air-Cooled all models '80 thru '83

VOLVO
97010 120, 130 Series & 1800 Sports '61 thru '73
97015 140 Series all models '66 thru '74
97020 240 Series all models '76 thru '93
97040 740 & 760 Series all models '82 thru '88
97050 850 Series all models '93 thru '97

TECHBOOK MANUALS
10205 Automotive Computer Codes
10206 OBD-II & Electronic Engine Management
10210 Automotive Emissions Control Manual
10215 Fuel Injection Manual '78 thru '85
10225 Holley Carburetor Manual
10230 Rochester Carburetor Manual
10305 Chevrolet Engine Overhaul Manual
10320 Ford Engine Overhaul Manual
10330 GM and Ford Diesel Engine Repair Manual
10331 Duramax Diesel Engines '01 thru '19
10332 Cummins Diesel Engine Performance
10333 GM, Ford & Chrysler Engine Performance
10334 GM Engine Performance
10340 Small Engine Repair Manual, 5 HP & Less
10341 Small Engine Repair Manual, 5.5 HP thru 20 HP
10345 Suspension, Steering & Driveline Manual
10355 Ford Automatic Transmission Overhaul
10360 GM Automatic Transmission Overhaul
10405 Automotive Body Repair & Painting
10410 Automotive Brake Manual
10411 Automotive Anti-lock Brake (ABS) Systems
10425 Automotive Heating & Air Conditioning
10435 Automotive Tools Manual
10445 Welding Manual

Over a 100 Haynes motorcycle manuals also available

10/22

Haynes North America, Inc. • (805) 498-6703 • www.haynes.com